International Association of Geodesy Symposia

Jeffrey T. Freymueller, Series Editor
Laura Sánchez, Assistant Editor

Series Editor
Jeffrey T. Freymueller
Endowed Chair for Geology of the Solid Earth
Department of Earth and Environmental Sciences
Michigan State University
East Lansing, MI, USA

Assistant Editor
Laura Sánchez
Deutsches Geodätisches Forschungsinstitut
Technische Universität München
Munich, Germany

International Association of Geodesy Symposia

Jeffrey T. Freymueller, Series Editor
Laura Sánchez, Assistant Editor

Symposium 110: From Mars to Greenland: Charting Gravity with Space and Airborne Instruments
Symposium 111: Recent Geodetic and Gravimetric Research in Latin America
Symposium 112: Geodesy and Physics of the Earth: Geodetic Contributions to Geodynamics
Symposium 113: Gravity and Geoid
Symposium 114: Geodetic Theory Today
Symposium 115: GPS Trends in Precise Terrestrial, Airborne, and Spaceborne Applications
Symposium 116: Global Gravity Field and Its Temporal Variations
Symposium 117: Gravity, Geoid and Marine Geodesy
Symposium 118: Advances in Positioning and Reference Frames
Symposium 119: Geodesy on the Move
Symposium 120: Towards an Integrated Global Geodetic Observation System (IGGOS)
Symposium 121: Geodesy Beyond 2000: The Challenges of the First Decade
Symposium 122: IV Hotine-Marussi Symposium on Mathematical Geodesy
Symposium 123: Gravity, Geoid and Geodynamics 2000
Symposium 124: Vertical Reference Systems
Symposium 125: Vistas for Geodesy in the New Millennium
Symposium 126: Satellite Altimetry for Geodesy, Geophysics and Oceanography
Symposium 127: V Hotine-Marussi Symposium on Mathematical Geodesy
Symposium 128: A Window on the Future of Geodesy
Symposium 129: Gravity, Geoid and Space Missions
Symposium 130: Dynamic Planet - Monitoring and Understanding...
Symposium 131: Geodetic Deformation Monitoring: From Geophysical to Engineering Roles
Symposium 132: VI Hotine-Marussi Symposium on Theoretical and Computational Geodesy
Symposium 133: Observing our Changing Earth
Symposium 134: Geodetic Reference Frames
Symposium 135: Gravity, Geoid and Earth Observation
Symposium 136: Geodesy for Planet Earth
Symposium 137: VII Hotine-Marussi Symposium on Mathematical Geodesy
Symposium 138: Reference Frames for Applications in Geosciences
Symposium 139: Earth on the Edge: Science for a sustainable Planet
Symposium 140: The 1st International Workshop on the Quality of Geodetic Observation and Monitoring Systems (GuQOMS'11)
Symposium 141: Gravity, Geoid and Height systems (GGHS2012)
Symposium 142: VIII Hotine-Marussi Symposium on Mathematical Geodesy
Symposium 143: Scientific Assembly of the International Association of Geodesy, 150 Years
Symposium 144: 3rd International Gravity Field Service (IGFS)
Symposium 145: International Symposium on Geodesy for Earthquake and Natural Hazards (GENAH)
Symposium 146: Reference Frames for Applications in Geosciences (REFAG2014)
Symposium 147: Earth and Environmental Sciences for Future Generations
Symposium 148: Gravity, Geoid and Height Systems 2016 (GGHS2016)
Symposium 149: Advancing Geodesy in a Changing World
Symposium 150: Fiducial Reference Measurements for Altimetry
Symposium 151: IX Hotine-Marussi Symposium on Mathematical Geodesy

More information about this series at http://www.springer.com/series/1345

IX Hotine-Marussi Symposium on Mathematical Geodesy

Proceedings of the Symposium in Rome, June 18 – 22, 2018

Edited by

Pavel Novák, Mattia Crespi, Nico Sneeuw, Fernando Sansò

Springer

Volume Editors
Pavel Novák
Department of Geomatics
University of West Bohemia
Pilsen, Czech Republic

Mattia Crespi
Geodesy and Geomatics Division
Department of Civil, Constructional and
Environmental Engineering
Sapienza University of Rome
Rome, Italy

Nico Sneeuw
Institute of Geodesy
University of Stuttgart
Stuttgart, Germany

Fernando Sansò
Dipartimento di Ingegneria Civile e Ambientale
Politecnico di Milano
Milano, Italy

Series Editor
Jeffrey T. Freymueller
Endowed Chair for Geology of the Solid Earth
Department of Earth and Environmental Sciences
Michigan State University
East Lansing, MI, USA

Assistant Editor
Laura Sánchez
Deutsches Geodätisches Forschungsinstitut
Technische Universität München
Munich, Germany

ISSN 0939-9585 ISSN 2197-9359 (electronic)
International Association of Geodesy Symposia
ISBN 978-3-030-54266-5 ISBN 978-3-030-54267-2 (eBook)
https://doi.org/10.1007/978-3-030-54267-2

This Springer imprint is published by the registered company Springer Nature Switzerland AG.
The registered company address is: Gewerbestrasse 11, 6330 Cham, Switzerland

Preface

This volume contains the proceedings of the IX Hotine-Marussi Symposium on Mathematical Geodesy which was held from 18 to 22 June 2018 in Rome, Italy. For a third time in a row, the Symposium took place at the Faculty of Civil and Industrial Engineering of the Sapienza University of Rome, Italy. As in 2009 and 2013, the venue of the Symposium was the beautiful cloister of the Basilica di San Pietro in Vincoli, worldwide known for the statue of Moses by Michelangelo.

A series of symposia focused on theoretical geodesy began in 1959 when Antonio Marussi organized the first Symposium on Three Dimensional Geodesy in Venice. The name of the symposia was changed in 1965 when the third Symposium on Mathematical Geodesy was held in Torino. The first three symposia were strongly influenced by the prominent British geodesist Martine Hotine. After his death in 1968, the series was renamed again and the first Hotine Symposium on Mathematical Geodesy was held in Trieste, 1969. This symposium and the following four symposia were organized by Antonio Marussi. After his death in 1984, the series was renamed to the Hotine-Marussi Symposia, the title still used today. The first five Hotine-Marussi Symposia (1985, 1989, 1994, 1998 and 2003) were organized by Fernando Sansò, the driving force behind the series of Hotine-Marussi symposia over more than three decades.

Since 2006, the series of the Hotine-Marussi Symposia has been under the responsibility of the Inter-Commission Committee on Theory (ICCT) within the International Association of Geodesy (IAG). The ICCT organized the last four Hotine-Marussi Symposia held in Wuhan (2006) and Rome (2009, 2013 and 2018). The overall goal of Hotine-Marussi Symposia is aligned with the main objective of the ICCT to advance geodetic theory in all branches of geodesy, reflecting developments in geodetic observing systems and interactions of geodesy with other Earth-related sciences. Thus, Hotine-Marussi Symposia on Mathematical Geodesy represent a main venue for theoretically oriented geodesists.

The Symposium attracted 119 participants from 30 countries who contributed 120 papers (83 oral and 37 poster presentations). The scientific program of the symposium was organized in ten regular sessions which were modelled thematically after ICCT study group topics. The regular sessions were convened by the chairs of the ICCT study groups who also constituted the Symposium's Scientific Committee:

1. *Geodetic Methods in Earth System Science*
   N. Sneeuw
2. *Theory of Multi-GNSS Parameter Estimation*
   A. Khodabandeh, M. Crespi
3. *Digital Terrain Modelling*
   R. Barzaghi
4. *Space Weather and Atmospheric Modelling*
   K. Börger, M. Schmidt
5. *Global Gravity Field Modelling and Heights Systems*
   D. Tsoulis, S. Claessens

6. *Theory of Modern Geodetic Reference Frames and Earth's Rotation*
   Z. Altamimi
7. *Deformation and Gravity Field Modelling at Regional Scales*
   J. Huang, Y. Tanaka
8. *Estimation Theory and Inverse Problems in Geodesy*
   A. Dermanis
9. *Advanced Numerical Methods in Geodesy*
   R. Čunderlík
10. *Multi-Sensor and Time Series Data Analysis*
    W. Kosek, K. Sosnica

Participants of the IX Hotine-Marussi Symposium in the cloister of the Basilica di San Pietro in Vincoli

Additionally, a special session was held at the Accademia Nazionale dei Lincei (the oldest scientific academy in the world, established in 1603 by Federico Cesi) on 19 June 2018. The session consisted of six invited talks focused on interactions of geodesy and selected Earth science components:

1. *Interaction Between Geodesy and Oceanography: Results and Open Problem*
   M. H. Rio
2. *Geodesy and Glaciology: Absolute Gravity and Surface Displacements in Greenland*
   O. Francis, T. van Dam
3. *Geodesy and Atmospheric Science: A Collaboration Mutually Beneficial*
   R. Pacione, J. Douša
4. *Geodesy and Mathematics: Acquisitions and Open Problems*
   W. Freeden, F. Sansò

5. *Solid Earth-System Structure from Satellites*
   R. Haagmans
6. *Geodesy and Seismology: A Key Synergy for the Understanding and Forecasting of Earthquakes*
   A. Peresan, M. Crespi, A. Mazzoni, G. Panza

The special session was organized by Fernando Sansò, Emeritus at the Politecnico di Milano, himself a member of the Accademia. One of the presentations within the special session is included into these proceedings in the form of a long paper by W. Freeden and F. Sansò, dealing with the interactions of geodesy and mathematics, while the combination of these two scientific disciplines forms the essence of the Hotine-Marussi symposia.

The scientific program of the symposium was complemented with a great social program including a night tour of the Vatican Museum and a social dinner at the restaurant Il Miraggio at the neighbourhood of Villa Farnesina and Palazzo Corsini, locations of the Accademia dei Lincei.

We would like to acknowledge all who contributed to the success of the IX Hotine-Marussi Symposium. The study group chairmen and the entire Scientific Committee (P. Novák, M. Crespi, N. Sneeuw, F. Sansò, G. Blewitt, R. Pail, M. Hashimoto, M. Santos, R. Gross, D. Tsoulis, R. Ćunderlík, M. Šprlák, K. Sośnica, J. Huang, R. Tenzer, A. Khodabandeh, S. Claessens, W. Kosek, K. Börger, Y. Tanaka, A. Dermanis, V. Michel and E. Grafarend) put much effort in organizing and convening sessions. They also served as the associated editors in a peer-review process lead by Jeffrey Freymueller and Laura Sánchez, the IAG Symposia Series editors. Although most of the reviewers remain anonymous for the authors, a complete list of reviewers is printed in this volume to express our gratitude for their dedication.

The Symposium was financially and promotionally supported by the Faculty of Civil and Industrial Engineering of the Sapienza University of Rome and by the Italian Space Agency. The IAG provided travel support to selected young participants of the Symposium.

However, most of our gratitude goes to the Local Organizing Committee of the Symposium. The team chaired by Mattia Crespi, the vice-president of the ICCT, consisted of members of the Area of Geodesy and Geomatics at the Faculty of Civil and Industrial Engineering of the Sapienza University of Rome: A. Mazzoni, F. Fratarcangeli, R. Ravanelli, A. Mascitelli, M. Ravanelli, M. Di Tullio, V. Belloni, G. Savastano, A. Nascetti, G. Colosimo, E. Benedetti, M. Branzanti, M. Di Rita, P. Capaldo and F. Pieralice. Through their effort and organization skills, Mattia Crespi and his team significantly contributed to the success of the Symposium.

Pilsen, Czech Republic — Pavel Novák
Rome, Italy — Mattia Crespi
Stuttgart, Germany — Nico Sneeuw
Milano, Italy — Fernando Sansò
May 2020

Contents

Part I

Gravity Field Modelling and Height Systems

Orbit Optimization for Future Satellite Gravity Field Missions: Influence of the Time Variable Gravity Field Models in a Genetic Algorithm Approach

Siavash Iran Pour, Nico Sneeuw, Matthias Weigelt, and Alireza Amiri-Simkooei

Abstract

Many studies in the past have discussed potential orbit configurations of future satellite gravity missions. Most of those works have targeted orbit optimization of the satellite missions of the next generation in the so-called Bender formation. The studies have investigated the impact of the Keplerian orbital parameters, especially the influence of the repeat orbits and mission altitude of both satellite pairs and the inclination of the second pair in Bender formation on the satellite configurations' gravity field recovery quality performance.

Obviously, the search space for the orbit optimization in the Bender formation is vast and, therefore, different approaches have been suggested for optimal orbit design. Among approaches, however, different assumptions about input geophysical models as well as the error models into the simulation software play a role. Our paper shows how different assumptions for input models change the orbit optimization results. For this purpose, the genetic algorithm has been utilized for orbit optimization of the Bender formation where different input models were considered. Those input models include (1) the updated ESA geophysical models, and (2) error models for the Ocean Tide (OT error) and Atmosphere-Ocean (AO error). Here, we focus on the impact of the models on relative difference of the longitude of ascending nodes between the two pairs in Bender formation. The results of the paper clearly state that our current and future knowledge about signal and error models can significantly affect the orbit optimization problem.

Keywords

Genetic algorithm · Gravity field recovery · Orbit optimization · Time-variable gravity field

S. Iran Pour (✉)
Department of Geomatics Engineering, Faculty of Civil Engineering and Transportation, University of Isfahan, Isfahan, Iran

Institute of Geodesy, University of Stuttgart, Stuttgart, Germany

N. Sneeuw
Institute of Geodesy, University of Stuttgart, Stuttgart, Germany

M. Weigelt
Institute of Geodesy, Leibniz University of Hannover, Hannover, Germany

A. Amiri-Simkooei
Department of Geomatics Engineering, Faculty of Civil Engineering and Transportation, University of Isfahan, Isfahan, Iran

Department of Control and Operations, Technical University of Delft, Delft, The Netherlands

1 Introduction

Soon after the launch of the Gravity Recovery and Climate Experiment (GRACE) mission (Tapley et al. 2004), several research studies have started to investigate the performance of next generation satellite gravity missions. Those researches range from replacing the GRACE by another inline formation (GRACE Follow-On) or alternative (advanced) formations to analyzing mission scenarios with two pairs, see (Sharifi et al. 2007; Bender et al. 2008; Wiese et al. 2009; Elsaka 2010; Wiese et al. 2012; Ellmer 2011; Elsaka et al. 2012; Iran Pour et al. 2013; Reubelt et al. 2014; Elsaka et al. 2014; Gruber et al. 2014; Klokočník

P. Novák et al. (eds.), *IX Hotine-Marussi Symposium on Mathematical Geodesy*, International Association of Geodesy Symposia 151, https://doi.org/10.1007/1345_2019_79

et al. 2015; Iran Pour et al. 2016; Iran-Pour et al. 2018). Moreover, ESA studies, among them the ESA projects “Assessment of a Next Generation Mission for Monitoring the Variations of Earth’s Gravity” (Anselmi et al. 2011) and “Assessment of Satellite Constellations for Monitoring the Variations in Earth’s Gravity Field” (Iran Pour et al. 2015) investigated the science requirements, performance criteria and design of future satellite gravity missions. These two ESA projects have looked into two advanced formations as well as Bender configurations, where the latter is the most probable candidate for the next generation of future gravity missions (according to the ESA proposal calls).

Most of the research works above did the selection of the individual constellation scenarios by rough assessment of sampling behavior of the missions, although the studies by (Wiese et al. 2012; Ellmer 2011; Iran Pour et al. 2013; Iran Pour et al. 2015) had a deeper look into some performance criteria for their double-pair mission scenario optimization search strategies. Furthermore, (Iran Pour et al. 2016) studied the groundtrack pattern evolution of double pair missions and discussed the impact of longitude of ascending nodes between the two satellites pairs ($\Delta\Omega$). They recognized three reasons for the quality variations of the gravity solutions, namely: (1) the different time evolution of sampling pattern of each satellite pair in a double pair mission with different repeat periods, (2) the impact of time-variation of $\Delta\Omega$ on the pattern change, and (3) the influence of the time-variable gravity signals on quality variations of the gravity field recovery.

The current research work tries to look into the influence of the time-variable gravity signals and error models on the gravity field recovery quality, where in particular the impact on the orbit optimization is studied. The focus of this paper is, however, the impact of the models on optimization of $\Delta\Omega$ which is shown to be a very influential parameter in the orbit optimization.

2 Methodology

2.1 Geophysical Models and Simulation Tool

For the simulation environment of this study, we employ the dominant mass variations of hydrology (H), ice (I) and solid Earth (S) of the Earth system by use of the updated time-variable gravity field generated from ESA-project “ESA Earth System Model for Gravity Mission Simulation Studies” (Dobslaw et al. 2015).

Time-span of 10-day for gravity field recovery, starting from January 1st 1996, is applied in the paper (Iran Pour et al. 2015). The forward and recovery simulations are both up to spherical harmonics degree $L_{max} = 90$.

The error models of our research work consist of:

- Ocean Tide (OT) error: The difference between the two OT models EOT08a (Savcenko and Bosch 2008) and GOT4.7 (Ray 2008).
- Atmosphere-Ocean (AO) error: The AO error product from IAPG (TU Munich), defined as two atmosphere models difference (ECMWF–NCEP) plus 10% of the ocean signal of the model OMCT (Iran Pour et al. 2015).

For the gravity field recovery simulation, a nominal circular orbit simulation software of so-called Reduced-Scale Tool, RST is employed. This software assumes nominal satellite orbits, i.e. the only orbit perturbation is secular J_2 effect by the Earth flattening, hence the semi-major axis, inclination angle and eccentricity of the satellites are not time-variable (Iran Pour et al. 2013). However, the equatorial bulge of the Earth (the Earth flattening effect) is responsible for variation of the Keplerian elements ω (argument of perigee), Ω (longitude of the ascending node) and M (mean anomaly). For circular orbits, argument of perigee is ill-defined, and therefore the variation of the other two elements are calculated by the following rates:

$$\dot{\Omega} = \frac{1}{na^2}\frac{\partial T}{\partial I}, \quad \dot{M} = n - \frac{2}{na}\frac{\partial T}{\partial a} \text{ with } n = \sqrt{\frac{GM}{a^3}} \tag{1}$$

where a and I are respectively the semi-major axis and inclination angle (the eccentricity here sets to zero for the circular orbits). GM is the gravitational constant times the Earth’s mass, and the parameter T stands for all the terms of the potential field beyond the central field term. We should also mention that the work by (Iran Pour et al. 2013) shows that other perturbing Earth’s gravity field terms rather than J_2 effect have negligible impact on the gravity field recovery performance in short time-spans (around one month and less).

2.2 Genetic Algorithm (GA)

Genetic algorithm (GA) is an evolutionary algorithm which generates solutions to optimization problems using techniques inspired by natural evolution such as inheritance, mutation, selection, and crossover. The algorithm is an implemented tool to find satisfactory solutions to non-linear search or optimization problems.

In a genetic algorithm, a problem is essentially solved by repeatedly guessing a solution, and then evaluating the quality of the solution. In this way, the trick is not guessing randomly, but rather on the basis of previously acquired knowledge about the solution space, where two main components are required: (1) a genome, which is a generalized representation of the solution space of a specific problem (the parameters space of the problem), and (2) a fitness function to evaluate the solution domain which in fact maps the attributes of a specific realization of the genome to a fitness value (Schmitt 2001; Ellmer 2011).

In this study, the global (accumulated) geoid height error RMS is chosen as the fitness function. The search space of our study for Bender configuration is identified as follows (Iran Pour et al. 2015):

- all the repeat modes with altitude between 340 and 500 km
- inclination of near-polar pair between 88° and 92° (minimizing polar gap) with step of 1°
- inclination of inclined pair within 65°–75° or 105°–115° with step of 1°
- free angle between orbital plane of the satellite pairs ($0^\circ \leq \Delta\Omega < 360^\circ$ with step of 1°)
- free mean anomaly difference between the pairs ($0^\circ \leq \Delta M < 360^\circ$ with step of 1°)

The GA software was run in MATLAB where the "TolFun" (for tolerance) was set to 10^{-6} with number of generations "StallGenLimit" over 50, i.e. the algorithm runs until the cumulative change in the fitness function value over 50 generations is less than 10^{-6}.

3 Results

In our study, the GA simulation tool has been run several times, every time with different initial conditions (input models). Since GA behaves in a stochastic way and in order to reduce the uncertainty of the results, the simulations with the same input models have been run several times.

The results of GA simulation runs consist of different input models, namely HIS (updated ESA model), only OT error model, only AO error model and full model (updated ESA HIS + OT error + AO error). The results of those runs are respectively depicted in Fig. 1 where the color bars show the impact of $\Delta\Omega$ on the orbit optimization in Bender configuration.

As, it can be seen from the figure, after around 300 simulations (approximately 6 generations), the first main convergence happens. It is also seen that the optimized

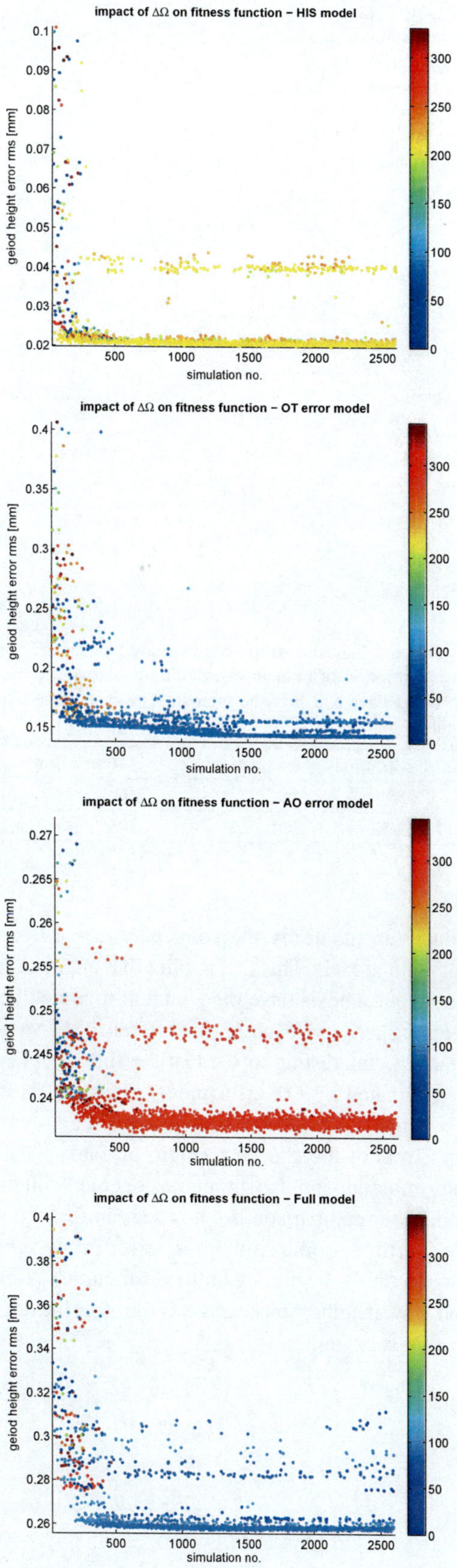

Fig. 1 Impact of $\Delta\Omega$ on fitness function of GA simulation runs with different input models. The colors stand for different $\Delta\Omega$ values (from 0° to 359°)

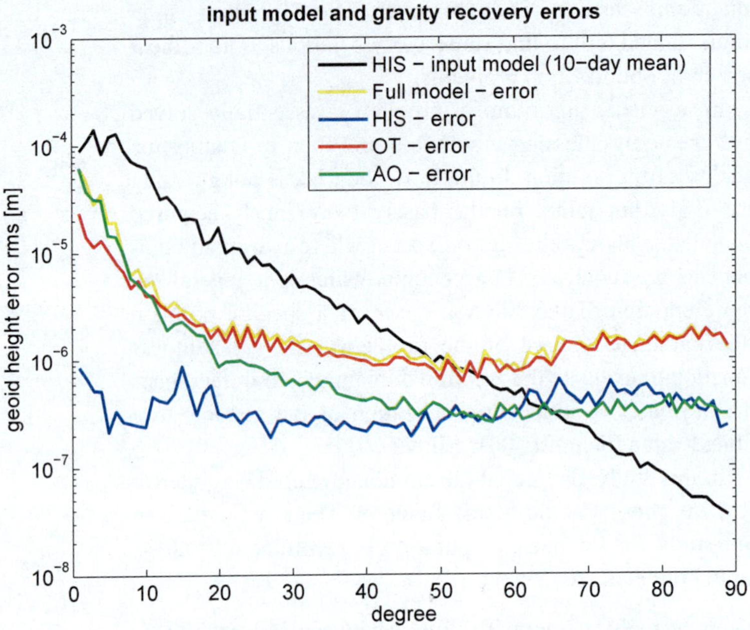

Fig. 2 Gravity field recovery errors of the chosen scenario of Table 1 by different simulation input assumptions together with the HIS input model (10-day mean of updated ESA HIS model) for comparison

Table 1 Selected Bender scenario of this study for detailed investigation (β/α is the ration of number of satellite revolutions to the repeat period in nodal days and ρ is the inter-satellite distance within each satellite pair)

Scenario	β/α (rev./day)	Inclination (deg.)	Altitude (km)	ρ (km)
	125/8	89.5°	360.7	100
	503/32	72°	305	100

$\Delta\Omega$ values (as shown by the color bars) are different for different input models. This fact implies that our assumptions about the input models have the potential to affect the orbit optimization, here by resulting in different $\Delta\Omega$ values. In particular, it is interesting to see that the full model error is largely dominated by OT error model. This fact is clearly shown by Fig. 2 as well, where it displays the gravity field recovery errors of the chosen scenario of Table 1 based on different simulation input assumptions together with the HIS input model for comparison. As it is seen in that figure, the full model error is dominated by AO error in the very long wavelengths (at the beginning of the error curve). At middle and short wavelengths, however, the OT error is the dominant error to a large extent.

We have also simulated the recovery error of 10-day solutions of the Table 1 scenario where the impact of different input models with the variation of the $\Delta\Omega$ angle values on the recovery error is in focus (Fig. 3). The results show different variation behaviors by the impact of different input models. One should also notice the significance of the variations' magnitude (y-axis of the graphs in Fig. 3). It is also important to mention that the geophysical signals are not distributed symmetrically on the globe (neither in space domain, nor in time domain), therefore one should not expect a perfect symmetry in the graphs. That is, in particular, a true statement for HIS signals. In addition, the results clearly illustrate that OT and AO errors are approximately one order of magnitude larger than HIS error, while the pattern of the full model error is almost dominated by OT error pattern.

Another interesting result is observed when we compare the outcome of the updated ESA HIS model by (Dobslaw et al. 2015) in Fig. 3 with that from the previous ESA HIS model by (Gruber et al. 2011) in Fig. 4. The comparison clearly shows slightly different variation behaviors by $\Delta\Omega$ impact on the recovery error where the magnitude of variations is also different. This fact, again, represents the influence of input models on orbit optimization.

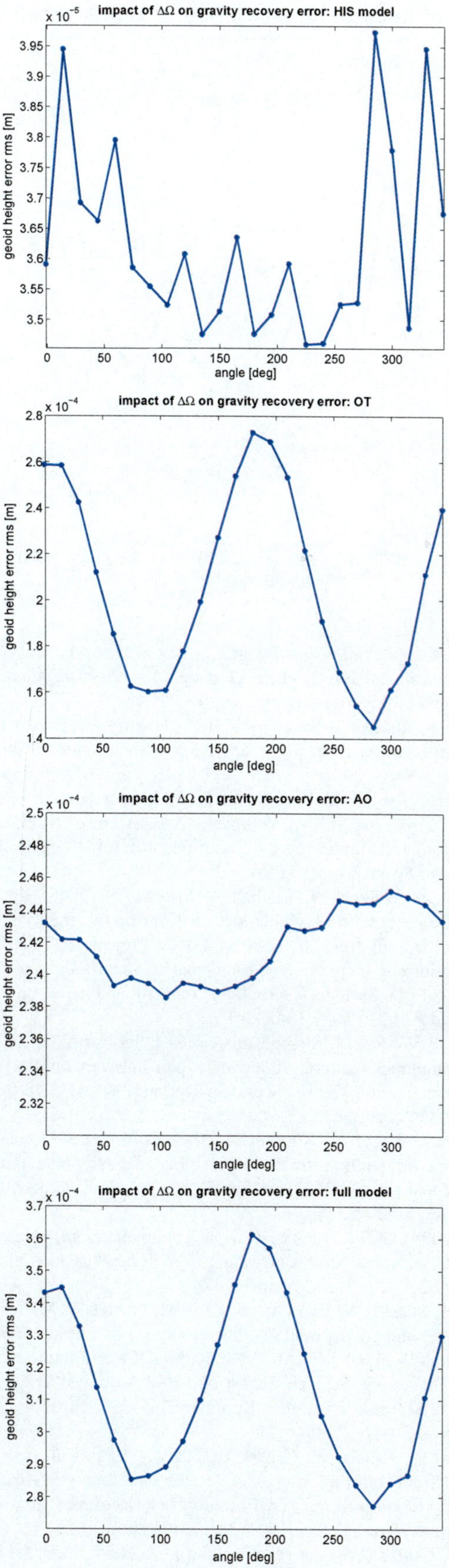

Fig. 3 Impact of $\Delta\Omega$ angle values on gravity field recovery error (in terms of geoid height RMS) for different input model assumptions

4 Discussion and Conclusion

This research work tried to look into the influence of the time-variable gravity signals and error models on the gravity field recovery quality, where the impact on the orbit optimization is studied. The focus of the paper was, however, the impact of the models on optimization of the orbital parameter "relative difference of the longitude of ascending nodes between the two pairs in Bender formation ($\Delta\Omega$)".

In this study, we have run a genetic algorithm simulation tool with different initial conditions (input models). We found that different input signal and error models result in different optimized $\Delta\Omega$, although the magnitude of impact and, hence, the importance in optimization is different. The OT error was found to be the most dominant model in the orbit optimization, where the magnitude of influence is approximately an order of magnitude larger than HIS signal. The AO error was also found influential, however, its effect is hidden by OT error impact in the full simulation run. The results of the paper clearly state that our current and future knowledge about signal and error models can significantly affect the orbit optimization problem. For example, improving the error models (i.e. AO and OT error models) has the potential to result in different values for optimized $\Delta\Omega$. That is also the case for HIS model which does not play an important role in the optimization problem at the time, but can be influential in the future if our OT and AO error models are significantly improved.

The focus of this paper was the impact of input models on parameter $\Delta\Omega$ optimization, which is in fact the most influential parameter in the general orbit optimization problem. However, in reality, this parameter is controlled by the orbit maintenance or even it is allowed to vary freely which might be of interest in some orbit designs. Therefore, it might be more interesting to keep the parameter $\Delta\Omega$ fixed and run the genetic algorithm for the other relative Keplerian parameters in Bender configuration. Furthermore, it would be of interest to study the impact of individual ocean tide constituents and investigate their relative importance in orbit optimization in more detail.

Fig. 4 Impact of $\Delta\Omega$ angle values on gravity field recovery error (in terms of geoid height RMS) for the old ESA HIS input model by (Gruber et al. 2011)

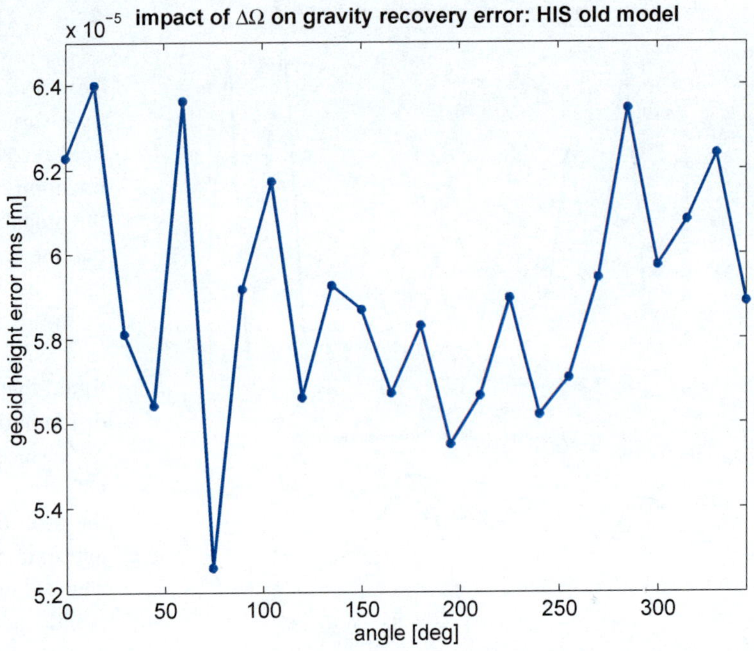

References

Anselmi A, Visser P, van Dam T, Sneeuw N, Gruber T, Altès B et al (2011) Assessment of a next generation gravity mission to monitor the variations of Earth's gravity field. European Space Agency (ESA)

Bender P, Wiese D, Nerem R (2008) A possible dual-GRACE mission with 90 degree and 63 degree inclination orbits. In: Proceedings of the third international Symposium on formation flying, missions and technologies, ESA/ESTEC, Noordwijk. ESA, ESTEC, Noordwijk, The Netherlands, pp 1–6. Retrieved April 23–25, 2008, from http://sci.esa.int/lisa/40135-3rd-international-symposium-on-formation-flying-missions-and-technologies/

Dobslaw H, Bergmann-Wolf I, Dill R, Forootan E, Klemann V, Kusche J, Sasgen I (2015) The updated ESA earth system model for future gravity mission simulation studies. J Geod 89:505–513. https://doi.org/10.1007/s00190-014-0787-8

Ellmer M (2011) Optimization of the orbit parameters of future gravity missions using genetic algorithms. University of Stuttgart, Stuttgart. Retrieved from http://elib.uni-stuttgart.de/opus/volltexte/2012/7122/pdf/Ellmer.pdf

Elsaka B (2010) Simulated satellite formation flights for detecting temporal variations of the Earth's gravity field. University of Bonn, Bonn. Retrieved from http://hss.ulb.uni-bonn.de/2010/2151/2213.pdf

Elsaka B, Kusche J, Ilk K (2012) Recovery of the Earth's gravity field from formation-flying satellites: temporal aliasing issues. Adv Space Res 50(11):1534–1552. https://doi.org/10.1016/j.asr.2012.07.016

Elsaka B, Raimondo J-C, Brieden P, Reubelt T, Kusche J, Flechtner F et al (2014) Comparing seven candidate mission configurations for temporal gravity field retrieval through full-scale numerical simulation. J Geod 88(2):31–43

Gruber T, Bamber J, Bierkens M, Dobslaw H, Murböck M, Thomas M et al (2011) Simulation of the time-variable gravity field by means of coupled geophysical models. Earth Syst Sci Data 3(1):19–35. https://doi.org/10.5194/essd-3-19-2011

Gruber T, Baldesarra M, Brieden P, Danzmann K, Daras I, Doll B et al (2014) e2motion - earth system mass transport mission - concept for a next generation gravity field mission. Deutsche Geodätische Kommission der Bayerischen Akademie der Wissenschaften, Reihe B, Angewandte Geodäsie, Munich, p 318

Iran Pour S, Reubelt T, Sneeuw N (2013) Quality assessment of sub-Nyquist recovery from future gravity satellite missions. Adv Space Res *52*(5):916–929. https://doi.org/10.1016/j.asr.2013.05.026

Iran Pour S, Reubelt T, Sneeuw N, Daras I, Murböck M, Gruber T et al (2015) ESA SC4MGV Project "Assessment of Satellite Constellations for Monitoring the Variations in Earth's Gravity Field". European Space Agency (ESA)

Iran Pour S, Weigelt M, Reubelt T, Sneeuw N (2016) Impact of groundtrack pattern of double pair missions on the gravity recovery quality: lessons from the ESA SC4MGV Project. In: International symposium on earth and environmental sciences for future generations, vol 147. Springer International Publishing, Prague, pp 97–101. https://doi.org/10.1007/1345_2016_228

Iran-Pour S, Weigelt M, Amiri-Simkooei A-R, Sneeuw N (2018) Impact of groundtrack pattern of a single pair mission on the gravity recovery quality. Geosciences (Switzerland) 8(9):315. https://doi.org/10.3390/geosciences8090315

Klokočník J, Wagner C, Kostelecký J, Bezděk A (2015) Ground track density considerations on the resolvability of gravity field harmonics in a repeat orbit. Adv Space Res 56(6):1146–1160. https://doi.org/10.1016/j.asr.2015.06.020

Ray R (2008) GOT4.7 (2008) (Private communication), extension of Ray R., A global ocean tide model from Topex/Poseidon altimetry. GOT99.2 NASA Tech Memo 209478

Reubelt T, Sneeuw N, Iran Pour S, Hirth M, Fichter W, Müller J et al (2014) Future gravity field satellite missions. In: Flechtner F, Sneeuw N, Schuh W-D (eds) CHAMP, GRACE, GOCE and future missions; Geotechnologien Science Report No. 20, "Advanced Technologies in Earth Sciences". Springer, Berlin, pp 165–230. https://doi.org/10.1007/978-3-642-32135-1_21

Savcenko R, Bosch W (2008) EOT08a – empirical ocean tide model from multi-mission satellite altimetry. Deutsches Geodätisches Forschungsinstitut (DGFI), München. Retrieved from https://mediatum.ub.tum.de/doc/1304928/1304928.pdf

Schmitt L (2001) Theory of genetic algorithms. Theor Comp Sci 259(1–2):1–61. https://doi.org/10.1016/S0304-3975(00)00406-0

Sharifi M, Sneeuw N, Keller W (2007) Gravity recovery capability of four generic satellite formations. In: Kilicoglu A, Forsberg R (eds) Gravity field of the Earth. General Command of Mapping, ISSN 1300-5790, Special Issue 18, pp 211–216

Tapley B, Bettadpur S, Watkins M, Reigber C (2004) The gravity recovery and climate experiment: mission overview and early results. Geophys Res Lett 31(9). https://doi.org/10.1029/2004GL019920

Wiese D, Folkner W, Nerem R (2009) Alternative mission architectures for a gravity recovery satellite mission. J Geod 83(6):569–581. https://doi.org/10.1007/s00190-008-0274-1

Wiese D, Nerem R, Lemoine F (2012) Design considerations for a dedicated gravity recovery satellite mission consisting of two pairs of satellites. J Geod 86(2):81–98. https://doi.org/10.1007/s00190-011-0493-8

Comparison of Criteria for the Identification of Correlated Orders in GRACE Spherical Harmonic Coefficients

Dimitrios Piretzidis, Michael G. Sideris, and Dimitrios Tsoulis

Abstract

The study of the Earth's time-varying gravity field using GRACE data requires the removal of correlated errors using filtering techniques in the spherical harmonic domain. The empirical decorrelation filter is an effective method of decorrelating order-wise series of spherical harmonic coefficients, although its improper implementation can lead to signal attenuation. To reduce geophysical signal over-filtering, decorrelation should be performed only for orders that show evidence of high correlation. In this paper we investigate and compare the behavior of three criteria, i.e., the root mean square ratio, the angle distribution of phase spectrum and the geometric properties of order-wise coefficient series, that can be used for the identification of correlated orders in GRACE data. Our analysis indicates that the root mean square ratio is the most reliable criterion, due to its simple implementation and for providing averaged time series of equivalent water height with smaller root mean square error, based on a simulation.

Keywords

Empirical decorrelation · Filtering · GRACE · Spherical harmonic coefficients

1 Introduction

Monthly gravity field solutions in the form of spherical harmonic coefficients derived from data of the Gravity Recovery and Climate Experiment (GRACE; Tapley et al. 2004) satellite mission are routinely used for monitoring mass variations in the Earth system. These solutions contain correlated errors, which manifest when month-to-month differences or monthly differences from a static gravity field solution are calculated. In the spatial domain, these errors produce longitudinal artifacts, commonly referred to as "stripes". The removal of correlated errors from GRACE coefficient changes is usually performed using an empirical decorrelation filter (EDF), also known as destriping filter. This type of filter removes the contribution of a smoothing polynomial from the even- and odd-degree coefficient series of a specific order (Swenson and Wahr 2006). The EDF is usually implemented from an order $m_{\min}$, where the correlated errors approximately start to appear, up to the maximum order of each monthly set of coefficient changes. A different and not so commonly used approach is the selective implementation of the EDF to the coefficient series that appear to be heavily influenced by correlated errors. This idea originates from Huang et al. (2012), who used a selective EDF scheme to study the groundwater storage variability in the Great Lakes and in Alberta, Canada. More recently, Piretzidis et al. (2018) accessed the capabilities of selective EDF by comparing it with conventional

Electronic Supplementary Material The online version of this chapter (https://doi.org/10.1007/1345_2019_83) contains supplementary material, which is available to authorized users.

D. Piretzidis (✉) · M. G. Sideris
Department of Geomatics Engineering, University of Calgary, Calgary, AB, Canada
e-mail: dimitrios.piretzidis@ucalgary.ca; sideris@ucalgary.ca

D. Tsoulis
Department of Geodesy and Surveying, Aristotle University of Thessaloniki, Thessaloniki, Greece
e-mail: tsoulis@topo.auth.gr

P. Novák et al. (eds.), *IX Hotine-Marussi Symposium on Mathematical Geodesy*,
International Association of Geodesy Symposia 151, https://doi.org/10.1007/1345_2019_83

decorrelation methods in North America and Greenland. Although different criteria are used to identify correlated coefficient series by these two studies, their findings suggest that the selective EDF ensures the removal of the majority of correlated errors while reducing the attenuation of useful geophysical signal. Piretzidis et al. (2018) also highlight the need to search for reliable criteria that can be used as proxies for the identification of correlated coefficient series in GRACE data. The objective of this study is to compare different criteria and provide insights on their performance. We examine three criteria, two of them already used in the studies of Huang et al. (2012) and Piretzidis et al. (2018), and one new, described in Sect. 3. We test them on monthly GRACE and Global Land Data Assimilation System (GLDAS; Rodell et al. 2004) data. We also study their similarities and differences, and indicate their advantages and disadvantages. Finally, we assess their performance with a simulation study.

Although the information about correlated errors is also contained in the full variance-covariance matrix of monthly spherical harmonic coefficients, this information is not used here because (a) it is not always available to the users and (b) it cannot be directly incorporated in the EDF algorithm. Alternatively, full variance-covariance matrices can be used in other decorrelation filters, such as the ones described in Kusche (2007), Klees et al. (2008), Kusche et al. (2009) and Horvath et al. (2018), that are not the focus of this study.

2 Data

All the experiments are conducted using the Center for Space Research (CSR) RL05 and RL06 of GRACE Level 2 data, spanning a period from 08/2002 to 08/2016. These data, also known as GSM products, are monthly sets of spherical harmonic coefficients $C_{n,m}$ and $S_{n,m}$ of degree n and order m up to maximum degree and order $N = 96$. The GLDAS 1.0 Noah monthly data are obtained for the same time period. Months with no available GRACE solutions are also removed from the GLDAS data. Their format consists of spatial grids with global land coverage. The terrestrial water storage is calculated for each month and transformed into spherical harmonic 'mass' coefficients $\hat{C}_{n,m}$ and $\hat{S}_{n,m}$. These coefficients are converted into dimensionless (or 'gravity') coefficients using eq. (12) from Wahr et al. (1998). Monthly coefficient changes $\Delta C_{n,m}$ and $\Delta S_{n,m}$ are derived by removing a long-term mean gravity field from both the GRACE and GLDAS monthly coefficients. For the GRACE data, the geopotential model GGM05C (Ries et al. 2016) is used as mean gravity field, whereas, for the GLDAS data a simple average of all monthly coefficients is used.

3 Criteria for the Identification of Correlated Orders

The three criteria selected for the identification of correlated order-wise coefficient series are: (1) the root mean square (RMS) ratio before and after decorrelation, (2) the angle distribution of phase spectrum and (3) the geometric shape properties. Criteria (1) and (3) are designed by previous studies explicitly for the identification of correlated coefficient series. Criterion (2) is only studied with respect to its behavior when different filters are implemented to the GRACE coefficients and never tested in the context of selective EDF.

3.1 RMS Ratio (RMSR) Before and After Decorrelation

The RMSR was firstly used as a criterion for the selective decorrelation of GRACE coefficient changes in Huang et al. (2012), and more recently in Huang et al. (2016). It is given by:

$$R(m) = \sqrt{\frac{\sum_{i=m}^{N} (\Delta C_{i,m})^2}{\sum_{i=m}^{N} (\Delta C^{d}_{i,m})^2}}, \tag{1}$$

where $\Delta C_{:,m}$ denotes the original and $\Delta C^{d}_{:,m}$ the decorrelated series of order m. Although Huang et al. (2012) used this criterion in the even- and odd-degree series separately, here we use it in the entire order-wise coefficient series for consistency with the other criteria examined. The removal of a polynomial contribution from a series suppresses constituents that correspond to either geophysical signal or correlated errors, and results in a decorrelated series with smaller RMS value. $R(m)$ is greater than one and its value increases with the magnitude of correlated errors, i.e., highly correlated series return a larger $R(m)$ value. The identification of correlated series using the RMSR criterion is performed by introducing a critical value R_C. Coefficient series with $R(m) \geq R_C$ are considered correlated and with $R(m) < R_C$ uncorrelated. The critical RMS value R_C should be selected in order to maximize the removal of correlated errors while minimizing the loss of geophysical signal. We follow the study of Huang et al. (2012), where a value of 2 was selected based on a simulation.

3.2 Angle Distribution of Phase Spectrum (ADPS)

The phase spectrum $\phi_{n,m}$ of a function with spherical harmonic representation $\{\Delta C_{n,m}, \Delta S_{n,m}\}$ is given by Devaraju and Sneeuw (2017):

$$\phi_{n,m} = \begin{cases} \arctan\left(-\dfrac{(-1)^m \Delta S_{n,m}}{(-1)^m \Delta C_{n,m}}\right), & m \neq 0 \\ \arctan\left(\dfrac{0}{\Delta C_{n,m}}\right), & m = 0. \end{cases} \quad (2)$$

Devaraju and Sneeuw (2017) examined the phase spectrum of a geopotential model of the Earth's static gravity field and showed that the order-wise angle distribution $\phi_{:,m}$ is nearly uniform, which is also likely to be the case for the time-varying gravity field. They also demonstrated that filtering GRACE coefficients with the EDF produces an angle distribution that tends towards uniformity. Taking these properties into account, the ADPS can be a potential criterion for the identification of correlated coefficient series. In order to assess how different (in a statistical sense) is the angle distribution from a uniform one, we follow Devaraju and Sneeuw (2017) and use Rao's spacing test, implemented by the CircStat toolbox (Berens 2009). Rao's spacing test calculates the value of the test-statistic U by comparing the distances between the phase angles to those expected from a uniform distribution, as follows (Russell and Levitin 1995):

$$U = \frac{1}{2}\sum_{i=1}^{i_{\max}} |T_i - \lambda|, \quad (3)$$

where $i_{\max} = N - m + 1$, $\lambda = 360°/i_{\max}$ and T_i is given by:

$$T_i = \begin{cases} \phi_{m+i,m} - \phi_{m+i-1,m}, & i \neq i_{\max} \\ 360° - \phi_{m+i_{\max}-1,m} + \phi_{m,m}, & i = i_{\max}. \end{cases} \quad (4)$$

The value U is then compared to a critical value U_C at a given level of significance. If $U > U_C$ then the angle distribution is significantly different from uniform, otherwise for $U \leq U_C$ the angle distribution is uniform. Due to the complex computations involving the evaluation of the probability density function of U, the U_C values are taken directly from published statistical tables.

3.3 Geometric Properties of Coefficient Series Shape (GPCS)

The shape of a coefficient series was firstly used as a criterion for the identification of correlated errors by Piretzidis et al. (2018). Their approach uses the rationale of Swenson and Wahr (2006), who noticed that correlated errors depend on the degree parity (i.e., even/odd) of a coefficient. This behavior, in the majority of cases, leads to an order-wise coefficient series with a distinct 'zigzagged' pattern that can be used to identify presence of correlated errors. For that purpose, the original coefficient series $\Delta C_{:,m}$ is separated into an even-order part $\Delta C_{:,m_e}$ and an odd-order part $\Delta C_{:,m_o}$. Piretzidis et al. (2018) examined several geometric features, denoted by f, such as: the sign change, the derivative sign change, the length, the convex hull area and the interior angles distribution. They also developed geometric indicators X_f, given by:

$$X_f(m) = \frac{f(\Delta C_{:,m_e}) + f(\Delta C_{:,m_o})}{f(\Delta C_{:,m})}. \quad (5)$$

For example, if the geometric property under study is the length, then $f(\Delta C_{:,m_e})$, $f(\Delta C_{:,m_o})$ and $f(\Delta C_{:,m})$ represent the total length of $\Delta C_{:,m_e}$, $\Delta C_{:,m_o}$ and $\Delta C_{:,m}$, respectively. Finally, a machine learning algorithm (MLA) was trained and used for predicting if a coefficient series is correlated, based on all X_f indicators. In this study, the MLA used is an artificial neural network (ANN) and the procedure followed is the same as in Piretzidis et al. (2018).

4 Results

For all the experiments described in Sects. 4.1 and 4.2, the EDF is implemented using a second-degree polynomial and a window with an order-dependent length w, given by Duan et al. (2009):

$$w(m) = \max\left(30e^{-\frac{m}{10}} + 1, 5\right). \quad (6)$$

For a maximum degree $N = 96$, the EDF implementation is restricted up to order 88 due to the small number of even/odd coefficients for greater orders.

4.1 Identification Results

We use the RMSR, ADPS and GPCS criteria to identify correlated series in both the GRACE and GLDAS coefficient changes. As the magnitude of GRACE correlated errors increases with degree and order, it is expected that the majority of low-order coefficient series will be identified as uncorrelated, and the high-order series as correlated. The CSR RL06 reprocessing strategy uses refined parameterization techniques and updated AOD1B, tide and force models, compared to RL05. These improvements resulted in the reduction of correlated errors in RL06 (Save et al. 2018). It is therefore expected that fewer coefficient series will be identified as correlated in RL06 than in RL05. The GLDAS coefficients do not contain the same type of correlated errors as the GRACE coefficients, i.e., errors of increasing

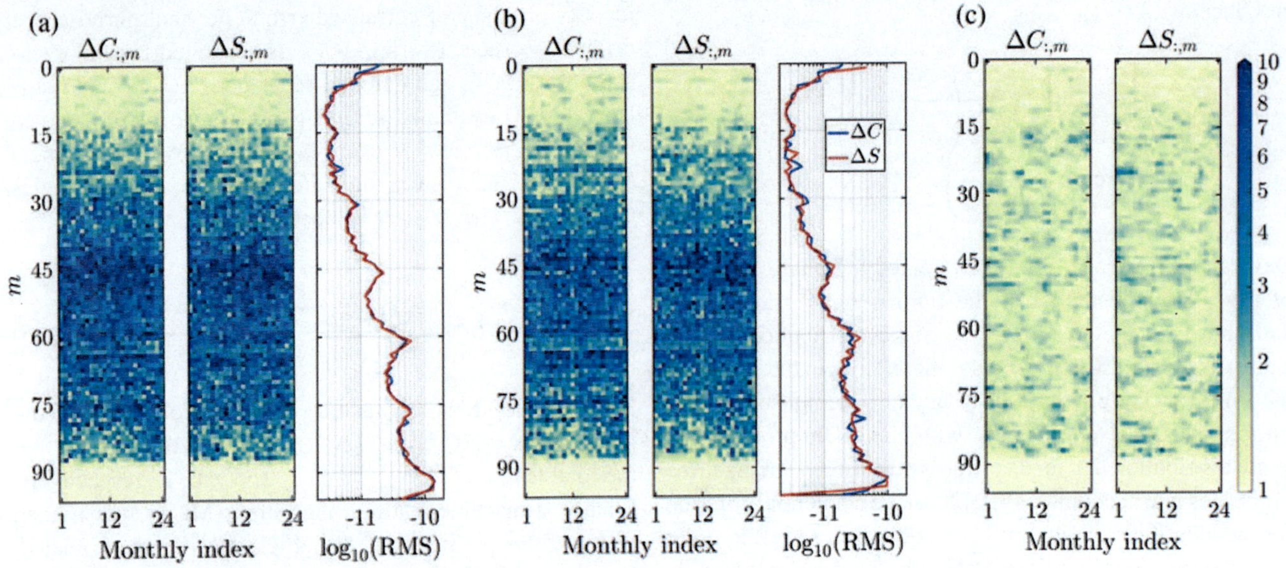

Fig. 1 $R(m)$ evaluated for (**a**) GRACE RL05, (**b**) GRACE RL06 and (**c**) GLDAS data

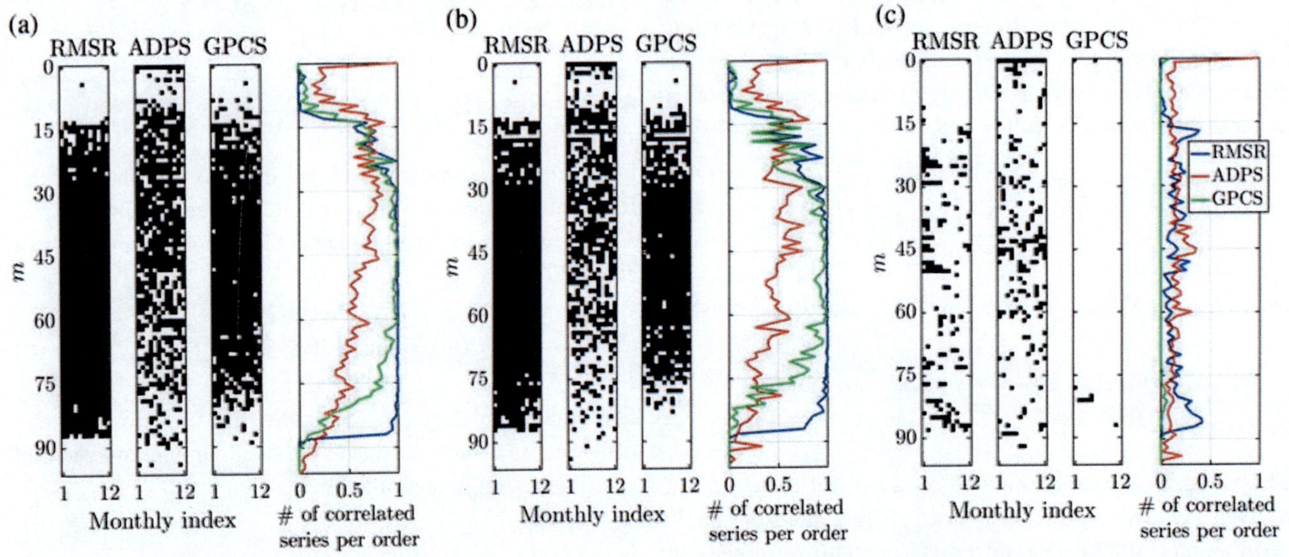

Fig. 2 Correlated orders in (**a**) GRACE RL05, (**b**) GRACE RL06 and (**c**) GLDAS data, identified using the three criteria

magnitude that appear as north-south stripes in the spatial domain, and should be identified as uncorrelated.

Equation (1) is evaluated for both the GRACE and GLDAS data (Fig. 1). The RMSR of GRACE $\Delta C_{:,m}$ and $\Delta S_{:,m}$ coefficient series shows a very similar behavior (Fig. 1a, b). There is also an increase of the RMSR around all orders multiples of 15 (i.e., 15, 30, 45, 60 and 75), forming a band-like pattern. An increase around these orders is also evident in the averaged order-wise RMS given in Fig. 1 for the GRACE data. This increase is related to temporal aliasing errors coming from both tidal and non-tidal geophysical signals. Murböck et al. (2014) showed that these errors appear mainly in orders multiples of the resonance orders, which are close to 15 and 16 for GRACE-like orbits. Seo et al. (2008) showed that higher correlated errors around the same orders are also due to the spatial aliasing coming from non-tidal geophysical model errors. The GRACE RMSR also rapidly increases around order 15 probably for the same reasons. An overall comparison between RL05 and RL06 shows that RL06 coefficient series result in a smaller RMSR due to the reduction of correlated errors. The RMSR of GLDAS coefficients shows an almost random behavior for both $\Delta C_{:,m}$ and $\Delta S_{:,m}$ series (Fig. 1c).

After selecting a critical value of $R_C = 2$, the GRACE and GLDAS identification results are given in the first panel of Fig. 2a, b, c. Orders having an $R(m)$ value greater than R_C are marked with black color and are considered correlated. The identification results using the ADPS criterion and Rao's

Table 1 Number of correlated and uncorrelated orders in GRACE and GLDAS data

	RMSR		ADPS		GPCS	
	GRACE	GLDAS	GRACE	GLDAS	GRACE	GLDAS
Correlated	(10,502) [10,096]	1,756	(7,434) [5,833]	2,095	(9,253) [8,109]	72
Uncorrelated	(2,670) [3,096]	11,416	(5,738) [7,339]	11,077	(3,919) [5,063]	13,100

spacing test are given in the second panel of Fig. 2a, b, c. The angle distributions of order-wise coefficient series that correspond to black pixels are significantly different from uniform. All zero-order series are identified as correlated. This is because all the coefficients in the $\Delta S_{:,0}$ series are by definition equal to zero, resulting in a phase angle of either 0° or 180° (depending on the sign of the corresponding coefficients in the $\Delta C_{:,0}$ series) and a highly non-uniform angle distribution. The results from the GPCS criterion are provided in the third panel of Fig. 2a, b, c. Black pixels denote orders that are classified as correlated by the trained ANN. These results correspond to an ensemble solution coming from 500 independent experiments, in order to account for biases due to the selection of an ANN training dataset. For each dataset, the number of correlated series per order, normalized with respect to the total number of monthly solutions, is provided in the fourth panel of Fig. 2a, b, c for the three criteria. The results of Figs. 1 and 2 for the complete GRACE and GLDAS data set are provided in the supplementary information.

The RMSR criterion shows that the majority of GRACE correlated series start at order 14 and their number gradually increases until order 37. Almost all series from order 38 to 75 appear correlated. A small decrease in the number of correlated series after order 75 is also evident. Differences in the identification results between RL05 and RL06 appear mainly from orders 14 to 37 and from 72 to 87. RL06 have less correlated coefficients, as expected. A substantial number of GLDAS series are also identified as correlated, the majority of them greater than order 16. According to the ADPS criterion, the majority of GRACE correlated series start from order 10 with a maximum value at order 14. The number of correlated series fluctuates significantly for orders 15 to 30, and decreases for higher orders. A reason for this relatively constant decrease is probably the limited number of coefficients in a series, that results in a weak statistical testing. RL06 appears to have a significantly smaller number of correlated coefficients than RL05, starting from order 15. The number of correlated GLDAS series fluctuates at the same level (~12%) for all orders, with an increase around orders 45 and 60. The GPCS criterion results show that most of the GRACE correlated series start at order 14 and their number increases until order 30. The unrealistic decreases of correlated series for orders higher than 60 is due to the small number of coefficients in high-order series that leads to low-quality geometric features (Piretzidis et al. 2018). A decrease in the number of correlated series from RL05 to RL06 is evident for almost all orders. The vast majority of GLDAS series are identified as uncorrelated with some exceptions for orders less than 11 and greater than 60. The results of Fig. 2 are also summarized in Table 1 for an overall quantitative comparison. Values inside round brackets denote RL05 and inside square brackets RL06.

Overall, the RMSR and GPCS criteria show similar behavior for GRACE data, especially up to order 60. The ADPS criterion identifies more series as correlated for orders up to 14 and less afterwards, compared to the GPCS and RMSR criteria. Table 1 also corroborates that the RMSR and GPCS results are closer to each other for GRACE data, with the former identifying the most number of correlated series and showing less differences between RL05 and RL06. On the contrary, the ADPS criterion identifies the least number of correlated series and results in the highest differences between RL05 and RL06. The examination of the GLDAS identification results shows that the GPCS criterion identifies by far the least number of correlated series. The RMSR and ADPS results are close, with the latter to identify slightly more GLDAS series as correlated. Although the GLDAS coefficients do not contain GRACE-type correlated errors, they contain errors from the global spherical harmonic analysis of equiangular data that do not preserve the orthogonality of Legendre functions towards the latitude direction (Sneeuw 1994). The reason that the GPCS criterion does not recognize this type of errors is due to the use of GRACE samples for the ANN training.

4.2 Simulation Results

We test the three criteria in a simulation environment to quantitatively assess their performance. The simulation is designed as follows. First, we use the GLDAS-derived gravity change coefficients, described in Sect. 2, and we assume it represents the true geophysical signal. We then add correlated errors to the true signal in order to produce a GRACE-like noisy geophysical signal. The correlated errors are simulated using the methodology of Vishwakarma et al. (2016), i.e., by filtering the GRACE coefficient changes with the EDF and a 400 km Gaussian smoothing filter, and subtracting the filtered coefficient changes from the original ones. The three criteria are then used to identify correlated orders in the noisy coefficient changes (some examples are provided in

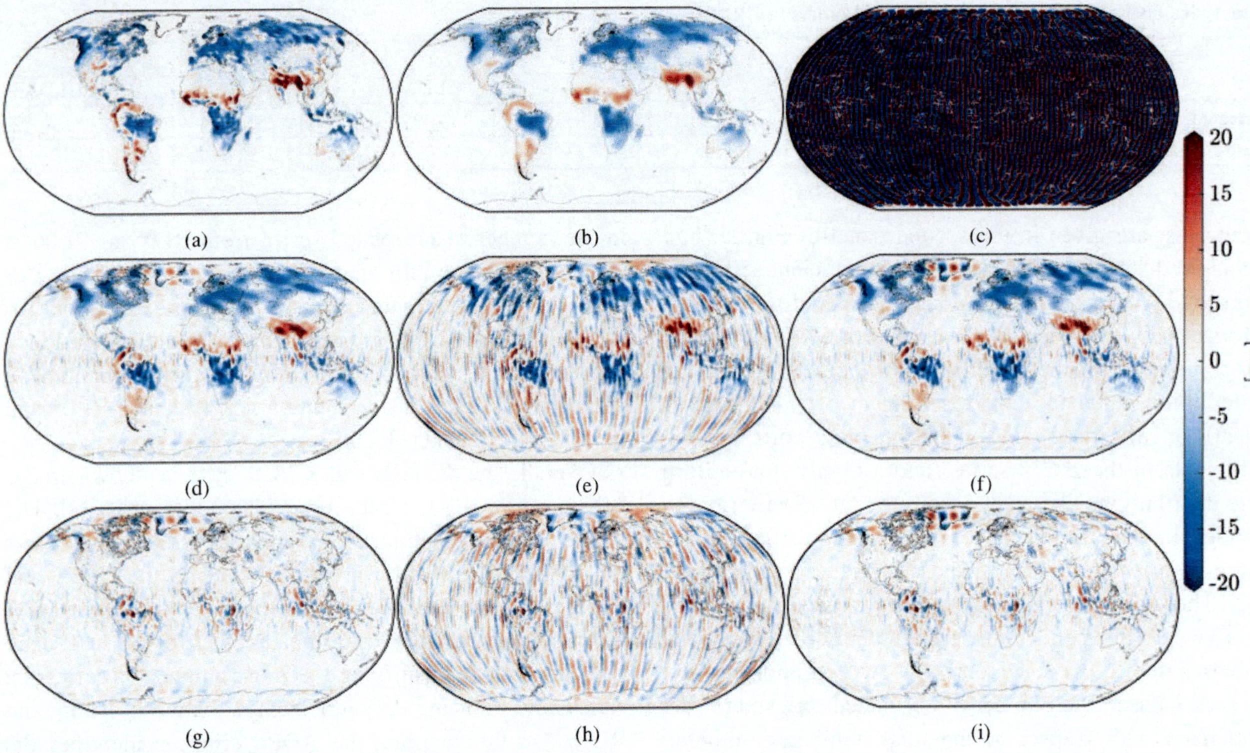

Fig. 3 (**a**) True, (**b**) filtered true and (**c**) noisy geophysical signal. (**d**) Selectively decorrelated and smoothed signal using RMSR, (**e**) ADPS and (**f**) GPCS. (**g**), (**h**), (**i**) Differences between filtered true signal and (**d**), (**e**), (**f**), respectively

the supplementary information), and the EDF is selectively applied to the orders identified as correlated. An additional 400 km Gaussian filter is also used to remove remaining errors. The results from the decorrelated coefficients in the spatial domain, in terms of equivalent water height (EWH), are provided in Fig. 3 for 1 month, where a 250 km Gaussian filter is applied to emphasize the remaining correlated errors and the differences amongst the three criteria. The examination of more monthly EWH changes shows that the RMSR and GPCS criteria remove most of the stripes and result in decorrelated spatial fields closer to each other, with the former one providing the least noisy fields. The ADPS criterion results in decorrelated fields that are still heavily influenced by stripes.

The comparison of the selectively decorrelated and smoothed signal with the true signal is performed in terms of basin averages of EWH. Fifteen basins are used, located in different geographic regions and having an area ranging from 4,672,876 km^2 (Amazon basin) to 521,829 km^2 (Brahmaputra basin). The decorrelated and smoothed basin averages are compared with the true basin averages, also filtered with a 400 km Gaussian filter. The decorrelated and smoothed basin averages are then corrected for filter-induced leakage effects using the data-driven method of deviations (Vishwakarma et al. 2017), and the leakage-repaired basin averages are compared with the true basin averages. For both cases, the RMS errors are provided in Table 2. The underlined values in Table 2 denote the smallest RMS error for each basin. The detailed time series of absolute differences between the true and the selectively decorrelated EWH signal are given in the supplementary information. In the vast majority of cases where filtered fields are compared, the RMSR criterion results in time series with smaller RMS error, with the GPCS criterion having a slightly higher RMS error and the ADPS criterion having the highest RMS error. When the leakage-repaired time series are compared, the RMSR and ADPS criteria again produce the smallest and highest RMS error, respectively, for the majority of cases. An increase in the RMS error is also noticed for both experiments as the basin area decreases.

5 Conclusions

We test three criteria (RMSR, ADPS and GPCS) for the identification of correlated order-wise coefficient series in GRACE monthly coefficient changes that can assist the selective implementation of the EDF. Regarding the computational complexity, the ADPS is the most efficient criterion that does not require the prior use of the EDF for its evaluation. The GPCS is by far the most complex and computationally demanding criterion, as it requires the evaluation of all geometric properties, the selection of a training dataset, and the design and training of a MLA. The RMSR criterion

Table 2 RMS error (in cm) of filtered and leakage-repaired EWH basin averages

	Filtered			Leakage-repaired		
Basin	RMSR	ADPS	GPCS	RMSR	ADPS	GPCS
Amazon	0.22	0.25	0.23	0.38	0.42	0.38
Congo	0.21	0.32	0.25	0.40	0.49	0.41
Mississippi	0.09	0.24	0.12	0.29	0.36	0.29
Ob	0.12	0.36	0.14	0.27	0.43	0.28
Yangtze	0.22	0.32	0.24	0.44	0.48	0.43
Mackenzie	0.20	0.49	0.23	0.49	0.71	0.56
Nelson	0.19	0.33	0.20	0.57	0.66	0.54
Indus	0.47	0.57	0.49	0.98	0.82	0.90
Zambezi	0.52	0.47	0.48	0.77	0.68	0.73
St. Lawrence	0.36	0.49	0.38	0.97	0.90	0.99
Ganges	0.74	0.93	0.82	1.32	1.64	1.47
Orange	0.53	0.67	0.52	0.90	0.94	0.85
Danube	0.37	0.52	0.36	0.80	1.08	0.87
Columbia	0.50	0.71	0.53	1.05	1.44	1.15
Brahmaputra	0.92	1.10	1.00	1.68	2.03	1.87

The basins are sorted in descending order with respect to their area

is simple and does not have any inherent biases. The only drawback could potentially be the selection of a wrong value for R_C and the EDF parameters.

The identification of correlated orders in GRACE data shows that the RMSR and GPCS criteria provide results in close agreement with each other for orders up to 60, and with the latter to identify less correlated series for orders greater than 60. Given that correlated errors in the GRACE coefficients increase with degree and order, the RMSR criterion shows the most realistic behavior. The ADPS criterion seems to overestimate the number of correlated series in low orders and underestimate the number of correlated series in medium and high orders. All criteria identify fewer correlated orders in GRACE RL06, indicating the reduction of correlated errors and the improvement on the quality of harmonic coefficients since RL05.

The results of the simulation study show that the RMSR criterion provides time series of EWH averages with the smallest RMS error for most of the basins examined. The GPCS criterion results are again in close agreement with the RMSR criterion, with a slightly higher RMS error. The reduced performance of GPCS is probably due to the unsuccessful identification of correlated series for orders greater than 60. The ADPS criterion results in the highest RMS error.

We conclude that the ADPS, despite having superior computational advantages, is not a reliable criterion. A stronger statistical test for uniformity would make the ADPS criterion the most efficient choice. The examination of different statistical tests for this task is still an open area of research. For the identification of correlated orders in GRACE data prior to the EDF implementation, we recommend the RMSR criterion that strikes a good balance between computational efficiency and reliability. If complexity is not an issue and the GRACE analysis is restricted up to degree and order 60, the GPCS criterion is also recommended.

Acknowledgements The two anonymous reviewers are thanked for their valuable comments and suggestions. NASA's PODAAC and GES DISC services are thanked for making freely available the GRACE Level 2 data and the GLDAS-1 Noah models. This work is financially aided by a grant from Canada's Natural Sciences and Engineering Research Council (NSERC) to the second author.

References

Berens P (2009) CircStat: A MATLAB Toolbox for Circular Statistics. J Stat Softw 31(10). https://doi.org/10.18637/jss.v031.i10

Devaraju B, Sneeuw N (2017) The polar form of the spherical harmonic spectrum: implications for filtering GRACE data. J Geod 91(12):1475–1489. https://doi.org/10.1007/s00190-017-1037-7

Duan XJ, Guo JY, Shum CK, van der Wal W (2009) On the postprocessing removal of correlated errors in GRACE temporal gravity field solutions. J. Geod 83(11):1095–1106. https://doi.org/10.1007/s00190-009-0327-0

Horvath A, Murböck M, Pail R, Horwath M (2018) Decorrelation of GRACE time variable gravity field solutions using full covariance information. Geosciences 8(9):323. https://doi.org/10.3390/geosciences8090323

Huang J, Halpenny J, van der Wal W, Klatt C, James TS, Rivera A (2012) Detectability of groundwater storage change within the Great Lakes Water Basin using GRACE. J Geophys Res Solid Earth 117(B8):B08401. https://doi.org/10.1029/2011JB008876

Huang J, Pavlic G, Rivera A, Palombi D, Smerdon B (2016) Mapping groundwater storage variations with GRACE: a case study in Alberta, Canada. Hydrogeol J 24(7):1663–1680. https://doi.org/10.1007/s10040-016-1412-0

Klees R, Revtova EA, Gunter BC, Ditmar P, Oudman E, Winsemius HC, Savenije HHG (2008) The design of an optimal filter for monthly GRACE gravity models. Geophys J Int 175(2):417–432. https://doi.org/10.1111/j.1365-246X.2008.03922.x

Kusche J (2007) Approximate decorrelation and non-isotropic smoothing of time-variable GRACEtype gravity field models. J Geod 81(11):733–749. https://doi.org/10.1007/s00190-007-0143-3

Kusche J, Schmidt R, Petrovic S, Rietbroek R (2009) Decorrelated GRACE time-variable gravity solutions by GFZ, and their validation using a hydrological model. J Geod 83(10):903–913. https://doi.org/10.1007/s00190-009-0308-3

Murböck M, Pail R, Daras I, Gruber T (2014) Optimal orbits for temporal gravity recovery regarding temporal aliasing. J Geod 88(2):113–126. https://doi.org/10.1007/s00190-013-0671-y

Piretzidis D, Sra G, Karantaidis G, Sideris MG, Kabirzadeh H (2018) Identifying presence of correlated errors using machine learning algorithms for the selective de-correlation of GRACE harmonic coefficients. Geophys J Int 215(1):375–388. https://doi.org/10.1093/gji/ggy272

Ries J, Bettadpur S, Eanes R, Kang Z, Ko U, McCullough C, Nagel P, Pie N, Poole S, Richter T, Save H, Tapley B (2016) The Combined Gravity Model GGM05c. https://doi.org/10.5880/icgem.2016.002, type: dataset

Rodell M, Houser PR, Jambor U, Gottschalck J, Mitchell K, Meng CJ, Arsenault K, Cosgrove B, Radakovich J, Bosilovich M, Entin* JK, Walker JP, Lohmann D, Toll D (2004) The global land data assimilation system. Bull Am Meteorol Soc 85(3):381–394. https://doi.org/10.1175/BAMS-85-3-381

Russell GS, Levitin DJ (1995) An expanded table of probability values for rao's spacing test. Commun Stat Simul Comput 24(4):879–888. https://doi.org/10.1080/03610919508813281

Save H, Tapley B, Bettadpur S (2018) GRACE RL06 reprocessing and results from CSR. In: EGU General Assembly, Vienna, 8–13 Apr 2018

Seo KW, Wilson CR, Chen J, Waliser DE (2008) GRACE's spatial aliasing error. Geophys J Int 172(1):41–48. https://doi.org/10.1111/j.1365-246X.2007.03611.x

Sneeuw N (1994) Global spherical harmonic analysis by least-squares and numerical quadrature methods in historical perspective. Geophys J Int 118(3):707–716. https://doi.org/10.1111/j.1365-246X.1994.tb03995.x

Swenson S, Wahr J (2006) Post-processing removal of correlated errors in GRACE data. Geophys Res Lett 33(8):L08402. https://doi.org/10.1029/2005GL025285

Tapley BD, Bettadpur S, Ries JC, Thompson PF, Watkins MM (2004) GRACE measurements of mass variability in the Earth system. Science 305(5683):503–505. https://doi.org/10.1126/science.1099192

Vishwakarma BD, Devaraju B, Sneeuw N (2016) Minimizing the effects of filtering on catchment scale GRACE solutions. Water Resour Res 52(8):5868–5890. https://doi.org/10.1002/2016WR018960

Vishwakarma BD, Horwath M, Devaraju B, Groh A, Sneeuw N (2017) A data-driven approach for repairing the hydrological catchment signal damage due to filtering of GRACE products. Water Resour Res 53(11):9824–9844. https://doi.org/10.1002/2017WR021150

Wahr J, Molenaar M, Bryan F (1998) Time variability of the Earth's gravity field: Hydrological and oceanic effects and their possible detection using GRACE. J Geophys Res Solid Earth 103(B12):30205–30229. https://doi.org/10.1029/98JB02844

Second- and Third-Order Derivatives of the Somigliana-Pizzetti Reference Gravity Field

Sten Claessens

Abstract

The computation of second- and third-order derivatives of the Somigliana-Pizzetti reference gravity field (reference gravity gradients and reference gravity field curvature values) is investigated. Closed expressions for these second- and third-order derivatives are derived in spheroidal coordinates. Rigorous equations for the second-order derivatives in a local north-oriented frame are also given. It is shown that on the surface of the reference ellipsoid, these lengthy expressions can be reduced to simple elegant formulas, akin to Somigliana's formula for the first-order derivative. Numerical results provide insight into the curvature of the reference plumb lines and spheropotential surfaces. It is shown that spheropotential surfaces up to 10,000 m in altitude differ from an oblate ellipsoid of revolution by less than 0.04 m. It is also shown that this fact can be utilised to approximate the reference gravity gradients through simple formulas.

Keywords

Gravity field curvature · Gravity gradients · Reference potential

1 Introduction

Terrestrial, airborne and space-borne observation of gravity gradients has been commonplace for many years (e.g. Völgyesi 2015, DiFrancesco et al. 2009, Rummel et al. 2011). The direct observation of rate of change of gravity gradients (gravity field curvature) has become an area of active research in recent years (e.g. Rosi et al. 2015).

Gravity gradients are the second-order derivatives of the gravity potential, and the rates of change of gravity gradients are the third-order derivatives of the gravity potential. Since in geodesy the gravity potential is customarily separated into a reference potential and a disturbing potential, the computation of the second- and third-order derivatives of the reference potential is of interest. These reference potential derivatives can be computed through the spherical harmonic expansion of the reference potential (e.g. Petrovskaya and Vershkov 2010, Hamáčková et al. 2016). Manoussakis (2013) presents an exact method to compute the second-order derivatives of the reference potential on the surface of the reference ellipsoid, and approximately in the vicinity of the reference ellipsoid through a linear approximation.

In this paper, closed expressions for the second- and third-order derivatives of the Somigliana-Pizzetti reference gravity field are derived in spheroidal coordinates. Rigorous, closed expressions for the second-order derivatives in a local north-oriented reference frame are also provided. It is shown that on the surface of the reference ellipsoid, the lengthy expressions are reduced to simple elegant formulas. Numerical results show the approximation error in these formulas when applied at altitude, and provide insight into the curvature of the reference plumb lines and spheropotential surfaces.

S. Claessens (✉)
The Institute for Geoscience Research, School of Earth and Planetary Sciences, Curtin University, Perth, Australia
e-mail: s.claessens@curtin.edu.au

P. Novák et al. (eds.), *IX Hotine-Marussi Symposium on Mathematical Geodesy*,
International Association of Geodesy Symposia 151, https://doi.org/10.1007/1345_2019_70

2 Derivatives in Spheroidal Coordinates

The Somigliana-Pizzetti reference potential U at any point in space can be expressed in spheroidal (Jacobi ellipsoidal) coordinates (u, β, λ) (Heiskanen and Moritz 1967; u is the semi-minor axis of an oblate spheroid with linear eccentricity E through the computation point, β is the reduced latitude with respect to this spheroid, and λ is the geocentric longitude)

$$U(u,\beta) = \frac{GM}{E}\operatorname{atan}\frac{E}{u} + \frac{1}{2}\omega^2 a^2 \frac{q}{q^0}\left(\sin^2\beta - \frac{1}{3}\right) + \tfrac{1}{2}\omega^2\left(u^2+E^2\right)\cos^2\beta \qquad (1)$$

where a and E are the semi-major axis and the linear eccentricity of the reference ellipsoid, respectively, GM is the Earth's geocentric gravitational constant, and ω is its angular velocity. The parameter q is closely related to the second-degree Legendre function of the second kind and q^0 is the value of q on the surface of the reference ellipsoid ($u = b$), where b is the semi-minor axis of the reference ellipsoid

$$q = \frac{1}{2}\left[\left(1 + 3\frac{u^2}{E^2}\right)\operatorname{atan}\frac{E}{u} - 3\frac{u}{E}\right] \qquad (2)$$

$$q^0 = \frac{1}{2}\left[\left(1 + 3\frac{b^2}{E^2}\right)\operatorname{atan}\frac{E}{b} - 3\frac{b}{E}\right] \qquad (3)$$

First-, second- and third-order derivatives of the reference potential with respect to the spheroidal coordinates can be obtained by single, double and triple differentiation of Eq. (1). The first-order derivatives are:

$$U_u = \frac{\partial U}{\partial u} = -\frac{GM}{u^2+E^2} + \frac{\omega^2 a^2 q_u}{2q^0}\left(\sin^2\beta - \frac{1}{3}\right) + \omega^2 u\cos^2\beta \qquad (4)$$

$$U_\beta = \frac{\partial U}{\partial \beta} = \left(a^2\frac{q}{q^0} - u^2 - E^2\right)\omega^2 \sin\beta\cos\beta \qquad (5)$$

The second-order derivatives are:

$$U_{uu} = \frac{\partial^2 U}{\partial u^2} = \frac{2GMu}{\left(u^2+E^2\right)^2} + \frac{\omega^2 a^2 q_{uu}}{2q^0}\left(\sin^2\beta - \frac{1}{3}\right) + \omega^2\cos^2\beta \qquad (6)$$

$$U_{\beta\beta} = \frac{\partial^2 U}{\partial \beta^2} = \left(a^2\frac{q}{q^0} - u^2 - E^2\right)\omega^2\left(\cos^2\beta - \sin^2\beta\right) \qquad (7)$$

$$U_{u\beta} = \frac{\partial^2 U}{\partial u\partial\beta} = \frac{\partial^2 U}{\partial\beta\partial u} = \left(a^2\frac{q_u}{q^0} - 2u\right)\omega^2\sin\beta\cos\beta \qquad (8)$$

And the third-order derivatives are:

$$U_{uuu} = \frac{\partial^3 U}{\partial u^3} = -\frac{2GM\left(3u^2 - E^2\right)}{\left(u^2+E^2\right)^3} + \frac{\omega^2 a^2 q_{uuu}}{2q^0}\left(\sin^2\beta - \frac{1}{3}\right) \qquad (9)$$

$$U_{uu\beta} = \frac{\partial^3 U}{\partial u^2\partial\beta} = \left(a^2\frac{q_{uu}}{q^0} - 2\right)\omega^2\sin\beta\cos\beta \qquad (10)$$

$$U_{u\beta\beta} = \frac{\partial^3 U}{\partial u\partial\beta^2} = \left(a^2\frac{q_u}{q^0} - 2u\right)\omega^2\left(\cos^2\beta - \sin^2\beta\right) \qquad (11)$$

$$U_{\beta\beta\beta} = \frac{\partial^3 U}{\partial\beta^3} = -4\left(a^2\frac{q}{q^0} - u^2 - E^2\right)\omega^2\sin\beta\cos\beta \qquad (12)$$

In Eqs. (4)–(12), q_u, q_{uu} and q_{uuu} are the first-, second- and third-order derivatives of q (Eq. (2))

$$q_u = \frac{dq}{du} = \frac{E}{u^2+E^2} - \frac{3}{E}\left(1 - \frac{u}{E}\operatorname{atan}\frac{E}{u}\right) \qquad (13)$$

$$q_{uu} = \frac{d^2q}{du^2} = -\frac{u\left(3u^2+5E^2\right)}{E\left(u^2+E^2\right)^2} + \frac{3}{E^2}\operatorname{atan}\frac{E}{u} = \frac{2}{u^2+E^2}\left(3q - uq_u\right) \qquad (14)$$

$$q_{uuu} = \frac{d^3q}{du^3} = -\frac{8E^3}{\left(u^2+E^2\right)^3} \qquad (15)$$

Note that the reference potential is rotationally symmetric around the semi-minor axis, and therefore all derivatives with respect to longitude λ are zero.

3 Reference Gravity Gradient Tensor in Local Cartesian Coordinates

Gravity gradients are often represented in a local Cartesian north-oriented reference frame, with the x-axis pointing north, the y-axis pointing west, and the z-axis pointing up. We here define the direction of the z-axis more precisely as perpendicular to the coordinate surface u = constant. The second-order derivatives in the local reference frame are most easily obtained from the first- and second-order derivatives in spheroidal coordinates (Eqs. (4)–(8)) through relations by Koop (1993, p. 31) (see also Vershkov and Petrovskaya 2016, Eq. 10)

$$U_{xx} = \frac{1}{u^2+E^2\sin^2\beta} \times \left[U_{\beta\beta} + \frac{u\left(u^2+E^2\right)}{u^2+E^2\sin^2\beta}U_u - \frac{E^2\sin\beta\cos\beta}{u^2+E^2\sin^2\beta}U_\beta\right] \qquad (16)$$

$$U_{yy} = \frac{1}{u^2 + E^2\sin^2\beta}\left[uU_u - \tan\beta\, U_\beta\right] \tag{17}$$

$$U_{zz} = \frac{1}{u^2 + E^2\sin^2\beta} \times \left[\left(u^2 + E^2\right)U_{uu} - \frac{uE^2\cos^2\beta}{u^2 + E^2\sin^2\beta}U_u + \frac{E^2\sin\beta\cos\beta}{u^2 + E^2\sin^2\beta}U_\beta\right] \tag{18}$$

$$U_{xz} = \frac{\sqrt{u^2 + E^2}}{u^2 + E^2\sin^2\beta} \times \left[U_{u\beta} - \frac{u}{u^2 + E^2\sin^2\beta}U_\beta - \frac{E^2\sin\beta\cos\beta}{u^2 + E^2\sin^2\beta}U_u\right] \tag{19}$$

These four second-order derivatives are the elements of the Eötvös/Marussi tensor of the reference gravity field

$$M = \mathrm{grad}(\mathrm{grad}\, U) = \begin{pmatrix} U_{xx} & 0 & U_{xz} \\ 0 & U_{yy} & 0 \\ U_{xz} & 0 & U_{zz} \end{pmatrix} \tag{20}$$

For any point on the surface of the reference ellipsoid ($u = b$), Eqs. (16)–(19) can be simplified considerably. Note that U_β and $U_{\beta\beta}$ are equal to zero in this case, and the diagonal elements of the Eötvös/Marussi tensor become

$$U^0_{xx} = \frac{a^2 b}{\left(b^2 + E^2\sin^2\beta\right)^2}U^0_u = -\frac{\gamma^0}{M} \tag{21}$$

$$U^0_{yy} = \frac{b}{b^2 + E^2\sin^2\beta}U^0_u = -\frac{\gamma^0}{N} \tag{22}$$

$$\begin{aligned} U^0_{zz} &= \frac{1}{b^2 + E^2\sin^2\beta}\left[a^2 U^0_{uu} - \frac{bE^2\cos^2\beta}{b^2 + E^2\sin^2\beta}U^0_u\right] \\ &= 2\omega^2 - U^0_{xx} - U^0_{yy} \\ &= \gamma^0\left(\frac{1}{M} + \frac{1}{N}\right) + 2\omega^2 \end{aligned} \tag{23}$$

where M and N are the principal radii of curvature in the prime meridian and prime vertical direction, respectively, and γ^0 is the magnitude of reference gravity at the reference ellipsoid, which can be found from Somigliana's formula (Heiskanen and Moritz 1967, Eqs. 2–78). Equations (21)–(23) could alternatively have been derived from geometrical considerations, given that the surface of the reference ellipsoid is a spheropotential surface, i.e. an equipotential surface of the reference gravity field. They can be found as a direct result of application of formulas due to Bruns that relate the second-order derivatives of the gravity potential to the curvature of the level surface (e.g. Heiskanen and Moritz 1967, Sects. 2–3).

The cross-derivative U_{xz} (Eq. (19)) can also be simplified on the surface of the reference ellipsoid

$$\begin{aligned} U^0_{xz} &= \frac{a}{b^2 + E^2\sin^2\beta}\left[U^0_{u\beta} - \frac{E^2\sin\beta\cos\beta}{b^2 + E^2\sin^2\beta}U^0_u\right] \\ &= \frac{ab\sin\beta\cos\beta}{b^2 + E^2\sin^2\beta}\left[2\left(\frac{\gamma_a}{a} - \frac{\gamma_b}{b}\right) + \frac{E^2\gamma^0}{b^2 N}\right] \end{aligned} \tag{24}$$

where γ_a and γ_b are the magnitude of reference gravity at the equator and at the poles, respectively. In the derivation of Eq. (24), use was made of Clairaut's theorem (Heiskanen and Moritz 1967, Eqs. 2–75)

$$\frac{a-b}{a} + \frac{\gamma_b - \gamma_a}{\gamma_a} = \frac{\omega^2 b}{\gamma_a}\left(1 - \frac{e' q^0_u}{2q^0}\right) \tag{25}$$

where e' is the second numerical eccentricity of the reference ellipsoid. Equation (24) can easily be expressed in terms of the more commonly used geodetic coordinates (geodetic latitude ϕ and longitude λ)

$$U^0_{xz} = \sin\phi\cos\phi\left[2\left(\frac{\gamma_a}{a} - \frac{\gamma_b}{b}\right) + \frac{e'^2\gamma^0}{N}\right] \tag{26}$$

The well-known Somigliana's formula for reference gravity gives us the first-order derivative of the reference potential on the surface of the reference ellipsoid, and Eqs. (21)–(23), (26) can be considered its counterpart for the second-order derivatives. Manoussakis (2013) has also derived an expression for U^0_{xz}, but through a completely different derivation based on geometric considerations. In the current notation, his formula is

$$U^0_{xz} = -\left.\frac{\partial\gamma}{\partial\phi}\right|^0 \frac{1}{\frac{\partial N}{\partial\phi}\left(\frac{b^2}{2a^2} - \frac{1}{2}\right)\sin 2\phi + N\left(\frac{b^2}{a^2}\cos^2\phi + \sin^2\phi\right)} \tag{27}$$

This equation requires the derivatives of reference gravity and radius of curvature in the prime vertical with respect to geodetic latitude and is not as easily applicable as Eq. (26).

4 Geometrical Interpretation

The second-order derivatives of the normal potential are closely related to the curvature coefficients of the spheropotential surfaces and reference gravity plumb lines (e.g. Torge 2001). This can be expressed by

$$\begin{pmatrix} U_{xx} & 0 & U_{xz} \\ 0 & U_{yy} & 0 \\ U_{xz} & 0 & U_{zz} \end{pmatrix} \approx \begin{pmatrix} -\gamma k_\beta & 0 & -\gamma k_u \\ 0 & -\gamma k_\lambda & 0 \\ -\gamma k_u & 0 & 2\omega^2 + 2\gamma J \end{pmatrix} \tag{28}$$

where k_β is the curvature of the spheropotential surface in north-south direction, k_λ is the curvature of the spheropotential surface in east-west direction, k_u is the curvature of the reference gravity plumb line, and J is the mean curvature of the spheropotential surface

$$J = \frac{1}{2}\left(\frac{1}{k_\beta} + \frac{1}{k_\lambda}\right) \tag{29}$$

Therefore, second-order derivatives can reveal curvatures of the surfaces and plumb lines, and *vice versa*, known or approximate values of curvature can be used to compute second-order derivatives.

Equations (16)–(19) were here used to investigate by how much the spheropotential surfaces of the GRS80 reference gravity field (Moritz 2000) differ from ellipsoidal surfaces. Knowledge of the exact shape of the spheropotential surfaces is of use in the interpretation of data influenced by the Somigliana-Pizzetti reference field, and can aid the efficient computation of gravity gradients (Sect. 5). While computation of the shape of spheropotential surfaces has long been possible with existing formulas, the results below and in Fig. 1 provide a succinct and instructive view that to the best of our knowledge has not been published before.

Spheropotential surfaces are not ellipsoidal due to the inclusion of the centrifugal potential in the gravity potential. If the angular velocity of the Earth were zero, the spheropotential surfaces would equal the spheroidal coordinate surfaces (u = constant), which are concentric ellipsoids with equal linear eccentricity. However, the centrifugal potential due to the Earth's rotation causes a bulge in the spheropotential surface at the equator of 0.87% of altitude. Furthermore, compared to an ellipsoidal surface fitted through a spheropotential surface at the poles and equator, there is an additional bulge at mid-latitudes of 0.0004% of altitude. Figure 1 shows the example for the spheropotential surface with an altitude of 10,000 m above the poles. The sizes of these bulges are approximately (but not exactly) linear with altitude, at least up to 10,000 m above the reference ellipsoid, and they have therefore been expressed as percentages.

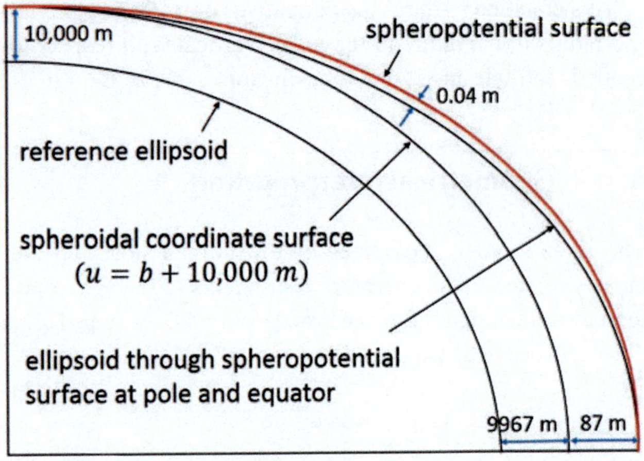

Fig. 1 The geometry of a spheropotential surface

5 Efficient Computation of Reference Gravity Gradients at Low Altitude

While Eqs. (16)–(19) can be used to compute the reference gravity gradients at any point, these gradients can also be computed in an approximate fashion through the simpler formulas (Eqs. (21)–(23), (26)). Just like reference gravity at low altitudes is customarily computed through a Taylor series expansion in terms of height (e.g. Heiskanen and Moritz 1967, Eqs. 2–123), the reference gravity gradients at low altitudes can be computed through a similar approach (Manoussakis 2013).

However, an alternative approximate method of computation is investigated here. Since the spheropotential surface can very closely be approximated by an ellipsoidal surface (see Fig. 1), it may be sufficiently accurate to use the geometry (a, E and derived parameters) of the ellipsoid fitted through the spheropotential surface at the poles and equator in Eqs. (21)–(23), (26). The main advantage of this approach is that the reference gravity gradients are computed using simple formulas in geodetic coordinates, avoiding the need for transformation to spheroidal coordinates. The approximation errors resulting from this approach are shown in Fig. 2 for an altitude of 10,000 m. It can be seen that the errors are largest in the U_{xz}-component, while the diagonal components are accurate to 0.1 mE ($10^{-12}\mathrm{s}^{-2}$). These errors increase approximately linearly with altitude, and they are thus one order of magnitude smaller at an altitude of 1000 m. The approximation errors shown in Fig. 2 are insignificant compared to the precision of common torsion balance gravity gradiometer observations, which is typically above 0.1 E (e.g. Rummel 2002, Hu et al. 2018).

6 Discussion and Conclusions

The second- and third order derivatives of the Somigliana-Pizzetti reference gravity potential in terms of spheroidal coordinates have been derived (Eqs. (6)–(12)). For the second-order derivatives, it is shown how these equations can be used to compute derivatives with respect to a local north-oriented reference frame using existing expressions (Eqs. (16)–(19)). In theory, this can also be done for third-order derivatives. Toth (2005) has derived the relations between third-order derivatives in the Cartesian and spherical coordinate systems, but explicit relations between the

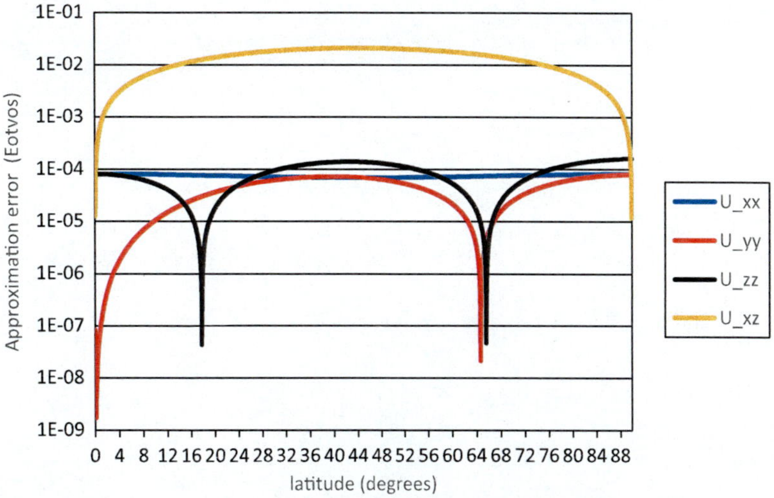

Fig. 2 Approximation error in Eqs. (21)–(23), (26) at 10,000 m altitude when using parameters for an ellipsoid fitted through the spheropotential surface at the poles and equator

Cartesian and spheroidal coordinate systems remain to be derived in future research.

Simple explicit relations for the second-order derivatives of the reference potential on the surface of the reference ellipsoid have been derived (Eqs. (21)–(23), (26)). As the well-known Somigliana's formula provides a simple equation for the first-order derivative of the reference potential (the magnitude of reference gravity), the equations derived here can be considered the equivalent for second-order derivatives (the magnitudes of the reference gravity gradients).

The derived equations were used to show the geometry of the spheropotential surfaces. The spheropotential surfaces up to 10,000 m in altitude differ from an oblate ellipsoid of revolution by less than 0.04 m. It has also been shown that this fact can be utilised to conveniently compute the gradients in an approximate way.

References

DiFrancesco D, Grierson A, Kaputa D, Meyer T (2009) Gravity gradiometer systems – advances and challenges. Geophys Prospect 57(4):615–623. https://doi.org/10.1111/j.1365-2478.2008.00764.x

Hamáčková E, Šprlak M, Pitoňak M, Novák P (2016) Non-singular expressions for the spherical harmonic synthesis of gravitational curvatures in a local north-oriented reference frame. Comput Geosci 88:152–162. https://doi.org/10.1016/j.cageo.2015.12.011

Heiskanen WA, Moritz H (1967) Physical geodesy. WH Freeman, San Francisco

Hu X, Yu Y, Liu J, Jiang R (2018) A torsion balance device for measuring the gravity gradient. J Phys Conf Ser 1065(04):2058. https://doi.org/10.1088/1742-6596/1065/4/042058

Koop R (1993) Global gravity modelling using satellite gravity gradiometry. Publications on Geodesy 38, Delft, The Netherlands

Manoussakis G (2013) Estimation of the normal Eötvös matrix for low geometric heights. Acta Geodaetica et Geophysica 48:179–189. https://doi.org/10.1007/s40328-013-0014-8

Moritz H (2000) Geodetic reference system 1980. J Geod 74(1):128–162. https://doi.org/10.1007/S001900050278

Petrovskaya MS, Vershkov AN (2010) Construction of spherical harmonic series for the potential derivatives of arbitrary orders in the geocentric Earth-fixed reference frame. J Geod 84:165–178. https://doi.org/10.1007/s00190-009-0353-y

Rosi G, Cacciapuoti L, Sorrentino F, Menchetti M, Prevedelli M, Tino GM (2015) Measurement of the gravity-field curvature by atom interferometry. Phys Rev Lett 114:013001. https://doi.org/10.1103/PhysRevLett.114.013001

Rummel R (2002) Gravity gradiometry: From Loránd Eötvös to modern space age. Acta Geodaetica et Geophysica Hungarica 37:435–444. https://doi.org/10.1556/AGeod.37.2002.4.7

Rummel R, Yi W, Stummer C (2011) GOCE gravitational gradiometry. J Geod 85:777–790. https://doi.org/10.1007/s00190-011-0500-0

Torge W (2001) Geodesy, 3rd edn. De Gruyter, Berlin

Toth G (2005) The gradiometric-geodynamic boundary value problem. In: Jekeli C, Bastos L, Fernandes J (eds) Gravity, geoid and space missions, IAG Symposia 129, pp 352–357, https://doi.org/10.1007/3-540-26932-0_61

Vershkov AN, Petrovskaya MS (2016) Nonsingular expansions of the gravity potential and its derivatives at satellite altitudes in the ellipsoidal coordinate system. Sol Syst Res 50(6):437–446. https://doi.org/10.1134/S0038094616060058

Völgyesi L (2015) Renaissance of torsion balance measurements in Hungary. Periodica Polytechnica Civil Eng 59(4):459–464. https://doi.org/10.3311/PPci7990

Check for updates

On the Advantage of Normal Heights

Once More on the Shape of Quasigeoid

Viktor V. Popadyev

In memory of Prof. Lyudmila V. Ogorodova

Abstract

This paper analyzes the arguments in the report "The shape of the quasigeoid" by Robert Kingdon, Petr Vaníček, Marcelo Santos presented in Rome (IX Hotin-Marussi Symposium on Theoretical Geodesy, Italy, Rome, June 18–June 22, 2018), which contains the criticisms of the basic concepts of Molodensky's theory: normal height and height anomaly of the point on the earth's surface, plotted on the reference ellipsoid surface and forming the surface of a quasigeoid. Also are presented the main advantages of the system of normal heights. They are closely related to the theory of determination of the external gravitational field and the Earth's surface, are presented.

Despite the fact that the main core of Molodensky's theory is the rigorous determining of the anomalous potential on the Earth's surface, the advantage of the normal heights system can be shown and proved separately. And this can be easily demonstrated by a simple hypothetical example of the spherical non-rotating Earth where the change of marks along the floor of a strictly horizontal tunnel in the spherical mountain massif serves as criterion for the convenience of the system. In this example, the difference in orthometric heights comes up to 3 cm per 1.5 km. It will require the same corrections to the measured elevations what with the effect of the orthometric heights system. Also the knowledge of the inner structure of the rock mass is necessary. In turn, the normal heights are constant along the tunnel and behave as dynamic ones and there is no need to introduce corrections.

Neither the ellipsoid nor the quasi-geoid is a reference surface for normal heights, because until now the heights are referenced to the initial tide gauge. The numerical values of heights are assigned to the physical surface. This is similar to the ideas of prof. L. V. Ogorodova about the excessive emphasis on the concept of quasigeoid itself. According to prof. V. V. Brovar the more general term is the "height anomaly" that exists both for points on the Earth's surface and at a distance from it and decreases together with an attenuation of the anomalous field.

Keywords

Geoid · Height systems · Modelling method · Normal height · Orthometric height · Quasigeoid

1 Introduction

The main disadvantage of the quasi-geoid boils down to the well-known fact that near the singular points of the earth's surface (conical cusps and faces) the gravitational field also

V. V. Popadyev (✉)
Federal Scientific-Technical Center of Geodesy, Cartography and Spatial Data Infrastructure, Moscow, Russia
e-mail: popadev_vv@nsdi.rosreestr.ru

P. Novák et al. (eds.), *IX Hotine-Marussi Symposium on Mathematical Geodesy*,
International Association of Geodesy Symposia 151, https://doi.org/10.1007/1345_2019_74

has a special feature. Since the normal field is smooth, the surface of the quasigeoid heights $\zeta = \frac{W-U}{\gamma}$ also has the peculiarity. Here we should remember that the solution of the boundary value problems of the Newton potential in all cases requires that the earth's surface should be smoothed out to the Lyapunov's conditions. The physical land surface can be smoothed out over gravimetric points. There are no features on the sea. The determination of the geoid does not become easier.

M. S. Molodensky showed that it is impossible to determine the form of the geoid from measurements taken on the earth's surface. Figure 1 shows the anomalous mass in the form of a sphere of radius $R_1 = 1$ and constant density $\delta_1 = 1$ centered on the geoid. If we replace this sphere by a concentric one which is equal in mass but having the radius $R_2 = 3R_1$, then from the equation

$$M_1 = \frac{4}{3}\pi R_1^3\delta_1 = M_2 = \frac{4}{3}\pi R_2^3\delta_2$$

we determine the new density $\delta_2 = \frac{1}{27}\delta_1$. Since the **external** field of such concentric spheres, equal in mass, is the same, the replacement will not be reflected in measurements on the earth's surface, but the **internal** potential of the sphere at the central point will be different:

$$V_1 = 2\pi G\delta_1 R_1^2 = 2\pi G, \quad V_2 = 2\pi G\delta_2 R_2^2 = \frac{2}{3}\pi G,$$

so that the geoid inside the sphere will change its shape. Thus, it is impossible to study the surface of the geoid from the measurements on the Earth's surface.

The impossibility of determining the geoid and the physical surface of the Earth was also known to F. Helmert:

> ...die Figur der physischen Erdoberfläche wird also richtig erhalten; nur die Lage des Geoids ist mehr oder weniger von einer unvermeidlichen Unsicherheit betroffen (Helmert 1900);
>
> Alle anderen Reductionsweisen haben das Gemeinsame, dass sie im Allgemeinen eine gewisse mehr oder weniger starke Veränderung (Idealisierung) der Massenvertheilung an der Erdoberfläche voraussetzen und deshalb mehr oder weniger fehlerhaft sind (Helmert 1902).

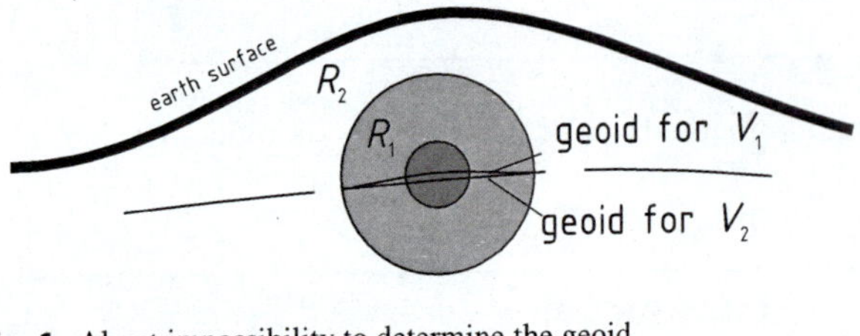

Fig. 1 About impossibility to determine the geoid

2 On the Quasigeoid

The report of Robert Kingdon, Petr Vaníček, Marcelo Santos *The shape of the quasigeoid* specifies the complexity of the quasi-geoid corresponding to the points of the earth's surface that forms "overhangs". The quasigeoid height related to the points on the earth's surface will also form an overhang. The correct arrangement of the quasigeoid is shown on Fig. 2 (see the report). The mean integral value g^m can be measured along the line 1–3 directly or we can use $\frac{\partial \Delta g}{\partial H}$ from the boundary value problem. Then the difference of the orthometric heights $H_3^g - H_1^g$ is just a real length of segment 1–3. This also is the difference of the geodetic height (neglecting the deflection of the plumb-line). Along line 2–4 the value g^m can be calculated by measuring the values of the gravity at the ends of a segment 4–5. So we obtain again the geometric length of the segment.

Shown at the beginning of the report *The shape of the quasigeoid* is the picture from the book of one of the authors (Vaníček and Krakiwsky 1986), where the quasi-geoid on land is located **under** the geoid and has a **smaller** curvature. This can be seen from the simple example in Fig. 3 while in own illustrations R. Kingdon, P. Vaníček and M. Santos place the surface of the quasi-geoid **over** the geoid and give it a **greater** curvature.

The above can be generalized as a more complex case showing how to calculate normal heights in tunnels crossing mountain massifs. Such a discussion was held back in the 1950–1960th when Burša (1959) and Yeremeyev (1965) discussed the system of normal heights in Czechoslovakia. A detailed investigation of the theory of orthometric, normal and dynamic heights is made in the monograph (Yeremeev and Yurkina 1972).

The advantage of the Molodensky theory is that it is possible to determine the anomalous potential T and the height anomaly $\zeta = T/\gamma$ with high accuracy on the earth's

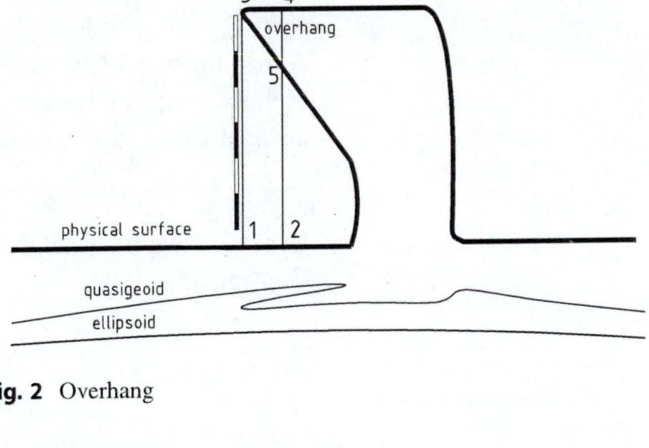

Fig. 2 Overhang

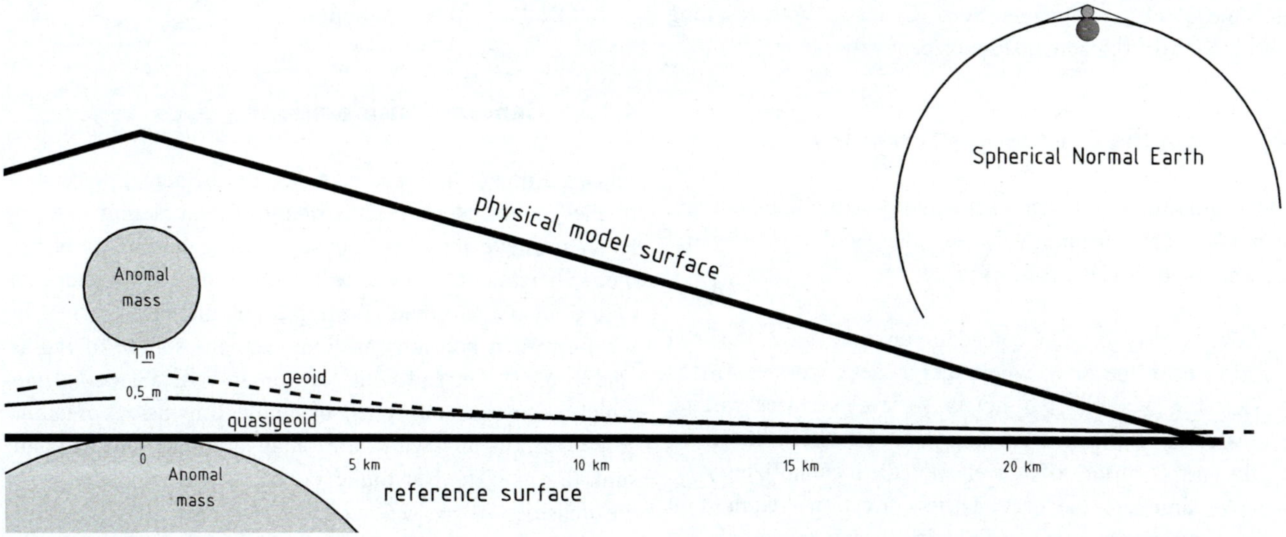

Fig. 3 Geoid and quasigeoid

surface (by the measurements taken on it). But the basic advantage of normal heights can also be proved without a connection with the solution of the geodetic boundary value problem.

3 Advantages of the Normal Heights

In the physical sense, the height position of the point is uniquely characterized by the geopotential number $W_0 - W = \int gdh$, equal to the difference of the gravity potential at the beginning of the height calculation W_0 and at the current point W.

The normal heights were first named by Molodensky as "auxiliary". Their introduction under condition[1] $W(H) - W_0 = U(H^\gamma) - U_0$ allowing him to strictly solve the geodetic boundary value problem which makes it possible to determine the geodetic heights of the points and the anomalous potential T on the earth's surface. Later it was noted that these auxiliary heights form a very convenient and harmonious system.

It is an interesting fact that in his draft of the article on the theory of heights, apparently dating back to 1950, V. F. Yeremeyev still uses the term "auxiliary" heights, and calls as the "normal" heights the approximate heights.[2] Responding to this work the head of the department of higher geodesy prof. V. Danilov recommended in his critical review to replace the "auxiliary" heights with another more appropriate term (as the impression is created "that these heights are secondary, not final"). He also noted:

> The advantages of normal heights are:
> 1. Normal heights can be precisely calculated, they are also uniquely related to the dynamic heights.
> 2. The discrepancies in the leveling polygons are reduced (in the limit of zero), so the adjustment of level networks in geopotential values and in normal heights gives identical results.
> 3. The level surfaces crossing the physical surface of the Earth are more important in practical terms than the geoid, since geodetic measurements have to deal with them but not with the geoid.

At the suggestion of M. S. Molodensky V. F. Eremeev introduced the term "normal height" (Yeremeev 1951):

> In this paper, instead of the term "auxiliary" altitude, the term "normal height" is adopted as being most appropriate to the physical meaning explained below.

Practically, it is difficult to obtain orthometric height from the geopotential number for a number of reasons: to determine the mean integral value g^m along the force line, it is required to know at least the first derivatives of the real gravity force (or the mass density distribution) up to the geoid surface, also unknown. To determine the orthometric height $H^g = 1$ km with an accuracy of 1 cm, it is required to know the average g^m with an accuracy of 10 mGal with the tolerances decreasing as the height grows (Yurkina 1998). In the meantime in the processing of high-precision leveling in the system of normal heights, the requirements will be less harsh to the accuracy of the points coordinates (up to 0.5 km) and gravimetric information (1 mGal). That will allow us to determine even if a small-scale topographic map is used, γ^m with an accuracy of 50 mGal to obtain a difference of normal heights with an accuracy of 0.1 mm. This fact permits us to calculate normal correction, see Eq. (1), even from the global field model. The normal corrections were calculated by D. A. Gulyyev and by O. V. Nefedova from EGM2008 for the leveling line 1,165 km long in Turkmenistan and for

[1]The strict condition is in spheroidal coordinate system.

[2]I.e. the heights determined from relation $\frac{\int \gamma dh}{\gamma^m}$, when there is no gravity map.

the line of 685 km long in North Caucasus with resulting absolute errors 0.6 and 1.0 mm respectively.

4 On the Calculation of Heights

In the question of choosing a heights system, the main thing is not how easy or intuitively clear the geoid or quasi-geoid is, what matters is the following:

- how the heights under consideration characterize the level surfaces at the place where the geodetic work is carried out (for geodesy, it is not those level surfaces passing inside the Earth that are important, but those that cross the earth's surface of a given spot, the local horizons).
- how large are the gravimetric corrections required in case when the measured elevations are converted into the height differences in the adopted system (if these corrections are significant, then they will have to be introduced into the measured elevations even in case of rough leveling at short distances which is one of well-known drawbacks of the dynamic heights system).

Let's take a simple example.

4.1 General Considerations

Since we do not have accurate data on the actual placement of the geoid and on the distribution of attracting masses, let us estimate the convenience of using normal or orthometric heights on a model with previously known elements of the anomalous field. We represent the Earth model in the form of a non-rotating homogeneous sphere of radius $R_0 = 6{,}378$ km and with Newtonian constant GM_0, into which a homogeneous sphere of smaller dimensions of radius $R_1 = 2$ km and with Newtonian constant GM_1 is half sunk (Fig. 4). The Newtonian constants are calculated from the densities $\delta_0 = 5.5$ g/cm^3, $\delta_1 = 2.3$ g/cm^3 and the volumes of spheres. The big sphere creates the potential V_0, the smaller one creates the potential V_1. Suppose now that in the obtained mountain massif it is required to build a strictly horizontal tunnel (so that the minimum fuel consumption may be achieved). So also let's assume that the removal of the soil does not change the internal field. Since all the elements

Fig. 4 Tunnel

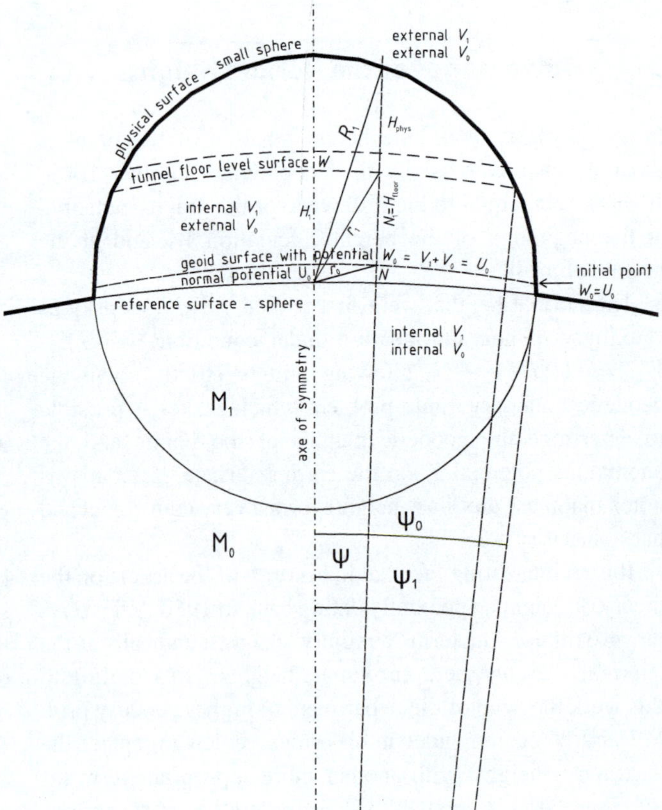

of the field are easily calculated with great precision, we can estimate the difficulties that arises when using normal or orthometric heights, namely **the difference in altitude when entering the tunnel and at other points**. It is clear that it will be more preferable to have a system of heights where the difference in altitude is less.

At once it is necessary to make a reservation that in case of strictly horizontal tunnel the choice of dynamic heights will be the most successful (without any correction). If there is a slope of the tunnel, the question remains open. This question can be solved in advance. The normal gravimetric correction consists of corrections for the nonparallelism of the level surfaces of the normal and real fields[3]:

$$\Delta H^{\gamma} = \sum \Delta h - \frac{1}{\gamma^m} \sum H^m \Delta\gamma_0 + \frac{1}{\gamma^m} \sum \Delta g \Delta h, \quad (1)$$

where H^m is mean height, Δg is gravity anomaly, Δh is elevation, γ_0 is normal gravity on the reference ellipsoid surface. Since there is no transition to another level surface and the normal field is spherical, then the gravimetric correction for the normal heights will be zero. In this particular case, the normal heights give a result equivalent to that of dynamic heights.

4.2 Numerical Experiment

Let us determine the heights of the points of the tunnel. Orthometric heights H^g can be found **from its definition** as the distance between the points on the tunnel floor and the geoid with the potential W_0. The latter is chosen from the value of the potential U_0 on the reference surface:

$$W_0 = U_0 = V_0(R_0) = \frac{GM_0}{R_0},$$

where the symbols for the gravity potential W, U are used traditionally, whereas in this case only the potential of attraction V makes sense (there is no centrifugal component).

At points of the geoid, we have the condition[4]

$$V_1^{in}(r_0) + V_0^{ex}(R_0 + N) = W_0, \quad (2)$$

here r_0 is the distance from the center of the small sphere to the point on the true geoid, N is the geoid height, the potential of the small sphere at the internal point[5] at a distance of r:

$$V_1^{in}(r) = \frac{GM_1}{R_1}\left[\frac{3}{2} - \frac{1}{2}\left(\frac{r}{R_1}\right)^2\right], \quad (3)$$

and where the potential of a large sphere to an external point is

$$V_0^{ex}(R_0 + N) = \frac{GM_0}{R_0 + N}.$$

The distance r_0 depends on ψ (angle between the symmetry axis of the model and the radius vector of the current point) and N:

$$r_0^2 = R_0^2 + (R_0 + N)^2 - 2R_0(R_0 + N)\cos\psi. \quad (4)$$

Substituting r_0 from (4) in (3) for r in (3) and in (2) we'll have

$$\frac{GM_1}{R_1}\left[\frac{3}{2} - \frac{R_0^2 + (R_0 + N)^2 - 2R_0(R_0 + N)\cos\psi}{2R_1^2}\right] + \\ + \frac{GM_0}{R_0 + N} = \frac{GM_0}{R_0}.$$

With $\kappa = M_1/M_0$ and $\beta = R_0 + N$ we have

$$\frac{\kappa}{R_1}\left[\frac{3}{2} - \frac{1}{2R_1^2}(R_0^2 + \beta^2 - 2R_0\beta\cos\psi)\right] + \frac{1}{\beta} = \frac{1}{R_0}.$$

The resulting cubic equation

$$-\frac{\kappa}{2R_1^3}\beta^3 + \frac{\kappa R_0}{R_1^3}\cos\psi \cdot \beta^2 + \left(\frac{3}{2}\frac{\kappa}{R_1} - \frac{\kappa R_0^2}{2R_1^3} - \frac{1}{R_0}\right)\beta + 1 = 0$$

can be solved numerically and we find the geoid height N at points with step $\Delta\psi$ from 0 to ψ_1 where the tunnel ends:

$$\cos\psi_1 \approx \frac{R_0^2 + (R_0 + H_1)^2 - R_1^2}{2R_0(R_0 + H_1)}.$$

At the same points, one can calculate the height of the level surface with the potential W_1 at the geodetic height $H_1 = 1$ km over the center of the small sphere:

$$W_1 = V_1^{in}(H_1) + V_0^{ex}(R_0 + H_1) = \\ = \frac{GM_1}{R_1}\left[\frac{3}{2} - \frac{1}{2}\left(\frac{H_1}{R_1}\right)^2\right] + \frac{GM_0}{R_0 + H_1}.$$

[3]The difference of the formula adopted by Yeremeyev from the corresponding formula (4–64) in the book "Physical geodesy" of Hofmann-Wellenhof B., Moritz H. (Springer, 2006) was noticed in the book review by M. Pick (Stud. Geophys. Geod., 50, 2006, 161–163).

[4]Hereafter the potentials V, U and W with parameters in brackets mean no multiplication.

[5]The adopted form $V^{in}(r) = 2\pi G\delta(R_1^2 - \frac{r^2}{3})$ can be easily expressed as next.

From the condition analogous to (2),

$$V_1^{in}(r_1) + V_0^{ex}(R_0 + N_1) = W_1,$$

one can obtain a cubic equation with respect to $\beta_1 = R_0 + N_1$ and numerically determine N_1, the geodetic height of the tunnel floor.

The normal heights H_1^γ of the floor surface can be calculated proceeding from Yeremeyev's formula (Yeremeev 1951):

$$H_1^\gamma = \frac{W_0 - W}{\gamma^m} = \frac{1}{\gamma^m}\left(\frac{GM_0}{R_0} - W_1\right),$$

where the mean integral value of the normal gravity can be determined directly:

$$\gamma^m = \frac{1}{H_1^\gamma}\int\limits_{\rho=R_0}^{\rho=R_0+H_1^\gamma} \frac{GM}{\rho^2}d\rho = \frac{1}{H_1^\gamma}\left(-\frac{GM}{\rho}\right)\Bigg|_{\rho=R_0}^{\rho=R_0+H_1^\gamma} =$$

$$= \frac{1}{H_1^\gamma}\left(\frac{GM}{R_0} - \frac{GM}{R_0 + H_1^\gamma}\right),$$

where in the right-hand side we can substitute H_1^γ for N_1, the geodetic height of the floor of the tunnel. For greater accuracy γ^m and H_1^γ can be calculated by successive approximations.

Normal height can be found strictly **from its definition** as such a height for which the real and normal geopotential numbers are equal:

$$W_0 - W_1 = U_0 - V_0(R_0 + H_1^\gamma) \tag{5}$$

or

$$\frac{GM_0}{R_0 + H_1^\gamma} = W_1.$$

Hence the normal heights of the points of the tunnel floor must be strictly constant. The quasigeoid height of the tunnel floor is $\zeta_{floor} = N_1 - H^\gamma$.

The orthometric height, by definition, is the height difference between the tunnel floor surfaces and the geoid $H^g = N_1 - N$. The results of calculations of the normal and orthometric height of the tunnel floor are in Table 1, the graphs of these values are shown in Fig. 5. The normal height varies within 9×10^{-6} mm, this change is due to rounding errors, while the orthometric height varies within 2.65 cm.

Table 1 Normal and orthometric heights of the tunnel floor points

ψ	0	5.6″	...	44.8″	50.4″
H^g, m	999,607	999,607	...	999,628	999,633
H^γ, m	999,639	999,639	...	999,639	999,639

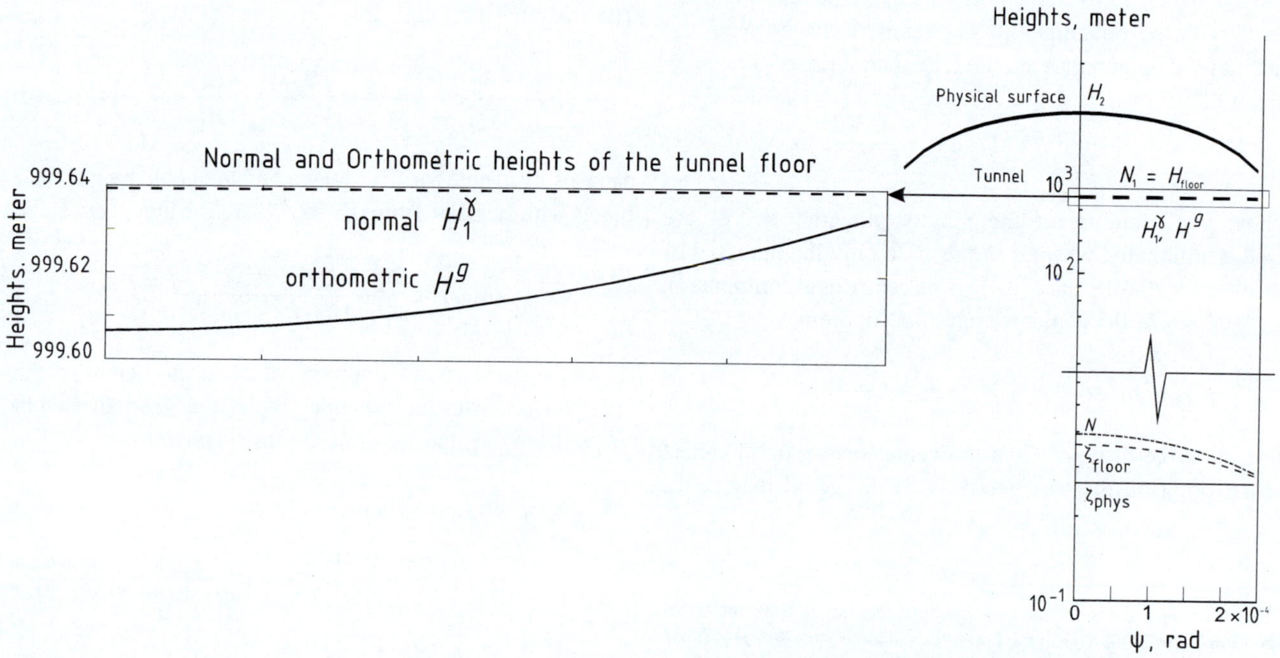

Fig. 5 Change in normal H_1^γ and orthometric H^g heights along the tunnel floor: N is the true geoid height; ζ_{floor} is the quasigeoid height of the tunnel floor; ζ_{phys} is the quasigeoid height of the external mountain surface; $N_1 = H_{floor}$ is the geodetic height of the tunnel floor; H_{phys} is the geodetic height of the external mountain surface

From the condition (5) one can also determine the normal heights H_2^γ and the quasigeoid heights ζ_{phys} for points on the outer surface of the mountain, whose real potential is

$$W_2 = \frac{GM_0}{R_0 + H_2} + \frac{GM_1}{R_1} = \frac{GM_0}{R_0 + H_2^\gamma},$$

where the geodetic height H_2 can be easily found from the cosine theorem (see Fig. 4)

$$H_2 = R_0(\cos\psi - 1) + \sqrt{R_0^2\cos^2\psi - R_0^2 + R_1^2}.$$

From the relation $\zeta_{phys} = \frac{T}{\gamma}$ where $T = \frac{GM_1}{R_1}$, $\gamma = \frac{GM_0}{(R_0+H_2^\gamma)^2}$ given on the outer surface we can control the quasigeoid height ζ_{phys} on the physical surface of the mountain.

5 Conclusions

Normal heights do not have a visual physical meaning, but they have important practical significance. They are strictly related to the solution of the Molodensky's geodetic boundary value problem, from where the anomalous potential is to be determined.

For points on the outer surface of the mountain and in the tunnel, the height anomalies will be different which reflects the decrease of the anomalous field with altitude. Such an idea of the height anomaly is justified at certain points above the earth's surface, for example, in the construction of large bridges (Pick 1970).

A quasigeoid is not a "vertical reference surface", what with the normal heights being counted from the reference point. After the precise leveling data are processed, to each **point** of the earth's surface, the height mark is assigned. The normal height is related to the segment from the ellipsoid to the point with the condition $W(H) - W_0 = U(H^\gamma) - U_0$. Along this segment the mean value γ^m is calculated, but the value of the normal height **refers to the point** of the earth's surface. Similarly the normal-orthometric height H^{no} is related to the segment from the Earth's surface *down* to the point, where the conditions $W(H) - W_0 = U(H) - U(H - H^{no})$. But the height value still refers to the point on the earth's surface. Between all the types of normal height (normal-orthometric, normal and possible intermediate variants) only the classical Molodensky's normal height doesn't require any geodetic height to know.

Acknowledgements I'm grateful to Prof. Elena Peneva and Tatiana Lambeva for their corrections and attention to this paper; I thank Ms. Asya N. Ananyeva for her help in translating.

References

Burša M (1959) Převod československých nivelačních základů do soustavy normálních výšek. Sbor. Výzk. prací. Ed. I VÚGTK, Praha

Helmert FR (1900) Zur Bestimmung kleiner Flächenstücke des Geoids aus Lotabweichhungen mit Rücksicht auf die Lotkrümmung. Sitzungsberichte der Königlich preussischen Akademie der Wissenschaften zu Berlin. N 42, 963–982; AkademieVorträge. 1993, 85–123. Frankfurt am Main. Verlag des Instituts für angewandte Geodäsie. S. 975

Helmert FR (1902) Über die Reduktion der auf der physischen Erdoberfläche beobachteten Schwerebeschleunigungen auf ein gemeinsames Niveau. Sitzungsberichte der Königlich Preussichen Akademie der Wissenschaften zu Berlin, 36, 843–855; Akademie Vorträge. 1993, 140–152. Frankfurt am Main Verlag des Instituts für angewandte Geodäsie. S. 844–845

Pick M (1970) Generalization of the system of normal heights. Studia Geophys Geod 14:156–159. In German: Pick M. Verallgemeinerung des Systems der Normalhöhen. Vermessungstechnik, Berlin, 1970, 11, pp. 446–448

Vaníček P, Krakiwsky EJ (1986) Geodesy. The concepts, 1st edn. Elsevier Science, Amsterdam, 714 p

Yeremeev VF (1951) Teorija ortometričeskih, dinamičeskih i normal'nyh vysot [Theory of orthometric, dynamic and normal heights]. Proceedings of the central scientific and research institute of geodesy and cartography, vol 86. In German: Jeremejew W. F. Theorie der orthometrischen, dynamischen und Normalhöhen. In: Sowjetische Arbeiten aus höherer Geodäsie u. Geophysik: Teil 2. Ausgewählte Übersetzungen fremdsprachlicher Fachliteratur. Geodätisches Institut d. Technischen Hochschule, 1958

Yeremeyev VF (1965) On the problem of determining normal heights. Studia Geophys Geod 9:1–13

Yeremeev VF, Yurkina MI (1972) Teorija vysot v gravitacionnom pole Zemli [Theory of heights in the Earth's gravity field]. Proceedings of the central scientific and research institute of geodesy and cartography, vol 191. In German: Eremeev V. F., Jurkina M. I. Theorie der Höhen im Gravitationsfeld der Erde. Arbeiten aus dem Vermessungs- und Kartenwesen der Deutschen Demokratischen Republik, Geodätischer Dienst, Leipzig, 1974, Band 32, Teil 1, 2, 282 S

Yurkina MI (1998) TsNIIGAiK i teorija figury Zemli [TsNIIGAiK and the Earth's figure theory]. Geodezija i kartografija 9:50–53

Green's Function Method Extended by Successive Approximations and Applied to Earth's Gravity Field Recovery

Petr Holota and Otakar Nesvadba

Abstract

The aim of the paper is to implement the Green's function method for the solution of the Linear Gravimetric Boundary Value Problem. The approach is iterative by nature. A transformation of spatial (ellipsoidal) coordinates is used that offers a possibility for an alternative between the boundary complexity and the complexity of the coefficients of Laplace's partial differential equation governing the solution. The solution domain is carried onto the exterior of an oblate ellipsoid of revolution. Obviously, the structure of Laplace's operator is more complex after the transformation. It was deduced by means of tensor calculus and in a sense it reflects the geometrical nature of the Earth's surface. Nevertheless, the construction of the respective Green's function is simpler for the solution domain transformed. It gives Neumann's function (Green's function of the second kind) for the exterior of an oblate ellipsoid of revolution. In combination with successive approximations it enables to meet also Laplace's partial differential equation expressed in the system of new (i.e. transformed) coordinates.

Keywords

Boundary value problems · Integral kernels · Laplace's operator · Method of successive approximations · Transformation of spatial coordinates

1 Introduction

Green's functions are an important tool in solving problems of mathematical physics. Equally this holds for applications in gravity field studies. The mathematical apparatus of classical physical geodesy is a typical example. Green's function is an integral kernel, which, convolved with input values, gives the solution of the particular problem considered. Regarding its construction, there exist elegant and powerful methods for one or two dimensional problems. However, only very few of these methods carried over to higher dimensions, indeed the higher the dimension of the Euclidean space the simpler the boundary of the region of interest had to be, see Roach (1982). In order to preserve the benefit of the Green's function method a suitable approximation procedure is discussed. The aim of the paper is to implement the procedure with the particular focus on the solution of the linear gravimetric boundary value problem. Two approaches immediately suggest themselves; either to approximate the boundary of the region of interest or approximate the domain functional (partial differential operator). We follow still another alternative that merges both of these approaches.

In this paper x_i, $i = 1, 2, 3$, mean rectangular Cartesian coordinates with the origin at the center of gravity of the Earth. We identify W and U with the gravity and a standard (or normal) potential of the Earth, respectively. Under this notation $\boldsymbol{g} = \boldsymbol{grad}\, W$ is the gravity vector and its

P. Holota (✉)
Research Institute of Geodesy, Topography and Cartography, Prague-East, Czech Republic
e-mail: petr.holota@pecny.cz

O. Nesvadba
Land Survey Office, Prague 8, Czech Republic
e-mail: nesvadba@sky.cz

P. Novák et al. (eds.), *IX Hotine-Marussi Symposium on Mathematical Geodesy*,
International Association of Geodesy Symposia 151, https://doi.org/10.1007/1345_2019_67

length $g = |\,\boldsymbol{grad}\, W\,|$ is the measured gravity. By analogy we put $\boldsymbol{\gamma} = \boldsymbol{grad}\, U$ and $\gamma = |\,\boldsymbol{grad}\, U\,|$ for the normal gravity. Finally, in the general point $\boldsymbol{x} = (x_1, x_2, x_3)$ we have $T(\boldsymbol{x}) = W(\boldsymbol{x}) - U(\boldsymbol{x})$ for the disturbing potential and $\delta g(\boldsymbol{x}) = g(\boldsymbol{x}) - \gamma(\boldsymbol{x})$ for the gravity disturbance.

We will discuss the Linear Gravimetric Boundary Value Problem (LGBVP). It is an oblique derivative problem. Its solution domain is the exterior of the Earth. We will denote it by Ω. The problem may be formulated as follows

$$\Delta T = div\, \boldsymbol{grad}\, T = 0 \quad \text{in} \quad \Omega, \tag{1}$$

$$\frac{\partial T}{\partial s} = \langle \boldsymbol{s}, \boldsymbol{grad}\, T \rangle = -\,\delta g \quad \text{on} \quad \partial\Omega, \tag{2}$$

where

$$\boldsymbol{s} = -\frac{1}{\gamma}\, \boldsymbol{grad}\, U, \tag{3}$$

$\langle\, ,\rangle$ is the inner product, Δ means Laplace's operator and $\partial\Omega$ represents the boundary of Ω, see Koch and Pope (1972), Bjerhammar and Svensson (1983), Grafarend (1989) and Holota (1997). Let us add in this connection that the vector $\boldsymbol{s}$ is assumed to be nowhere tangential to $\partial\Omega$.

Now we introduce ellipsoidal coordinates u, β, λ (β is the reduced latitude and λ is the geocentric longitude in the usual sense) related to Cartesian coordinates x_1, x_2, x_3 by the equations

$$x_1 = \sqrt{u^2 + E^2} \cos\beta \cos\lambda, \tag{4}$$

$$x_2 = \sqrt{u^2 + E^2} \cos\beta \sin\lambda, \tag{5}$$

$$x_3 = u \sin\beta, \tag{6}$$

where $E = \sqrt{a^2 - b^2}$ is the linear eccentricity of an ellipsoid of revolution with semiaxes a and b, $a \geq b$, whose center is in the origin of our Cartesian system and whose axis of rotation coincides with the x_3-axis.

In our considerations we will suppose that $h(\beta, \lambda)$ is a function that describes the boundary $\partial\Omega$ of our solution domain Ω with respect to the level ellipsoid $u = b$, i.e. $\partial\Omega$ is represented by

$$x_1 = \sqrt{[b + h(\beta, \lambda)]^2 + E^2} \cos\beta \cos\lambda, \tag{7}$$

$$x_2 = \sqrt{[b + h(\beta, \lambda)]^2 + E^2} \cos\beta \sin\lambda, \tag{8}$$

$$x_3 = [b + h(\beta, \lambda)] \sin\beta. \tag{9}$$

In addition, referring to Heiskanen and Moritz (1967), we can reproduce that $\partial U/\partial\lambda = 0$ for the normal (Somigliana-Pizzeti) potential U and that for $h = 0$ we have $\partial U/\partial\beta = 0$. Moreover, for $\partial\Omega$ close to the level ellipsoid, we can even adopt that with a high (sufficient) accuracy $\partial U/\partial\beta = 0$ is valid for a realistic range of h representing the boundary $\partial\Omega$ (surface of the Earth). In consequence the boundary condition above, Eq. (2), can be interpreted in terms of a derivative of T with respect to u, i.e.,

$$\frac{\partial T}{\partial u} = -\,w\,(b + h, \beta)\, \delta g \quad \text{on} \quad \partial\Omega, \tag{10}$$

where

$$w\,(u, \beta) = \sqrt{\frac{u^2 + E^2 \sin^2\beta}{u^2 + E^2}}. \tag{11}$$

In the following approach to the solution the LGBVP a transformation (small modification) of ellipsoidal coordinates will be applied together with an attenuation function. This will open a way for an alternative between the boundary complexity and the complexity of the coefficients of the partial differential equation governing the solution. The approach represents a generalization of the concept discussed in Holota (1985, 1986, 1989, 1992a, b, 2016) and Holota and Nesvadba (2016).

2 Transformation of Coordinates and an Attenuation Function

Our starting point will be the mapping given by Eqs. (4)–(6), but with

$$u = z + \omega(z) h\,(\beta, \lambda)\,, \tag{12}$$

where z is a new coordinate and $\omega(z)$ is a twice continuously differentiable attenuation function defined for $z \in [b, \infty)$, such that

$$\omega(z) h\,(\beta, \lambda) > -b, \tag{13}$$

$$\omega(b) = 1, \quad \frac{d\omega}{dz}(b) = 0 \tag{14}$$

and

$$\omega(z) = 0 \quad \text{for} \quad z \in [z_{ext}, \infty)\,, \text{ where } b < z_{ext}. \tag{15}$$

Stress that the assumption concerning the continuity of ω and its first and the second derivatives implies

$$\lim \omega(z) = 0, \quad \lim \frac{d\omega(z)}{dz} = 0, \quad \lim \frac{d^2\omega(z)}{dz^2} = 0 \tag{16}$$

for $z \to z_{ext}^-$, i.e. for z approaching z_{ext} from the left. Obviously, z, β, λ form a system of new curvilinear coordinates and in case that

$$\frac{du}{dz} = 1 + \frac{d\omega}{dz} h > 0 \tag{17}$$

the transformation given by Eqs. (4)–(6) with u as in Eq. (12) represents a one-to-one mapping between the original solution domain Ω and the outer space Ω_{ell} of our oblate ellipsoid of revolution.

The construction of the attenuation function $\omega(z)$ in the interval $[b, z_{ext})$, i.e. for $b \le z < z_{ext}$, deserves some attention. Here we give an example, which is also applied in this work. We put

$$\omega(z) = \exp\left[2 - \frac{2\,(\Delta z)^2}{(\Delta z)^2 - (z-b)^2}\right], \tag{18}$$

where $\Delta z = z_{ext} - b$. By direct computation we can verify that $\omega(b) = 1$ and $\lim_{z \to z_{ext}^-} \omega(z) = 0$. For the first derivative of $\omega(z)$ we obtain

$$\frac{d\omega(z)}{dz} = -\frac{4\,(\Delta z)^2\,(z-b)}{\left[(\Delta z)^2 - (z-b)^2\right]^2}\,\omega(z), \tag{19}$$

$$\frac{d\omega(b)}{dz} = 0 \quad \text{and} \quad \lim_{z \to z_{ext}^-} \frac{d\omega(z)}{dz} = 0. \tag{20}$$

Similarly for the second derivative of $\omega(z)$ we can verify that

$$\begin{aligned} \frac{d^2\omega(z)}{dz^2} &= -\frac{4\,(\Delta z)^2 (z-b)}{\left[(\Delta z)^2-(z-b)^2\right]^2}\,\frac{d\omega(z)}{dz} - \\ &- \left\{\frac{4\,(\Delta z)^2}{\left[(\Delta z)^2-(z-b)^2\right]^2} + \frac{16\,(\Delta z)^2 (z-b)^2}{\left[(\Delta z)^2-(z-b)^2\right]^3}\right\}\omega(z) \end{aligned} \tag{21}$$

and

$$\lim_{z \to z_{ext}^-} \frac{d^2\omega(z)}{d^2z} = 0. \tag{22}$$

3 Transformation of the Boundary Condition

In the coordinates z, β, λ the boundary $\partial\Omega$ is defined by $z = b$ and its image $\partial\Omega_{ell}$ coincides with our oblate ellipsoid of revolution. In addition the transformation changes the formal representation of the LGBVP. Indeed, the boundary condition turns into

$$\frac{\partial T}{\partial z} = -w\left[z + \omega(z)h(\beta,\lambda), \beta\right]\delta g \quad \text{for} \quad z = b. \tag{23}$$

Hence, denoting by $\partial/\partial n$ the derivative in the direction of the unit (outer) normal $\boldsymbol{n}$ of $\partial\Omega_{ell}$ and recalling $\partial T/\partial n = (\partial T/\partial z)\,(dz/dn)$, where $dz/dn = 1/w(z,\beta)$, which follows from differential geometry considerations, we obtain

$$\frac{\partial T}{\partial n} = -\sqrt{1+\varepsilon}\;\delta g \quad \text{on} \quad \partial\Omega_{ell}, \tag{24}$$

where

$$\varepsilon = \frac{E^2\left(2bh + h^2\right)\cos^2\beta}{\left(a^2\sin^2\beta + b^2\cos^2\beta\right)\left[(b+h)^2 + E^2\right]} \tag{25}$$

may practically be neglected (in our case). Using the values of the parameters a and b as, e.g., in the Geodetic Reference System 1980, see Moritz (1992), together with $h_{\max} = 8848$ m, we can deduce that $\varepsilon < 1.9 \times 10^{-5}\cos^2\beta$.

4 Metric Tensor

Expressing Laplace's operator of T in terms of the curvilinear coordinates z, β, λ, which do not form an orthogonal system, is somewhat more complicated. In the first step we approach the construction of the metric tensor. Putting

$$y_1 = z, \quad y_2 = \beta, \quad y_3 = \lambda, \tag{26}$$

we easily deduce that the Jacobian (Jacobian determinant)

$$J = \left|\frac{\partial x_i}{\partial y_j}\right| = -\left(1 + \frac{d\omega}{dz}h\right)\left[(z+\omega h)^2 + E^2\sin^2\beta\right]\cos\beta \tag{27}$$

of the transformation in Sect. 2 is negative (apart from its zero values for $\beta = -\pi/2$ and $\pi/2$). Thus, the transformation is a one-to-one mapping. Now we use the tensor calculus and by means of some algebra we obtain the components of the metric tensor

$$g_{ij}(y) = \frac{\partial x_k}{\partial y_i} \cdot \frac{\partial x_k}{\partial y_j} \tag{28}$$

in the coordinates y_i. In the original notation this means that

$$g_{11} = \left(1 + \frac{d\omega}{dz}h\right)^2 \alpha, \; g_{12} = \left(1 + \frac{d\omega}{dz}h\right)\alpha\,\omega\,\frac{\partial h}{\partial \beta}, \tag{29}$$

$$g_{13} = \left(1 + \frac{d\omega}{dz}h\right)\alpha\,\omega\,\frac{\partial h}{\partial \lambda}, \tag{30}$$

$$g_{22} = (z+\omega h)^2 + E^2\sin^2\beta + \alpha\,\omega^2\left(\frac{\partial h}{\partial \beta}\right)^2, \tag{31}$$

$$g_{23} = \alpha\,\omega^2 \frac{\partial h}{\partial \beta}\,\frac{\partial h}{\partial \lambda}, \tag{32}$$

$$g_{33} = \left[(z+\omega h)^2 + E^2\right]\cos^2\beta + \alpha\,\omega^2\left(\frac{\partial h}{\partial \lambda}\right)^2, \tag{33}$$

where $\alpha = w^2(z+\omega\,h,\beta)$.

5 Associated (Conjugate) Metric Tensor

Of similar importance is the associate (conjugate) metric tensor. For the determinant $g = |g_{ij}|$ we have $g = J^2$. Denoting the cofactor of g_{ij} in the determinant g by G^{ij} and putting $g^{ij} = G^{ij}/g$ for the components of the associated metric tensor, we get

$$g^{11} = \tfrac{1}{\alpha}\left(1+\tfrac{d\omega}{dz}h\right)^{-2} + \left(1+\tfrac{d\omega}{dz}h\right)^{-2}\times \\ \times\left\{\frac{\omega^2}{(z+\omega h)^2+E^2\sin^2\beta}\left(\frac{\partial h}{\partial\beta}\right)^2 + \frac{\omega^2}{\left[(z+\omega h)^2+E^2\right]\cos^2\beta}\left(\frac{\partial h}{\partial\lambda}\right)^2\right\}, \tag{34}$$

$$g^{12} = -\left(1+\frac{d\omega}{dz}h\right)^{-1}\frac{\omega}{(z+\omega h)^2+E^2\sin^2\beta}\,\frac{\partial h}{\partial\beta}, \tag{35}$$

$$g^{13} = -\left(1+\frac{d\omega}{dz}h\right)^{-1}\frac{\omega}{\left[(z+\omega h)^2+E^2\right]\cos^2\beta}\,\frac{\partial h}{\partial\lambda}, \tag{36}$$

$$g^{22} = \frac{1}{(z+\omega h)^2+E^2\sin^2\beta}, \tag{37}$$

$$g^{23} = 0 \quad\text{and}\quad g^{33} = \frac{1}{\left[(z+\omega h)^2+E^2\right]\cos^2\beta}. \tag{38}$$

6 Laplacian and Topography-Dependent Coefficients

Now we are ready to approach Laplace's operator applied on T. In terms of the curvilinear coordinates y_i (i.e. in z, β, λ) it has the following general form

$$\Delta T = \frac{1}{\sqrt{g}}\frac{\partial}{\partial y_i}\left(\sqrt{g}\,g^{ij}\,\frac{\partial T}{\partial y_j}\right) = g^{ij}\frac{\partial^2 T}{\partial y_i\partial y_j} + \frac{1}{\sqrt{g}}\frac{\partial\sqrt{g}\,g^{ij}}{\partial y_i}\frac{\partial T}{\partial y_j}, \tag{39}$$

see Sokolnikoff (1971). After some algebra and neglecting the difference

$$w^2(z+\omega h,\beta) - w^2(z,\beta) \le \frac{E^2}{z^2}\left[2\omega\frac{h}{z} + \left(\omega\frac{h}{z}\right)^2\right]\cos^2\beta, \tag{40}$$

which for $h_{\max} = 8848$ m and the values of E^2 and $z = b$ taken from the Geodetic Reference System 1980, see Moritz (1992), can be estimated from above by $1.9\times10^{-5}\cos^2\beta$, we can deduce that

$$\Delta T = \frac{z^2+E^2\sin^2\beta}{(z+\omega h)^2+E^2\sin^2\beta}\left[\Delta_{ell}T - \delta(T,h)\right], \tag{41}$$

where

$$\Delta_{ell}T = \frac{1}{z^2+E^2\sin^2\beta}\left[\left(z^2+E^2\right)\frac{\partial^2 T}{\partial z^2} + 2z\frac{\partial T}{\partial z} + \right. \\ \left. + \frac{\partial^2 T}{\partial\beta^2} - \frac{\sin\beta}{\cos\beta}\frac{\partial T}{\partial\beta} + \frac{z^2+E^2\sin^2\beta}{\left(z^2+E^2\right)\cos^2\beta}\frac{\partial^2 T}{\partial\lambda^2}\right], \tag{42}$$

$$\delta(T,h) = A_1\frac{\partial T}{\partial z} + A_2\frac{\partial^2 T}{\partial z^2} + \\ + A_3\frac{1}{\sqrt{z^2+E^2\sin^2\beta}}\frac{\partial^2 T}{\partial z\partial\beta} + A_4\frac{\sqrt{\alpha}}{\sqrt{z^2+E^2}\cos\beta}\frac{\partial^2 T}{\partial z\partial\lambda} \tag{43}$$

and A_i are topography dependent coefficients given by

$$A_1 = \left(1+\tfrac{d\omega}{dz}h\right)^{-1}\left[2\left(\tfrac{d\omega}{dz}-\tfrac{\omega}{z}\right)\frac{zh}{z^2+E^2\sin^2\beta} + \omega\,\Delta_E h\right] - \\ - 2\left(1+\tfrac{d\omega}{dz}h\right)^{-2}\omega\,\tfrac{d\omega}{dz}|\mathbf{grad}_E h|^2 + \\ + \left(1+\tfrac{d\omega}{dz}h\right)^{-3}\left[\frac{(z+\omega h)^2+E^2}{z^2+E^2\sin^2\beta} + \omega^2|\mathbf{grad}_E h|^2\right]\frac{d^2\omega}{dz^2}h, \tag{44}$$

$$A_2 = \left(1+\tfrac{d\omega}{dz}h\right)^{-2}\left\{2\left(\tfrac{d\omega}{dz} - \frac{\omega z}{z^2+E^2}\right)h + \right. \\ \left. + \left[\left(\tfrac{d\omega}{dz}\right)^2 - \frac{\omega^2}{z^2+E^2}\right]h^2\right\}\frac{z^2+E^2}{z^2+E^2\sin^2\beta} - \\ - \left(1+\tfrac{d\omega}{dz}h\right)^{-2}\omega^2|\mathbf{grad}_E h|^2, \tag{45}$$

$$A_3 = \left(1+\frac{d\omega}{dz}h\right)^{-1}\frac{2\omega}{\sqrt{z^2+E^2\sin^2\beta}}\,\frac{\partial h}{\partial\beta}, \tag{46}$$

$$A_4 = \left(1+\frac{d\omega}{dz}h\right)^{-1}\frac{2\,\omega\,\sqrt{\alpha}}{\sqrt{z^2+E^2}\cos\beta}\,\frac{\partial h}{\partial\lambda} \tag{47}$$

with

$$|\mathbf{grad}_E h|^2 = \frac{1}{z^2+E^2\sin^2\beta}\left[\left(\frac{\partial h}{\partial\beta}\right)^2 + \frac{z^2+E^2\sin^2\beta}{(z^2+E^2)\cos^2\beta}\left(\frac{\partial h}{\partial\lambda}\right)^2\right] \tag{48}$$

and

$$\Delta_E h = \frac{1}{z^2+E^2\sin^2\beta}\left[\frac{\partial^2 h}{\partial\beta^2} - \frac{\sin\beta}{\cos\beta}\frac{\partial h}{\partial\beta} + \frac{z^2+E^2\sin^2\beta}{(z^2+E^2)\cos^2\beta}\frac{\partial^2 h}{\partial\lambda^2}\right] \tag{49}$$

being the first and the second Beltrami differential operators.

7 Linear GBVP and Neumann's Function

The disturbing potential T is a harmonic function in the original solution domain Ω. In the space of the curvilinear coordinates z, β, λ, therefore, T satisfies Laplace's equation $\Delta T = 0$ for $z > b$, which in view of Eq. (41) yields

$$\Delta_{ell} T = \delta(T, h) \quad \text{for} \quad z > b, \tag{50}$$

where $\delta(T, h)$ is given by Eq. (43). Hence in combination with Eq. (24) the linear gravimetric boundary value problem in terms of the curvilinear coordinates z, β, λ attains the form

$$\Delta_{ell} T = f \quad \text{in} \quad \Omega_{ell}, \tag{51}$$

$$\frac{\partial T}{\partial n} = -\sqrt{1+\varepsilon}\,\delta g \quad \text{on} \quad \partial\Omega_{ell}, \tag{52}$$

where $f = \delta(T, h)$ and ε given by Eq. (25) is as small that it may be omitted.

Neglecting the fact that $f = \delta(T, h)$ depends on T, we can represent the solution of the problem formally by means of a classical apparatus of mathematical physics. The natural point of departure is Green's third identity (Green's representation formula)

$$\begin{aligned} T_P = \frac{1}{4\pi} \int_{\partial\Omega_{ell}} \left[T\frac{\partial}{\partial n}\left(\frac{1}{l}\right) - \frac{1}{l}\frac{\partial T}{\partial n}\right] dS \\ - \frac{1}{4\pi}\int_{\Omega_{ell}} \frac{1}{l}\,\Delta_{ell} T\, dV \end{aligned} \tag{53}$$

with l being the distance between the computation and the variable point of integration and dS and dV denoting the surface and the volume element, respectively. Similarly, the quantities with and without the subscript P are referred to the computation and the variable point of integration. We will generalize the formula a little. To do that, we take into consideration a function H harmonic in Ω_{ell}. Hence $\Delta H = 0$ in Ω_{ell} and by Green's second identity we have

$$\int_{\partial\Omega_{ell}} \left(T\frac{\partial H}{\partial n} - H\frac{\partial T}{\partial n}\right) dS = \int_{\Omega_{ell}} H \Delta_{ell} T\, dV. \tag{54}$$

Writing now

$$G = \frac{1}{l} - H \tag{55}$$

and combining Eqs. (53) and (54), we obtain the generalized Green representation formula

$$T_P = \frac{1}{4\pi}\int_{\partial\Omega_{ell}} \left(T\frac{\partial G}{\partial n} - G\frac{\partial T}{\partial n}\right) dS - \frac{1}{4\pi}\int_{\Omega_{ell}} G\Delta_{ell} T\, dV. \tag{56}$$

In the following we will use the function G constructed under Neumann's boundary condition, i.e.

$$\frac{\partial G}{\partial n} = 0 \quad \text{on} \quad \partial\Omega_{ell}, \tag{57}$$

which means that we have to look for a function $H = H(z, \beta, \lambda)$ such that

$$\frac{\partial H}{\partial n} = \frac{\partial}{\partial n}\left(\frac{1}{l}\right) \quad \text{for} \quad z = b. \tag{58}$$

In this case G represents Green's function of the second kind, usually called Neumann's function. We will denote the function G by N and from Eq. (56) we obtain that

$$T_P = \frac{1}{4\pi}\int_{z=b} N\,\delta g\, dS - \frac{1}{4\pi}\int_{b<z<z_{ext}} N\,\delta(T, h)\, dV, \tag{59}$$

where in addition we took into consideration Eq. (50) and the properties of the attenuation function $\omega(z)$, see Sect. 2. On the other hand the construction of Neumann's function itself for the exterior of an oblate ellipsoid of revolution is not routine as yet in contrast to problems formulated for a spherical boundary, as e.g. in Holota (2003). For an oblate ellipsoid of revolution the construction is discussed in Holota (2004, 2011), Holota and Nesvadba (2014, 2018b) and in particular in Holota and Nesvadba (2018a), equally as its relation to Green's function of the first kind and to the so-called reproducing kernel.

8 Iteration Process

The integral formula (59) represents an integro-differential equation for T. For clarity we put

$$F_P = \frac{1}{4\pi} \int_{z=b} N\, \delta g\; dS, \tag{60}$$

$$(KT)_P = -\frac{1}{4\pi} \int_{b<z<z_{ext}} N\, \delta\,(T,h)\; dV, \tag{61}$$

where F is a harmonic function and KT is an integro-differential operator applied on T, such that

$$\Delta_{ell} K\, T = \delta\,(T,h) \quad \text{in} \quad \Omega_{ell} \tag{62}$$

and

$$\frac{\partial K\, T}{\partial n} = 0 \quad \text{on} \quad \partial\Omega_{ell}, \tag{63}$$

which follows from general principles applied in constructing Neumann's function. Under this notation the problem is to find T from

$$T = F + K\, T. \tag{64}$$

Our aim is to apply the method of successive approximations, i.e.

$$T = \lim_{n} T_n, \qquad T_n = F + K\, T_{n-1}, \tag{65}$$

where $n = 1, 2, \ldots \infty$ and T_0 is the starting approximation, e.g. $T_0 = F$.

9 Operator with Reduced Degree of Derivatives

For practical use it is convenient to modify the operator K in order to reduce the degree of derivatives involved in $\delta(T,h)$ and to display the mutual interplay of individual terms in $\delta(T,h)$ more explicitly. Integrating by parts and neglecting terms multiplied by E^2/z^3, we get

$$(KT)_P = -\tfrac{1}{4\pi} \int_{z=b} N A_2\, \delta g dS - \tfrac{1}{4\pi} \int_{b<z<z_{ext}} N A_5 \tfrac{\partial T}{\partial z} dV +$$
$$+ \tfrac{1}{4\pi} \int_{b<z<z_{ext}} \left(A_2 \tfrac{\partial N}{\partial z} + \tfrac{A_3}{\sqrt{z^2+E^2 \sin^2\beta}} \tfrac{\partial N}{\partial \beta} + \tfrac{A_4 \sqrt{\alpha}}{\sqrt{z^2+E^2}\cos\beta} \tfrac{\partial N}{\partial \lambda} \right) \tfrac{\partial T}{\partial z} dV, \tag{66}$$

where

$$A_5 = A_1 - \tfrac{\partial A_2}{\partial z} - \tfrac{2}{\sqrt{z^2+E^2 \sin^2\beta}} A_2 -$$
$$- \tfrac{1}{\sqrt{z^2+E^2 \sin^2\beta}} \left(\tfrac{\partial A_3}{\partial \beta} - \tfrac{\sin\beta}{\cos\beta} A_3 \right) - \tfrac{\sqrt{\alpha}}{\sqrt{z^2+E^2}\cos\beta} \tfrac{\partial A_4}{\partial \lambda}. \tag{67}$$

Note. It may be interesting that for $\omega(z) = 1$, i.e. $z_{ext} = \infty$, we get $A_5 = -\,\Delta_E h$ directly from Eq. (49).

10 Conclusions

Loosely speaking, the operator K "consumes" derivatives. The question is how the operator transforms the differentiability of the function T or what is the range of the operator for an initially chosen function space, i.e. an initially chosen domain of the operator? This feature is of considerable importance. Its impact will take effect immediately in case that we try to proof the convergence of the iteration procedure as in Eq. (65) by means of tools of functional analysis. The key step is to show that K is a contraction mapping which (if proved) guarantees the convergence of the iteration procedure on the basis of Banach's fixed point theorem, see e.g. Lyusternik and Sobolev (1965). This approach was already discussed in Holota (1985, 1986, 1989, 1992a, b) for $E = 0$ and functions from Sobolev's space $W_2^{(2)}$ produced (roughly speaking) by functions which together with their (generalized) derivatives of the 1st and the 2nd order are square integrable on a spherical layer. In this case it was shown that K is as mapping from $W_2^{(2)}$ onto $W_2^{(2)}$ and its contractivity depends on essential supreme values of the topography dependent coefficients A_i, $i = 1,2,3,4$. The most intricate step to estimate the second order derivatives of KT has been done by means of the Calderon-Zygmund inequality (which belongs to L_p estimates for Poisson's equation), see Gilbarg and Trudinger (1983). As a result the convergence of the iteration procedure was proved for a realistic range of heights and relatively gentle slopes and curvatures of the topography, see Holota (1992b).

Nevertheless, by nature these are a priori estimates and the results concerning the solvability of the LGBVP may differ a bit. Indeed, studies on the existence, uniqueness and stability of the LGBVP, as e.g. in Holota (1997) and by Sansò in Sansò and Sideris (2013), show that the requirements on the topography may be considerably milder. In particular, in his proof Sansò shows that the inclination should be smaller than about 89°. In addition also the use of the ellipsoidal apparatus for the construction of the iteration procedure has its impact on the behavior and the speed of the convergence of the successive approximations.

For all these reasons it may be very instructive to use a numerical approach. The idea is given attention in the ongoing research. First step in this direction was the application of the integration by parts in Sect. 9 that decreases the order of derivatives in the operator K and keeps Lebesgue integrability at the same time. Considerable attention is also given to the investigation on how the successive approximations of the solution behave close to the boundary and how they attain the boundary values. Preference is given to the classical (pointwise) definition of these properties. These goals are challenging, but we believe they will enrich the solution of the problem.

Acknowledgements The work on this paper was supported by the Ministry of Education, Youth and Sports of the Czech Republic through Project No. LO1506. This support is gratefully acknowledged.

References

Bjerhammar A, Svensson L (1983) On the geodetic boundary-value problem for a fixed boundary surface - satellite approach. Bull Geod 57:382–393

Gilbarg D, Trudinger NS (1983) Elliptic partial differential equations of second order. Springer, Berlin

Grafarend EW (1989) The geoid and the gravimetric boundary-value problem. Rep 18 Dept Geod. The Royal Institute of Technology, Stockholm

Heiskanen WA, Moritz H (1967) Physical geodesy. W.H. Freeman and Company, San Francisco

Holota P (1985) A new approach to iteration solutions in solving geodetic boundary value problems for real topography. In: Proc. 5th Int. Symp. Geod. and Phys. of the Earth, GDR, Magdeburg, Sept. 23rd–29th, 1984, Part II. Veroff. d. Zentr. Inst. f. Phys. d. Erde, Nr. 81, Teil II, pp 4–15

Holota P (1986) Boundary value problems in physical geodesy: present state, boundary perturbation and the Green-Stokes representation. In: Proc. 1st Hotine-Marussi Symp. on Math. Geodesy, Rome, 3–5 June 1985, vol 2. Politecnico di Milano, pp 529–558

Holota P (1989) Laplacian versus topography in the solution of the Molodensky problem by means of successive approximations. In: Kejlso E, Poder K, Tscherning CC (eds) Festschrift to Torben Krarup, Geodaetisk Inst., Meddelelse No. 58, Kobenhavn, pp 213–227

Holota P (1992a) On the iteration solution of the geodetic boundary-value problem and some model refinements. Contribution to Geodetic Theory and Methodology. In: XXth General Assembly of the IUGG, IAG-Sect. IV, Vienna, 1991. Politecnico di Milano, 1991, pp 31–60; also in: Travaux de l'Association Internationale de Geodesie, Tome 29, Paris: 260–289

Holota P (1992b) Integral representation of the disturbing potential: effects involved, iteration technique and its convergence. In: Holota P, Vermeer M (eds) Proc. First continental workshop on the geoid in Europe, Prague, May 11–14, 1992. Research Inst. of Geod., Topog. and Cartog., Prague, in co-operation with IAG-Subcommis. for the Geoid in Europe, Prague, pp 402–419

Holota P (1997) Coerciveness of the linear gravimetric boundary value problem and geometrical interpretation. J Geod 71:640–651

Holota P (2003) Green's function and external masses in the solution of geodetic boundary-value problems. In: Tziavos IN (ed) Gravity and Geoid, 3rd Meeting of the Intl. Gravity and Geoid Commission, Thessaloniki, Greece, August 26–30, 2002. Ziti Editions, Thessaloniki, pp 108–113

Holota P (2004) Some topics related to the solution of boundary-value problems in geodesy. In: Sansò F (ed) V Hotine-Marussi Symposium on Mathematical Geodesy, Matera, Italy, June 17–21, 2002. International Association of Geodesy Symposia, vol 127. Springer, Berlin, pp 189–200

Holota P (2011) Reproducing kernel and Galerkin's matrix for the exterior of an ellipsoid: application in gravity field studies. Studia geophysica et geodaetica 55(3):397–413

Holota P (2016) Domain transformation and the iteration solution of the linear gravimetric boundary value problem. In: Freymueller J, Sánchez L (eds) International symposium on earth and environmental sciences for future generations. Proceedings of the IAG General Assembly, Prague, Czech Republic, June 22–July 2, 2015. International Association of Geodesy Symposia, vol 147. Springer, Cham, pp 47–52. https://doi.org/10.1007/1345_2016_236

Holota P, Nesvadba O (2014) Reproducing kernel and Neumann's function for the exterior of an oblate ellipsoid of revolution: application in gravity field studies. Studia geophysica et geodaetica 58(4):505–535

Holota P, Nesvadba O (2016) Small modifications of curvilinear coordinates and successive approximations applied in geopotential determination. 2016 AGU Fall Meeting, Session G21B (Scientific and Practical Challenges of Replacing NAD 83, NAVD 88, and IGLD 85), San Francisco, USA, 12–16 December 2016, poster. https://agu.confex.com/agu/fm16/meetingapp.cgi/Paper/189936

Holota P, Nesvadba O (2018a) Neumann's function and its derivatives constructed for the exterior of an ellipsoid and adapted to an iteration solution of the linear gravimetric boundary value problem. In: Geophysical Research Abstracts, vol 20, EGU2018-18558

Holota P, Nesvadba O (2018b) Boundary complexity and kernel functions in classical and variational concepts of solving geodetic boundary value problems. In: Freymueller J, Sánchez L (eds) International symposium on advancing geodesy in a changing world. International Association of Geodesy Symposia, vol 149. Springer, Cham, pp 31–41. https://doi.org/10.1007/1345_2018_34

Koch KR, Pope AJ (1972) Uniqueness and existence for the geodetic boundary-value problem using the known surface of the Earth. Bull Geod 106:467–476

Lyusternik LA, Sobolev VI (1965) Foundations of functional analysis. Nauka Publishers, Moscow. (in Russian)

Moritz H (1992) Geodetic reference system 1980. In: Tscherning CC (ed) The Geodesist's Handbook 1992. Bulletin Géodésique, vol 66, no 2, pp 187–192

Roach GF (1982) Green's functions, 2nd edn. Cambridge University Press, Cambridge

Sansò F, Sideris MG (2013) Geoid determination - theory and methods. Springer, Berlin

Sokolnikoff IS (1971) Tensor analysis. Theory and applications to geometry and mechanics of continua. Nauka Publishers, Moscow. (in Russian)

On Combining the Directional Solutions of the Gravitational Curvature Boundary-Value Problem

Martin Pitoňák, Pavel Novák, Michal Šprlák, and Robert Tenzer

Abstract

In global studies, the Earth's gravitational field is conveniently described in terms of spherical harmonics. Four integral-based solutions to a gravitational curvature boundary-value problem can formally be formulated for the vertical-vertical-vertical, vertical-vertical-horizontal, vertical-horizontal-horizontal and horizontal-horizontal-horizontal components of the third-order gravitational tensor. Each integral equation provides an independent set of spherical harmonic coefficients because each component of the third-order gravitational tensor is sensitive to gravitational changes in the different directions. In this contribution, estimations of spherical harmonic coefficients of the gravitational potential are carried out by combining four solutions of the gravitational curvature boundary-value problem using three methods, namely an arithmetic mean, a weighted mean and a conditional adjustment model. Since the third-order gradients of the gravitational potential are not yet observed by satellite sensors, we synthesise them at the satellite altitude of 250 km from a global gravitational model up to the degree 360 while adding a Gaussian noise with the standard deviation of 6.3×10^{-19} m^{-1} s^{-2}. Results of the numerical analysis reveal that the arithmetic mean model provides the best solution in terms of the RMS fit between predicted and reference values. We explain this result by the facts that the conditions only create additional stochastic bindings between estimated parameters and that more complex numerical schemes for the error propagation are unnecessary in the presence of only a random noise.

Keywords

Conditional adjustment · Gravitational curvature · Spherical harmonics

1 Introduction

Solutions to a spherical boundary-value problem lead to spherical harmonic series or surface integrals with Green's kernel functions (e.g., Jekeli 2009). When solving this problem for higher-order gradients of the gravitational potential as boundary conditions, more than one solution is obtained. The solutions to the gravimetric, gradiometric and gravitational curvature boundary-value problems (Martinec 2003; Šprlák and Novák 2016) lead to two, three and four formulas, respectively. From a theoretical point of view, all formulas should provide the same solution, but practically, when discrete noisy observations are exploited, they do not.

M. Pitoňák (✉) · P. Novák
NTIS – New Technologies for the Information Society, Faculty of Applied Sciences, University of West Bohemia, Plzeň, Czech Republic
e-mail: pitonakm@ntis.zcu.cz

M. Šprlák
School of Engineering and Built Environment, University of Newcastle, Callaghan, NSW, Australia

R. Tenzer
Department of Land Surveying and Geo-informatics, The Hong Kong Polytechnic University, Kowloon, Hong Kong

P. Novák et al. (eds.), *IX Hotine-Marussi Symposium on Mathematical Geodesy*,
International Association of Geodesy Symposia 151, https://doi.org/10.1007/1345_2019_68

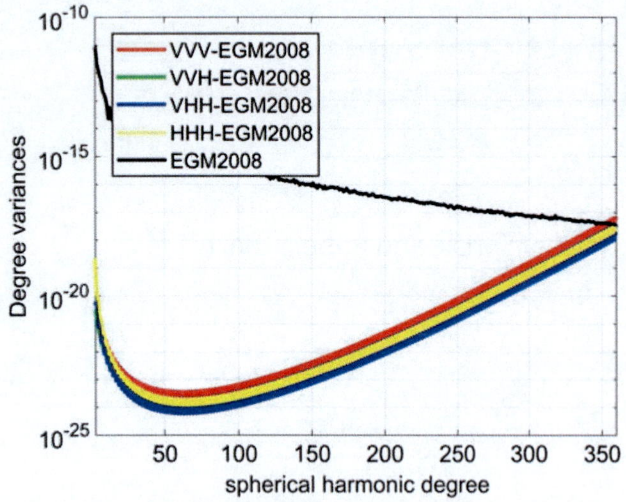

Fig. 1 Differences between vertical-vertical-vertical (VVV), vertical-vertical-horizontal (VVH), vertical-horizontal-horizontal (VHH) and horizontal-horizontal-horizontal (HHH) solutions and EGM2008 up to the degree 360

This is illustrated by Fig. 1, where differences between the vertical-vertical-vertical (VVV), vertical-vertical-horizontal (VVH), vertical-horizontal-horizontal (VHH) and horizontal-horizontal-horizontal (HHH) solutions to the gravitational curvature boundary-value problem are plotted and compared with the EGM2008 gravitational spectrum (Pavlis et al. 2012, 2013) complete up to the degree 360.

Since the four solutions of the gravitational curvature boundary-value problem are not equal, their combination is required. In this study, we investigate three different methods for combining the VVV, VVH, VHH and HHH solutions by applying an arithmetic mean, a weighted mean and a conditional adjustment model (CAM) in order to obtain the unique solution. To do so, we modify the CAM to fit the combination of the four solutions and then derive corresponding expressions for estimating the adjusted errors. A similar study was conducted for the spherical gradiometric boundary-value problem by Eshagh (2010), who applied the variance component estimation technique to the CAM in order to obtain an improved combined solution.

Currently, sensors for measuring components of the gravitational curvature tensor at the Earth's surface or at the satellite altitude are not yet available. However, the VVV component was already observed in laboratory conditions, see, e.g., (Balakin et al. 1997; Rosi et al. 2015). Hence, we use synthetic data that are compiled from an existing global gravitational model.

The paper is organized into five sections, beginning with a brief overview of the spherical gravitational curvature boundary-value problem in Sect. 2, which is followed by the description of the combination strategies in Sect. 3 and the numerical experiment in Sect. 4. The study is concluded in Sect. 5.

2 Gravitational Curvature Boundary-Value Problem

The solutions to the gravitational curvature boundary-value problem in the spectral form were presented by Šprlák et al. (2016):

$$C^{(0)}_{n,m} = \frac{B^0_n}{4\pi GM t_n} \int_\Omega T_{zzz}(r,\Omega)\, \overline{Y}_{n,m}(\Omega)\, d\Omega, \tag{1a}$$

$$C^{(1)}_{n,m} = \frac{B^1_n}{4\pi GM t_n} \int_\Omega \Big[T_{xzz}(r,\Omega) \frac{\partial}{\partial\varphi} \overline{Y}_{n,m}(\Omega) - T_{yzz}(r,\Omega) \frac{1}{\cos\varphi} \frac{\partial}{\partial\lambda} \overline{Y}_{n,m}(\Omega) \Big]\, d\Omega, \tag{1b}$$

$$C^{(2)}_{n,m} = \frac{B^2_n}{4\pi GM t_n} \int_\Omega \left\{ \begin{array}{l} \left[T_{xxz}(r,\Omega) - T_{yyz}(r,\Omega)\right] \\ \times \left(\tan\varphi \frac{\partial}{\partial\varphi} + \frac{\partial^2}{\partial\varphi^2} - \frac{1}{\cos^2\varphi} \frac{\partial^2}{\partial\lambda^2} \right) \overline{Y}_{n,m}(\Omega) \\ - 4T_{xyz}(r,\Omega) \left(\frac{\tan\varphi}{\cos\varphi} \frac{\partial}{\partial\lambda} + \frac{1}{\cos\varphi} \frac{\partial^2}{\partial\varphi\partial\lambda} \right) \overline{Y}_{n,m}(\Omega) \end{array} \right\} d\Omega, \tag{1c}$$

$$C^{(3)}_{n,m} = \frac{B^3_n}{4\pi GM t_n} \int_\Omega \left\{ \begin{array}{l} \left[T_{xxx}(r,\Omega) - 3T_{xyy}(r,\Omega)\right] \\ \times \left(\frac{3 - 2\cos^2\varphi}{\cos^2\varphi} \frac{\partial}{\partial\varphi} + 3\tan\varphi \frac{\partial^2}{\partial\varphi^2} - \frac{6\tan\varphi}{\cos^2\varphi} \frac{\partial^2}{\partial\lambda^2} \right. \\ \left. + \frac{\partial^3}{\partial\varphi^3} - \frac{3}{\cos^2\varphi} \frac{\partial^3}{\partial\varphi\partial\lambda^2} \right) \overline{Y}_{n,m}(\Omega) \\ - \left[T_{yyy}(r,\Omega) - 3T_{xxy}(r,\Omega)\right] \\ \times \left(\frac{2\left(1 + 3\sin^2\varphi\right)}{\cos^3\varphi} \frac{\partial}{\partial\lambda'} + \frac{9\tan\varphi}{\cos\varphi} \frac{\partial^2}{\partial\varphi\partial\lambda} \right. \\ \left. + \frac{3}{\cos\varphi} \frac{\partial^3}{\partial\varphi^2\partial\lambda} - \frac{1}{\cos^3\varphi} \frac{\partial^3}{\partial\lambda^3} \right) \overline{Y}_{n,m}(\Omega) \end{array} \right\} d\Omega, \tag{1d}$$

where GM denotes the geocentric gravitational constant, $t_n = (R/r)^{n+4}$ is the degree-dependent attenuation factor, R is the Earth's mean radius and $\overline{Y}_{n,m}(\Omega)$ is the (fully-normalized) spherical harmonic function of degree n and order m. The 3-D position in Eqs. (1a)–(1d) and thereafter is defined by the spherical coordinates (r, Ω); where r is the geocentric radius, and $\Omega = (\varphi, \lambda)$ is the geocentric direction with the spherical latitude φ and longitude λ. The parameter B_n^i in Eqs. (1a)–(1d) is

$$B_n^i = (-1)^{(i+1)} \frac{R^4\,(n-i)!}{(n+3)!}, i = 0, 1, 2, 3, \tag{2}$$

where the index i = 0, 1, 2, 3 specifies respectively the VVV, VVH, VHH and HHH solutions (cf. Šprlák et al. 2016; Šprlák and Novák 2016). Symbols $T_{\mu\nu\tau}$, μ, ν, $\tau \in \{x, y, z\}$ in Eqs. (1a)–(1d) stand for ten components of the gravitational curvature tensor. We note that only the VVV solution can be used for the recovery of the full Earth's gravitational spectrum (up to a certain degree) because the VVH solution is restricted to non-zero degree spherical harmonics. Similarly, the solutions to the VHH and HHH components comprise the spherical harmonics above the degree one and two, respectively. Our combined solution is thus carried out for higher than second-degree spherical harmonics (i.e., for $n > 2$).

After vectorization of Eqs. (1a)–(1d), we arrive at

$$\begin{aligned}
C_{n,m}^{(0)} &= \mathbf{b}_{n,m}^{zzz}\mathbf{t}_{zzz},\\
C_{n,m}^{(1)} &= \mathbf{b}_{n,m}^{xzz}\mathbf{t}_{xzz} + \mathbf{b}_{n,m}^{yzz}\mathbf{t}_{yzz},\\
C_{n,m}^{(2)} &= \mathbf{b}_{n,m}^{xxz}\mathbf{t}_{xxz} + \mathbf{b}_{n,m}^{yyz}\mathbf{t}_{yyz} + \mathbf{b}_{n,m}^{xyz}\mathbf{t}_{xyz},\\
C_{n,m}^{(3)} &= \mathbf{b}_{n,m}^{xxx}\mathbf{t}_{xxx} + \mathbf{b}_{n,m}^{xyy}\mathbf{t}_{xyy} + \mathbf{b}_{n,m}^{yyy}\mathbf{t}_{yyy} + \mathbf{b}_{n,m}^{xxy}\mathbf{t}_{xxy},
\end{aligned} \tag{3}$$

where $\mathbf{b}_{n,m}^{xxx}$, $\mathbf{b}_{n,m}^{xxy}$, $\mathbf{b}_{n,m}^{xxz}$, $\mathbf{b}_{n,m}^{xyy}$, $\mathbf{b}_{n,m}^{xyz}$, $\mathbf{b}_{n,m}^{xzz}$, $\mathbf{b}_{n,m}^{yyy}$, $\mathbf{b}_{n,m}^{yyz}$, $\mathbf{b}_{n,m}^{yzz}$ and $\mathbf{b}_{n,m}^{zzz}$ are the row vectors obtained from discretization of Eqs. (1a)–(1d); and $\mathbf{t}_{xxx}$, $\mathbf{t}_{xxy}$, $\mathbf{t}_{xxz}$, $\mathbf{t}_{xyy}$, $\mathbf{t}_{xyz}$, $\mathbf{t}_{xzz}$, $\mathbf{t}_{yyy}$, $\mathbf{t}_{yyz}$, $\mathbf{t}_{yzz}$ and $\mathbf{t}_{zzz}$ are the column vectors of T_{xxx}, T_{xxy}, T_{xxz}, T_{xyy}, T_{xyz}, T_{xzz}, T_{yyy}, T_{yyz}, T_{yzz} and T_{zzz} observations.

3 Combination Strategies

We apply three different strategies to combine the solutions based on Eqs. (1a)–(1d), namely the arithmetic and weighted means as well as the CAM. Details are given in the following subsections.

3.1 Arithmetic Mean

The simplest strategy to combine the VVV, VVH, VHH and HHH solutions is based on applying a simple arithmetic mean. Each of the four solutions of the gravitational curvature boundary-value problem contributes to the combination with the equal weight. We then write for the arithmetic mean solution

$$C_{n,m}^{SM} = \frac{C_{n,m}^{(0)} + C_{n,m}^{(1)} + C_{n,m}^{(2)} + C_{n,m}^{(3)}}{4}. \tag{4}$$

3.2 Weighted Mean

Alternatively, we can apply a weighted mean, so that the solution reads

$$C_{n,m}^{WM} = \frac{C_{n,m}^{(0)}p_{n,m}^{(0)} + C_{n,m}^{(1)}p_{n,m}^{(1)} + C_{n,m}^{(2)}p_{n,m}^{(2)} + C_{n,m}^{(3)}p_{n,m}^{(3)}}{p_{n,m}^{(0)} + p_{n,m}^{(1)} + p_{n,m}^{(2)} + p_{n,m}^{(3)}}, \tag{5}$$

where the individual weights are defined by

$$p_{n,m}^{(0)} = \left(1/\varepsilon_{n,m}^{(0)}\right)^2, p_{n,m}^{(1)} = \left(1/\varepsilon_{n,m}^{(1)}\right)^2, p_{n,m}^{(2)} = \left(1/\varepsilon_{n,m}^{(2)}\right)^2, p_{n,m}^{(3)} = \left(1/\varepsilon_{n,m}^{(3)}\right)^2. \tag{6}$$

Mean errors $\varepsilon_{n,m}^{(i)}$, $i = 0, 1, 2, 3$, respectively, correspond to the VVV, VVH, VHH and HHH solutions.

3.3 Conditional Adjustment

Before describing the CAM, we set six conditions of the form

$$\begin{aligned}
C_{n,m}^{(0)} - C_{n,m}^{(1)} = 0, C_{n,m}^{(0)} - C_{n,m}^{(2)} = 0, C_{n,m}^{(0)} - C_{n,m}^{(3)} = 0,\\
C_{n,m}^{(1)} - C_{n,m}^{(2)} = 0, C_{n,m}^{(1)} - C_{n,m}^{(3)} = 0, C_{n,m}^{(2)} - C_{n,m}^{(3)} = 0.
\end{aligned} \tag{7}$$

These conditions postulate that the differences between the four solutions (in terms of spherical harmonics) equal zero. We then define the CAM as follows (Koch 1999)

$$\mathbf{B}\,(\mathbf{L} - \boldsymbol{\varepsilon}) = \mathbf{w} \text{ with } E\left\{\boldsymbol{\varepsilon}\boldsymbol{\varepsilon}^T\right\} = \sigma_0^2\mathbf{Q}, \tag{8}$$

where $\mathbf{B}$ is the coefficient matrix of observations, $\mathbf{L}$ is the observation vector, $\boldsymbol{\varepsilon}$ stands for the vector of the observation noise, $\mathbf{w}$ is the misclosure vector, $E\{\cdot\}$ denotes the expectation operator, σ_0^2 is an *a priori* variance factor, and $\mathbf{Q}$ is the co-factor matrix. The solution to the system of conditional equations in Eq. (8) in terms of the estimated residuals reads

$$\hat{\boldsymbol{\varepsilon}}_{n,m} = \mathbf{Q}(\mathbf{B}_{n,m})^T \left[\mathbf{B}_{n,m}\mathbf{Q}(\mathbf{B}_{n,m})^T\right]^{-1} \mathbf{w}_{n,m} = \mathbf{Q}(\mathbf{B}_{n,m})^T [\mathbf{C}_{n,m}]^{-1} \mathbf{w}^{n,m}. \tag{9}$$

From Eqs. (1a)–(1d), we write six conditions but only arbitrary five of them are independent and the matrix $\mathbf{C}_{n,m}$ is regular, i.e., an inversion of $\mathbf{C}_{n,m}$ exists. We chose the first five conditions defined in Eq. (7) in our numerical experiment. The conditional equations have the following vector-matrix form

$$\mathbf{B}_{n,m}\,(\mathbf{L} - \boldsymbol{\varepsilon}) = \mathbf{0}, \tag{10a}$$

where

$$\mathbf{L} = \left[\mathbf{t}_{zzz}\ \mathbf{t}_{xzz}\ \mathbf{t}_{yzz}\ \mathbf{t}_{xxz}\ \mathbf{t}_{yyz}\ \mathbf{t}_{xyz}\ \mathbf{t}_{xxx}\ \mathbf{t}_{xyy}\ \mathbf{t}_{yyy}\ \mathbf{t}_{xxy}\right]^T, \tag{10b}$$

and

$$\mathbf{B}_{n,m} = \begin{bmatrix} \mathbf{b}^{zzz}_{n,m} & -\mathbf{b}^{xzz}_{n,m} & -\mathbf{b}^{yzz}_{n,m} & 0 & 0 & 0 & 0 & 0 & 0 & 0 \\ \mathbf{b}^{zzz}_{n,m} & 0 & 0 & -\mathbf{b}^{xxz}_{n,m} & -\mathbf{b}^{yyz}_{n,m} & -\mathbf{b}^{xyz}_{n,m} & 0 & 0 & 0 & 0 \\ \mathbf{b}^{zzz}_{n,m} & 0 & 0 & 0 & 0 & 0 & -\mathbf{b}^{xxx}_{n,m} & -\mathbf{b}^{xyy}_{n,m} & -\mathbf{b}^{yyy}_{n,m} & -\mathbf{b}^{xyy}_{n,m} \\ 0 & \mathbf{b}^{xzz}_{n,m} & \mathbf{b}^{yzz}_{n,m} & -\mathbf{b}^{xxz}_{n,m} & -\mathbf{b}^{yyz}_{n,m} & -\mathbf{b}^{xyz}_{n,m} & 0 & 0 & 0 & 0 \\ 0 & \mathbf{b}^{xzz}_{n,m} & \mathbf{b}^{yzz}_{n,m} & 0 & 0 & 0 & -\mathbf{b}^{xxx}_{n,m} & -\mathbf{b}^{xyy}_{n,m} & -\mathbf{b}^{yyy}_{n,m} & -\mathbf{b}^{xyy}_{n,m} \end{bmatrix}. \tag{10c}$$

The misclosure vector comprises five elements, particularly

$$\mathbf{w}_{n,m} = \begin{bmatrix} (w_1)_{n,m} \\ (w_2)_{n,m} \\ (w_3)_{n,m} \\ (w_4)_{n,m} \\ (w_5)_{n,m} \end{bmatrix} = \begin{bmatrix} -\mathbf{b}^{zzz}_{n,m}\mathbf{t}^T_{zzz} + \mathbf{b}^{xzz}_{n,m}\mathbf{t}^T_{xzz} + \mathbf{b}^{yzz}_{n,m}\mathbf{t}^T_{yzz} \\ -\mathbf{b}^{zzz}_{n,m}\mathbf{t}^T_{zzz} + \mathbf{b}^{xxz}_{n,m}\mathbf{t}^T_{xxz} + \mathbf{b}^{yyz}_{n,m}\mathbf{t}^T_{yyz} + \mathbf{b}^{xyz}_{n,m}\mathbf{t}^T_{xyz} \\ -\mathbf{b}^{zzz}_{n,m}\mathbf{t}^T_{zzz} + \mathbf{b}^{xxx}_{n,m}\mathbf{t}^T_{xxx} + \mathbf{b}^{xyy}_{n,m}\mathbf{t}^T_{xyy} + \mathbf{b}^{yyy}_{n,m}\mathbf{t}^T_{yyy} + \mathbf{b}^{xxy}_{n,m}\mathbf{t}^T_{xxy} \\ -\mathbf{b}^{xzz}_{n,m}\mathbf{t}^T_{xzz} - \mathbf{b}^{yzz}_{n,m}\mathbf{t}^T_{yzz} + \mathbf{b}^{xxz}_{n,m}\mathbf{t}^T_{xxz} + \mathbf{b}^{yyz}_{n,m}\mathbf{t}^T_{yyz} + \mathbf{b}^{xyz}_{n,m}\mathbf{t}^T_{xyz} \\ -\mathbf{b}^{xzz}_{n,m}\mathbf{t}^T_{xzz} - \mathbf{b}^{yzz}_{n,m}\mathbf{t}^T_{yzz} + \mathbf{b}^{xxx}_{n,m}\mathbf{t}^T_{xxx} + \mathbf{b}^{xyy}_{n,m}\mathbf{t}^T_{xyy} + \mathbf{b}^{yyy}_{n,m}\mathbf{t}^T_{yyy} + \mathbf{b}^{xxy}_{n,m}\mathbf{t}^T_{xxy} \end{bmatrix}. \tag{11}$$

If we disregard correlations between components of the third-order disturbing gravitational tensor, the co-factor matrix simplifies to the following diagonal form

$$\mathbf{Q} = \mathrm{diag}\left[\mathbf{Q}_{zzz}\ \mathbf{Q}_{xzz}\ \mathbf{Q}_{yzz}\ \mathbf{Q}_{xxz}\ \mathbf{Q}_{yyz}\ \mathbf{Q}_{xyz}\ \mathbf{Q}_{xxx}\ \mathbf{Q}_{xyy}\ \mathbf{Q}_{yyy}\ \mathbf{Q}_{xxy}\right], \tag{12}$$

where $\mathbf{Q}_{ijk}$, $i, j, k = x, y, z$ denote the individual co-factor matrices for $\mathbf{t}_{ijk}$, $i, j, k = x, y, z$.

The elements of the matrix $\mathbf{C}_{n,m}$ in Eq. (9) read

$$
\begin{aligned}
(c_{1,1})_{n,m} &= \mathbf{b}^{xzz}_{n,m}\mathbf{Q}_{xzz}\left(\mathbf{b}^{xzz}_{n,m}\right)^T + \mathbf{b}^{yzz}_{n,m}\mathbf{Q}_{yzz}\left(\mathbf{b}^{yzz}_{n,m}\right)^T + \mathbf{b}^{zzz}_{n,m}\mathbf{Q}_{zzz}\left(\mathbf{b}^{zzz}_{n,m}\right)^T,\\
(c_{2,1})_{n,m} &= \mathbf{b}^{zzz}_{n,m}\mathbf{Q}_{zzz}\left(\mathbf{b}^{zzz}_{n,m}\right)^T, (c_{1,3})_{n,m} = (c_{3,1})_{n,m} = (c_{2,3})_{n,m} = (c_{3,2})_{n,m} = (c_{1,2})_{n,m} = (c_{2,1})_{n,m},\\
(c_{4,1})_{n,m} &= -\mathbf{b}^{xzz}_{n,m}\mathbf{Q}_{xzz}\left(\mathbf{b}^{xzz}_{n,m}\right)^T - \mathbf{b}^{yzz}_{n,m}\mathbf{Q}_{yzz}\left(\mathbf{b}^{yzz}_{n,m}\right)^T,\\
(c_{1,4})_{n,m} &= -(c_{4,5})_{n,m} = -(c_{5,4})_{n,m} = (c_{1,5})_{n,m} = (c_{5,1})_{n,m} = (c_{4,1})_{n,m},\\
(c_{2,2})_{n,m} &= \mathbf{b}^{xxz}_{n,m}\mathbf{Q}_{xxz}\left(\mathbf{b}^{xxz}_{n,m}\right)^T + \mathbf{b}^{xyz}_{n,m}\mathbf{Q}_{xyz}\left(\mathbf{b}^{xyz}_{n,m}\right)^T + \mathbf{b}^{yyz}_{n,m}\mathbf{Q}_{yyz}\left(\mathbf{b}^{yyz}_{n,m}\right)^T + \mathbf{b}^{zzz}_{n,m}\mathbf{Q}_{zzz}\left(\mathbf{b}^{zzz}_{n,m}\right)^T,\\
(c_{4,2})_{n,m} &= (c_{2,4})_{n,m} = \mathbf{b}^{xxz}_{n,m}\mathbf{Q}_{xxz}\left(\mathbf{b}^{xxz}_{n,m}\right)^T + \mathbf{b}^{xyz}_{n,m}\mathbf{Q}_{xyz}\left(\mathbf{b}^{xyz}_{n,m}\right)^T + \mathbf{b}^{yyz}_{n,m}\mathbf{Q}_{yyz}\left(\mathbf{b}^{yyz}_{n,m}\right)^T,\\
(c_{2,5})_{n,m} &= (c_{5,2})_{n,m} = (c_{4,3})_{n,m} = (c_{3,4})_{n,m} = 0,\\
(c_{3,3})_{n,m} &= \mathbf{b}^{xxx}_{n,m}\mathbf{Q}_{xxx}\left(\mathbf{b}^{xxx}_{n,m}\right)^T + \mathbf{b}^{xxy}_{n,m}\mathbf{Q}_{xyy}\left(\mathbf{b}^{xxy}_{n,m}\right)^T + \mathbf{b}^{xyy}_{n,m}\mathbf{Q}_{xxy}\left(\mathbf{b}^{xyy}_{n,m}\right)^T + \mathbf{b}^{yyy}_{n,m}\mathbf{Q}_{yyy}\left(\mathbf{b}^{yyy}_{n,m}\right)^T\\
&\quad + \mathbf{b}^{zzz}_{n,m}\mathbf{Q}_{zzz}\left(\mathbf{b}^{zzz}_{n,m}\right)^T,\\
(c_{3,5})_{n,m} &= (c_{5,3})_{n,m} = \mathbf{b}^{xxx}_{n,m}\mathbf{Q}_{xxx}\left(\mathbf{b}^{xxx}_{n,m}\right)^T + \mathbf{b}^{xxy}_{n,m}\mathbf{Q}_{xyy}\left(\mathbf{b}^{xxy}_{n,m}\right)^T + \mathbf{b}^{xyy}_{n,m}\mathbf{Q}_{xxy}\left(\mathbf{b}^{xyy}_{n,m}\right)^T\\
&\quad + \mathbf{b}^{yyy}_{n,m}\mathbf{Q}_{yyy}\left(\mathbf{b}^{yyy}_{n,m}\right)^T,\\
(c_{4,4})_{n,m} &= \mathbf{b}^{xxz}_{n,m}\mathbf{Q}_{xxz}\left(\mathbf{b}^{xxz}_{n,m}\right)^T + \mathbf{b}^{xyz}_{n,m}\mathbf{Q}_{xyz}\left(\mathbf{b}^{xyz}_{n,m}\right)^T + \mathbf{b}^{xzz}_{n,m}\mathbf{Q}_{xzz}\left(\mathbf{b}^{xzz}_{n,m}\right)^T + \mathbf{b}^{yyz}_{n,m}\mathbf{Q}_{yyz}\left(\mathbf{b}^{yyz}_{n,m}\right)^T\\
&\quad + \mathbf{b}^{yzz}_{n,m}\mathbf{Q}_{yzz}\left(\mathbf{b}^{yzz}_{n,m}\right)^T,\\
(c_{5,5})_{n,m} &= \mathbf{b}^{xxx}_{n,m}\mathbf{Q}_{xxx}\left(\mathbf{b}^{xxx}_{n,m}\right)^T + \mathbf{b}^{xxy}_{n,m}\mathbf{Q}_{xyy}\left(\mathbf{b}^{xxy}_{n,m}\right)^T + \mathbf{b}^{xyy}_{n,m}\mathbf{Q}_{xxy}\left(\mathbf{b}^{xyy}_{n,m}\right)^T + \mathbf{b}^{yyy}_{n,m}\mathbf{Q}_{yyy}\left(\mathbf{b}^{yyy}_{n,m}\right)^T\\
&\quad + \mathbf{b}^{xzz}_{n,m}\mathbf{Q}_{xzz}\left(\mathbf{b}^{xzz}_{n,m}\right)^T + \mathbf{b}^{yzz}_{n,m}\mathbf{Q}_{yzz}\left(\mathbf{b}^{yzz}_{n,m}\right)^T.
\end{aligned}
\tag{13}
$$

Finally, after correcting observed values (i.e., components of the third-order gravitational tensor) by applying the adjusted errors,

$$\hat{\mathbf{L}} = \mathbf{L} - \hat{\boldsymbol{\varepsilon}}, \tag{14}$$

all four solutions must provide the same spherical harmonic coefficients. The resulting spherical harmonic coefficients can then conveniently be computed from the simplest solution according to Eq. (1a).

4 Numerical Experiment

The three methods of combining the solutions to the spherical gravitational curvature boundary-value problem presented in Sect. 3 are compared here.

4.1 Data Preparation

We synthetize ten components of the third-order gravitational tensor by expressions derived by Hamáčková et al. (2016) at a satellite altitude of 250 km (above the mean sphere of the radius $R = 6378136.3$ m). We use EGM2008 coefficients up to the spherical harmonic degree 360. The gravitational curvatures are generated in a global equiangular grid with the sampling interval of 0.25 arc-deg to avoid aliasing (Rexer 2017). We then generate the Gaussian noise with the standard deviation of 6.3×10^{-19} m^{-1} s^{-2}. This theoretical value characterizes an analytical error of the differential accelerometry (Šprlák et al. 2016) by assuming the following parameters (of a hypothetical gravity-dedicated satellite mission): the mutual separation of two adjacent accelerometers $\Delta\mathbf{x} = \mathbf{x}_3 - \mathbf{x}_2 = \mathbf{x}_2 - \mathbf{x}_1$, $\mathbf{x} = \{x, y, z\}$ is 0.5 m, the mission duration is 1,270 days, the satellite altitude is 250 km, the maximum spherical harmonic degree is 360 and the data-sampling interval is 1 s.

4.2 Design of the Numerical Experiment

We perform the spherical harmonic analysis according to the expressions in Eqs. (1a)–(1d) to calculate the spherical harmonic coefficients (up to the harmonic degree 360) individually for the VVV, VVH, VHH and HHH solutions. As inputs, we used ten defining components of the gravitational curvature tensor polluted by the Gaussian noise. Note that the noise of the spherical harmonics obtained from the VVV, VVH, VHH and HHH solutions respects the ratio $\varepsilon^{(0)}_{n,m} : \varepsilon^{(1)}_{n,m} : \varepsilon^{(2)}_{n,m} : \varepsilon^{(3)}_{n,m} \approx 1 : 1 : 3 : 10$ as derived in Šprlák et al. (2016). In other words, the VVV and VVH solutions contain the least noise, while the HHH solution is the most polluted. We removed from spherical harmonic coefficients the discretization error, before we calculated combined solutions. The discretization error was estimated as the difference between spherical harmonic coefficients from EGM2008 up to the degree 360 and spherical harmonic coefficients calculated from Eqs. (1a)–(1d). Note that we used gravitational curvatures synthetized directly from EGM2008 up to the degree 360, i.e., input values contained no noise. We then compute the combined solutions $C_{n,m}^{SM}$ by applying the arithmetic mean (Eq. 4), the weighted mean (Eq. 5) and the CAM (Sect. 3.3).

4.3 Results

To compare the three solutions, we present results in two different ways. First, we plot the number of common digits in the combined solutions obtained from the simple mean, the weighted mean and the CAM, see Fig. 2. As it can be seen, all three solutions have a relatively good fit with the EGM2008 reference values at low degrees roughly up to the degree 100. By increasing the spherical harmonics, the number of common digits decreases.

In addition, we plot the degree variances of all three solutions, see Fig. 3. As seen, all combined solutions fit the EGM2008 gravitational spectrum up to the degree 240. We obtained the largest differences with respect to EGM2008 for the CAM solution. The differences are about five orders of magnitude larger than those attributed to the arithmetic- and weighted-mean solutions. Moreover, the arithmetic mean solution has the best fit with the EGM2008 spectrum.

The results can be explained in a very simple way. As we discuss in Sect. 4.1, each of the ten components of the gravitational curvature tensor was polluted by the Gaussian noise with the standard deviation of 6.3×10^{-19} m^{-1} s^{-2}. Thus, the random errors in the four solutions of the gravitational curvature boundary-value problem are uncorrelated and reach the ratio $\varepsilon_{n,m}^{(0)} : \varepsilon_{n,m}^{(1)} : \varepsilon_{n,m}^{(2)} : \varepsilon_{n,m}^{(3)} \approx 1:1:3:10$, see Šprlák et al. (2016) for more information. In the solution based on the arithmetic mean each input had the same weight while in the other two solutions weights of inputs were based on the ratio defined above. Different weights applied to the four solutions of the gravitational curvature boundary-value problem have been reflected in the estimated values of spherical harmonics. Five conditions in the CAM only form additional stochastic bindings between the unknown spherical harmonics and the model is solvable also without conditions. As we could see the conditional equations make the solution even worse and do not improve the fit achieved by the weighted mean.

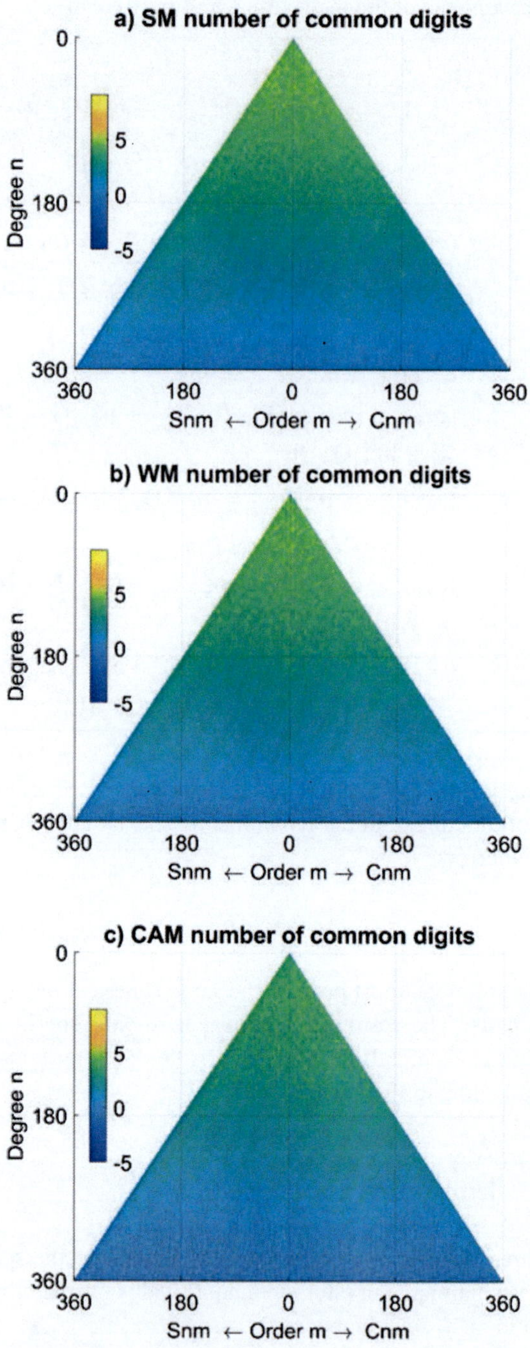

Fig. 2 Number of common digits of combined solution based on the simple mean (SM), **a**), the weighted mean (WM), **b**) and the condition adjustment model (CAM), **c**)

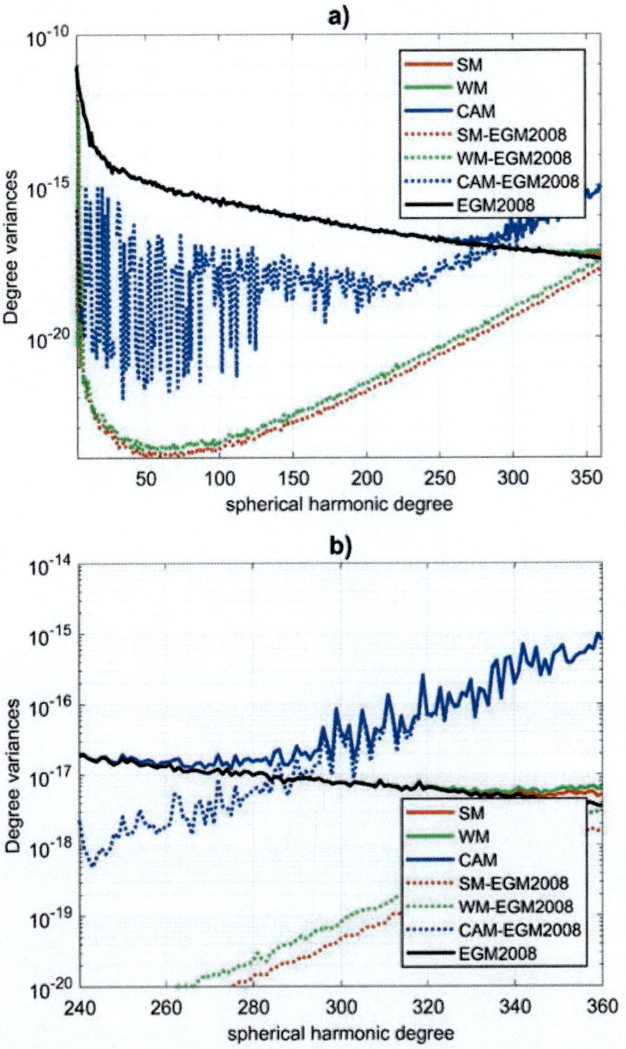

Fig. 3 Differences of degree variances between combinations obtained by the simple mean (SM), weighted mean (WM) and the condition adjustment model (CAM) with respect to EGM2008 up to the degree 360 (**a**) and the detail for the bandwidth 240–360 (**b**)

5 Conclusions

We test three different methods for combining the four solutions to the spherical gravitational curvature boundary-value problem, i.e., the simple and weighted means, and the conditional least-squares adjustment.

Despite the highest complexity of the conditional least-squares adjustment, this method does not provide the best result. On the contrary, the simple mean provides the best solution in terms of the fit of the predicted values to the EGM2008 reference model. Our results also confirm findings of Eshagh (2010), who conducted a similar study for combining the three solutions of the spherical gradiometric boundary-value problem and concluded that the simple mean method provides superior results.

The results also reveal that in the presence of an uncorrelated random noise equal for all ten components of the gravitational curvature tensor, the arithmetic mean provides the best solution and the more complex estimation models as the weighted mean or the conditional adjustment are unnecessary.

Acknowledgements Martin Pitoňák acknowledges the Czech Ministry of Education, Youth and Sport for a financial support of this research by the project No. LO1506. Pavel Novák was supported by the project No. 18-06943S of the Czech Science Foundation. Robert Tenzer was supported by the HK science project 1-ZE8F: Remote-sensing data for studying the Earth's and planetary inner structure. Thoughtful and constructive comments of two anonymous reviewers, the Assistant Editor-in-Chief and Series Editor are gratefully acknowledged.

References

Balakin AB, Daishev RA, Murzakhanov ZG, Skochilov AF (1997) Laser-interferometric detector of the first, second and third derivatives of the potential of the Earth gravitational field. Izvestiya vysshikh uchebnykh zavedenii, seriya Geologiya i Razvedka 1:101–107

Eshagh M (2010) Optimal combination of integral solutions of gradiometric boundary-value problem using variance component estimation in Earth gravitational modelling. Earth Planet Space 62:437–448

Hamáčková E, Šprlák M, Pitoňák M, Novák P (2016) Non-singular expressions for the spherical harmonic synthesis of gravitational curvatures in a local north-oriented reference frame. Comput Geosci 88:152–162

Jekeli C (2009) Potential theory and static gravity field of the earth. In: Herring T (ed) Treatise on geophysics: geodesy. Elsevier, Amsterdam, pp 12–41

Koch K-R (1999) Parameter estimation and hypothesis testing in linear models. Springer, Berlin

Martinec Z (2003) Green's function solution to spherical gradiometric boundary-value problem. J Geodesy 77:41–49

Pavlis NK, Holmes SA, Kenyon SC, Factor JK (2012) The development and evaluation of the Earth Gravitational Model 2008 (EGM2008). J Geophys Res 117:B04406. https://doi.org/10.1029/2011JB008916

Pavlis NK, Holmes SA, Kenyon SC, Factor JK (2013) Correction to "The development of the Earth Gravitational Model 2008 (EGM2008)". J Geophys Res 118:2633. https://doi.org/10.1002/jgrb.50167

Rexer M (2017) Spectral solutions to the topographic potential in the context of high-resolution global gravity field modelling. PhD thesis, Technische Universität München

Rosi G, Cacciapuoti L, Sorrentino F, Menchetti M, Prevedelli M, Tino GM (2015) Measurements of the gravity-field curvature by atom interferometry. Phys Rev Lett 114:013001

Šprlák M, Novák P (2016) Spherical gravitational curvature boundary-value problem. J Geodesy 90:727–739

Šprlák M, Novák P, Pitoňák M (2016) Spherical harmonic analysis of gravitational curvatures and implications for future satellite mission. Surv Geophys 37:681–700

Part II

Theory of Modern Geodetic Reference Frames

Review of Reference Frame Representations for a Deformable Earth

Zuheir Altamimi, Paul Rebischung, Xavier Collilieux, Laurent Métivier, and Kristel Chanard

Abstract

Our planet Earth is constantly deforming under the effects of geophysical processes that cause linear and nonlinear displacements of the geodetic stations upon which the International Terrestrial Reference Frame (ITRF) is established. The ITRF has traditionally been defined as a secular (linear) frame in which station coordinates are described by piecewise linear functions of time. Nowadays, some particularly demanding applications however require more elaborate reference frame representations that can accommodate non-linear displacements of the reference stations. Two such types of reference frame representations are reviewed: the usual linear frame enhanced with additional parametric functions such as seasonal sine waves, and non-parametric time series of quasi-instantaneous reference frames. After introducing those two reference frame representations, we briefly review the systematic errors present in geodetic station position time series. We finally discuss the practical issues raised by the existence of these systematic errors for the implementation of both types of non-linear reference frames.

Keywords

ITRF · Nonlinear motions · Reference frames · Reference systems

1 Introduction

Where am I and how to accurately navigate between places on Earth, oceans and in space? How to plan for territory and land management (construction, mining, civil engineering, national boundaries delimitation)? How to ensure that geospatial data are inter-operable within a country, a region and globally? How to locate areas and people at risk (natural disasters: earthquakes, tsunamis and flooding)? How to accurately determine orbits of artificial satellites? How to measure self-consistent sea level rise over several decades, through the usage of satellite altimetry data and tide gauges? How to accurately determine point positions on the Earth surface that is constantly deforming? In order to answer these crucial questions and to enable operational geodesy and Earth science applications, a unified terrestrial reference system and its materialization by an accurate terrestrial reference frame are needed. This is the purpose of the International Terrestrial Reference System (ITRS; see Chapter 4 in Petit and Luzum 2010) and of its materialization by the International Terrestrial Reference Frame (ITRF).

The successive releases of the ITRF are provided in the form of reference regularized coordinates $X(t)$ for a set of geodetic stations, described by mathematical functions of time and obtained from the adjustment of data from the four contributing space geodetic techniques (VLBI, SLR, GNSS, DORIS). From ITRF91 (Altamimi et al. 1993; Boucher et al. 1992) to ITRF2008 (Altamimi et al. 2011), reference station coordinates have been described by piecewise linear functions of time, able to capture linear station motions

Z. Altamimi (✉) · P. Rebischung · L. Métivier · K. Chanard
Institut National de l'Information Géographique et Forestière (IGN), Université Paris Diderot, Paris, France
e-mail: zuheir.altamimi@ign.fr; paul.rebischung@ign.fr; laurent.metivier@ign.fr; kristel.chanard@ign.fr

X. Collilieux
Ecole Nationale des Sciences Géographiques, Champs sur Marne, France
e-mail: xavier.collilieux@ensg.eu

P. Novák et al. (eds.), *IX Hotine-Marussi Symposium on Mathematical Geodesy*, International Association of Geodesy Symposia 151, https://doi.org/10.1007/1345_2019_66

(e.g., tectonic motions, post-glacial rebound) and abrupt position changes (e.g., co-seismic displacements, equipment changes). The latest ITRF2014 solution (Altamimi et al. 2016) additionally includes logarithmic and exponential functions that describe the post-seismic displacements of stations affected by large earthquakes.

Other non-linear crustal motions, such as non-tidal loading deformation, are however not accounted for by the ITRF station coordinates. This implies that the ITRF station coordinates do not represent the instantaneous shape of the Earth, but only a linearly varying approximation. This also has for consequence that the ITRF origin can indeed not coincide with the instantaneous Earth's center of mass (CM), but can only follow CM linearly with time.

The needs of most Earth science and operational geodesy applications can be met with a linear frame such as the ITRF. Some demanding applications such as satellite precise orbit determination (POD) however require precise instantaneous station coordinates (i.e., including non-linear station motions) expressed with respect to the instantaneous CM. This paper therefore discusses possible alternative reference frame representations able to capture non-linear station motions, and issues related to their practical implementation.

2 Non-linear Reference Frame Representations

Two main types of reference frame representations can be considered in order to capture non-linear station motions. The first possibility is to enhance the piecewise linear model of ITRF station coordinates with additional parametric functions. The second possibility is to represent a reference frame as a (non-parametric) time series of quasi-instantaneous frames. These two possible representations are briefly introduced in the next subsections.

2.1 Augmented Parametric Reference Frame

The classical piecewise linear model of ITRF station coordinates can in principle be augmented with additional parametric functions of time in order to describe non-linear station motions. This is already partly the case with the ITRF2014 solution, which includes logarithmic and exponential functions that describe the post-seismic displacements of stations affected by large earthquakes. In such an augmented parametric representation, the reference coordinates $X(t)$ of a station would be given by:

$$X(t) = X(t_0) + (t - t_0) \cdot \dot{X} + \sum \delta X(t)_{PSD} + \sum \delta X(t)_S \tag{1}$$

where $X(t_0) + (t - t_0) \cdot \dot{X}$ is the classical linear model of the station coordinates, $\sum \delta X(t)_{PSD}$ is a sum of parametric functions of time describing post-seismic displacements of the station (if any), and $\sum \delta X(t)_S$ is a sum of additional parametric functions of time describing the other non-linear displacements of the station with respect to CM.

Various choices could in principle be made for these additional parametric functions, such as polynomials, Fourier series or splines (Dermanis 2008). However, a large fraction of the non-linear variability in observed geodetic station position time series arises from seasonal variations, and the still unmodeled geophysical phenomena that induce non-linear deformation of the Earth's crust (e.g., non-tidal loading, thermal expansion) are also dominated by seasonal variations (Collilieux et al. 2007; Altamimi and Collilieux 2010). We therefore argue that sine waves at the annual frequency (and its few first harmonics) are likely the most beneficial choice of additional parametric functions $\sum \delta X(t)_S$. We will hence limit our discussion about augmented parametric reference frames to the augmentation of the ITRF piecewise linear model with annual, semi-annual, etc. sine waves.

2.2 Non-parametric Reference Frame

A quasi-instantaneous reference frame consists of reference coordinates of a network of stations valid at a given epoch only. A time series of such quasi-instantaneous CM-centered reference frames can in principle embed the non-linear motions of the stations with respect to CM, and defines what we refer to as a "non-parametric reference frame". Compared to a linear reference frame augmented with seasonal sine waves, such a non-parametric reference frame has the theoretical advantage that it can capture non-linear station motions at all frequencies, and not only at the annual harmonic. An example of such a non-parametric reference frame is the JTRF2014 (Abbondanza et al. 2017): a time series of weekly reference frames determined using the ITRF2014 input data of the four techniques and a Kalman filter and smoother approach. Another example of such frame is an SLR time series of station coordinates that naturally follow the instantaneous CM (as sensed by SLR) and the SLR intrinsic scale, but aligned in orientation to an external frame such as the ITRF.

3 Technique Systematic Errors

Whether obtained from the adjustment of parametric functions or from the sequential adjustment of quasi-instantaneous coordinates, non-linear reference frames would in any case be based on station position time series provided by the four contributing space geodetic techniques. The non-linear variations in those time series are however known to comprise various systematic errors, which, as we shall see in Sect. 4, raise serious practical issues for the implementation of non-linear reference frames.

Non-linear variations in GNSS station position time series are for instance known to reflect:

- Real geophysical crustal motions. Modeled loading deformation explains in particular about half of the observed vertical annual variations; but only 15–20% in horizontal (Ray et al. 2011; Xu et al. 2017), as well as a modest part of the observed aperiodic variations in vertical (Rebischung et al. 2018). Thermoelastic surface deformation has been shown to explain an additional 7–9% of the observed annual variations, both in horizontal and vertical (Yan et al. 2009; Xu et al. 2017).
- Artificial variations, such as the spurious periodic signals at harmonics of the GPS draconitic year evidenced by Ray et al. (2008), due to errors in GNSS observations or in their modelling.
- Unexplained variations: a significant fraction of the observed seasonal variations, as well as most of the observed aperiodic variations, remain to be precisely understood, but likely result from the superposition of multiple sources such as local (non-loading) ground deformation, thermal deformation of the monuments, GNSS systematic errors

Non-linear variations in SLR, VLBI and DORIS station position time series are generally noisier than in GNSS station position time series, hence not as well characterized, but must similarly result from the superposition of real ground deformation, monument deformation and technique systematic errors. Systematic errors are not only present in individual station position time series, but also in the technique determinations of the location of CM (i.e., geocenter motion) and of the terrestrial scale:

- Although they have benefited from various modelling improvements, recent DORIS determinations of geocenter motion and of the terrestrial scale still exhibit spurious non-linear variations (Moreaux et al. 2016; Altamimi et al. 2016).
- Geocenter motion time series derived from the GNSS contribution to ITRF2014 similarly show unreliable non-linear variations (Rebischung et al. 2016).
- SLR determinations of non-linear geocenter motion are considered as the most reliable, but are likely not free of systematic errors. Besides, SLR determinations of the terrestrial scale are known to be significantly affected by station range biases (Appleby et al. 2016).
- Finally, VLBI is insensitive to the location of CM, and VLBI determinations of the terrestrial scale are prone to errors due to thermal and gravitational deformation of the antennas (Sarti et al. 2009, 2011; Gipson 2018).

4 Consequences for the Implementation of Non-linear Reference Frames

This last section discusses several practical issues raised by the existence of the technique systematic errors summarized in Sect. 3 for the implementation of non-linear reference frames.

4.1 Augmented Parametric Reference Frame

The purpose of implementing non-linear reference frames is to provide demanding users with instantaneous reference station coordinates, which describe the instantaneous shape of the Earth, and are expressed with respect to the instantaneous CM. In the simplest possible case of a linear reference frame augmented with annual sine waves only, the objective is thus to determine reference annual motion of geodetic sites with respect to CM.

As mentioned in the previous section, DORIS and GNSS determinations of non-linear geocenter motion are currently still unreliable, so there is no other practical choice but to refer those reference annual site motions to the origin of SLR-derived annual displacements, even if it is not perfectly CM. The situation is similar as for the linear ITRF coordinates, which are referred, by default, to the linear CM as sensed by SLR.

The question then arises how to transfer the origin of SLR-derived annual displacements to the annual displacements of the other techniques. This could be done reliably if technique-specific annual displacements could be assumed to be similar (in the sense of a geometrical similarity transformation) over a set of co-located stations. The annual displacements from the different techniques could then be combined into unique reference annual site motions, just like station velocities from the different techniques are combined into unique site velocities in the ITRF computation.

Given the existence of the various systematic errors that affect technique-specific annual displacements (Sect. 3), the assumption of their similarity can however be questioned. What can indeed be the meaning of a combination of annual displacements from the different techniques when significant

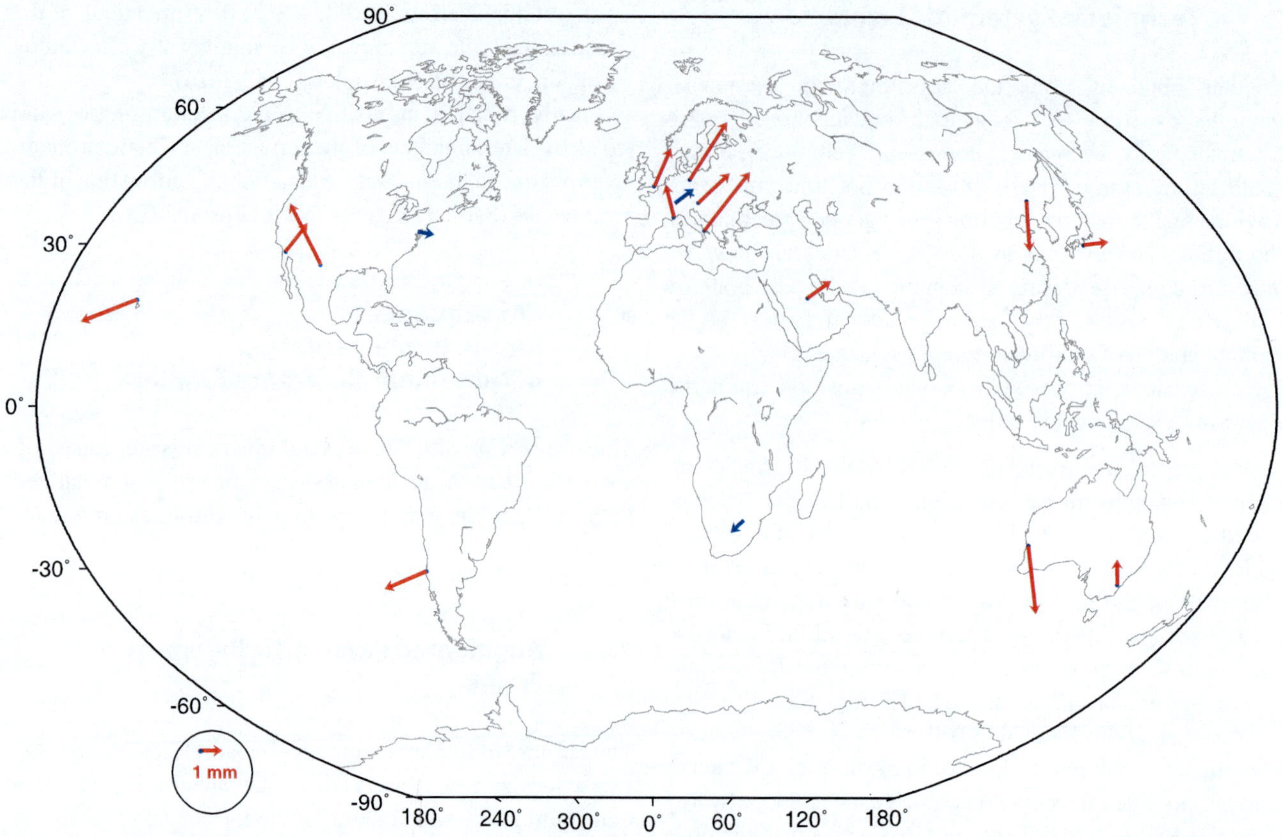

Fig. 1 Differences between vertical annual displacements at SLR stations derived, on one hand, from SLR solutions only, and on the other, from a combination of solutions from the four techniques. The amplitudes and phases of the differences are represented by the lengths and orientations of the arrows, respectively. Blue (resp. red) arrows indicate differences with amplitudes smaller (resp. larger) than 1 mm

fractions of the observed displacements do not reflect real ground motion, but technique systematic errors? In order to illustrate this issue, Fig. 1 shows the differences between vertical annual displacements at SLR stations derived, on one hand, from SLR solutions only, and on the other, from a combination of solutions from the four techniques. The data used are in both cases the technique inputs to ITRF2014, and the annual displacements are in both cases referred to the origin of SLR-derived annual displacements. The large differences visible in this figure for the majority of sites are the consequence of combining discrepant SLR- and GNSS-derived annual signals at co-location sites. Only 4 co-location sites, out of 18, show an agreement between SLR and GNSS annual signals better than 1 mm in amplitude.

A detailed study about the feasibility of combining annual displacements from the four techniques was carried out by Collilieux et al. (2018). They highlighted a number of co-location sites where the annual displacements from the different techniques are in clear disagreement. Apart from these "outliers", they found an overall level of agreement between annual displacements from GNSS, SLR and VLBI of the order of 1 mm in horizontal and 2 mm in vertical, DORIS-derived annual signals being clearly less consistent with the other techniques. These numbers are an indication of the best level of consistency (precision) that can be reached with current geodetic solutions. While they might appear satisfactory, the question nevertheless remains of how accurately the obtained combined annual displacements describe real crustal motion with respect to CM. This question can only be answered when:

- systematic errors in the annual signals from the different techniques are better understood,
- the accuracy of the annual geocenter motion sensed by SLR is externally assessed.

4.2 Non-parametric Reference Frame

Regarding the implementation of a non-parametric reference frame (i.e., a time series of quasi-instantaneous reference frames), the same issues as raised in the previous section for a linear reference frame augmented with annual variations basically hold, but over all frequencies instead of at the annual frequency only. The additional questions to be considered are thus:

- How accurate is the non-linear, non-annual geocenter motion sensed by SLR?
- How similar (i.e., combinable) are the non-linear, non-annual displacements from the different techniques at co-location sites?
- How representative of real ground motion are those non-linear, non-annual displacements?

These are to our knowledge open research questions which require dedicated investigations. A partial answer to the last question can however already be given: part of the observed non-linear, non-annual displacements, such as draconitic signals in GNSS station position time series, do indeed definitely not represent real ground motion. Such known systematic errors would consequently need to be taken into account when implementing of a non-parametric reference frame which aims at describing the actual shape of the Earth, rather than a mix of technique errors. More generally, the implementation of a non-parametric reference frame would require the technique systematic and random errors to be well characterized over all frequencies, so that they could be either filtered out, or accounted for statistically, when combining the non-linear displacements from the different techniques.

5 Conclusion

Two types of non-linear terrestrial reference frame representations have been considered in this paper: the usual linear frame enhanced with additional parametric functions such as seasonal sine waves, and non-parametric time series of quasi-instantaneous reference frames. The difficulties raised by the existence of technique systematic errors for the implementation of both types of non-linear reference frames have been discussed. The key issue lies in the fact that current knowledge does not allow to separate technique systematic errors from real non-linear ground motion/geocenter motion. In view of establishing reliable and accurate non-linear terrestrial reference frames, a better understanding and characterization of the technique systematic errors is therefore essential.

Acknowledgements The ITRF activities are funded by the Institut national de l'information géographique et forestière (IGN), France, and partly by Centre National d'Etudes Spatiales (CNES).

References

Abbondanza C, Chin TM, Gross RS, Heflin MB, Parker J, Soja BS, van Dam T, Wu X (2017) JTRF2014, the JPL Kalman filter, and smoother realization of the international terrestrial reference system. J Geophys Res Solid Earth 122(10):8474–8510. https://doi.org/10.1002/2017JB014360

Altamimi Z, Collilieux X (2010) Quality assessment of the IDS contribution to ITRF2008. Adv Space Res 45(12):1500–1509. https://doi.org/10.1016/j.asr.2010.03.010

Altamimi Z, Boucher C, Duhem L (1993) The worldwide centimetric terrestrial reference frame and its associated velocity field. Adv Space Res 13(11):151–160

Altamimi Z, Collilieux X, Métivier L (2011) ITRF2008: an improved solution of the international terrestrial reference frame. J Geod 85(8):457–473. https://doi.org/10.1007/s00190-011-0444-4

Altamimi Z, Rebischung P, Métivier L, Collilieux X (2016) ITRF2014: a new release of the international terrestrial reference frame modeling non-linear station motions. J Geophys Res Solid Earth 121(8):6109–6131. https://doi.org/10.1002/2016JB013098

Appleby G, Rodriguez R, Altamimi Z (2016) Assessment of the accuracy of global geodetic satellite laser ranging observations and estimated impact on ITRF scale: estimation of systematic errors in LAGEOS observations 1993–2014. J Geod. 90(12):1371–1388. https://doi.org/10.1007/s00190-016-0929-2

Boucher C, Altamimi Z, Duhem L (1992) ITRF 91 and its associated velocity field. IERS Tech Note 12:1–142

Collilieux X, Altamimi Z, Coulot D, Ray J, Sillard P (2007) Comparison of very long baseline interferometry, GPS, and satellite laser ranging height residuals from ITRF2005 using spectral and correlation methods. J Geophys Res 112(B12403). https://doi.org/10.1029/2007JB004933

Collilieux X, Altamimi Z, Métivier L, Rebischung P, Chanard K, Ray J, Coulot D (2018) Comparison of the seasonal displacement parameters estimated in the ITRF2014 processing, what can we learn? In: 42nd COSPAR scientific assembly, Pasadena, California, 14–22 July

Dermanis A (2008) The ITRF beyond the "linear" model. Choices and challenges. In: VI Hotine-Marussi symposium on theoretical and computational geodesy. Springer, Berlin/Heidelberg, pp. 111–118

Gipson J (2018) Impact of gravitational deformation in VLBI analysis on the reference frame. In: Abstract G42A-04 presented at AGU fall meeting 2018, Washington, DC, 10–14 Dec

Moreaux G, Lemoine FG, Capdeville H, Kuzin S, Otten M, Stepanek P, Willis P, Ferrage P (2016) The international DORIS service contribution to the 2014 realization of the international terrestrial reference frame. Adv Space Res 58(12):2479–2504. https://doi.org/10.1016/j.asr.2015.12.021

Petit G, Luzum B (2010) IERS conventions (2010), IERS Technical Note 36. Verlag des BundesamtsfürKartographie und Geodäsie, Frankfurt am Main, 179 pp.

Ray J, Altamimi Z, Collilieux X, van Dam T (2008) Anomalous harmonics in the spectra of GPS position estimates. GPS Solutions 12(1):55–64. https://doi.org/10.1007/s10291-007-0067-7

Ray J, Collilieux X, Rebischung P, van Dam TM, Altamimi Z (2011) Consistency of crustal loading signals derived from models and GPS: inferences for GPS positioning errors. In: Abstract G51B-06 presented at AGU fall meeting 2011, San Francisco, CA, 5–9 Dec

Rebischung P, Altamimi Z, Ray J, Garayt B (2016) The IGS contribution to ITRF2014. J Geod 90(7):611–630. https://doi.org/10.1007/s00190-016-0897-6

Rebischung P, Altamimi Z, Chanard K, Métivier L (2018) Aperiodic surface mass transport observed in GRACE and GNSS time series. In: Presented at IGS workshop 2018, Wuhan, China, 29 Oct.–2 Nov

Sarti P, Abbondanza C, Vittuari L (2009) Gravity-dependent signal path variation in a large VLBI telescope modelled with a combination of surveying methods. J Geod 83(11):1115–1126. https://doi.org/10.1007/s00190-009-0331-4

Sarti P, Abbondanza C, Petrov L, Negusini M (2011) Height bias and scale effect induced by antenna gravity deformations in geodetic VLBI data analysis. J Geod 85(1):1–8. https://doi.org/10.1007/s00190-010-0410-6

Xu X, Dong D, Fang M, Zhou T, Wei N, Zhou F (2017) Contributions of thermoelastic deformation to seasonal variations in GPS station position. GPS Solutions 21(3):1265–1274. https://doi.org/10.1007/s10291-017-0609-6

Yan H, Chen W, Zhu Y, Zhang W, Zhong M (2009) Contributions of thermal expansion of monuments and nearby bedrock to observed GPS height changes. Geophys Res Lett 36(13):L13301. https://doi.org/10.1029/2009GL038152

Impacts of the LARES and LARES-2 Satellite Missions on the SLR Terrestrial Reference Frame

Rolf König, Susanne Glaser, Ignazio Ciufolini, and Antonio Paolozzi

Abstract

LARES, an Italian satellite launched in 2012, and its successor LARES-2 approved by the Italian Space Agency, aim at the precise measurement of frame dragging predicted by General Relativity and other tests of fundamental physics. Both satellites are equipped with Laser retro-reflectors for Satellite Laser Ranging (SLR). Both satellites are also the most dense particles ever placed in an orbit around the Earth thus being nearly undisturbed by nuisance forces as atmospheric drag or solar radiation pressure. They are, therefore, ideally suited to contribute to the terrestrial reference frame (TRF). At GFZ we have implemented a tool to realistically simulate observations of all four space-geodetic techniques and to generate a TRF from that. Here we augment the LAGEOS based SLR simulation by LARES and LARES-2 simulations. It turns out that LARES and LARES-2, alone or in combination, can not deliver TRFs that meet the quality of the LAGEOS based TRF. However, once the LARES are combined with the LAGEOS satellites the formal errors of the estimated ground station coordinates and velocities and the co-estimated Earth Rotation Parameters are considerably reduced. The improvement is beyond what is expected from error propagation due to the increased number of observations. Also importantly, the improvement concerns in particular origin and scale of the TRF of about 25% w.r.t. the LAGEOS-combined TRF. Furthermore, we find that co-estimation of weekly average range biases for all stations does not change the resulting TRFs in this simulation scenario free of systematic errors.

Keywords

LAGEOS · LAGEOS-2 · LARES · LARES-2 · Terrestrial reference frame

1 Introduction

The project GGOS-SIM (Schuh et al. 2015) resulted in a powerful tool that enables the simulation of the space-geodetic techniques Very Long Baseline Interferometry (VLBI), Satellite Laser Ranging (SLR), Global Navigation Satellite Systems (GNSS), and Doppler Orbitography and Radiopositioning Integrated by Satellite (DORIS) in order to test various effects on the Terrestrial Reference Frame (TRF). The requirements set by the Global Geodetic Observing System (GGOS) on accuracy and stability of the TRF are 1 mm and 0.1 mm/year (Gross et al. 2009). In a first attempt, the observations of the 2008 to 2014 (inclusive) ground networks of all the space-geodetic techniques have

R. König (✉)
German Research Centre for Geosciences GFZ, DLR Oberpfaffenhofen, Wessling, Germany
e-mail: koenigr@gfz-potsdam.de

S. Glaser
German Research Centre for Geosciences GFZ, Potsdam, Germany

I. Ciufolini
Università del Salento, Lecce, Italy

A. Paolozzi
Sapienza Università di Roma, Roma, Italy

P. Novák et al. (eds.), *IX Hotine-Marussi Symposium on Mathematical Geodesy*, International Association of Geodesy Symposia 151, https://doi.org/10.1007/1345_2019_84

been simulated close to reality. Eventually the individual techniques are evaluated for the derivation of technique-specific and combined TRFs. The simulation of VLBI observations and VLBI-only TRFs is described in Glaser et al. (2016). The combination of the VLBI and SLR techniques based on so-called global and local ties and the extension of the global VLBI network by new stations is discussed in Glaser et al. (2017). The extension by new stations in case of SLR is discussed in Otsubo et al. (2016), Kehm et al. (2018) and Glaser et al. (2019b). The impact of different local tie scenarios on the combined GPS, SLR, and VLBI TRF was investigated in Glaser et al. (2019a). Simulations of LARES and LARES-2 regarding their main purpose to test General Relativity were performed by e.g., Ciufolini et al. (2013, 2017b)

For recent global TRFs, f.i. the ITRF2014 (Altamimi et al. 2016), SLR provides the fundamental datum parameters origin and, together with VLBI, the scale. The input from SLR to the ITRF2014 is provided by the analysis and combination centers of the International Laser Ranging Service (ILRS, Pearlman et al. 2002) where the solution is mainly based on LAGEOS and LAGEOS-2 observations. Also involved are observations to the ETALON and ETALON-2 satellites, however their amount is so small that they hardly play any role. Therefore, the GGOS-SIM SLR base is composed of LAGEOS mission data only. Currently, the Analysis Standing Committee (ASC) of the ILRS has pilot projects on the way to also include LARES observations for the contribution to the next generation ITRF.

In the following, the GGOS-SIM base of LAGEOS and LAGEOS-2 SLR simulated observations is augmented by simulated observations to the satellites LARES and LARES-2. With the augmented data base we evaluate their impact on the resulting TRF with a particular view on origin and scale.

2 The Satellite Missions and Data Used

The characteristics of the satellite missions involved here are listed in Table 1. Where the LAGEOS satellites have been designed for geophysical applications, the LARES satellites serve the measurement of frame-dragging, a phenomenon predicted by General Relativity (GR) (Ciufolini et al. 2017a). However, both objectives can be assigned to each mission due to the cannon ball shape of the satellites and their favorable area-to-mass ratio minimizing nuisance forces, e.g., solar radiation pressure. Indeed LARES obeys the lowest value of area-to-mass ratio, making it the densest object ever sent into orbit and therefore makes it together with the large eccentricity of the orbit a nearly ideal particle for testing effects of GR (Paolozzi et al. 2015).

For GGOS-SIM the LAGEOS and LAGEOS-2 SLR observations are simulated close to reality in terms of time of operation of a station, and in terms of number and accuracy. For this the real SLR observations of 51 ground stations were analyzed first. The simulations followed then assuming no systematic errors, just white noise, with no leaps in the coordinate time series. Figure 1 shows the number of observations for each station for each arc over the analysis period for the real and the simulated data at the example of LAGEOS-2. Slight differences can be found where stations observing in reality with different eccentricities (and therefore with different occupation numbers) are simulated as one site only. Also one station with very few passes was left out.

LARES has been tracked by SLR since its launch in 2012, therefore the simulations are as in case of the LAGEOS satellites simulated close to reality in terms of time of operations, and number and accuracy of the observations. Figure 2 shows the orbital fits for each station for each arc at the example of LARES and LARES-2 in the simulation. It has to be noted that the observation period of LARES starts due to its launch in 2012 only which yields an overlap with the LAGEOS analysis period of three years only. In order to get a longer analysis period to properly solve for the velocities of new stations in the evolving network, the analysis for LARES is prolonged to 2017 (inclusive). LARES-2 is not yet in orbit, however already approved as mission by the Italian Space Agency (Agenzia Spaziale Italiana – ASI) and scheduled for a launch around 2019 to 2020. The simulation of the LARES-2 observations follows the real world scenario of LARES in the years 2012 to 2017 in terms of time of operations, and number and accuracy of observed ranges. Potentially the number of LARES-2 observations could be higher than that of LARES due to its higher orbital altitude. In fact, in spite of the expected much lower intensity of laser returns from LARES-2 with respect to LARES, the coverage for a higher altitude satellite is more favourable (see Fig. 3).

Table 1 Characteristics of the SLR missions

Satellite	Launch (year)	Altitude (km)	Eccentricity	Inclination (deg)	Area-to-mass ratio (m^2/kg)
LAGEOS	1976	5,900	0.004	109.8	0.000695
LAGEOS-2	1992	5,800	0.014	52.6	0.000697
LARES	2012	1,440	0.001	69.5	0.000269
LARES-2	2019–2020	5,900	0.001	70.2	0.000269

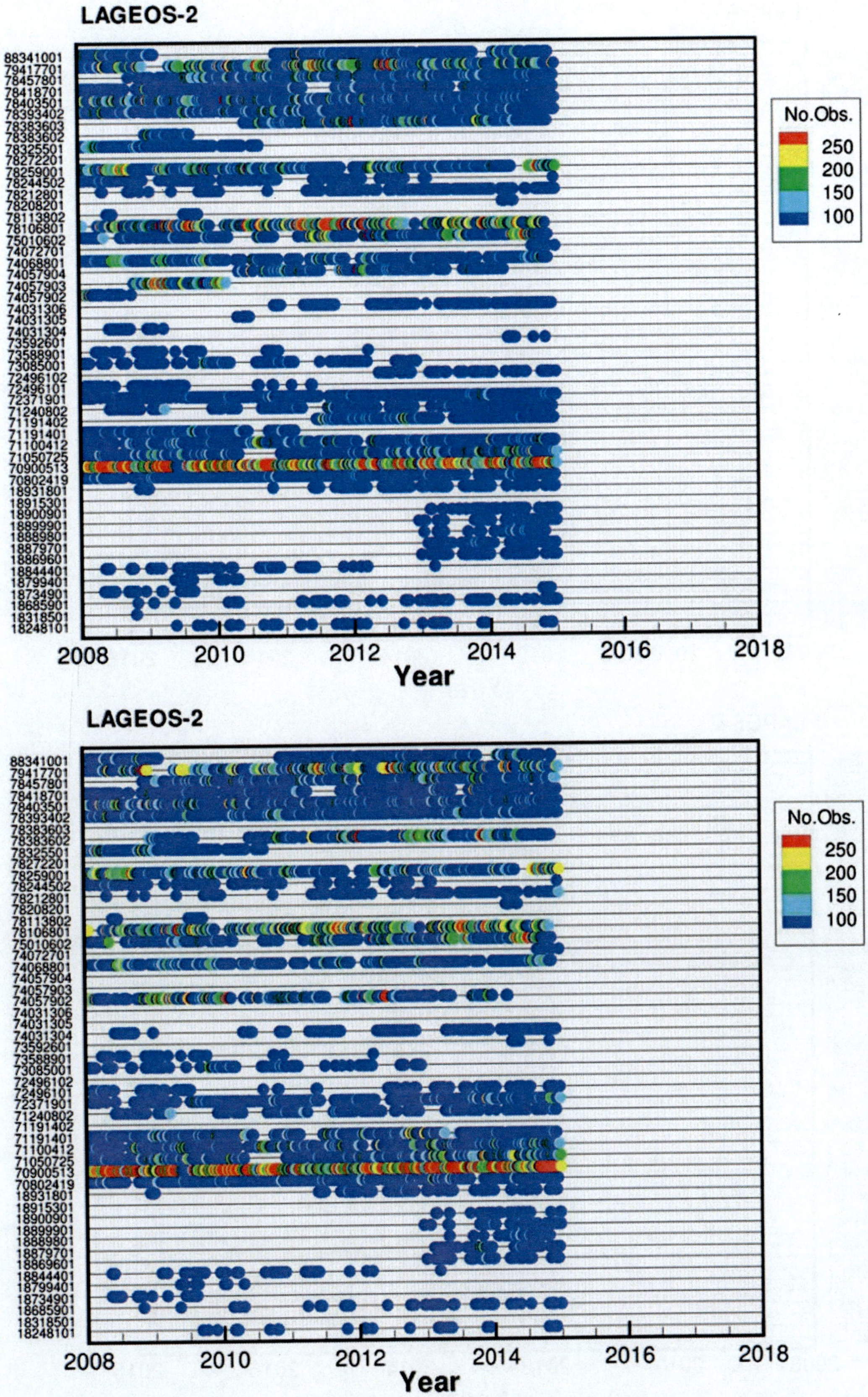

Fig. 1 Timely distribution and number of observations for LAGEOS-2, (top) for the real data, (bottom) for the simulated data

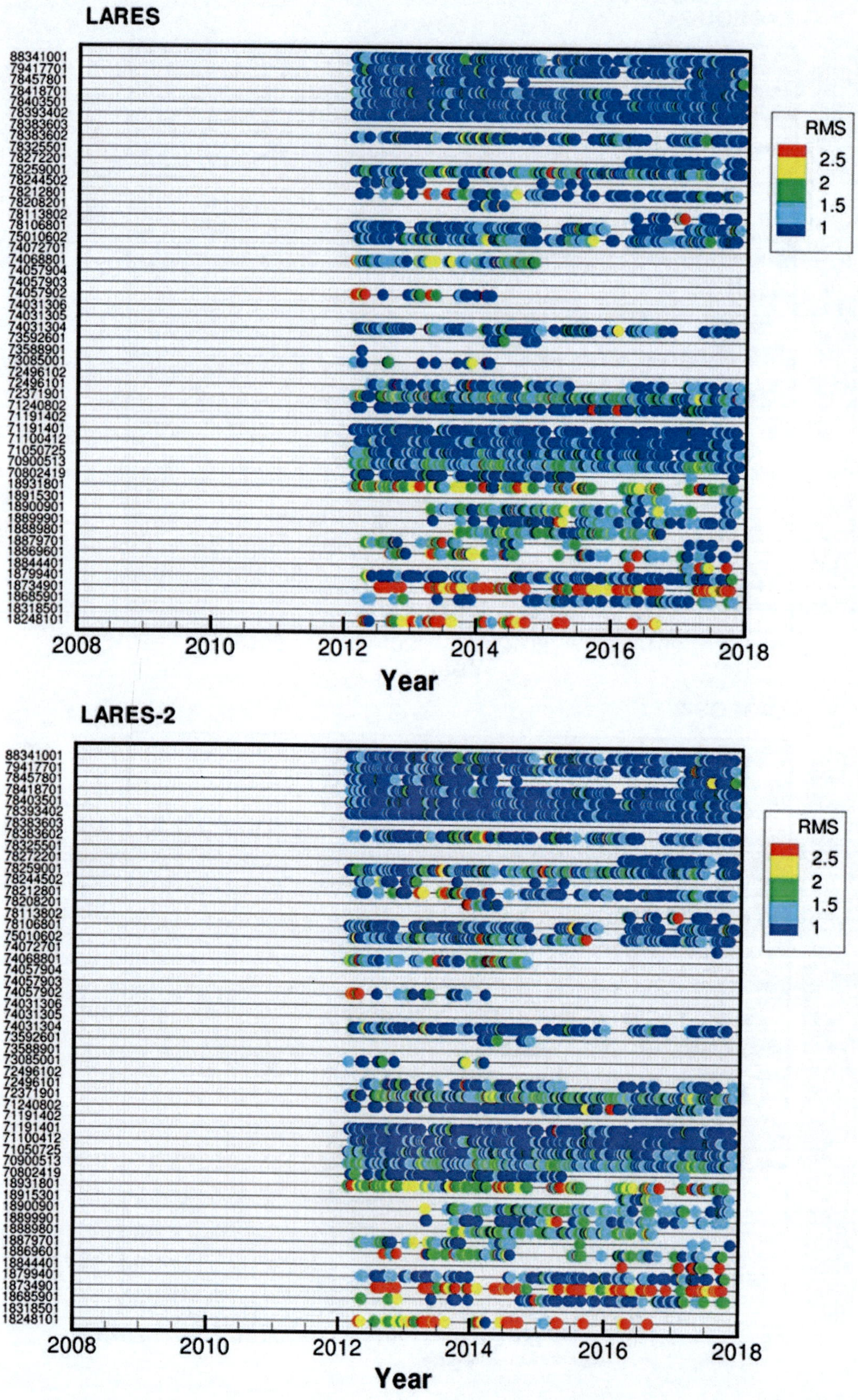

Fig. 2 Timely distribution and accuracy of the simulated observations for LARES and LARES-2

Fig. 3 Location and number of the simulated observations

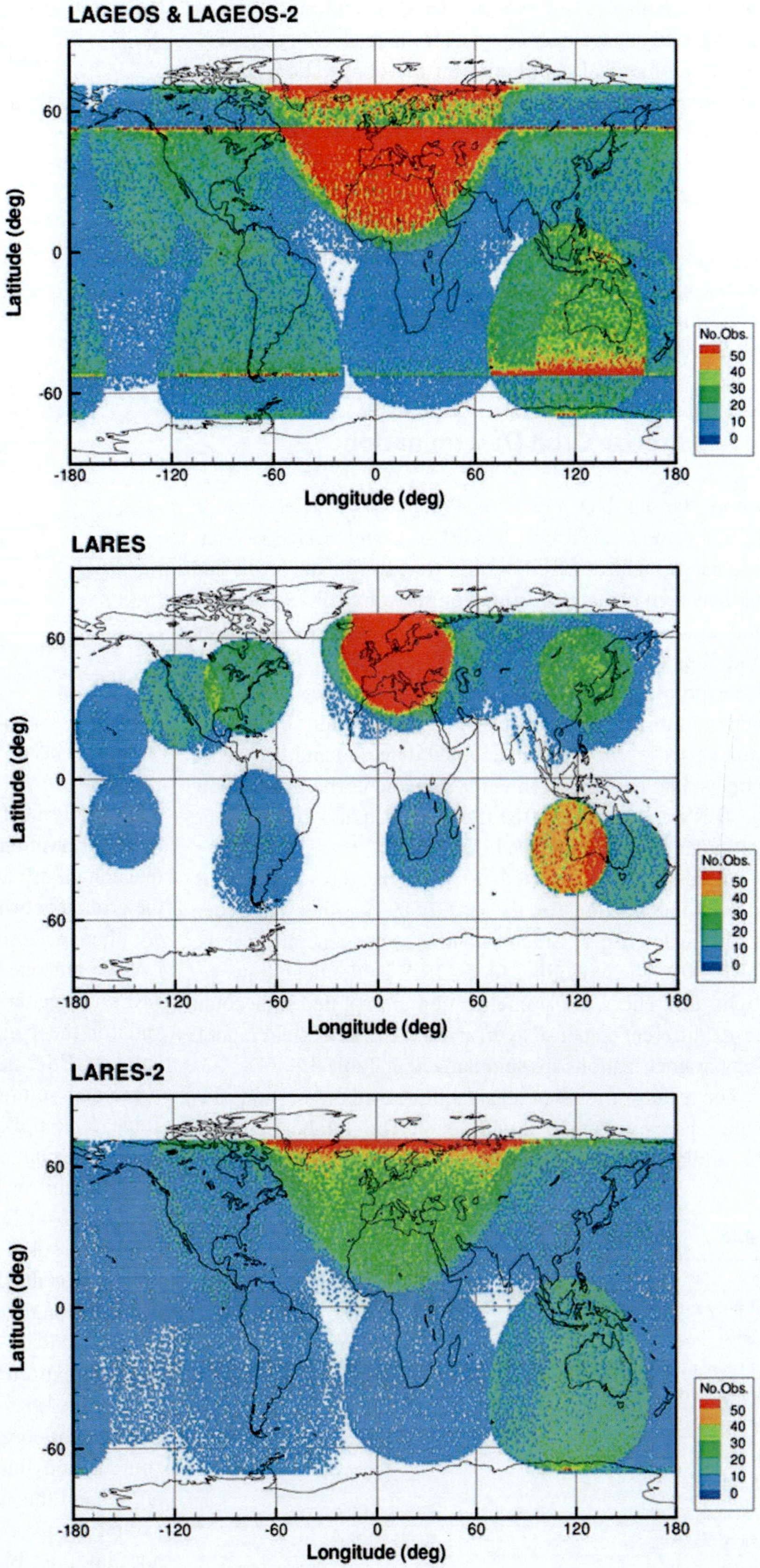

For the simulations however the geometry will not suffer as the observations are distributed over the respective arcs.

The geometrical distribution of the observations can be seen in Fig. 3 where the footprints of all observations in the analysis periods are sampled for number of occurrence in 1 × 1 degree bins over the Earth's surface. It becomes clear that large parts of the orbit of LARES are not covered with observations. But LARES-2 covers about the same geographical area as LAGEOS due to its identical orbital altitude and its inclination supplementary to that of LAGEOS delimiting the geographical distribution towards Northern and Southern latitudes the same way.

3 Precise Orbit Determination

Before starting the simulations, Precise Orbit Determination (POD) of real LAGEOS, LAGEOS-2, and LARES observations was performed. For this we rely on our orbit and Earth system parameter estimation software EPOS-OC (Zhu et al. 2004). EPOS-OC uses the dynamic approach, based on modelling the forces acting on the satellite. The highly nonlinear problem is solved by differential parameter improvement minimizing the residuals of the observations in the least squares sense. Most of the adopted dynamic and geometric models, and the measurement systematic corrections follow the IERS conventions 2010 (Petit and Luzum 2010), some particular choices are given in Table 2.

Processing is conducted in seven day arcs. Modelling and parametrization for the LAGEOS satellites is chosen according to GFZ's SLR contribution to the generation of the ITRF2008. The modelling for LARES is identical, however slight differences in parametrization are applied to account for its different response to errors of the gravity field model. The parametrizations are summarized in Table 3.

The results of POD of all satellites based on real and simulated data are compared in Table 4. The numbers confirm the similarity of the simulations with reality.

Table 2 Models for POD

Type	Model
Gravity model	EIGEN-6C
ERPs	IERS C04 08 Bizouard and Gambis (2011)
Ephemerides	JPL421
Solar radiation	Cannon ball
Albedo	Heurtel
Ocean tides	Not modelled
Ocean pole tides	Desai (2002)
Coordinates	SLRF2008
Ocean loading	Chalmers feat. FES2004
Atmospheric loading	Not applied
Troposphere	Mendes and Pavlis (2004)

Table 3 Parametrization

Type	LAGEOS	LARES
Initial states	6/arc	6/arc
Albedo global scaling factor	1/arc	1/arc
Atmosph. drag global scaling factor	–	1/arc
Empirical accelerations	1 const. acc./4d in T 1cpr/4d in T	1 const. acc./4d in T 1cpr/4d in T
ERPs	xp, yp, LOD /d	xp, yp, LOD /d
Coordinates	X, Y, Z /station/arc	X, Y, Z /station/arc
Velocities	$\dot{X}$ $\dot{Y}$, $\dot{Z}$ /station/arc	$\dot{X}$ $\dot{Y}$, $\dot{Z}$ /station/arc

Table 4 Orbital fits

	Real data		Simulated data	
Satellite	RMS (cm)	No.	RMS (cm)	No.
LAGEOS	0.88	528,742	0.86	529,600
LAGEOS-2	0.91	468,869	0.89	469,994
LARES	1.23	476,270	1.20	477,505
LARES-2	–	–	1.20	474,453

4 Terrestrial Reference Frame

After verifying the simulated observations in POD, the simulated observations are further processed to yield weekly normal equations containing position and velocity parameters of the ground network and ERPs. The weekly normal equations are then accumulated to yield one normal equation for LARES and one for LARES-2. From these normal equations, TRFs are generated for LARES and LARES-2 separately and for the combination of both. Eventually the LARES and LARES-2 normal equations are added to the LAGEOS-combined solution either one by one or in combination. Table 5 compiles the mean percentages of improvement of the formal errors for positions and velocities of the stations, and of the ERPs, i.e. the two polar motions and the Length-of-Day (LOD) parameter, w.r.t. the LAGEOS-combined solution. The improvement expected from error propagation due to the increased number of observations are also given in the last column denoted by "Exp.".

The LARES-only and the LARES-2-only TRFs do not meet the expected precision of the station position and ERP parameters, so does not the LARES-combined TRF featuring a number of observations comparable to that of the LAGEOS combination due to a less favourable geometry. On the other hand the station velocities from the LARES solutions exhibit the expected precision meaning that geometry does not play that role here. At a profit the improvement of the LAGEOS-combined TRF by adding LARES and LARES-2 can be seen in pronounced smaller formal errors of the

Table 5 Improvement of formal errors w.r.t. LAGEOS-combined (LC)

Satellite	Pos. (%)	Vel. (%)	ERPs (%)	Exp. (%)
LARES	−95	−38	−143	−45
LARES-2	−103	−43	−149	−46
LARES+LARES-2	−37	3	−64	−2
LC+LARES	38	53	10	18
LC+LARES-2	36	51	9	18
LC+LARES+LARES-2	43	57	16	28

Table 6 Improvement in origin and scale w.r.t. LAGEOS-combined (LC)

Satellite	Tx (%)	Ty (%)	Tz (%)	Scale (%)
LARES	−79	−73	−46	−157
LARES-2	−79	−71	−72	−140
LARES+LARES-2	−19	−15	−5	−68
LC+LARES	12	13	21	4
LC+LARES-2	12	14	14	5
LC+LARES+LARES-2	23	24	29	10

estimated positions and velocities of the ground stations. This improvement goes beyond the expectation due to the increased number of observations and can be attributed to a better observation geometry in case of the positions, the station velocities benefit from the longer analysis period. The increase in precision of the estimated ERPs however stays behind expectation and is owned to the small overlap of three years only between the analysis periods of the LAGEOS and LARES satellites.

The improvement of the TRF in its defining parameters origin and scale is computed according to the approach by Sillard and Boucher (2001) where the variance-covariance matrix of the solution is divided into a datum dependent part and an independent one. The dependent part shows the reference system effect in the standard deviations of the Helmert parameters. As SLR provides origin and scale in international TRF solutions where the space-geodetic techniques are combined, we compile in Table 6 the improvement of origin and scale w.r.t. the LAGEOS-combined solution of the LARES-only, the LARES-2-only, and of the LARES-combined TRF and the impact of the addition of LARES or LARES-2 or of both to the LAGEOS-combined solution.

The LARES-only and the LARES-2-only TRFs can not compete with the LAGEOS-combined TRF in terms of origin and scale definition. The LARES-combined TRF shows just a slight degradation in origin but a large deficiency in scale. The latter one might come from the relatively low altitude of LARES coming along with smaller ranges between the ground stations and the satellite and therefore resulting in less favorable ratios between the observed ranges and their errors. However once the LARES and LARES-2 observations are combined with the LAGEOS-combined solution, indeed considerable improvements of up to 29% can be expected for the core contribution of SLR to the TRF, i.e. origin and scale.

Apart from the improvements in the stochastic characteristics of the estimated station positions and velocities, the question arises whether adding of LARES and LARES-2 to the TRF solution leads to any systematic changes of the TRF. Therefore 14-parameter Helmert transformations are carried out where the LAGEOS+LARES, LAGEOS+LARES-2, and LAGEOS+LARES+LARES-2 TRFs are transformed w.r.t. the LAGEOS-combined TRF. All Helmert parameters are in the sub-millimeter range and are statistically not significant. This means that the addition of the new missions to the LAGEOS-combined TRF does not lead to a systematic change in the definition of the TRF.

5 On the Estimation of Range Biases

Appleby et al. (2016) advertised to estimate weekly average range biases for all SLR stations in the network in order to reduce the scale difference between SLR and very long baseline interferometry (VLBI) in the recent ITRFs. In preparation of the next generation ITRF the ILRS is running a pilot project where the estimation of range biases is analyzed. Here we follow these recommendations and estimate range biases besides station positions and velocities and ERPs for all solution types. The range biases are set up per station per satellite per arc (week) and endowed with an a priori sigma of 1 m. From all solutions with range biases being estimated adjacent TRFs are generated that can be compared to their counterparts with no range biases estimated. The comparison is done via 14-parameter Helmert transformations, the results are compiled in Table 7.

In all cases, i.e. LAGEOS-combined, LARES-only, LARES-2-only, and LAGEOS-combined plus LARES plus LARES-2, all Helmert parameters turn out with values for translations, rotations and scale and their derivatives below statistical significance. This means that estimation of range biases from simulated observations (with known a priori values) as described does not lead to a significant change of the TRF-defining parameters. In particular one should take note that the scale is not destroyed by estimating range biases in this simulation scenario where no systematic errors have been introduced a priori. In the real world, where systematic errors can not be ruled out, the conclusion might be different. To find out if systematic errors would change the above findings, extensive analyzes will be needed that are beyond the scope of this paper and are left therefore for future studies.

Table 7 14-Parameter Helmert transformations between solutions with range biases being estimated yes and no

Parameter	Value (mm)	St.Dev. (mm)	Deriv. (mm/a)	St.Dev. (mm/a)
LAGEOS-comb				
Tx	−0.09	0.48	−0.30	0.30
Ty	−0.12	0.49	0.08	0.30
Tz	−0.29	0.47	−0.10	0.29
Rx	−0.09	0.59	−0.09	0.36
Ry	−0.16	0.58	−0.19	0.36
Rz	0.05	0.56	−0.02	0.35
Sc	−0.35	0.46	−0.21	0.29
LARES				
Tx	−0.46	0.80	0.03	0.17
Ty	1.11	0.81	−0.25	0.17
Tz	−0.58	0.77	0.04	0.17
Rx	−0.20	0.97	0.02	0.21
Ry	0.00	0.95	0.02	0.20
Rz	0.59	0.93	−0.10	0.20
Sc	−0.58	0.77	0.19	0.17
LARES-2				
Tx	−0.99	1.77	0.20	0.30
Ty	0.37	1.79	−0.20	0.30
Tz	0.11	1.71	−0.14	0.29
Rx	0.38	2.16	0.00	0.36
Ry	−1.04	2.10	0.15	0.35
Rz	0.15	2.05	−0.06	0.35
Sc	−1.78	1.70	0.29	0.29
LAGEOS-comb+LARES+LARES-2				
Tx	0.12	0.19	−0.02	0.04
Ty	−0.08	0.19	0.04	0.04
Tz	−0.01	0.18	−0.01	0.04
Rx	0.26	0.23	−0.02	0.05
Ry	0.04	0.22	0.00	0.05
Rz	0.09	0.21	−0.03	0.05
Sc	−0.17	0.18	0.07	0.04

6 Summary and Conclusions

The project GGOS-SIM has provided a tool to simulate the space-geodetic techniques for the generation of global TRFs. Available are realistic, representative solutions for the years 2008 to 2014 where the SLR solutions are based on the LAGEOS and LAGEOS-2 missions. Here we simulated SLR observations to the LARES satellite over the years 2012 to 2017 following closely the analysis of the real world data. In addition we simulated SLR observations to the planned LARES-2 satellite relying on the LARES scenario in terms of accuracy and number of observations. It turns out that both LARES missions, either alone or in combination, can hardly compete with the LAGEOS combined TRF. However in combination with the LAGEOS they considerably improve the resulting coordinates and velocities of the SLR stations in terms of lower formal errors. The improvement is beyond what is expected from error propagation by the increased number of observations. The ERPs are also improved in the formal errors however at a lesser amount as in case of the coordinates because of the shorter overlap in the parameter space. Also origin and scale of the resulting TRFs are improved by about 25% when the LAGEOS and LARES missions are combined. Systematic changes of the TRF defining parameters identified by 14-parameter Helmert transformations were not found when adding the LARES missions. An attempt was made to assess the effect of estimating weekly average range biases for all stations for all satellites besides station positions and velocities and EOPs. The resulting TRFs are statistically not different from their counterparts where the said range biases are not estimated.

Acknowledgements SLR data and a priori station coordinates are provided by the ILRS. EOPs are provided by IERS (IERS 2018). The basic LAGEOS results have been achieved within project GGOS-SIM (SCHU 1103/8-1) of the German Research Foundation (Deutsche Forschungsgemeinschaft, DFG). I.C. and A.P. acknowledge ASI for supporting both LARES and LARES 2 missions under agreements No. 2015-021-R.0 and No. 2017-23-H.0. The authors would like to thank two anonymous reviewers for their valuable comments on the manuscript.

References

Altamimi Z, Rebischung P, Métivier L, Collilieux X (2016) ITRF2014: a new release of the international terrestrial reference frame modeling nonlinear station motions. J Geophys Res Solid Earth 121(8). https://doi.org/10.1002/2016JB013098

Appleby G, Rodriguez J, Altamimi Z (2016) Assessment of the accuracy of global geodetic satellite laser ranging observations and estimated impact on ITRF scale: estimation of systematic errors in LAGEOS observations 1993–2014. J Geod 90:12. https://doi.org/10.1007/s00190-016-0929-2

Bizouard C, Gambis, D (2011) The combined solution C04 for earth orientation parameters consistent with international terrestrial reference frame 2008. http://hpiers.obspm.fr/iers/eop/eopc04/C04.guide.-pdf

Ciufolini I, Moreno Monge B, Paolozzi A, Koenig R, Sindoni G, Michalak G, Pavlis EC (2013) Monte Carlo simulations of the LARES space experiment to test general relativity and fundamental physics. Classical Quantum Gravity 30:23. https://doi.org/10.1088/0264-9381/30/23/235009

Ciufolini I, Paolozzi A, Pavlis EC, Sindoni G, Koenig R, Ries JC, Matzner R, Gurzadyan V, Penrose R, Rubincam D, Paris C (2017a) A new laser-ranged satellite for general relativity and space geodesy: I. An introduction to the LARES2 space experiment. Eur Phys J Plus. https://doi.org/10.1140/epjp/i2017-11635-1

Ciufolini I, Pavlis EC, Sindoni G, Ries JC, Paolozzi A, Matzner R, Koenig R, Paris C (2017b) A new laser-ranged satellite for general relativity and space geodesy: II. Monte Carlo simulations and covariance analyses of the LARES 2 experiment. Eur Phys J Plus. https://doi.org/10.1140/epjp/i2017-11636-0

Desai S (2002) Observing the pole tide with satellite altimetry. J Geophys Res 107(C11). https://doi.org/10.1029/2001JC001224

Glaser S, Ampatzidis D, König R, Nilsson T, Heinkelmann R, Flechner F, Schuh H (2016) Simulation of VLBI observations to determine a global TRF for GGOS. IAG symposia series. Springer, Berlin. https://doi.org/10.1007/1345_2016_256

Glaser S, König R, Ampatzidis D, Nilsson T, Heinkelmann R, Flechner F, Schuh H (2017) A global terrestrial reference frame from simulated VLBI and SLR data in view of GGOS. J Geod. https://doi.org/10.1007/s00190-017-1021-2

Glaser S, König R, Neumayer KH, Nilsson T, Heinkelmann R, Flechtner F, Schuh H (2019a) On the impact of local ties on the datum realization of global terrestrial reference frames. J Geod. https://doi.org/10.1007/s00190-018-1189-0

Glaser S, König R, Neumayer KH, Balidakis K, Schuh H (2019b) Future SLR station networks in the framework of simulated multi-technique terrestrial reference frames. J Geod https://doi.org/10.1007/s00190-019-01256-8

Gross R, Beutler G, Plag HP (2009) Integrated scientific and societal user requirements and functional specifications for the GGOS. In: Global geodetic observing system: meeting the requirements of a global society on a changing planet in 2020. Springer, Berlin, pp 209–224. https://doi.org/10.1007/978-3-642-02687-4_7

IERS (2018) International earth rotation and reference systems service. http://www.iers.org. Accessed 30 Oct 2018

Kehm A, Blossfeld M, Pavlis EC, Seitz F (2017) Future global SLR network evolution and its impact on the terrestrial reference frame. J Geod 92:625. https://doi.org/10.1007/s00190-017-1083-1

Mendes VB, Pavlis EC (2004) High accuracy zenith delay prediction at optical wavelengths. Geophys Res Lett 31:L14602. https://doi.org/10.1029/2004GL020308

Otsubo T, Matsuo K, Aoyama Y, Yamamoto K, Hobiger T, Kubo-oka T, Sekido M (2016) Effective expansion of satellite laser ranging network to improve global geodetic parameters. Earth Planets Space 68:65. https://doi.org/10.1186/s40623-016-0447-8

Paolozzi A, Ciufolini I, Paris C, Sindoni G (2015) LARES: a new satellite specifically designed for testing general relativity. Int J Aerosp Eng 2015, Article ID 341384. https://doi.org/10.1155/2015/341384

Pearlman MR, Degnan JJ, and Bosworth JM (2002) The international laser ranging service. Adv Space Res 30(2):135–143

Petit G, Luzum B (2010) IERS conventions (2010). Bundesamts für Kartographie und Geodäsie, Frankfurt am Main

Schuh H, König R, Ampatzidis D, Glaser S, Flechner F, Heinkelmann R, Nilsson T (2015) GGOS-SIM: simulation of the reference frame for the global geodetic observing system. In: van Dam T (eds) REFAG 2014, International association of geodesy symposia, vol 146, pp 95–100. https://doi.org/10.1007/1345_2015_217

Sillard P, Boucher C (2001) A review of algebraic constraints in terrestrial reference frame datum definition. J Geod 75(2–3):63–73

Zhu S, Reigber Ch, König R (2004) Integrated adjustment of CHAMP, GRACE, and GPS data. J Geod 78(1–2):103–108

Permanent GNSS Network Optimization Considering Tectonic Motions

Lisa Pertusini, Giulio Tagliaferro, Rosa Pacione, and Giuseppe Bianco

Abstract

The contribution discusses the optimal design of a Global Navigation Satellite System (GNSS) network compromising between the estimation of the tectonic motion with other geodetic criteria. It considers the case of a pre-existing network to be densified by the addition of new stations. An optimization principle that minimizes the error of the estimated background motion and maximizes the spatial uniformity of the stations is formulated. A means to solve approximately the proposed target function is presented. The proposed procedure is preliminary tested for the case of the densification of the Agenzia Spaziale Italiana (ASI) GNSS network in Italy.

Keywords

Background motion · GNSS · Network · Optimization · Tectonic motions

1 Introduction

The densification of a GNSS permanent network is a common problem for both national and international permanent networks. The paper considers the problem of the design of a permanent Global Navigation Satellite System (GNSS) network that best allows to estimate the velocities of its stations. The optimization principle will account for the presence of already installed N_0 stations, and will seek an optimum for the position for the remaining $N - N_0$ stations. This has been designed to solve a practical problem for the Italian context. Two different criteria are going to be considered for the creation of a target function. The first one aims to best estimate the tectonic background motion, the second one is the spatial uniformity of the stations. As we will see in order to define the best placement of the new stations we need a prior knowledge of what we will call the background (or smooth) motion/pattern. Since this can be generally achieved by monitoring a dense network of already existing GNSS stations, it might seem that the purpose of the paper is in contradiction with the existing prior information. However this is not the case, because we are talking of two different types of networks; one is a reference network that must satisfy all the criteria of uniform high-level accuracy, continuity, archiving and availability of the data, robustness of the station monumentation, etc., as stated by International agreements on standards (Bruyninx 2013); among other things this reference network is generally analyzed in a consistent and continuous way by some authority that can also guarantee the correctness of the results. This is the net for which the problem of an optimal increase from N_0 to N stations is studied. The second type of network is not in reality a unique net, but rather a congeries of GNSS networks implemented by different public or private entities for completely different purposes, sometimes giving data over different periods of time and certainly with a non uniform accuracy. This is at least the situation in Italy and

L. Pertusini (✉)
Geomatics Research and Development srl, Lomazzo, Italy
e-mail: lisa.pertusini@g-red.eu

G. Tagliaferro
Geomatics Research and Development srl, Lomazzo, Italy

Politecnico di Milano, Dipartimento di Ingegneria Civile e Ambientale, Milano, Italy

R. Pacione
e-GEOS S.p.A., Roma, Italy

G. Bianco
Agenzia Spaziale Italiana, Roma, Italy

P. Novák et al. (eds.), *IX Hotine-Marussi Symposium on Mathematical Geodesy*,
International Association of Geodesy Symposia 151, https://doi.org/10.1007/1345_2020_89

it is thanks to the work of INGV researchers (Devoti et al. 2014) that a large amount of such in-homogeneous data has been analyzed and reduced to a unique picture, as we will better describe in Sect. 3. The resulting velocity field has been further analyzed and split into regions of homogeneous background motion (Biagi and Pertusini 2018). So in the end we have the following frame for the problem we want to treat: we have an area where points are moving along a certain velocity field, of which we have prior knowledge as for its functional (or correlation) characteristics. In this area there is an already existing net of N_0 stations and we would like to increase it to N stations, trying to compromise between two targets: one is to make the new net as uniform as possible because we want to use it as reference network for positioning and the other is that the new network could be used at best to improve (and monitor) our knowledge of the velocity field, for the sake of geophysical studies. Both purposes leads to its own optimal criterion. The two criteria are discussed and derived in Sects. 2.1 and 2.2 and then combined as customary in multitarget problems. Due to the high non-linearity of the resulting compound principle, practical means to compute an approximate solution are discussed in Sect. 2.3. Finally, the case study of the Agenzia Spaziale Italiana (ASI) permanent GNSS network is presented in Sect. 3.

2 Optimal Placement of the Stations

In this section a target function implementing the optimization principle for the placement of $N - N_0$ stations is formulated. The analytical formulation of the two criteria mentioned in the introduction is presented. Then an approximate way to find the minimum is discussed.

2.1 Optimal Estimation of Background Motion

Let's assume that the velocity field is generally smooth apart from outliers and localized geophysical phenomena (e.g. volcanic activity) that would prevent the background motion from being described using a smooth function. The problem to identify such outlier points has been addressed and studied in depth for the Italian case in Biagi and Pertusini (2018). The background motion (u_0 East and v_0 North components) has to be optimally estimated by interpolation using the empirical velocity field that can be estimated from the finally placed N permanent stations. The need of predicting the background motion at points other than the reference stations is urged when monitoring the position of a new point: one would like to distinguish its variations when due to either tectonic motion or local phenomena. Notice that here interpolation does not mean "exact" interpolation, but rather it refers to the fact that we are estimating the velocity field only inside the area where we have observations too. Someone would just call it approximation. A first choice has to be made with respect to the representation of the background field. We assume that (u_0, v_0) can be represented separately by a spline model, namely:

$$u_0(P) = \sum_{k=1}^{n} \phi_k(P) p_k \tag{1}$$

$$v_0(P) = \sum_{k=1}^{n} \phi_k(P) q_k \tag{2}$$

where $n < N$ and ϕ is the spline base function and P is a generic point in the area. In principle the number of splines could be different between the two components. However, since this does not affect the approach we are going to present we will assume them to be equal. This is the case because we are estimating the two velocity component separately. The suitability of this choice will be discussed later in the present section. We start by writing the velocities at the N station points using the spline model described in (1) and (2) and assuming a white noise error, i.e. a diagonal covariance proportional to identity:

$$u_0(\boldsymbol{r}_i) = \sum_{k=1}^{n} \phi_k(\boldsymbol{r}_i) p_k + \nu_i \tag{3}$$

$$v_0(\boldsymbol{r}_i) = \sum_{k=1}^{n} \phi_k(\boldsymbol{r}_i) q_k + \eta_i \tag{4}$$

where $\boldsymbol{r}_i, i = 1, \cdots, N$.

Let us notice that in general ν and η might also be correlated one another. Yet such a correlation can be often decreased, if not cancelled, by suitably choosing two axes rotated with respect to the North, East directions (Biagi and Pertusini 2018). Indeed a more general stochastic model should be used without too many changes of the approach.

For the component u_0, (3) can be written in vector form:

$$\boldsymbol{u}_0 = \Phi \boldsymbol{p} + \boldsymbol{\nu} \tag{5}$$

and we assume that

$$C_{\nu} = \sigma_{\nu}^2 I;$$

yet the formulation can be easily adapted to the case that ν has a known covariance matrix.

We go to the least squares estimates of $\boldsymbol{p}$ denoted as $\hat{\boldsymbol{p}}$:

$$\hat{\boldsymbol{p}} = W^{-1} \Phi^{\top} \boldsymbol{u_0} \quad W = \Phi^{\top} \Phi. \tag{6}$$

This implies that at any point P the interpolated $\hat{u_0}(P)$ is given by:

$$\hat{u}(P) = \boldsymbol{\phi}^{\top}(P) \hat{\boldsymbol{p}} = \sum_{k=1}^{n} \phi_k(P) \hat{p}_k \tag{7}$$

and its propagated error $\delta\hat{u}$ is:

$$\delta\hat{u} = \boldsymbol{\phi}^{\top}(P)\,(\hat{\boldsymbol{p}} - \boldsymbol{p}) \tag{8}$$

with variance:

$$\sigma^2(\delta\hat{u}(P)) = \sigma_v^2 \boldsymbol{\phi}^{\top}(P) W^{-1} \boldsymbol{\phi}(P). \tag{9}$$

To avoid confusion it might be worth to mention that $\boldsymbol{\Phi}$ refers the base functions evaluated at the station points while $\boldsymbol{\phi}$ refers to the base functions evaluated at a generic point P. So a first term we want to include into our target function is the mean quadratic residual of the error over the area of interest (A), namely

$$\begin{aligned}\Xi_u^2 &= \int_A \sigma^2(\delta\hat{u}) dS_P = \sigma_v^2 \int_A \boldsymbol{\phi}^{\top}(P) W^{-1} \boldsymbol{\phi}(P) dS_P \\ &= \sigma_v^2 Tr\{W^{-1}K\}\end{aligned} \tag{10}$$

where:

$$K_{hk} = \int_A \boldsymbol{\phi}_k(P)\boldsymbol{\phi}_h(P) dS_P \tag{11}$$

being h, k the indexes of our base functions. Similarly we can construct a similar target function for the v component:

$$\Xi_v^2 = \sigma_\eta^2 Tr\{W^{-1}K\}. \tag{12}$$

The total quadratic interpolation error will then be:

$$\Xi_{tot}^2 = (\sigma_\eta^2 + \sigma_v^2) Tr\{W^{-1}K\}. \tag{13}$$

One might object that the chosen model is too simple, to the extent that no attempt of including a model error in (1) and (2) has been done. This is because the optimization in the end has to be done with respect to positions $\boldsymbol{r}_i$ and the dependency of the total error Ξ_{tot}^2 on such variables is generally quite complicated as one can already see from (13), where by the way the points $\boldsymbol{r}_i$ enter only in W^{-1}. Nevertheless we can shortly outline how our target function modifies when for instance we assume that each component u, v of the velocity vector can be split into a signal, with known covariance structure and a noise. Referring to u for instance, and assuming that $C(P, Q)$ is its covariance, the interpolator (7) would become in this case (Moritz 1980):

$$\hat{u}(P) = \sum_{i,k} C(P, P_i)\{C(P_i, P_k) + \sigma^2\delta_{i,k}\}^{(-1)} u_0(P_k) \tag{14}$$

and the corresponding prediction error would be

$$\begin{aligned}E\{e^2(P)\} = C(P, P) - \sum_{i,k} C(P, P_i)\{C(P_i, P_k) \\ + \sigma^2\delta_{i,k}\}^{(-1)} C(P_k, P).\end{aligned} \tag{15}$$

Therefore the total error in this case has the form

$$\Xi_{tot}^2 = \int_A C(P, P) dS_P - Tr\{(\mathscr{C} + \sigma_v^2 I)^{-1} K\} \tag{16}$$

where $\mathscr{C}_{i,k} = C(P_i, P_k)$ and $K_{i,k} = \int_A C(P, P_i) C(P_k, P)\, dS_P$.

We can notice that in (16) the first term to the right is constant, while the second depends on the position $\{P_i\}$ through both $(\mathscr{C} + \sigma_v^2 I)^{-1}$ and K. So the dependence of this function on the configuration of the network is indeed more complicated than that of (13), where positions enter only in W.

It is maybe worth mentioning that a similar analytical problem is met if instead of considering the background motion as a stochastic signal in general, we would rather require that the unknown parameters would agree with the prior values that one could derive from the prior knowledge of the velocity field. In any event we shall restrict our attention to the total error function (13).

Remark 1 There are two objections of general character that a careful reader might raise; they are both somehow related to the choice of the reference system. The first point is that it is known that velocities are generally not estimable unless we fix the reference system e.g. by a suitable minimal constraint (Dermanis 2019). This however is a problem that has been solved by the authors of Devoti et al. (2014) for the estimation of the prior velocity field. A second point is that indeed in this case the covariance matrix of $\boldsymbol{u}_0$, $\boldsymbol{v}_0$ will depend on such a choice. So in our simplified solution we have avoided the problem by assigning to v and η a white noise covariance. As we have already claimed it would not be difficult though to include a given covariance Σ or its pseudoinverse Σ^+ for $\boldsymbol{u}_0$ or $\boldsymbol{v}_0$ into the target function, where the $W = \Phi^T \Phi$ matrix should be substituted rather by $W = \Phi^T \Sigma^+ \Phi$.

2.2 Uniformity of the Network

As previously mentioned, the idea of minimizing Ξ_{tot}^2 could be balanced by the concept that we should not move the position of the station too much from the one that we would derive from a simple geodetic setting. This criterion is to have network stations as uniformly distributed as possible in space. This is important for instance in case the network has to be used for a geodetic service for positioning. Since we have already N_0 stations in the area we would like to place the remaining $N - N_0$ stations in a way that the overall distribution of N station be as uniform as possible. This could be done simply identifying $N - N_0$ optimal positions e.g. by visual inspection; $\bar{\boldsymbol{r}}_{0_i}$ $\quad i = N_0 + 1, \cdots, N$

selected optimal positions. Therefore to balance Ξ^2_{tot} we would add a term into the optimization criterion, assuming that also

$$\sum_{i=N_0+1}^{N} |\boldsymbol{r}_i - \bar{\boldsymbol{r}}_{0i}|^2 = G^2(\boldsymbol{r}) \tag{17}$$

should be small. It is important to underline that the difference $\boldsymbol{r}_i - \bar{\boldsymbol{r}}_{0i}$ should be taken using the closest $\bar{\boldsymbol{r}}_{0i}$ to $\boldsymbol{r}_i$. The selected criterion is not an uniformity criterion itself but rather a proximity to the uniform solution. Indeed more rigorous criteria could be applied for the two choice of $\bar{\boldsymbol{r}}_{0i}$ in (17); for instance it would be possible to choose the point than minimize the variance of areas of the Voronoi polygons (Okabe et al. 2009) constructed using the location of the stations. This is the choice made for the case study described in Sect. 3.

2.3 The Target Function and Its Minimization

In order to combine the two criteria we can simply take a weighted sum (*lambda* being the weight) of the target functions, so arriving at the principle

$$\{\boldsymbol{r}_i\ i = N_0 + 1, \cdots, N\} = argMin\ F(\boldsymbol{r}),$$

$$F(\boldsymbol{r}) = \Xi^2_{tot} + \lambda G^2(\boldsymbol{r}) = \sigma_0^2 Tr\{W^{-1}K\} + \lambda \sum_{i=N_0+1}^{N} |\boldsymbol{\xi}_i|^2 \tag{18}$$

where:

$$\boldsymbol{\xi}_i = \boldsymbol{r}_i - \bar{\boldsymbol{r}}_{0_i} \tag{19}$$

$$\sigma_0^2 = \sigma_\eta^2 + \sigma_\nu^2 \tag{20}$$

The minimization problem (18) can be discussed from different points of view. Here we write standard minimum conditions. As we can observe by inspecting (18) where the unknowns $\boldsymbol{r}_i$ enter in both terms and in particular in a complicated way into the matrix W (see (6)), the present optimization problem differs significantly from traditional network optimization problems studied long ago in geodetic literature (Grafarend and Sansò 1985). Formally the minimizing principle has the normal equation described from differencing F with respect to $\delta\boldsymbol{\xi}$:

$$\delta F = 2\lambda \delta\boldsymbol{\xi}_i^{\top}\boldsymbol{\xi}_i - \sigma_0^2 Tr\{W^{-1}\delta W W^{-1}K\} \tag{21}$$

where, taking into account that $\delta\boldsymbol{r}_i = \delta\boldsymbol{\xi}_i$, we have:

$$\delta W_{h,k} = \delta\boldsymbol{\xi}_i^{\top}\nabla\phi_h(\boldsymbol{r}_i)\phi_k(\boldsymbol{r}_i) + \phi_h(\boldsymbol{r}_i)\delta\boldsymbol{\xi}_i^{\top}\nabla\phi_k(\boldsymbol{r}_i). \tag{22}$$

Coming back to (21), we can put:

$$H = W^{-1}KW^{-1}, \tag{23}$$

to find:

$$Tr\{W^{-1}\delta W W^{-1}K\} = 2\sum_{k,h,i} \delta\boldsymbol{\xi}_i^{\top}\nabla\phi_h(\boldsymbol{r}_i)H_{k,h}\phi_k(\boldsymbol{r}_i). \tag{24}$$

Therefore the equation for the minimum is:

$$\boldsymbol{\xi}_i = \frac{\sigma_0^2}{\lambda}\sum_{k,h,i} \nabla\phi_h(\boldsymbol{r}_i)H_{k,h}\phi_k(\boldsymbol{r}_i). \tag{25}$$

As we see in (25) the minimum principle is translated into a highly non linear normal equation that we might solve only iteratively.

Remark 2 Apart from convergence considerations, the solution (25) depends heavily on the choice of the weight λ. In fact, we can see from (18) that if $\lambda \to 0$ the solution $\boldsymbol{r}_i$ will tend to minimize Ξ^2_{tot} only, while if we let $\lambda \to \infty$ the solution will tend to minimize $G^2(\boldsymbol{r}_i)$ only, namely $\boldsymbol{\xi}_i = 0$. This last consideration is well known in Tikhonov regularization theory (Tikhonov and Arsenin 1977). Since the criterion for the choice of λ is in this case quite vague, we could conceive instead of finding one specific solution (one for every λ) to give the two positions $\bar{\boldsymbol{r}}_{0i}$ and $\boldsymbol{r}'_i$ that minimize respectively G^2 and Ξ^2_{tot}, and then look for a feasible implementation of the station at a point along the straight line connecting the two points. This is certainly not rigorous, however it is sufficient for a practical solution where many other factors have to be taken into account, for instance the availability at the stations of electricity or of an internet connection etc.

3 Case Study

The Italian Space Agency (ASI) intends to densify the current Italian GNSS reference frame network. At present it consists of 15 permanent stations throughout the country, 3 of them being co-located at the Space Geodesy Center in Matera (Fig. 1). Thirty-one new permanent stations have to be installed, that will contribute to the EPN – EUREF European GNSS permanent station network densification over the Italian area (Fig. 1).

As previously discussed, the new GNSS network has to be as homogeneously distributed as possible (static criterion) and best identify the background motion (kinematic criterion) insisting over the Italian peninsula. In fact it is well known that Italy is continuously and strongly moving

Fig. 1 GNSS Reference frame network currently powered by ASI (in red) and Italian EPN – EUREF network (in black)

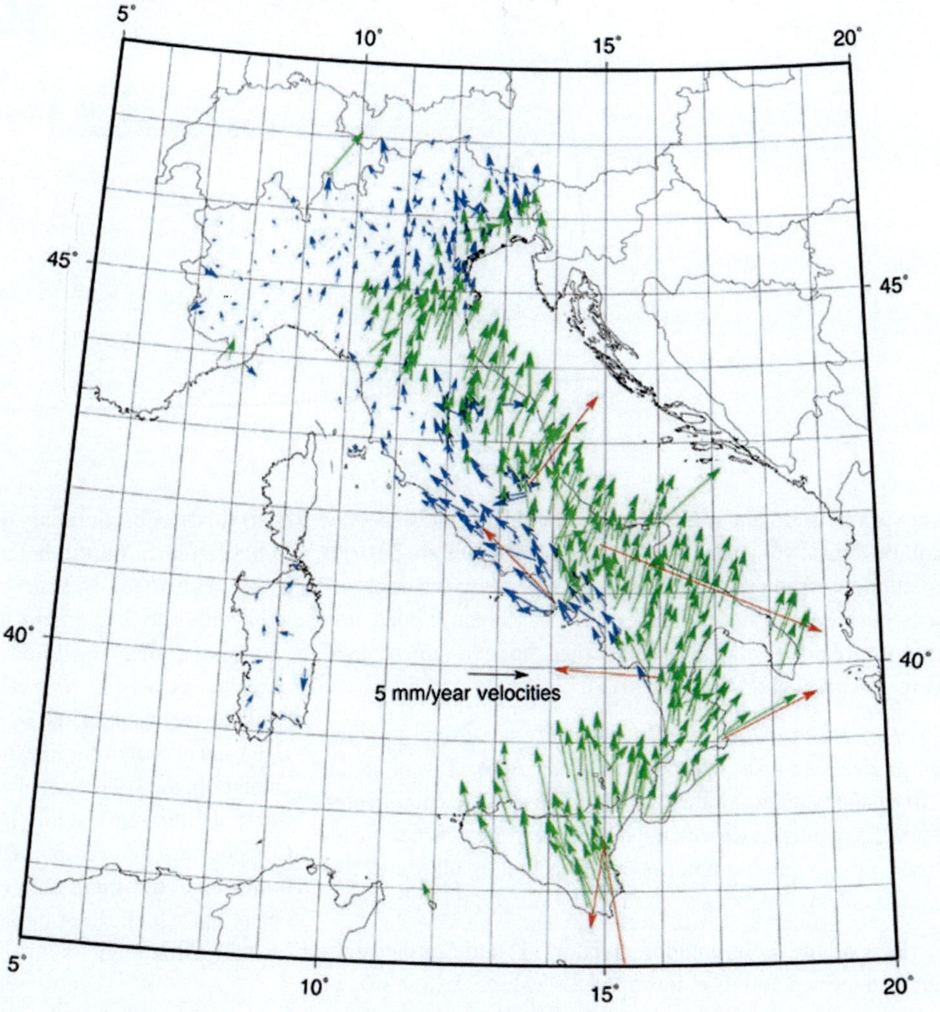

Fig. 2 Velocity vectors of the GNSS permanent stations estimated by the INGV. Plot based on velocity estimated from Devoti et al. (2014) and taken from Biagi and Pertusini (2018). Blue and green clusters. In red: outliers

within the European Plate (Devoti et al. 2011). Actually, Italy seems clearly divided into different sub-regions that move in different mean directions but have quite smooth internal behaviors, except for a few discussed outliers (plotted in red) (Fig. 2). For several years, the Istituto Nazionale di Geofisica e Vulcanologia (INGV) has been continuously monitoring a

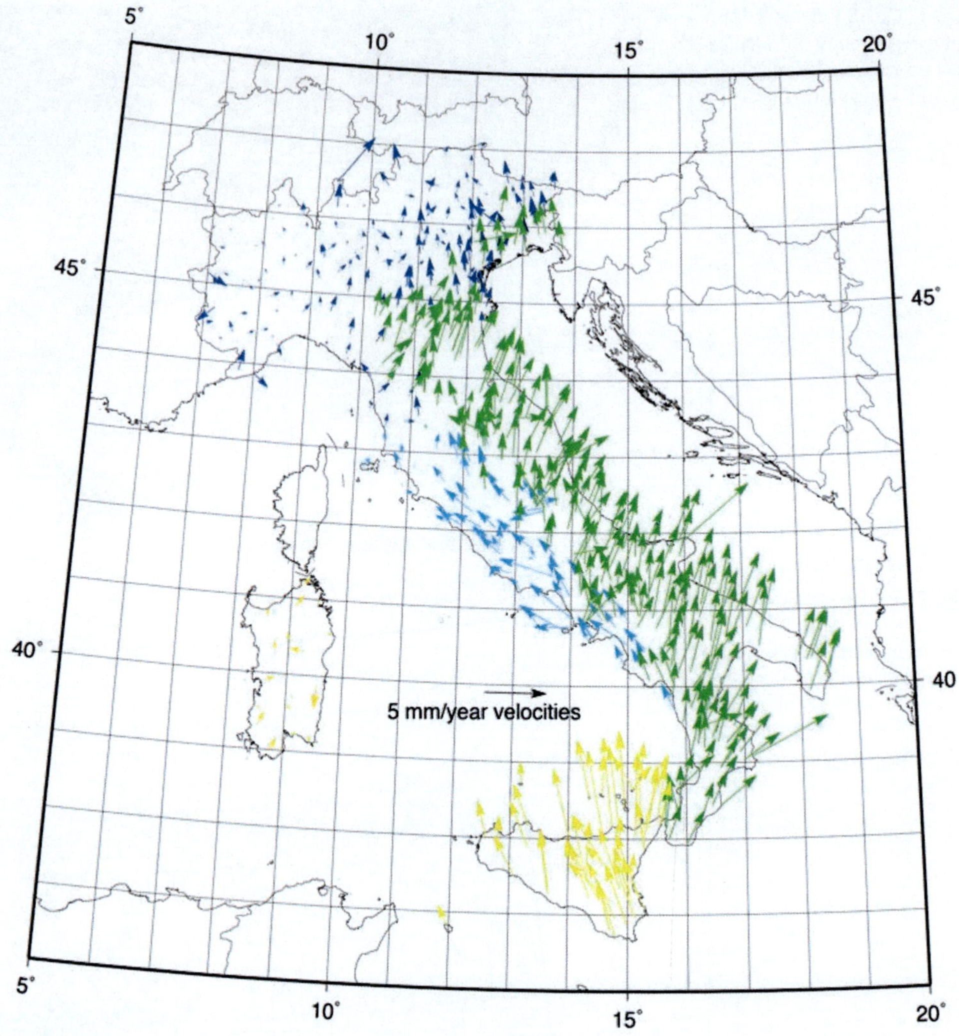

Fig. 3 Five areas with homogeneous background motion derived in Italy: blue cluster, light blue cluster, green cluster, Sicily, Sardinia (image taken from Biagi and Pertusini (2018))

network of permanent GNSS stations. Based on a specific solution that was estimated in 2013 (Devoti et al. 2014) a method to predict at any point the "background motion" was developed. The Italian territory has been divided into five sub-regions with an internal homogeneous velocity field (Fig. 3) (Biagi and Pertusini 2018).

Remark 3 Let us observe that in reality according to some geological analysis (Serpelloni et al. 2016; Livani et al. 2018), the behavior of the Italian region is more complicated than the one here described. Nevertheless the practical solution of using smaller patches described below alleviates this inconvenient.

Beyond the representation given by (1) and (2), the background motion could be described for instance by the sum of a deterministic part and a stochastic part; the latter has been found to be much smaller than the first one, with a correlation length of the signal shorter than 20 km (Biagi and Pertusini 2018). This is in fact the critical datum we wanted to get from this stochastic analysis. Even more, since we wanted to have control on the behaviour of Ξ^2_{tot}, which is difficult if we want to move together the positions of 31 stations, we decided to implement the optimization by dividing the whole area analyzed into smaller patches where only one new station was to be placed and the background motion could be represented just by a linear function. Indeed we have also tried higher order interpolations which gave insignificant changes in the solution, so we preferred to stay with the linear interpolation, more stable in terms of the spline parameters. Considering the vector of known horizontal velocities at points $\boldsymbol{r}_i$ in the small area of interest, an a priori model can be set up, which describes a regular motion, see (7), that in East direction reads

$$u_0(\boldsymbol{r}) = \sum p_k \phi_k(\boldsymbol{r}) = \boldsymbol{\Phi}^\top(\boldsymbol{r})\hat{\boldsymbol{p}}. \tag{26}$$

With $\phi_k(\boldsymbol{r})$ spanning the space of first degree polynomials, and where $\hat{p}_k$ have been least squares estimated.

Using (10) and (11) we can write, identifying $\boldsymbol{r}$ with the corresponding point P:

$$\int_A \sigma^2[e_{0E}(P)]dS = \sigma_{0\,E}^2 Tr\{(\boldsymbol{\Phi}^\top \boldsymbol{\Phi})^{-1} K\} = \sigma_{0\,E}^2 \Xi^2(\boldsymbol{r}_i) \tag{27}$$

where $K = \int_A \boldsymbol{\phi}(P)\boldsymbol{\phi}(P)^\top d_2 S$.

The integral formula for K (11) has been computed numerically discretizing the area using a regular grid. Following the simplified procedure previously presented, the target function becomes:

$$H(\boldsymbol{r}_i) = \lambda \sum |\boldsymbol{r}_i - \bar{\boldsymbol{r}}_{0i}|^2 + \sigma_0^2 F(\boldsymbol{r}_i). \tag{28}$$

As previously observed changing λ, the optimized solution $\hat{\boldsymbol{r}}_i$ runs between the point following the uniformity criterion and the point which minimizes $F(\boldsymbol{r}_i)$. Since finding the final place of the stations is also a practical problem where considerations of access to electricity, internet and of security also enter, we decided to adopt the already discussed approach, consisting in providing the two positions and searching for a station that could be operationally monumented along the line connecting them, with a tolerance buffer of 20 km, suggested by the above discussed covariance analysis. In Fig. 4 an example of this solution for the Italian North-West area is shown. The point that minimizes the variance on the areas of the Voronoi polygons over the area of interest was chosen as point that maximizes the uniformity of the network (see Fig. 5).

Remark 4 One remark on Fig. 4 is in order at this point. The fact that the minimization of Ξ^2 falls at the boundary of the area is by no means accidental: in fact this depends from the choice of a linear local interpolator. Nevertheless this means also that, considering Fig. 4, the direction between the point of maximal uniformity $\bar{\boldsymbol{r}}_0$ and the "pink star" $\boldsymbol{r}'$ point is that of maximum inclination for Ξ^2 so placing a station

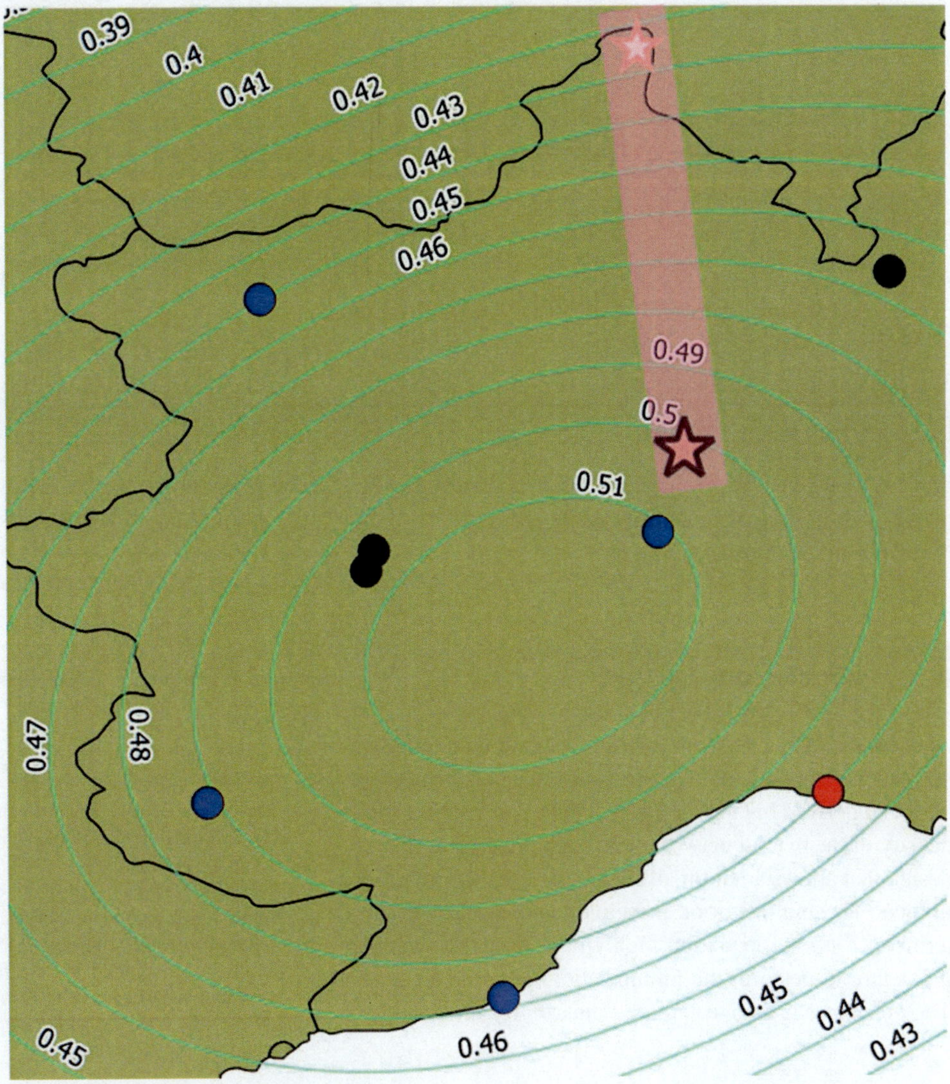

Fig. 4 Proposed new station position in the Italian NorthWest region. Red dot represents the already existing ASI station in Genoa, black dots are the already existing EPN – EUREF network stations, blue dots are the already fixed new stations by convenience criteria of co-location with other instruments networks, such as tide gauges or automated weather stations. In pink star with black borderline the maximal uniformity position and in white star with pink borderline the optimal trend estimation solution. Contour lines represent function (27) normalized by the area, unit in $m^4/1e18$

Fig. 5 Contour lines representing the variance of the areas of the Voronoi polygon over the area of interest, unit in $m^4/1e18$

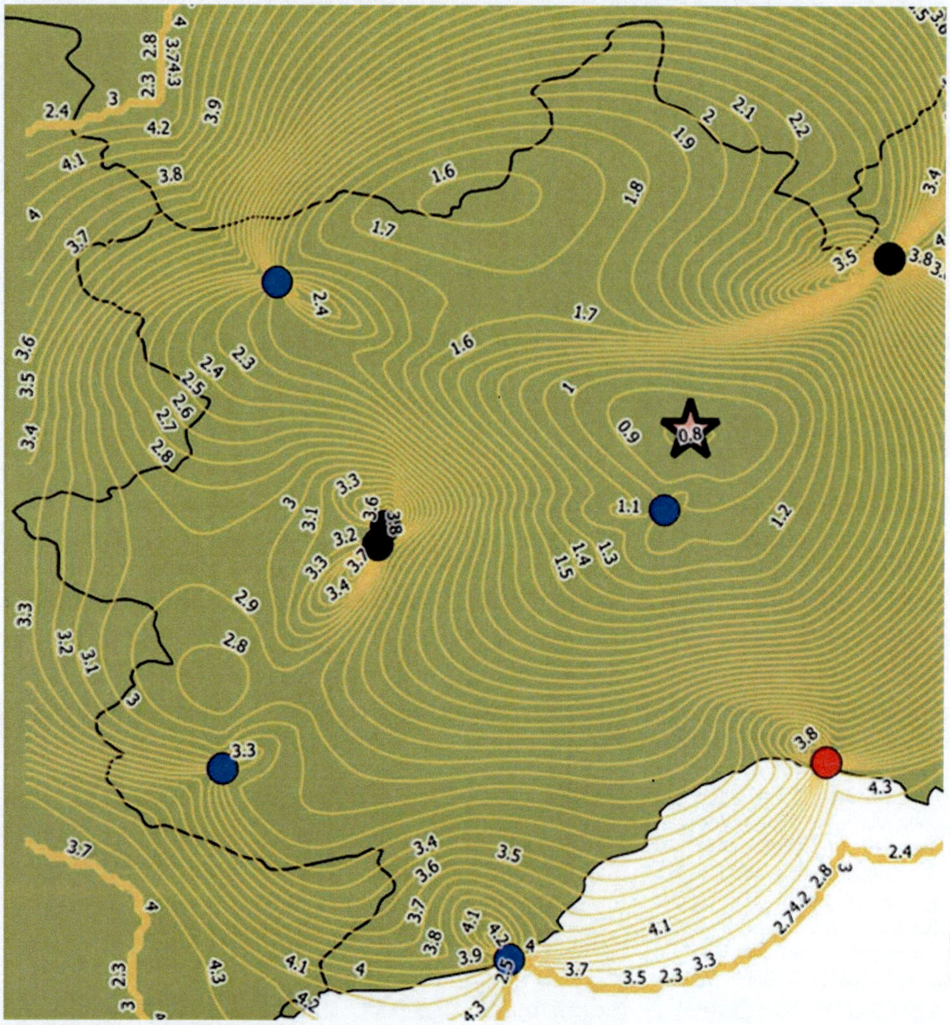

along the segment between the two is really in some sense minimizing the total signal estimation error. The mentioned discretization computation of the Ξ^2 function has as matter of fact allowed us to evaluate its shape that resulted in a convex cup in all cases, confirming the above interpretation of its minimum point.

4 Conclusion

An optimization criterion for the design of an improved permanent network, augmenting an already existing network, has been derived balancing the geodetic criterion of uniform density of the overall network with a kinematic criterion of a design that allows an optimally estimation of the background motion. The minimization principle has been expressed in terms of a normal equation highly non linear. A simplified procedure to perform the minimization has been presented. The procedure is then tested preliminarily for the case of the densification of the ASI permanent GNSS station. Additional work is needed to refine the proposed procedure and to evaluate the impact of the arbitrary choices still present by the procedure. In particular the sensitivity of the solution to the partitioning of the network is of main concern.

References

Biagi L, Pertusini L (2018) A model of the deformation pattern in the Italian area and its relationship with the national reference frame. Rend Lincei Sci Fis Natur 29(1):59–67

Bruyninx C (2013) Guidelines for EPN Stations & Operational Centres. EPN Central Bureau

Dermanis A (2019) Geodetic methods for monitoring crustal motion and deformation. In: Handbuch der Geodäsie: 6 Bände. Springer, Berlin, pp 1–71

Devoti R, Esposito A, Pietrantonio G, Pisani AR, Riguzzi F (2011) Evidence of large scale deformation patterns from GPS data in the Italian subduction boundary. Earth Planet Sci Lett 311(3–4):230–241

Devoti R, Pietrantonio G, Riguzzi F (2014) GNSS networks for geodynamics in Italy. Física de la Tierra 26(2014):11–24

Grafarend EW, Sansò F (1985) Optimization and design of geodetic networks. Springer, Berlin

Livani M, Scrocca D, Arecco P, Doglioni C (2018) Structural and stratigraphic control on salient and recess development along a thrust belt front: the Northern Apennines (Po Plain, Italy). J Geophys Res Solid Earth 123:4360

Moritz H (1980) Advanced physical geodesy. Advances in planetary geology. Herbert Wichmann, Karlsruhe

Okabe A, Boots B, Sugihara K, Chiu SN (2009) Spatial tessellations: concepts and applications of Voronoi diagrams, vol 501. Wiley, Chichester

Serpelloni E, Vannucci G, Anderlini L, Bennett R (2016) Kinematics, seismotectonics and seismic potential of the eastern sector of the European Alps from GPS and seismic deformation data. Tectonophysics 688:157–181

Tikhonov A, Arsenin VY (1977) Methods for solving ill-posed problems. Wiley, New York

Part III

Estimation Theory and Inverse Problems in Geodesy

Adjustment of Gauss-Helmert Models with Autoregressive and Student Errors

Boris Kargoll, Mohammad Omidalizarandi, and Hamza Alkhatib

Abstract

In this contribution, we extend the Gauss-Helmert model (GHM) with t-distributed errors (previously established by K.R. Koch) by including autoregressive (AR) random deviations. This model allows us to take into account unknown forms of colored noise as well as heavy-tailed white noise components within observed time series. We show that this GHM can be adjusted in principle through constrained maximum likelihood (ML) estimation, and also conveniently via an expectation maximization (EM) algorithm. The resulting estimator is self-tuning in the sense that the tuning constant, which occurs here as the degree of freedom of the underlying scaled t-distribution and which controls the thickness of the tails of that distribution's probability distribution function, is adapted optimally to the actual data characteristics. We use this model and algorithm to adjust 2D measurements of a circle within a closed-loop Monte Carlo simulation and subsequently within an application involving GNSS measurements.

Keywords

Autoregressive process · Circle fitting · Constrained maximum likelihood estimation · Expectation maximization algorithm · Gauss-Helmert model · Scaled t-distribution · Self-tuning robust estimator

1 Introduction

When a deterministic model used to approximate observations are characterized by condition equations in which multiple observations and unknown parameters are linked with each other, an adjustment by means of the Gauss-Helmert model (GHM) is often the procedure of choice. The classical formulation of that method, being based on the method of least squares, does not require the specification of a probability density function (pdf) for the random deviations or the observables. When the observables are outlier-afflicted or heavy-tailed, this least-squares approach can be expected to break down. However, it can be turned into an outlier-resistant ('robust') procedure by including a re-weighting or variance-inflation scheme based on a heavy-tailed error law such as Student's t-distribution (Koch 2014a,b). This procedure is implemented as an expectation maximization (EM) algorithm, which allows for the estimation not only of the parameters within the condition equations, but also of the scale factor and degree of freedom of the underlying t-distribution. The latter feature turns the method into a *self-tuning* robust estimator in the sense of Parzen (1979).

An additional common characteristic of observables that complicates their adjustment is given by autocorrelations or colored noise, which phenomena frequently occur with electronic instruments measuring at a high sampling rate (cf. Kuhlmann 2003). When the data covariance matrix is unknown or too large, autoregressive (AR) or AR moving average (ARMA) processes enable a parsimonious modeling of correlations in situations when the measurements can be treated as a time series (e.g., Schuh 2003). Such processes are also attractive as they may be easily transformed into the easy-to-interpret autocovariance or spectral density function

B. Kargoll (✉) · M. Omidalizarandi · H. Alkhatib
Geodetic Institute, Leibniz University Hannover, Hannover, Germany
e-mail: kargoll@gih.uni-hannover.de

P. Novák et al. (eds.), *IX Hotine-Marussi Symposium on Mathematical Geodesy*,
International Association of Geodesy Symposia 151, https://doi.org/10.1007/1345_2019_82

(cf. Krasbutter et al. 2015; Loth et al. 2019). The adjustment of multiple (possibly nonlinear) regression time series with AR and Student errors within the framework of the Gauss-Markov model was investigated by Alkhatib et al. (2018). A similar approach is undertaken in the current contribution now in the framework of the Gauss-Helmert model, thereby extending K.R. Koch's aforementioned model by the inclusion of estimable AR processes.

In Sect. 2, we specify the observation model, the correlation model, and the stochastic model, which jointly define the GHM with AR and t-distributed errors. In this model we allow for different groups of observations characterized by individual correlation and stochastic models. In doing so, time series that stem from different sensors or that form a multivariate measurement process, such as measured three-dimensional coordinate time series, can be modeled flexibly to take the heterogeneous data characteristics usually present in such time series into account. In Sect. 3, we formulate the optimization principle employed to adjust this model, and we derive for the purpose of convenient computation an EM algorithm, deriving in detail the normal equations to be solved. The Monte Carlo simulation results in Sect. 4 demonstrate the biases to be expected in the practical situation of data approximation by a circle. Section 5 contains results stemming from the adjustment of a real data set.

2 A Gauss-Helmert Model with Autoregressive and t-Distributed Errors

We intend to adjust N groups of observables $\mathcal{L}_k = [\mathcal{L}_{k,1}, \ldots, \mathcal{L}_{k,n_k}]^T$ for $k = 1, \ldots, N$, with the kth group consisting of n_k observables. Each random variable $\mathcal{L}_{k,t}$ is modeled by an individual location parameter $\mu_{k,t}$ and an additive random deviation $\mathcal{E}_{k,t}$, that is,

$$\mathcal{L}_{k,t} = \mu_{k,t} + \mathcal{E}_{k,t} \quad (t = 1, \ldots, n_k). \tag{1}$$

In this and the following equations the unknown parameters are generally denoted by Greek letters, random variables by calligraphic letters, and real-valued constants by Roman letters. Thus, a random variable (e.g., $\mathcal{E}_{k,t}$) and its realization ($e_{k,t}$) can be distinguished. Moreover, matrices and vectors are symbolized by bold letters. The index values of t represent equidistant time instances, so that the sequence $\boldsymbol{\mathcal{L}}_k$ and correspondingly $\boldsymbol{\mathcal{E}}_k$ constitute time series. Defining the vector $\boldsymbol{\mu}_k = [\mu_{k,1}, \ldots, \mu_{k,n_k}]^T$ for $k = 1, \ldots, N$, we mean by $\mathbf{I}_{k,t}$ the tth row of the $(n_k \times n_k)$-identity matrix. Consequently, the observation model (1) can also be written as

$$\mathcal{L}_{k,t} = \mathbf{I}_{k,t}\boldsymbol{\mu}_k + \mathcal{E}_{k,t}. \tag{2}$$

Whereas each location parameter $\mu_{k,t}$ is treated as an unknown fixed parameter (to be estimated), each random deviation $\mathcal{E}_{k,t}$ is a random variable that follows a group-specific AR(p_k) process

$$\mathcal{E}_{k,t} = \alpha_{k,1}\mathcal{E}_{k,t-1} + \cdots + \alpha_{k,p_k}\mathcal{E}_{k,t-p_k} + \mathcal{U}_{k,t}. \tag{3}$$

To fix the initial values of this recursive model for all of the time instances $t-1, \ldots, t-p_k$ with values less than 1, the associated random variables are assumed to take the constant value 0. The AR coefficients $\boldsymbol{\alpha} = [\boldsymbol{\alpha}_1^T, \ldots, \boldsymbol{\alpha}_N^T]^T$ with $\boldsymbol{\alpha}_k = [\alpha_{k,1}, \ldots, \alpha_{k,p_k}]^T$ for $k = 1, \ldots, N$ are taken to be unknown parameters (also to be estimated). Depending on their values, the time series $\boldsymbol{\mathcal{E}}_1, \ldots, \boldsymbol{\mathcal{E}}_N$ exhibit group-specific auto-correlation patterns or colored noise characteristics. Using the notation $L^j\mathbf{Z}_{k,t} := \mathbf{Z}_{k,t-j}$ with $\boldsymbol{\alpha}_k(L) := 1 - \alpha_{k,1}L - \ldots - \alpha_{k,p_k}L^p$ and $\overline{\mathbf{Z}}_{k,t} = \boldsymbol{\alpha}_{k,t}(L)\mathbf{Z}_{k,t}$ for an arbitrary family of matrices $(\mathbf{Z}_{k,t} \mid t \in \{1, \ldots, n_k\})$, we also have

$$\begin{aligned} u_{k,t} = \overline{e}_{k,t} &= \boldsymbol{\alpha}_k(L)e_{k,t} = \boldsymbol{\alpha}_k(L)(\ell_{k,t} - \mathbf{I}_{k,t}\boldsymbol{\mu}_k) \\ &= \boldsymbol{\alpha}_k(L)\ell_{k,t} - \boldsymbol{\alpha}_k(L)\mathbf{I}_{k,t}\boldsymbol{\mu}_k = \overline{\ell}_{k,t} - \overline{\mathbf{I}}_{k,t}\boldsymbol{\mu}_k. \end{aligned} \tag{4}$$

This enables an interpretation of the quantities $\overline{e}_{k,t}$, $\overline{\ell}_{k,t}$ and $\overline{\mathbf{I}}_{k,t}$ as the outputs of the digital filter $\boldsymbol{\alpha}_k(L)$, applied respectively to the $k$the group of random deviations $\mathbf{e}_k$ as well as observations $\boldsymbol{\ell}_k$, and to the identity matrix $\mathbf{I}_k$. Thus, $\boldsymbol{\alpha}_k(L)$ may be viewed as turning the colored noise sequence $\mathbf{e}_k$ into white noise $\mathbf{u}_k$, in the sense of a *decorrelation filter*.

In each time series $\boldsymbol{\mathcal{U}}_k = [\mathcal{U}_{k,1}, \ldots, \mathcal{U}_{k,n_k}]^T$, we assume the individual components $\mathcal{U}_{k,1}, \ldots, \mathcal{U}_{k,n_k}$ to independently follow a scaled t-distribution (cf. Lange et al. 1989) with expectation 0 and group-specific degree of freedom ν_k as well as scale factor σ_k, that is,

$$\mathcal{U}_{k,t} \overset{\text{ind.}}{\sim} t_{\nu_k}(0, \sigma_k^2). \tag{5}$$

Due to the previous independence assumption the joint pdf of each white noise series $\boldsymbol{\mathcal{U}}_k$ can be factorized into

$$f(\mathbf{u}_k) = \prod_{t=1}^{n_k} \frac{\Gamma\left(\frac{\nu_k+1}{2}\right)}{\sqrt{\nu_k \pi \sigma_k^2}\,\Gamma\left(\frac{\nu_k}{2}\right)} \left[1 + \left(\frac{u_{k,t}}{\sigma_k}\right)^2 / \nu_k\right]^{-\frac{\nu_k+1}{2}}, \tag{6}$$

where Γ is the gamma function and $\mathbf{u}_k = [u_{k,1}, \ldots, u_{k,n_k}]$ the vector of possible realizations of $\boldsymbol{\mathcal{U}}_k$. Assuming moreover that there are no cross-correlations between the white noise series, the joint pdf of $\boldsymbol{\mathcal{U}} = [\boldsymbol{\mathcal{U}}_1^T, \ldots, \boldsymbol{\mathcal{U}}_N^T]^T$ with possible realizations $\mathbf{u} = [\mathbf{u}_1^T, \ldots, \mathbf{u}_N^T]^T$ can be factorized into $f(\mathbf{u}) = f(\mathbf{u}_1)\cdots f(\mathbf{u}_N)$. Taking the natural logarithm yields

$$\log f(\mathbf{u}) = \sum_{k=1}^{N}\left(n \log\left[\frac{\Gamma\left(\frac{\nu_k+1}{2}\right)}{\sqrt{\nu_k \pi \sigma_k^2}\,\Gamma\left(\frac{\nu_k}{2}\right)}\right] - \frac{\nu_k+1}{2}\sum_{t=1}^{n_k}\log\left[1+\left(\frac{u_{k,t}}{\sigma_k}\right)^2/\nu_k\right]\right), \tag{7}$$

similar to the stochastic model considered in Alkhatib et al. (2018). As this function includes also the observations, the location parameters and the AR coefficients through (4), it could be used to define a log-likelihood function for the purpose of maximum likelihood (ML) estimation of all model parameters. However, being based on the intricate t-distribution model, this function cannot be maximized conveniently, which problem is solved as follows. As shown for instance in Koch and Kargoll (2013), the t-distribution model (5) is equivalent to the conditional Gaussian variance-inflation model

$$\mathcal{U}_{k,t}|p_{k,t} \overset{\text{ind.}}{\sim} N(0, \sigma_k^2/p_{k,t}) \tag{8}$$

involving independently, rescaled chi-square-distributed latent variables

$$\mathcal{P}_{k,t} \overset{\text{ind.}}{\sim} \frac{\chi^2_{\nu_k}}{\nu_k}. \tag{9}$$

In practice, the values $p_{k,t}$ (which will later take the role of weights within an iteratively reweighted least squares algorithm) of the random variables $\mathcal{P}_{k,t}$ are unobservable, so that the latter are considered as latent variables. The distributions of the random variables (8)–(9) are defined by the corresponding pdfs

$$f(u_{k,t}|p_{k,t}) = \frac{1}{\sqrt{2\pi\sigma_k^2/p_{k,t}}}\exp\left\{-\frac{u_{k,t}^2}{2\sigma_k^2/p_{k,t}}\right\}.$$

and

$$f(p_{k,t}) = \begin{cases} \frac{\left(\frac{\nu_k}{2}\right)^{\frac{\nu_k}{2}}}{\Gamma\left(\frac{\nu_k}{2}\right)}\cdot(p_{k,t})^{\frac{\nu_k}{2}-1}\cdot e^{-\frac{\nu_k}{2}\cdot p_{k,t}} & \text{if } p_{k,t} > 0, \\ 0 & \text{if } p_{k,t} \le 0 \end{cases}.$$

The previous independence assumptions allow then for the formation of the joint, factorized pdf

$$f(\mathbf{u},\mathbf{p}) = \prod_{t=1}^{n_k} f(u_{1,t}, p_{1,t})\cdots\prod_{t=1}^{n_k} f(u_{N,t}, p_{N,t})$$
$$= \prod_{t=1}^{n_k} f(p_{1,t})\, f(u_{1,t}|p_{1,t},)\cdots\prod_{t=1}^{n_k} f(p_{N,t})\, f(u_{N,t}|p_{N,t}),$$

where the vector $\mathbf{p} = [\mathbf{p}_1^T, \ldots, \mathbf{p}_N^T]^T$ contains all of the unknown "weights" as realizations of $\boldsymbol{\mathcal{P}} = [\boldsymbol{\mathcal{P}}_1^T, \ldots, \boldsymbol{\mathcal{P}}_N^T]^T$ with $\boldsymbol{\mathcal{P}}_k = [\mathcal{P}_{k,1}, \ldots, \mathcal{P}_{k,n_k}]^T$.

Viewing the observation equations (2) as a linear regression where the design matrix is given by an identity matrix, we thus have that the entire model constitutes a special case of the regression model adjusted in Alkhatib et al. (2018), which is a multivariate extension of the model considered in Kargoll et al. (2018). Now, in the general situation of a Gauss-Helmert model (described in detail in Koch 2014a), the location parameters $\boldsymbol{\mu} = [\boldsymbol{\mu}_1^T, \ldots, \boldsymbol{\mu}_N^T]^T$ alongside additional parameters $\boldsymbol{\xi} = [\xi_1, \ldots, \xi_u]^T$ have to satisfy r (possibly nonlinear) condition equations $\mathbf{h}(\boldsymbol{\xi}, \boldsymbol{\mu}) = \mathbf{0}_{[r\times 1]}$.

3 The Adjustment Procedure

Let us define $\boldsymbol{\sigma} = [\sigma_1, \ldots, \sigma_N]^T$ as well as $\boldsymbol{\nu} = [\nu_1, \ldots, \nu_N]^T$. To estimate the unknown model parameters $\boldsymbol{\phi} = [\boldsymbol{\xi}^T, \boldsymbol{\mu}^T, \boldsymbol{\alpha}^T, \boldsymbol{\sigma}^T, \boldsymbol{\nu}^T]^T$, we apply the method of constrained ML estimation. In analogy to the model in Alkhatib et al. (2018), the model of Sect. 2 gives rise to the logarithmized pdf

$$\log f(\mathbf{u},\mathbf{p}) = \text{const.} - \sum_{k=1}^{N}\frac{n_k}{2}\log(\sigma_k^2) + \sum_{k=1}^{N}\frac{n_k}{2}\nu_k\log\left(\frac{\nu_k}{2}\right) - \sum_{k=1}^{N}\sum_{t=1}^{n_k}\frac{1}{2}\left[\nu_k + \left(\frac{\boldsymbol{\alpha}_k(L)(\boldsymbol{\ell}_{k,t} - \mathbf{I}_{k,t}\boldsymbol{\mu}_k)}{\sigma_k}\right)^2\right]p_{k,t} - \sum_{k=1}^{N} n_k\log\Gamma\left(\frac{\nu_k}{2}\right) + \sum_{k=1}^{N}\sum_{t=1}^{n_k}\frac{1}{2}(\nu_k - 1)\log p_{k,t}. \tag{10}$$

Note that (10) does not depend on all of the parameters in $\boldsymbol{\phi}$, but only on the parameters $\boldsymbol{\theta} = [\boldsymbol{\mu}^T, \boldsymbol{\alpha}^T, \boldsymbol{\sigma}^T, \boldsymbol{\nu}^T]^T$. Next, we define that logarithmic pdf to be the log-likelihood function $\log L(\boldsymbol{\theta}; \boldsymbol{\ell}, \mathbf{p})$, to be maximized under the constraints. Here, $\boldsymbol{\ell} = [\boldsymbol{\ell}_1^T, \ldots, \boldsymbol{\ell}_N^T]^T$ is the vector of given numerical observations for the observables $\boldsymbol{\mathcal{L}} = [\boldsymbol{\mathcal{L}}_1^T, \ldots, \boldsymbol{\mathcal{L}}_N^T]^T$. To deal with the missing data $\mathbf{p}$, we establish an (iterative) EM algorithm in which the E-step determines the conditional expectation

$$Q(\boldsymbol{\theta}|\boldsymbol{\theta}^{(s)}) = E_{\boldsymbol{\mathcal{P}}|\boldsymbol{\ell};\boldsymbol{\theta}^{(s)}}\{\log L(\boldsymbol{\theta}; \boldsymbol{\ell}, \boldsymbol{\mathcal{P}})\}$$

of the log-likelihood function with respect to the given stochastic model for the latent variables $\boldsymbol{\mathcal{P}}$, using the given observations $\boldsymbol{\ell}$ alongside initial parameter values $\boldsymbol{\theta}^{(s)}$ known from the preceding iteration step s. As the likelihood function, being defined by (10), is actually a function of $\mathbf{u}$ rather than $\boldsymbol{\ell}$, we condition directly on $\mathbf{u}$ besides $\boldsymbol{\theta}^{(s)}$, which values fix the dependent observations $\boldsymbol{\ell}$ through the Eqs. (2) and (3). Thus, the Q-function takes the form

$$Q(\boldsymbol{\theta}|\boldsymbol{\theta}^{(s)}) = \text{const.} - \sum_{k=1}^{N} \frac{n_k}{2} \log(\sigma_k^2) + \sum_{k=1}^{N} \frac{n_k}{2} \nu_k \log\left(\frac{\nu_k}{2}\right)$$
$$- \sum_{k=1}^{N}\sum_{t=1}^{n_k} \frac{1}{2}\left[\nu_k + \left(\frac{\boldsymbol{\alpha}_k(L)(\ell_{k,t} - \mathbf{I}_{k,t}\boldsymbol{\mu}_k)}{\sigma_k}\right)^2\right] E_{\mathcal{P}|\mathbf{u};\boldsymbol{\theta}^{(s)}}\{\mathcal{P}_{k,t}\}$$
$$- \sum_{k=1}^{N} n_k \log\Gamma\left(\frac{\nu_k}{2}\right) + \sum_{k=1}^{N}\sum_{t=1}^{n_k} \frac{1}{2}(\nu_k - 1) E_{\mathcal{P}|\mathbf{u};\boldsymbol{\theta}^{(s)}}\{\log \mathcal{P}_{k,t}\},$$

where

$$E_{\mathcal{P}|\mathbf{u};\boldsymbol{\theta}^{(s)}}\{\mathcal{P}_{k,t}\} = E_{\mathcal{P}_{k,t}|u_{k,t};\boldsymbol{\theta}^{(s)}}\{\mathcal{P}_{k,t}\},$$
$$E_{\mathcal{P}|\mathbf{u};\boldsymbol{\theta}^{(s)}}\{\log \mathcal{P}_{k,t}\} = E_{\mathcal{P}_{k,t}|u_{k,t};\boldsymbol{\theta}^{(s)}}\{\log \mathcal{P}_{k,t}\}.$$

The Q-function can then be shown to be

$$Q(\boldsymbol{\theta}|\boldsymbol{\theta}^{(s)}) = \text{const.} - \frac{1}{2}\sum_{k=1}^{N} n_k \log(\sigma_k^2)$$
$$- \frac{1}{2}\sum_{k=1}^{N} \frac{1}{\sigma_k^2} \sum_{t=1}^{n_k} p_{k,t}^{(s)} [\boldsymbol{\alpha}_k(L)(\ell_{k,t} - \mathbf{I}_{k,t}\boldsymbol{\mu}_k)]^2$$
$$+ \frac{1}{2}\sum_{k=1}^{N} n_k \nu_k \log \nu_k - \sum_{k=1}^{N} n_k \log\Gamma\left(\frac{\nu_k}{2}\right)$$
$$+ \frac{1}{2}\sum_{k=1}^{N} n_k \nu_k \left[\psi\left(\frac{\nu_k^{(s)}+1}{2}\right) - \log\left(\nu_k^{(s)}+1\right)\right.$$
$$\left. + \frac{1}{n_k}\sum_{t=1}^{n_k}\left(\log p_{k,t}^{(s)} - p_{k,t}^{(s)}\right)\right] \tag{11}$$

where ψ is the digamma function and where each imputed weight $p_{k,t}^{(s)}$, as the conditional expectation of the latent variable $\mathcal{P}_{k,t}$, is determined by

$$p_{k,t}^{(s)} = E_{\mathcal{P}_{k,t}|u_{k,t};\boldsymbol{\theta}^{(s)}}\{\mathcal{P}_{k,t}\} = \frac{\nu_k^{(s)} + 1}{\nu_k^{(s)} + \left(\frac{\boldsymbol{\alpha}_k^{(s)}(L)(\ell_{k,t} - \mathbf{I}_{k,t}\boldsymbol{\mu}_k^{(s)})}{\sigma_k^{(s)}}\right)^2}. \tag{12}$$

The proof is the same as for the Q-function in Alkhatib et al. (2018) if one replaces the parameters $\boldsymbol{\xi}$ there by $\boldsymbol{\mu}$ and if one considers the specific linear functional model $h_{k,t}(\boldsymbol{\mu}) = \mathbf{I}_{k,t}\boldsymbol{\mu}_k$. For shorter expressions, we define the diagonal matrices

$$\operatorname{diag}(\mathbf{W}_k^{(s)}) = \left[p_{k,1}^{(s)}/\sigma_k^2, \ldots, p_{k,n_k}^{(s)}/\sigma_k^2\right]. \tag{13}$$

The subsequent M-step maximizes this Q-function under the constraints, which step we solve by maximizing the Lagrangian

$$F(\boldsymbol{\phi}, \boldsymbol{\lambda}|\boldsymbol{\theta}^{(s)}) = Q(\boldsymbol{\theta}|\boldsymbol{\theta}^{(s)}) - \boldsymbol{\lambda}^T \mathbf{h}(\boldsymbol{\xi}, \boldsymbol{\mu}), \tag{14}$$

where $\boldsymbol{\lambda}$ is the $(r \times 1)$-vector of unknown Lagrange multipliers. If the function $\mathbf{h}$ is nonlinear, we linearize it as shown generally in Koch (2014a, Sect. 2); as we adjust N groups of observations, we may introduce summations over these groups and write

$$\mathbf{h}(\boldsymbol{\xi}, \boldsymbol{\mu}) \approx \mathbf{A}\boldsymbol{\Delta\xi} + \sum_{k=1}^{N} \mathbf{B}_k(\boldsymbol{\mu}_k - \boldsymbol{\ell}_k) + \mathbf{m}_p \tag{15}$$

with $\boldsymbol{\Delta\xi} = \boldsymbol{\xi} - \boldsymbol{\xi}^{(s)}$ and pseudo-misclosures

$$\mathbf{m}_p = \mathbf{m} + \sum_{k=1}^{N} \mathbf{B}_k(\boldsymbol{\ell}_k - \boldsymbol{\mu}_k^{(s)}), \tag{16}$$

where we define $\mathbf{m} = \mathbf{h}(\boldsymbol{\xi}^{(s)}, \boldsymbol{\mu}^{(s)})$ to be the vector of misclosures at the most recent estimates $\boldsymbol{\xi}^{(s)}$ and $\boldsymbol{\mu}^{(s)}$. Here, $\mathbf{A} = \partial\mathbf{h}(\boldsymbol{\xi}^{(s)}, \boldsymbol{\mu}^{(s)})/\partial\boldsymbol{\xi}$ is the $(r \times u)$-matrix of partial derivatives of $\mathbf{h}(\boldsymbol{\xi}, \boldsymbol{\mu})$ with respect to $\boldsymbol{\xi}$ evaluated at $\boldsymbol{\xi}^{(s)}$ and $\boldsymbol{\mu}^{(s)}$, and $\mathbf{B}_k = \partial\mathbf{h}(\boldsymbol{\xi}^{(s)}, \boldsymbol{\mu}^{(s)})/\partial\boldsymbol{\mu}_k$ is the $(r \times n_k)$-matrix of partial derivatives of $\mathbf{h}(\boldsymbol{\xi}, \boldsymbol{\mu})$ with respect to $\boldsymbol{\mu}_k$ (also evaluated at $\boldsymbol{\xi}^{(s)}$ and $\boldsymbol{\mu}^{(s)}$). The first-order conditions for the maximization of the Lagrangian (including linearization) read

$$\mathbf{0}_{[u\times 1]} = \frac{\partial F(\boldsymbol{\phi}, \boldsymbol{\lambda})}{\partial \boldsymbol{\Delta\xi}} = -\mathbf{A}^T\boldsymbol{\lambda}, \tag{17}$$

$$\mathbf{0}_{[n_k\times 1]} = \frac{\partial F(\boldsymbol{\phi}, \boldsymbol{\lambda})}{\partial \boldsymbol{\mu}_k} = -\bar{\mathbf{I}}_k \mathbf{W}_k^{(s)}(\bar{\boldsymbol{\ell}}_k - \bar{\mathbf{I}}_k\boldsymbol{\mu}_k) - \mathbf{B}_k^T\boldsymbol{\lambda}, \tag{18}$$

$$\mathbf{0}_{[p_k\times 1]} = \frac{\partial F(\boldsymbol{\phi}, \boldsymbol{\lambda})}{\partial \boldsymbol{\alpha}_k} = -\mathbf{E}_k^T \mathbf{W}_k^{(s)}(\mathbf{e}_k - \mathbf{E}_k\boldsymbol{\alpha}_k), \tag{19}$$

$$0 = \frac{\partial F(\boldsymbol{\phi}, \boldsymbol{\lambda})}{\partial \sigma_k^2} = -\frac{n_k}{2\sigma_k^2} - \frac{1}{2\sigma_k^4} \times \sum_{t=1}^{n_k} p_{k,t}^{(s)} [\boldsymbol{\alpha}_k(L)(\ell_{k,t} - \mu_{k,t})]^2, \tag{20}$$

$$0 = \frac{\partial F(\boldsymbol{\phi}, \boldsymbol{\lambda})}{\partial \nu_k} = 1 + \log\left(\nu_k^{(s+1)}\right) - \log\left(\nu_k^{(s+1)} + 1\right) - \psi\left(\nu_k^{(s+1)}/2\right) + \psi\left([\nu_k^{(s+1)} + 1]/2\right) + \frac{1}{n_k}\sum_{t=1}^{n_k}\left[\log p_{k,t}^{(s)} - p_{k,t}^{(s)}\right], \tag{21}$$

$$\mathbf{0}_{[r\times 1]} = \frac{\partial F(\boldsymbol{\phi}, \boldsymbol{\lambda})}{\partial \boldsymbol{\lambda}} = -\left(\mathbf{A}\boldsymbol{\Delta\xi} + \sum_{k=1}^{N} \mathbf{B}_k(\boldsymbol{\mu}_k - \boldsymbol{\ell}_k) + \mathbf{m}_p\right). \tag{22}$$

with $\mathbf{e}_k = \boldsymbol{\ell}_k - \mathbf{I}_k\boldsymbol{\mu}_k$ and

$$\mathbf{E}_k = \begin{bmatrix} e_{k,0} & \cdots & e_{k,1-p_k} \\ \vdots & & \vdots \\ e_{k,n_k-1} & \cdots & e_{k,n_k-p_k} \end{bmatrix}. \tag{23}$$

Combining (18) with (22) yields

$$\sum_{k=1}^{N} \mathbf{B}_k \left(\bar{\mathbf{I}}_k^T \mathbf{W}_k^{(s)} \bar{\mathbf{I}}_k\right)^{-1} \mathbf{B}_k^T \boldsymbol{\lambda} + \mathbf{A}\boldsymbol{\Delta\xi} =$$

$$\sum_{k=1}^{N} \mathbf{B}_k \boldsymbol{\ell}_k - \mathbf{m}_p - \sum_{k=1}^{N} \mathbf{B}_k \left(\bar{\mathbf{I}}_k^T \mathbf{W}_k^{(s)} \bar{\mathbf{I}}_k\right)^{-1} \bar{\mathbf{I}}_k^T \mathbf{W}_k^{(s)} \bar{\boldsymbol{\ell}}_k$$

and thus, in conjunction with (17),

$$\begin{bmatrix} \boldsymbol{\lambda} \\ \boldsymbol{\Delta\xi} \end{bmatrix} = \begin{bmatrix} \sum_{k=1}^{N} \mathbf{B}_k \left(\bar{\mathbf{I}}_k^T \mathbf{W}_k^{(s)} \bar{\mathbf{I}}_k\right)^{-1} \mathbf{B}_k^T & \mathbf{A} \\ \mathbf{A}^T & \mathbf{0} \end{bmatrix}^{-1} \tag{24}$$

$$\times \begin{bmatrix} \sum_{k=1}^{N} \mathbf{B}_k \boldsymbol{\ell}_k - \mathbf{m}_p - \sum_{k=1}^{N} \mathbf{B}_k \left(\bar{\mathbf{I}}_k^T \mathbf{W}_k^{(s)} \bar{\mathbf{I}}_k\right)^{-1} \bar{\mathbf{I}}_k^T \mathbf{W}_k^{(s)} \bar{\boldsymbol{\ell}}_k \\ \mathbf{0} \end{bmatrix}.$$

Note that the decorrelation filter $\boldsymbol{\alpha}_k(L)$ transforms the diagonal matrix $\mathbf{I}_k$ into the matrix $\bar{\mathbf{I}}_k$ having an additional diagonal for each AR coefficient. Now, (18) gives us

$$\boldsymbol{\mu}_k = \left(\bar{\mathbf{I}}_k^T \mathbf{W}_k^{(s)} \bar{\mathbf{I}}_k\right)^{-1} \left(\mathbf{B}_k^T \boldsymbol{\lambda} + \bar{\mathbf{I}}_k^T \mathbf{W}_k^{(s)} \bar{\boldsymbol{\ell}}_k\right), \tag{25}$$

(19) can be written as the equation system

$$\boldsymbol{\alpha}_k = \left(\mathbf{E}_k^T \mathbf{W}_k^{(s)} \mathbf{E}_k\right)^{-1} \mathbf{E}_k^T \mathbf{W}_k^{(s)} \mathbf{e}_k, \tag{26}$$

and (20) yields with (4)

$$\sigma_k^2 = \frac{1}{n_k} \cdot \sum_{t=1}^{n_k} p_{k,t}^{(s)} u_{k,t}^2. \tag{27}$$

The solution of (21) is obtained by means of a zero search algorithm, using for instance MATLAB's fzero.m routine. Since each of the five parameter groups (a) $(\boldsymbol{\lambda}, \boldsymbol{\Delta\xi})$, (b) $\boldsymbol{\mu}$, (c) $\boldsymbol{\alpha}$, (d) $\boldsymbol{\sigma}^2$ and (e) $\boldsymbol{\nu}$ can evidently be solved for from (a) to (e) after determining values for the preceding group, we may employ conditional maximization within the M-step; together with the E-step, we thus have a so-called expectation conditional maximization (ECM) algorithm (cf. McLachlan and Krishnan 2008). To make sure that the parameters $\boldsymbol{\xi}$ and $\boldsymbol{\mu}$ satisfy the (linearized) constraints (15), we adapt the constrained EM algorithm proposed by Takai (2012) and iterate the solution of (24)–(25) within the current M-iteration step $s+1$ until the misclosures $\mathbf{m}$ become sufficiently small. After achieving this, that M-step is completed by computing the solutions (26) using $\boldsymbol{\mu}_k^{(s+1)}$ to determine $\mathbf{e}^{(s+1)}$ and $\mathbf{E}^{(s+1)}$, then the solutions (27) using $\boldsymbol{\alpha}_k^{(s+1)}$ to determine $\mathbf{u}^{(s+1)}$, and finally the solutions (21). Having thus obtained the solution $\boldsymbol{\phi}^{(s+1)}$, the next E-iteration step is executed by computing new weights $p_{k,t}^{(s+1)}$. As indicated in McLachlan and Krishnan (2008), it is more efficient to solve for ν_k by maximizing the log-likelihood function (7) with respect to ν_k, leading to the first-order condition

$$0 = \log \nu_k^{(s+1)} + 1 - \psi\left(\frac{\nu_k^{(s+1)}}{2}\right) + \psi\left(\frac{\nu_k^{(s+1)} + 1}{2}\right)$$
$$- \log\left(\nu_k^{(s+1)} + 1\right) + \frac{1}{n_k} \sum_{t=1}^{n_k} \left(\log p_{k,t}^{(s+1)} - p_{k,t}^{(s+1)}\right) \tag{28}$$

with

$$p_{k,t}^{(s+1)} = \frac{\nu_k^{(s+1)} + 1}{\nu_k^{(s+1)} + \left(u_{k,t}^{(s+1)} / \sigma_k^{(s+1)}\right)^2}. \tag{29}$$

With this modification, the previous ECM algorithm turns into an expectation conditional maximization *either* (ECME) algorithm, which is constrained in our case. The E- and M-steps are iterated until a preset maximum number of iterations is reached or the maximum difference between subsequent solutions is sufficiently small, say $\varepsilon_\nu = 10^{-4}$ concerning $\nu_1, \ldots, \nu_N$ and $\varepsilon = 10^{-8}$ for the other parameters. In the sequel, the estimates of the final EM step are denoted by $\hat{\boldsymbol{\xi}}$, $\hat{\boldsymbol{\mu}}_k$, $\hat{\boldsymbol{\alpha}}_k$, $\hat{\sigma}_k^2$ and $\hat{\nu}_k$. Note that $\hat{\boldsymbol{\mu}}_k$ may be viewed as the adjusted observations $\hat{\boldsymbol{\ell}}_k$.

4 Monte Carlo Results

The purpose of the following Monte Carlo simulation is to evaluate the estimation bias within a closed-loop simulation based on a true model with fixed, known parameter values.

4.1 Simulation Setup

To generate n observations, we first determined $n/2$ true observation points with coordinates $\mathbf{x} = [x_1, \ldots, x_{n/2}]^T$ and $\mathbf{y} = [y_1, \ldots, y_{n/2}]^T$ on a 2D circle with center $(x_c, y_c) = (1, 1)$ and radius $r = 2$ by means of the equations

$$x_t = x_c + r\cos(\phi_t), \tag{30}$$

$$y_t = y_c + r\sin(\phi_t), \tag{31}$$

with equidistant radians values

- $\phi_t = 0.2, \ldots, 10$ with step size 0.2 for $n/2 = 50$,
- $\phi_t = 0.02, \ldots, 10$ with step size 0.02 for $n/2 = 500$,
- $\phi_t = 0.002, \ldots, 10$ with step size 0.002 for $n/2 = 5{,}000$.

Regarding the noise, we generated $5n$ random numbers (using MATLAB's trnd.m routine) from the t-distribution with scale factor $\sigma_1^2 = 0.001^2$ and degree of freedom $\nu_1 = 2.5$ for the x-coordinates, as well as from the t-distribution with scale factor $\sigma_2^2 = 0.001^2$ and degree of freedom $\nu_2 = 3$ for the y-coordinates. The resulting white noise series $\mathbf{u}_1$ and $\mathbf{u}_2$ were then correlated by means of an AR(1) process with coefficient $\alpha_1 = -0.6$ and an AR(1) process with coefficient $\alpha_2 = 0.5$, respectively, using (3). We extracted the last $n/2$ values from each of the resulting colored noise series, discarding the beginnings in order to minimize the warm-up effect caused by the initial process values $e_{1,0} = e_{2,0} = 0$. The resulting vectors $\mathbf{e}_1$ and $\mathbf{e}_2$ were then added to $\mathbf{x}$ and $\mathbf{y}$, giving the observation vectors $\boldsymbol{\ell}_1$ and $\boldsymbol{\ell}_2$. This generation procedure was repeated for 1,000 times, as part of a Monte Carlo closed-loop simulation.

4.2 Estimation Results

Combining (30) and (31) results in the well-known circle equation

$$(x_t - x_c)^2 + (y_t - y_c)^2 = r^2, \tag{32}$$

which, however, is generally not satisfied by the generated or measured coordinates x_t and y_t due to the noise that they contain. Therefore, these coordinates are modeled as realizations of random variables

$$\mathcal{X}_t = \mathcal{L}_{1,t} = \mu_{1,t} + \mathcal{E}_{1,t}, \tag{33}$$

$$\mathcal{Y}_t = \mathcal{L}_{2,t} = \mu_{2,t} + \mathcal{E}_{2,t}, \tag{34}$$

according to (1), and the values x_t and y_t in (32) are replaced by the unknown location parameters $\mu_{1,t}$ and $\mu_{2,t}$, respectively. These substitutions give us the constraints

$$\mathbf{h}(\boldsymbol{\xi}, \boldsymbol{\mu}) = \begin{bmatrix} h_1(\boldsymbol{\xi}, \boldsymbol{\mu}) \\ \vdots \\ h_{n/2}(\boldsymbol{\xi}, \boldsymbol{\mu}) \end{bmatrix} = \begin{bmatrix} (\mu_{1,1} - x_c)^2 + (\mu_{2,1} - y_c)^2 - r^2 \\ \vdots \\ (\mu_{1,n/2} - x_c)^2 + (\mu_{2,n/2} - y_c)^2 - r^2 \end{bmatrix} = \begin{bmatrix} 0 \\ \vdots \\ 0 \end{bmatrix} \tag{35}$$

with $\boldsymbol{\xi} = [x_c, y_c, r]^T$. The model parameters $\boldsymbol{\xi}$, $\boldsymbol{\mu}$, α_1, α_2, σ_1^2, σ_2^2, ν_1 and ν_2 can be estimated by means of the algorithm described in Sect. 3, using the initial values $x_c^{(0)} = y_c^{(0)} = 0.7$, $r^{(0)} = 1.7$, $\mu_{1,t}^{(0)} = x_t$, and $\mu_{2,t}^{(0)} = y_t$ $(t = 1, \ldots, n/2)$. Consequently, the linearization at the corresponding initial Taylor point is based on the Jacobi matrices

$$\mathbf{A} = \begin{bmatrix} -2(x_1 - x_c^{(0)}) & -2(y_1 - y_c^{(0)}) & -2r^{(0)} \\ \vdots & \vdots & \vdots \\ -2(x_{n/2} - x_c^{(0)}) & -2(y_{n/2} - y_c^{(0)}) & -2r^{(0)} \end{bmatrix}, \tag{36}$$

$$\mathbf{B}_1 = \begin{bmatrix} 2(x_1 - x_c^{(0)}) & & 0 \\ & \ddots & \\ 0 & & 2(x_{n/2} - x_c^{(0)}) \end{bmatrix}, \tag{37}$$

$$\mathbf{B}_2 = \begin{bmatrix} 2(y_1 - y_c^{(0)}) & & 0 \\ & \ddots & \\ 0 & & 2(y_{n/2} - y_c^{(0)}) \end{bmatrix}. \tag{38}$$

The results of these adjustments, alongside the number of EM iteration steps and the mean of the root mean square error

$$\text{RMSE} = \sqrt{\left(\begin{bmatrix} \hat{\boldsymbol{\mu}}_1 \\ \hat{\boldsymbol{\mu}}_2 \end{bmatrix} - \begin{bmatrix} \mathbf{x} \\ \mathbf{y} \end{bmatrix}\right)^T \left(\begin{bmatrix} \hat{\boldsymbol{\mu}}_1 \\ \hat{\boldsymbol{\mu}}_2 \end{bmatrix} - \begin{bmatrix} \mathbf{x} \\ \mathbf{y} \end{bmatrix}\right)} \tag{39}$$

are summarized in Table 1. The parameter values of the three estimated circle parameters are correct up to three digits already for 100 observations, and their biases are very small for 1,000 or more observations. Figure 1 shows that the

Table 1 Results of the 1,000 adjustments of a 2D circle with center $x_c = 1$, $y_c = 1$ and radius $r = 2$ with AR(1) errors in both dimensions ($\alpha_1 = -0.6$, $\alpha_2 = 0.5$) as well as Student errors in both dimensions ($\nu_1 = 2.5$, $\nu_2 = 3$, $\sigma_1^2 = \sigma_2^2 = 0.001^2$) for $n = 100/1{,}000/10{,}000$ observations

n	100	1,000	10,000
Mean(EM steps)	82	76	68
Mean($\hat{x}_c$)	1.00000061	0.99999987	1.00000064
Mean($\hat{y}_c$)	1.00001679	1.00000949	0.99999974
Mean($\hat{r}$)	1.99928616	2.00000233	1.99999850
Mean($\hat{\alpha}_1$)	−0.5102	−0.5761	−0.5848
Mean($\hat{\alpha}_2$)	0.2196	0.3867	0.4175
Mean($\hat{\sigma}_1$)	0.00066	0.00068	0.00069
Mean($\hat{\sigma}_2$)	0.00052	0.00053	0.00053
Median($\hat{\nu}_1$)	2.9804	2.0504	2.0309
Median($\hat{\nu}_2$)	2.3854	1.7531	1.7384
Mean(RMSE)	0.00114	0.00169	0.00173

The shown quantities are: (1) The number of EM steps carried out, (2) Mean($\hat{x}_c$), (3) Mean($\hat{y}_c$) and (4) Mean($\hat{r}$) are, respectively, the (arithmetic) means of the estimated coordinates of the circle center and of the radius, (5) Mean($\hat{\alpha}_1$) and (6) Mean($\hat{\alpha}_2$) are the means of the estimated AR coefficients, (7) Mean($\hat{\sigma}_1$) and (8) Mean($\hat{\sigma}_2$) are the means of the square roots of estimated scale factors, (9) Median($\hat{\nu}_1$) and (10) Median($\hat{\nu}_2$) are the medians of the estimated degrees of freedom, and (11) Mean(RMSE) is the mean of the root mean square error defined by (39)

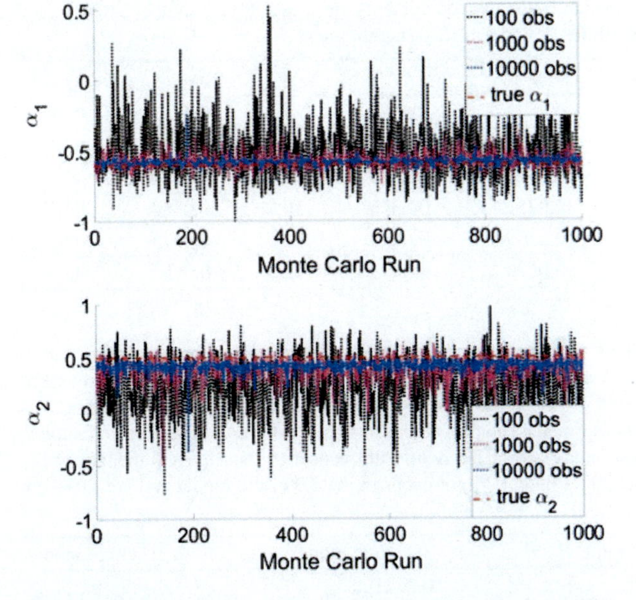

Fig. 1 Comparison of the estimated AR parameters α_1 and α_2 for different sample sizes: 100 (black), 1,000 (magenta) and 10,000 (blue)

variability of the two estimated AR parameters decreases strongly with increasing sample size n, whereas their biases decrease rather slowly and are still quite large for $10,000$ observations (see Table 1). However, the two qualitatively and quantitatively very different AR(1) processes are separated well by the algorithm. The biases of the estimated scale factors and of the estimated degrees of freedom could not be decreased by increasing the number of observations. For about 5% of the Monte Carlo samples the estimated degrees of freedom turned out to be rather large. As such large values would greatly distort the mean value, the median of the estimates throughout all Monte Carlo runs was taken as a measure of location. Figure 2 shows for the Monte Carlo simulation based on 10,000 observations that the medians of the two estimated degrees of freedom indeed closely match the peaks of the histogram plots, justifying this choice of measure. The mean RMSE values indicate that an overall good approximation of the data by the adjusted circle is achieved already for $n = 100$ observations.

4.3 Outlier Identification

The weights imputed within the adjustment by the EM algorithm can be used to identify outliers. In the following we demonstrate a graphical procedure for a part of a simulated time series. Figure 3 (top) shows the 500 x-coordinates generated within a single Monte Carlo run based precisely on the parameter values of the previous simulation setup, with the exception that the larger scale factors $\sigma_1^2 = \sigma_2^2 = 0.1^2$ were used in order to visually bring out the outliers more clearly.

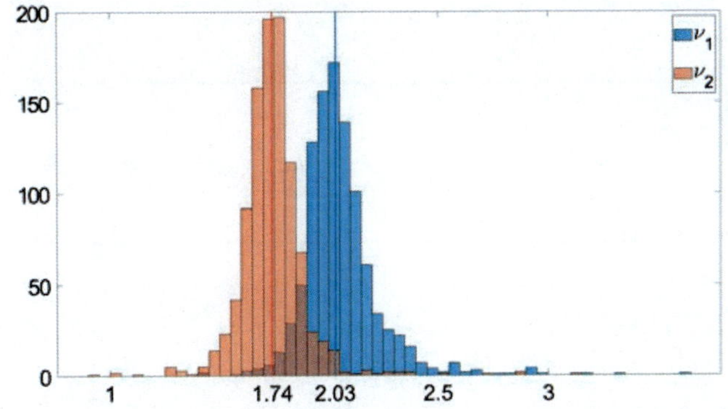

Fig. 2 Histogram plots of the estimated degrees of freedom for the Monte Carlo simulation based on 10,000 observations. The vertical lines indicate the location of the medians (see also Table 1), which are near the peaks of the histograms

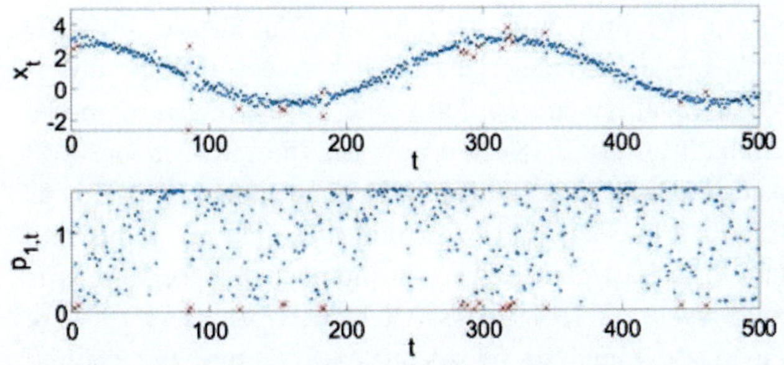

Fig. 3 Display of the generated x-coordinates within a single Monte Carlo run and the corresponding weights after the adjustment; the red crosses in the upper subplot indicate the observations associated with the 20 least weights (red crosses in the lower subplot)

After adjusting the entire data set of x- and y-coordinates the weights $p_{1,t}$ corresponding to this first group of observations were also plotted (see Fig. 3, bottom). All weights less than 0.1 were marked by red crosses, and the observations with the same indexes t were then also marked by red crosses. It is seen that all of these small weights correspond to visibly outlying x-coordinates. Since the transition from small to large weights is rather smooth, the separation of inliers from outliers could be achieved via the definition of a threshold, (in analogy to the "3σ rule", cf. Lehmann 2013) or by means of an additional classification procedure (such as the method suggested by Koch 2012). As the algorithm works reasonably well in a controlled numerical experiment, we apply it in the following to a more challenging real data set.

5 A Numerical Example Based on Real Data

We adjust GNSS observations of the circularly and constantly rotating multi-sensor system (MSS) described in Paffenholz (2012), given in an North-East-Up (NEU) coordinate

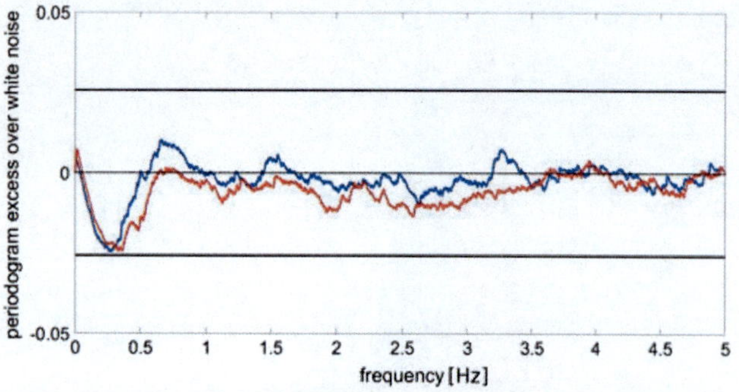

Fig. 4 Excess of the estimated periodograms of the decorrelated residuals for the North (red) and the East (blue) component based on the AR(12) model over the theoretical white noise periodogram (constant zero); approximate 99% significance bounds are displayed as the two heavy black lines

Table 2 Estimated coefficients of the AR(12) models fitted to the North ($\hat{\alpha}_{1,j}$) and East ($\hat{\alpha}_{2,j}$) component

j	$\hat{\alpha}_{1,j}$	$\hat{\alpha}_{2,j}$	j	$\hat{\alpha}_{1,j}$	$\hat{\alpha}_{2,j}$
1	0.2522	0.0284	7	0.0272	0.0844
2	0.0991	0.1327	8	0.0235	0.0368
3	0.0544	0.0677	9	0.0739	0.0488
4	0.0751	0.0326	10	0.0380	0.0358
5	0.0209	0.0952	11	0.0478	0.0802
6	0.0670	−0.0013	12	0.0965	0.0565

Table 3 Estimated parameters of the 3D circle model and t-distributions fitted in Alkhatib et al. (2018) that can be directly compared with the estimates of the current 2D model and t-distributions (the estimated rotation angles and the center coordinate, scale factor as well as degree of freedom with respect to the Up/z-component of the 3D circle model do not occur in the 2D circle model and are therefore omitted)

Parameter	3D circle model	2D circle model
$\hat{r}$	0.2971 m	0.2973 m
$\hat{x}_c$	12.2340 m	12.2337 m
$\hat{y}_c$	−16.6317 m	−16.6319 m
$\hat{\sigma}_1^2$	$7.3 \cdot 10^{-7}$ m^2	$1.6 \cdot 10^{-7}$ m^2
$\hat{\sigma}_2^2$	$1.4 \cdot 10^{-6}$ m^2	$3.3 \cdot 10^{-7}$ m^2
$\hat{\nu}_1$	120	3
$\hat{\nu}_2$	88	3

system. We previously approximated this dataset by a 3D circle model parameterized in such a way that the observation equations are conform with a Gauss-Markov model (Alkhatib et al. 2018). We now use the circle model (32) instead and employ the Gauss-Helmert model (33)–(35) as described in Sects. 2–4, after projecting the 3D points into the North-East-plane. The total number of observations is given by $n = 15{,}792$. As AR models of orders 15 were previously found to be adequate to capture the colored measurement noise in the North and East component, we tried out orders between 10–20 and selected the smallest order ($p = 12$) for which Bartlett's periodogram-based white noise test (described in Kargoll et al. 2018) is accepted for both components upon convergence of the EM algorithm of Sect. 3 (see Fig. 4). For this AR model order, the algorithm converged after 60 EM steps. The estimated AR coefficients are shown in Table 2; the rather small values indicate that the processes are stable and do not cause numerical problems during the filtering steps. Concerning the 3D circle model fitted in Alkhatib et al. (2018), the optimal AR model orders for the comparable North and East components were $p = 15$ each, but the periodogram excesses of the corresponding estimated AR models exceeded the 99 % significance bounds, so that these models were not fully satisfactory. Thus, the estimated AR models with respect to the current 2D model are both adequate in view of the accepted white noise tests and more parsimonious due to the smaller model orders. The estimated degrees of freedom $\hat{\nu}_1 = 3$ and $\hat{\nu}_2 = 3$ indicate substantial heavy-tailedness of the white noise components. Table 3 shows that the estimated radius $\hat{r}$ and circle center coordinates in the North-East (x–y) plane differ only slightly ($\approx$0.3 mm) between the 3D and 2D circle model. In contrast, the estimated scale factors and degrees of freedom with respect to these two coordinate components differ greatly between the two circle models. As the fitted AR models and thus the colored noise characteristics have been found to be quite different for the two adjustment models, differences of the white noise models in terms of scale factors and degrees of freedom appear to be a reasonable consequence. As the results of the previous Monte Carlo simulation showed a quite large variability of the estimates for these parameters, it is currently unclear how significant the observed numerical differences actually are.

6 Summary, Conclusions, and Outlook

We extended the Gauss-Helmert model with Student errors to situations, where the random deviations of different observation groups follow independent AR processes, where the white noise components of each process are modeled by a Student distribution with individual scale factor and degree of freedom. We formulated the optimization problem within the framework of constrained ML estimation, leading to a computationally convenient EM algorithm. The linearization of the condition equations was shown to be simply nested within the conditional maximization step with respect to the affected parameters. The conditional maximization steps with respect to the AR coefficients, the scale factors and the degrees of freedoms are similar to those of a comparable EM algorithm for the previously established multivariate regression time series models with AR and Student errors.

We showed that the proposed algorithm produces reasonable estimates in a numerical simulation. The biases of the three estimated circle parameters are very small already for a a number of observations as small as 1,000. The finding that the biases of the estimated AR coefficients decrease rather slowly, being still quite large for 10,000 observations, suggests that the number of observations should be much larger than that (say, $n > 1{,}000{,}000$) in practice. The modeling of AR processes with higher orders should then also be based on larger numbers of observations, as observed in Kargoll et al. (2018). As to the reason for the persistent bias it could be investigated in the future whether the warm-up effect due to the simple choice of zeros for the initial values of the AR processes plays an important role. The focus of the presented closed-loop simulation was to evaluate the bias of the estimator but not the robustness of the estimator against AR model misspecification. As the AR model order is usually unknown in practical situations it seems to be a worthwhile task, as part of future work, to study the effect of a misspecified AR model order on the estimation of the other model parameters. Such research related to the problem of model selection could include the application of information criteria (besides the white noise tests), which were also beyond the scope of the current contribution. The biases of the estimated scale factors and of the estimated degrees of freedom could not be decreased by increasing the number of observations, which problem should also be investigated in the future. Despite these shortcomings it could be demonstrated by a numerical example that small weights imputed by the EM algorithm correspond to visible outliers in the given time series. The application of the proposed adjustment procedure to the fitting of a circle to a data set consisting of real GNSS observations led to more adequate and more parsimonious AR models than the previous approximation of that data by means of a fundamentally different adjustment procedure.

The investigated Gauss-Helmert model with AR and Student errors appears to be a useful framework in the context of multivariate time series analysis of sensor data affected by auto-correlations and outliers. In case such phenomena do not play a significant role, the Student distribution tends towards a Gaussian distribution (since the estimated degree of freedom takes a large value), and the estimated AR models are eliminated through the aforementioned model selection procedure. Thus, the classical Gauss-Helmert model may be viewed as a special case of the presented model when the observations can be grouped in a certain way.

Acknowledgements Funded by the Deutsche Forschungsgemeinschaft (DFG, German Research Foundation) – 386369985.

References

Alkhatib H, Kargoll B, Paffenholz JA (2018) Further results on robust multivariate time series analysis in nonlinear models with autoregressive and t-distributed errors. In: Rojas I, Pomares H, Valenzuela O (eds) Time series analysis and forecasting. ITISE 2017, pp 25–38. Contributions to Statistics. Springer, Cham. https://doi.org/10.1007/978-3-319-96944-2_3

Kargoll B, Omidalizarandi M, Loth I, Paffenholz JA, Alkhatib H (2018) An iteratively reweighted least-squares approach to adaptive robust adjustment of parameters in linear regression models with autoregressive and t-distributed deviations. J Geod 92(3):271–297. https://doi.org/10.1007/s00190-017-1062-6

Koch KR (2012) Robust estimation by expectation maximization algorithm. J Geod 87:(2)107–116. https://doi.org/10.1007/s00190-012-0582-3

Koch KR (2014) Robust estimations for the nonlinear Gauss Helmert model by the expectation maximization algorithm. J Geod 88(3):263–271. https://doi.org/10.1007/s00190-013-0681-9

Koch KR (2014) Outlier detection for the nonlinear Gauss Helmert model with variance components by the expectation maximization algorithm. J Appl Geod 8(3):185–194. https://doi.org/10.1515/jag-2014-0004

Koch KR, Kargoll B (2013) Expectation-maximization algorithm for the variance-inflation model by applying the t-distribution. J Appl Geod 7:217–225. https://doi.org/10.1515/jag-2013-0007

Krasbutter I, Kargoll B, Schuh W-D (2015) Magic square of real spectral and time series analysis with an application to moving average processes. In: Kutterer H, Seitz F, Alkhatib H, Schmidt M (eds) The 1st international workshop on the quality of geodetic observation and monitoring systems (QuGOMS'11). International Association of Geodesy Symposia, vol 140. Springer International Publishing, Berlin, pp 9–14. https://doi.org/10.1007/978-3-319-10828-5_2

Kuhlmann H (2003) Kalman-filtering with coloured measurement noise for deformation analysis. In: Proceedings of the 11th FIG International Symposium on Deformation Measurements, FIG

Lange KL, Little RJA, Taylor JMG (1989) Robust statistical modeling using the t-distribution. J. Am. Stat. Assoc. 84:881–896. https://doi.org/10.2307/2290063

Lehmann R (2013) 3σ-rule for outlier detection from the viewpoint of geodetic adjustment. J. Surv. Eng. 139(4):157–165. https://doi.org/10.1061/(ASCE)SU.1943-5428.0000112

Loth I, Schuh W-D, Kargoll B (2019) Non-recursive representation of an autoregressive process within the Magic Square, IAG Symposia (First Online), Springer. https://doi.org/10.1007/1345_2019_60

McLachlan GJ, Krishnan T (2008) The EM algorithm and extensions, 2nd edn. Wiley, Hoboken

Paffenholz JA (2012) Direct geo-referencing of 3D point clouds with 3D positioning sensors. Committee for Geodesy (DGK) of the Bavarian Academy of Sciences and Humanities, Series C: Dissertations, No. 689, Munich

Parzen E (1979) A density-quantile function perspective on robust estimation. In: Launer L, Wilkinson GN (eds) Robustness in statistics, pp. 237–258. Academic Press, New York. https://doi.org/10.1016/B978-0-12-438150-6.50019-4

Schuh WD (2003) The processing of band-limited measurements; filtering techniques in the least squares context and in the presence of data gaps. Space Sci Rev 108(1):67–78. https://doi.org/10.1023/A:1026121814042

Takai K (2012) Constrained EM algorithm with projection method. Comput Stat 27:701–714. https://doi.org/10.1007/s00180-011-0285-x

How Abnormal Are the PDFs of the DIA Method: A Quality Description in the Context of GNSS

Safoora Zaminpardaz and Peter J. G. Teunissen

Abstract

The DIA-method, for the detection, identification and adaptation of modeling errors, has been widely used in a broad range of applications including the quality control of geodetic networks and the integrity monitoring of GNSS models. The DIA-method combines two key statistical inference tools, estimation and testing. Through the former, one seeks estimates of the parameters of interest, whereas through the latter, one validates these estimates and corrects them for biases that may be present. As a result of this intimate link between estimation and testing, the quality of the DIA outcome $\bar{x}$ must also be driven by the probabilistic characteristics of both estimation and testing. In practice however, the evaluation of the quality of $\bar{x}$ is never carried out as such. Instead, use is made of the probability density function (PDF) of the estimator under the identified hypothesis, say $\hat{x}_i$, thereby thus neglecting the conditioning process that led to the decision to accept the ith hypothesis. In this contribution, we conduct a comparative study of the probabilistic properties of $\bar{x}$ and $\hat{x}_i$. Our analysis will be carried out in the framework of GNSS-based positioning. We will also elaborate on the circumstances under which the distribution of the estimator $\hat{x}_i$ provides either poor or reasonable approximations to that of the DIA-estimator $\bar{x}$.

Keywords

Detection, identification and adaptation (DIA) · DIA-estimator · Global Navigation Satellite System (GNSS) · Probability density function (PDF) · Statistical testing

1 Introduction

In the DIA-method for the detection, identification and adaptation of mismodelling errors, next to estimation of parameters of interest, a statistical testing is also exercised to check the validation of underlying model. The actual DIA outcome is then the one which rigorously captures this combination of estimation and testing, and was introduced as the *DIA estimator* in Teunissen (2017b). The DIA-method has been widely used in a variety of applications, including the quality control of geodetic networks and the integrity monitoring of GNSS models, see e.g. DGCC (1982), Teunissen (1990), Salzmann (1995), Tiberius (1998), Perfetti (2006), Khodabandeh and Teunissen (2016), Zaminpardaz et al. (2015). As a result of the combined estimation-testing scheme of the DIA-method, the DIA outcome $\bar{x}$ must also be evaluated on the basis of characteristics of both estimation and testing. In practice however, the evaluation of the quality of $\bar{x}$ is carried out based upon the probability density function (PDF) of the estimator under the identified hypothesis, say

S. Zaminpardaz (✉)
Geospatial Sciences, School of Science, RMIT University, Melbourne, VIC, Australia
e-mail: safoora.zaminpardaz@rmit.edu.au

P. J. G. Teunissen
Department of Geoscience and Remote Sensing, Delft University of Technology, Delft, The Netherlands

GNSS Research Centre, School of Earth and Planetary Sciences, Curtin University, Perth, WA, Australia
e-mail: p.j.g.teunissen@tudelft.nl

P. Novák et al. (eds.), *IX Hotine-Marussi Symposium on Mathematical Geodesy*,
International Association of Geodesy Symposia 151, https://doi.org/10.1007/1345_2019_57

$\hat{x}_i$, without regard to the conditioning process that led to the decision of accepting the hypothesis $\mathcal{H}_i$. In this contribution, a comparative study of the probabilistic properties of $\bar{x}$ and $\hat{x}_i$ is conducted to highlight the impact of neglecting the estimation-testing link on follow-on quality evaluations and to elaborate on the circumstances under which such negligence may still be considered acceptable.

This contribution is organized as follows. We first give a brief overview of the Detection, Identification, and Adaptation procedure in Sect. 2. Using a partitioning of the misclosure space, the DIA-estimator and its statistical distribution are then presented in Sect. 3. The difference between the PDF of the DIA-estimator and that of $\hat{x}_i$ is discussed and shown to be driven by the DIA-method decision probabilities which can be categorized as probability of correct acceptance (CA), of false alarm (FA), of correct/missed detection (CD/MD) and of correct/wrong identification (CI/WI). In Sect. 4, we outline the estimation and testing strategies that we use for our analyses. Section 5 contains our numerical evaluations of the distribution of the DIA-estimator $\bar{x}$ and its normally-distributed individual components $\hat{x}_i$ $(i = 0, 1, \ldots, k)$. We graphically demonstrate, for binary hypothesis testing, i.e. $\mathcal{H}_0$ and $\mathcal{H}_1$, applied to a single-unknown, single-redundancy observational model, the PDF of $\bar{x}$, $\hat{x}_0$ and $\hat{x}_1$ under both $\mathcal{H}_0$ and $\mathcal{H}_1$. The distributional comparison is then continued for a Global Navigation Satellite System (GNSS) single point positioning (SPP) model where multiple-hypothesis testing is involved. Finally a summary with conclusions are presented in Sect. 6.

2 DIA Overview

As our point of departure, we first formulate our statistical hypotheses. The hypothesis believed to be true under nominal working conditions is referred to as the *null hypothesis*. Denoted by $\mathcal{H}_0$, the null hypothesis is assumed to be given as

$$\mathcal{H}_0: \quad \mathsf{E}(y) = Ax \quad , \quad \mathsf{D}(y) = Q_{yy} \tag{1}$$

with $\mathsf{E}(\cdot)$ and $\mathsf{D}(\cdot)$ denoting the expectation and dispersion operators, respectively. According to (1), under $\mathcal{H}_0$, the expectation of the normally-distributed random vector of observables $y \in \mathbb{R}^m$ is characterized through the unknown parameter vector $x \in \mathbb{R}^n$ and the full-rank design matrix $A \in \mathbb{R}^{m\times n}$ ($\text{rank}(A) = n$), while the dispersion of the observables y is described by the positive-definite variance-covariance matrix $Q_{yy} \in \mathbb{R}^{m\times m}$. The redundancy of $\mathcal{H}_0$ is $r = m - \text{rank}(A) = m - n$. The corresponding estimator of x on the basis of (1) is denoted by $\hat{x}_0$.

The observational model in (1) could be misspecified in several ways like, for example, $\mathsf{E}(y) \neq Ax$ and/or $\mathsf{D}(y) \neq Q_{yy}$. Here we assume that a misspecification is restricted to an underparametrization of the mean of y, which is the most common error that occurs when formulating the model (Teunissen 2017a). Thus, the alternative hypothesis $\mathcal{H}_i$ is formulated as

$$\mathcal{H}_i: \quad \mathsf{E}(y) = Ax + C_i b_i \quad , \quad \mathsf{D}(y) = Q_{yy} \tag{2}$$

where $b_i \in \mathbb{R}^q$ is the unknown bias vector while $C_i \in \mathbb{R}^{m\times q}$ is known which together with the design matrix A form a full-rank matrix, i.e. $\text{rank}([A,\ C_i]) = n + q$ with $q \leq m - n$. The corresponding estimator of x on the basis of (2) is denoted by $\hat{x}_i$.

In practical applications, we usually have to consider several alternative hypotheses about the physical reality at hand. For example when modeling GNSS observations, we may need to take into account hypotheses describing code outliers, phase cycle slips, ionospheric gradients, etc. The statistical validity of $\mathcal{H}_0$ and the multiple, say k, alternatives $\mathcal{H}_i$ $(i = 1, \ldots, k)$ is usually checked through the following three steps of detection, identification and adaptation (DIA) (Baarda 1968; Teunissen 1990).

1. *Detection* The null hypothesis undergoes a validity check using an overall model test, without considering a particular set of alternatives. If $\mathcal{H}_0$ is accepted, then $\hat{x}_0$ is provided as the estimate of x.
2. *Identification* In case $\mathcal{H}_0$ is rejected, a search is carried out among the specified alternative hypotheses $\mathcal{H}_i$ $(i = 1, \ldots, k)$ with the purpose of pinpointing the potential source of model error. In doing so, one of the alternative hypotheses, say $\mathcal{H}_i$, is identified as the suspected model error.
3. *Adaptation* The identified hypothesis $\mathcal{H}_i$ becomes the new null hypothesis. The $\mathcal{H}_0$-based inferences are then accordingly corrected and $\hat{x}_i$ is provided as the estimate of x.

The required information to realize the above steps of the DIA-method is contained in the *misclosure* vector $t \in \mathbb{R}^r$ given as

$$t = B^T y; \qquad Q_{tt} = B^T Q_{yy} B \tag{3}$$

where $B \in \mathbb{R}^{m\times r}$ is a full-rank matrix, with $\text{rank}(B) = r$, such that $[A,\ B] \in \mathbb{R}^{m\times m}$ is invertible and $A^T B = 0$. With $y \overset{\mathcal{H}_i}{\sim} \mathcal{N}(Ax + C_i b_i, Q_{yy})$ for $i = 0, 1, \ldots, k$ and $C_0 b_0 = 0$, the misclosure vector is then distributed as

$$t \overset{\mathcal{H}_i}{\sim} \mathcal{N}(\mu_{t_i} = B^T C_i b_i, Q_{tt}), \qquad \text{for } i = 0, 1, \ldots, k \tag{4}$$

An unambiguous testing procedure can be established through unambiguously assigning the outcomes of t to the statistical hypotheses $\mathcal{H}_i$ for $i = 0, 1, \ldots, k$, which can be realized through a partitioning of the misclosure space (Teunissen 2017b). As such, let $\mathcal{P}_i \in \mathbb{R}^r$ $(i = 0, 1, \ldots, k)$ be a partitioning of the misclosure space $\mathbb{R}^r$, i.e. $\cup_{i=0}^{k}\mathcal{P}_i = \mathbb{R}^r$ and $\mathcal{P}_i \cap \mathcal{P}_j = \emptyset$ for $i \neq j$, then the unambiguous testing procedure is defined as follows (Teunissen 2017b)

$$\text{Select } \mathcal{H}_i \iff t \in \mathcal{P}_i, \qquad \text{for } i = 0, 1, \ldots, k \tag{5}$$

3 On the Outcome of the DIA Method

Looking at the three steps of the DIA-method presented in Sect. 2, it can be realized that estimation and testing are combined in the DIA procedure. To gain a better appreciation of this combination, the DIA procedure is schematically visualized in Fig. 1. One can then find out that this is indeed the testing decision which determines how to estimate the unknown parameter vector x. Therefore, the actual DIA outcome, denoted by $\bar{x}$, will inherit the characteristics of not only the estimation scheme but also the testing scheme as well.

3.1 DIA Estimator

The combined estimation-testing scheme of the DIA-method can be captured in one single *DIA estimator* which was introduced in Teunissen (2017b) and is formulated as

$$\bar{x} = \sum_{i=0}^{k} \hat{x}_i \, p_i(t) \tag{6}$$

in which the contribution to $\bar{x}$ from the estimation scheme is captured by the individual estimators $\hat{x}_i$ $(i = 0, 1, \ldots, k)$, and from the testing scheme by the indicator functions $p_i(t)$ $(i = 0, 1, \ldots, k)$ defined as $p_i(t) = 1$ if $t \in \mathcal{P}_i$ and $p_i(t) = 0$ elsewhere. The DIA outcome $\bar{x}$ is therefore a binary weighted average of all the solutions corresponding with the hypotheses at hand. We note that although $\bar{x}$ is linear in the estimators $\hat{x}_i$ $(i = 0, 1, \ldots, k)$, it is nonlinear in t as the indicator functions $p_i(t)$ $(i = 0, 1, \ldots, k)$ are nonlinear functions of t. As a consequence, even if all the individual estimators $\hat{x}_i$ $(i = 0, 1, \ldots, k)$ are normally distributed, $\bar{x}$ does *not* have a normal distribution.

3.2 Abnormality of the PDF of $\bar{x}$

A general probabilistic evaluation of the DIA-estimator is presented in Teunissen (2017b), see also Teunissen et al. (2017). With (6), the probability density function (PDF) of $\bar{x}$, under an arbitrary hypothesis like $\mathcal{H}_j$, can be expressed in terms of the probabilistic properties of the estimators $\hat{x}_i$ $(i = 0, 1, \ldots, k)$ and t as

$$f_{\bar{x}}(\theta|\mathcal{H}_j) = \sum_{i=0}^{k} \int_{\mathcal{P}_i} f_{\hat{x}_i,t}(\theta, \tau|\mathcal{H}_j)\, d\tau \tag{7}$$

where $f_{\hat{x}_i,t}(\theta, \tau|\mathcal{H}_j)$ is the joint PDF of $\hat{x}_i$ and t under $\mathcal{H}_j$. The abnormality of the DIA-estimator PDF can clearly be seen in the above equation. It is important to note, upon application of the DIA-method, that all the follow-on evaluations and inferences must be derived from the probabilistic properties of $\bar{x}$ which are captured by its PDF in (7). In practice however, if a certain hypothesis, say $\mathcal{H}_i$, is selected through the testing procedure, use is made of the PDF of the estimator under the selected hypothesis, i.e. $\hat{x}_i$, neglecting the conditioning process that led to the decision to accept this hypothesis, see e.g. Salzmann (1995), Klein et al. (2018).

To get a better insight into what such negligence could incur, we highlight the difference between the PDF of the DIA-estimator $\bar{x}$ and that of the estimator $\hat{x}_i$, which, under $\mathcal{H}_j$, can be expressed in

$$\begin{aligned} &f_{\bar{x}}(\theta|\mathcal{H}_j) - f_{\hat{x}_i}(\theta|\mathcal{H}_j) = \\ &\left\{ f_{\bar{x}|t\notin\mathcal{P}_i}(\theta|t \notin \mathcal{P}_i,\, \mathcal{H}_j) - f_{\hat{x}_i|t\notin\mathcal{P}_i}(\theta|t \notin \mathcal{P}_i\,, \mathcal{H}_j)\right\} \times \\ &\left\{1 - \mathrm{P}(t \in \mathcal{P}_i|\mathcal{H}_j)\right\} \end{aligned} \tag{8}$$

in which $\mathrm{P}(\cdot)$ denotes the probability of occurrence of the event within parentheses. The above expression results from an application of the conditional probability rule and the fact that the event $(\bar{x}|t \in \mathcal{P}_i)$ is equivalent to $(\hat{x}_i|t \in \mathcal{P}_i)$. As (8) shows, the difference between $f_{\bar{x}}(\theta|\mathcal{H}_j)$ and $f_{\hat{x}_i}(\theta|\mathcal{H}_j)$ is governed by the difference between the conditional PDFs $f_{\bar{x}|t\notin\mathcal{P}_i}(\theta|t \notin \mathcal{P}_i,\, \mathcal{H}_j)$ and $f_{\hat{x}_i|t\notin\mathcal{P}_i}(\theta|t \notin \mathcal{P}_i\,, \mathcal{H}_j)$ as well as the probability $\mathrm{P}(t \in \mathcal{P}_i|\mathcal{H}_j)$. If, for instance, the probability $\mathrm{P}(t \in \mathcal{P}_i|\mathcal{H}_j)$ gets close to one, the non-normal PDF $f_{\bar{x}}(\theta|\mathcal{H}_j)$ gets closer to the normal PDF $f_{\hat{x}_i}(\theta|\mathcal{H}_j)$. Depending on the values of i and j, probabilities $\mathrm{P}(t \in$

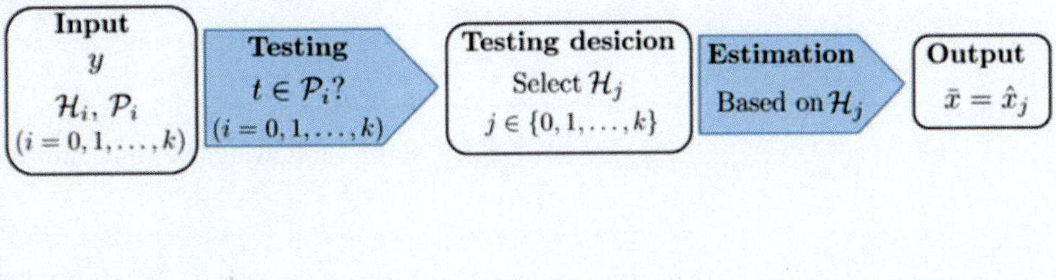

Fig. 1 Schematic illustration of the DIA-method

$\mathcal{P}_i|\mathcal{H}_j)$ $(i, j = 0, 1, \ldots, k)$ can be categorized as

$$\begin{array}{lll} \mathrm{P_{CA}} & = \mathrm{P}(t \in \mathcal{P}_0|\mathcal{H}_0) & \text{Pro. correct acceptance} \\ \mathrm{P_{FA}} & = 1 - \mathrm{P_{CA}} & \text{Pro. false alarm} \\ \mathrm{P_{MD_j}} & = \mathrm{P}(t \in \mathcal{P}_0|\mathcal{H}_{j\neq 0}) & \text{Pro. missed detection} \\ \mathrm{P_{CD_j}} & = 1 - \mathrm{P_{MD_j}} & \text{Pro. correct detection} \\ \mathrm{P_{CI_j}} & = \mathrm{P}(t \in \mathcal{P}_{j\neq 0}|\mathcal{H}_{j\neq 0}) & \text{Pro. correct identification} \\ \mathrm{P_{WI_j}} & = \mathrm{P_{CD_j}} - \mathrm{P_{CI_j}} & \text{Pro. wrong identification} \end{array} \tag{9}$$

where 'Pro.' stands for 'Probability of'. Distinguished by index j, the last four probabilities are different from alternative to alternative. Also note that the last two probabilities become of importance when more than one alternative hypothesis need to be considered. For the single alternative case, say $\mathcal{H}_1$, we have $\mathrm{P_{CI_1}} = \mathrm{P_{CD_1}}$ and $\mathrm{P_{WI_1}} = 0$.

4 Estimation and Testing Strategy

Here, we outline the estimation and testing method as employed in our numerical analysis of the following section. We also remark that our evaluations will be carried out for scalar biases, i.e. $b_i \in \mathbb{R}$, revealing that C_i will take the form of a vector $c_i \in \mathbb{R}^m$.

Estimation To estimate the unknown parameters, use is made of the Best Linear Unbiased Estimation (BLUE) method. As such, $\hat{x}_0$ corresponding with (1) and $\hat{x}_i$ corresponding with (2) are given by

$$\begin{aligned} \hat{x}_0 &= A^+ y = (A^T Q_{yy}^{-1} A)^{-1} A^T Q_{yy}^{-1} y \\ \hat{x}_i &= \bar{A}_i^+ y = (\bar{A}_i^T Q_{yy}^{-1} \bar{A}_i)^{-1} \bar{A}_i^T Q_{yy}^{-1} y \end{aligned} \tag{10}$$

where the superscript '+' denotes the BLUE-inverse, $\bar{A}_i = P_{c_i}^{\perp} A$ and $P_{c_i}^{\perp} = I_m - c_i(c_i^T Q_{yy}^{-1} c_i)^{-1} c_i^T Q_{yy}^{-1}$. Assuming that the observation vector y is normally distributed, $\hat{x}_0$ and $\hat{x}_i$ in (10), as linear functions of y, have also normal distributions. It can be shown, through the Tienstra transformation (Tienstra 1956; Teunissen 2017b), that all the information in the observation vector y is contained in the two independent vectors $\hat{x}_0$ and t (cf. 3). The estimator $\hat{x}_i$, as a linear function of y, can then be expressed as a linear function of $\hat{x}_0$ and t as

$$\hat{x}_i = \hat{x}_0 - L_i t \tag{11}$$

in which $L_i = A^+ c_i (c_{t_i}^T Q_{tt}^{-1} c_{t_i})^{-1} c_{t_i}^T Q_{tt}^{-1}$ with $c_{t_i} = B^T c_i$.

Testing Our testing procedure is specified through defining the regions $\mathcal{P}_0$ and $\mathcal{P}_{i\neq 0}$ (cf. 5) as follows

$$\begin{aligned} \mathcal{P}_0 &= \left\{ t \in \mathbb{R}^r \middle| \, \|t\|^2_{Q_{tt}} \leq k_{\alpha,r} \right\} \\ \mathcal{P}_{i\neq 0} &= \left\{ t \in \mathbb{R}^r / \mathcal{P}_0 \middle| \, |w_i| = \max_{j\in\{1,\ldots,k\}} |w_j| \right\}, \quad i = 1, \ldots, k \end{aligned} \tag{12}$$

in which $\|.\|^2_{Q_{tt}} = (.)^T Q_{tt}^{-1}(.)$, α is the user-defined false alarm probability $\mathrm{P_{FA}}$, $k_{\alpha,r}$ is the α-percentage of the central Chi-square distribution with r degrees of freedom, and w_i is Baarda's test statistic computed as (Baarda 1967; Teunissen 2000)

$$w_i = \frac{c_{t_i}^T Q_{tt}^{-1} t}{\sqrt{c_{t_i}^T Q_{tt}^{-1} c_{t_i}}}; \quad i = 1, \ldots, k \tag{13}$$

5 Numerical Evaluations

In this section, we emphasize the discrepancies between the non-normal PDF of $\bar{x}$ and the normal PDFs of its individual components, i.e. $\hat{x}_i$ $(i = 0, 1, \ldots, k)$. In addition, we investigate situations in which the abnormality of the PDF of $\bar{x}$ gets mitigated. In doing so, we first consider a simple observational model with only a single alternative hypothesis $\mathcal{H}_1$, and then continue with a multiple-hypothesis example in the context of GNSS single point positioning.

5.1 Single-Alternative Case

Suppose that under $\mathcal{H}_0$, the observational model in (1) contains only one unknown parameter $(n = 1)$ with one redundancy $(r = 1)$, i.e. $x \in \mathbb{R}$ and $t \in \mathbb{R}$. We furthermore assume that there is only one single alternative hypothesis, say $\mathcal{H}_1$, against which the null hypothesis $\mathcal{H}_0$ is to be tested. For this binary example, the partitioning of the misclosure space is formed by two regions, $\mathcal{P}_0$ and its complement $\mathcal{P}_0^c$. The DIA-estimator is then constructed by the two estimators $\hat{x}_0$ and $\hat{x}_1$ and the misclosure t as

$$\bar{x} = \hat{x}_0 \, p_0(t) + \hat{x}_1(1 - p_0(t)) \tag{14}$$

To compute the PDF of $\bar{x}$, we assume that t, $\hat{x}_0$ and $\hat{x}_1$ are distributed as

$$\begin{array}{ll} t \overset{\mathcal{H}_0}{\sim} \mathcal{N}(0, \sigma_t^2) & , \; t \overset{\mathcal{H}_1}{\sim} \mathcal{N}(\mu_{t_1}, \sigma_t^2) \\ \hat{x}_0 \overset{\mathcal{H}_0}{\sim} \mathcal{N}(0, \sigma_{\hat{x}_0}^2) & , \; \hat{x}_0 \overset{\mathcal{H}_1}{\sim} \mathcal{N}(L_1\mu_{t_1}, \sigma_{\hat{x}_0}^2) \\ \hat{x}_1 \overset{\mathcal{H}_0}{\sim} \mathcal{N}(0, \sigma_{\hat{x}_0}^2 + L_1^2\sigma_t^2) & , \; \hat{x}_1 \overset{\mathcal{H}_1}{\sim} \mathcal{N}(0, \sigma_{\hat{x}_0}^2 + L_1^2\sigma_t^2) \end{array} \tag{15}$$

for some non-zero scalar μ_{t_1}. Note that $\hat{x}_1$ is unbiased both under $\mathcal{H}_0$ and $\mathcal{H}_1$. Under $\mathcal{H}_0$, there is no bias to be considered, and under $\mathcal{H}_1$, $\hat{x}_1$ is obtained based on a model in which the bias b_1 has already been taken into account. With (8), (14) and (15), the difference between the PDF of the DIA-estimator $\bar{x}$ and the normal PDFs of $\hat{x}_0$ and $\hat{x}_1$ is driven by σ_t, $\sigma_{\hat{x}_0}$, L_1, $\mathcal{P}_0$ and the value of μ_{t_1} which comes into play under the alternative hypothesis $\mathcal{H}_1$. In the following, we show how the PDF differences in (8) behave

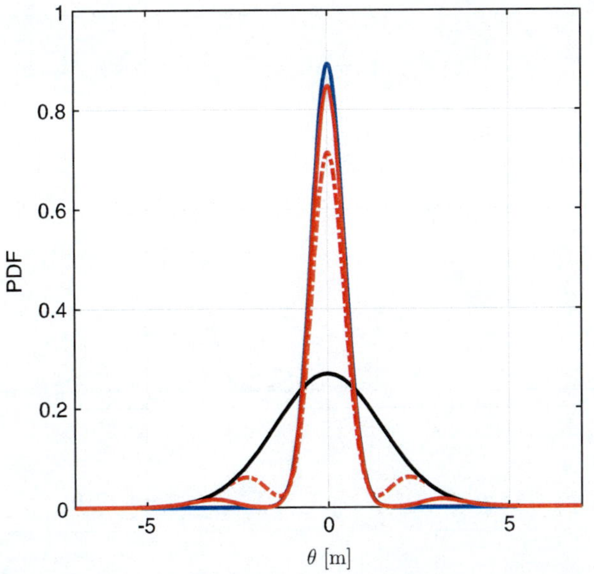

Fig. 2 PDF of the DIA-estimator $\bar{x}$ versus those of $\hat{x}_0$ and $\hat{x}_1$ under $\mathcal{H}_0$ in (14), given $\sigma_t = \sqrt{2}\,\text{m}$, $\sigma_{\hat{x}_0} = \sqrt{0.2}\,\text{m}$ and $L_1 = 1$. *blue*: $f_{\hat{x}_0}(\theta|\mathcal{H}_0)$; *black*: $f_{\hat{x}_1}(\theta|\mathcal{H}_0)$; *red dashed-dotted*: $f_{\bar{x}}(\theta|\mathcal{H}_0)$ for $\text{P}_{\text{FA}} = 0.2$; *red solid*: $f_{\bar{x}}(\theta|\mathcal{H}_0)$ for $\text{P}_{\text{FA}} = 0.05$

as function of some of these parameters under both $\mathcal{H}_0$ and $\mathcal{H}_1$. Note that instead of $\mathcal{P}_0$, we equivalently work with the probability of false alarm $\text{P}_{\text{FA}} = \text{P}(t \in \mathcal{P}_0^c|\mathcal{H}_0)$, which is usually a priori set by the user.

Evaluation Under $\mathcal{H}_0$ Given $\sigma_t = \sqrt{2}\,\text{m}$, $\sigma_{\hat{x}_0} = \sqrt{0.2}\,\text{m}$ and $L_1 = 1$, Fig. 2 shows the normal PDFs of $\hat{x}_0$ (blue) and $\hat{x}_1$ (black) as well as the PDF of $\bar{x}$ (red) under $\mathcal{H}_0$. The DIA-estimator PDF is illustrated for two different values of P_{FA} distinguished by their line style; *dashed-dotted:* $\text{P}_{\text{FA}} = 0.2$, *solid:* $\text{P}_{\text{FA}} = 0.05$. As it can be seen, the PDF of the DIA-estimator does *not* resemble a normal distribution, but in fact a multi-modal distribution. Like the shown two normal PDFs, the red graphs are symmetric w.r.t. the center, which means that the DIA-estimator is unbiased under the null hypothesis, i.e. $\mathsf{E}(\bar{x}|\mathcal{H}_0) = 0$. It is observed that the PDF of the DIA-estimator gets close to the normal PDF of $\hat{x}_0$ as the false alarm probability decreases. It indeed makes sense as decreasing the false alarm probability means that it is getting more likely that the testing procedure leads to the decision to correctly accept the null hypothesis. This in turn will result in the contribution of $\hat{x}_0$ to the construction of $\bar{x}$ getting larger. In the extreme case of $\text{P}_{\text{FA}} = 0$ (no testing), the DIA-estimator PDF becomes identical to the normal PDF of $\hat{x}_0$.

If the data precision Q_{yy} gets scaled by a factor of $\kappa \in \mathbb{R}^+$, then the precision of $\hat{x}_0$, $\hat{x}_1$ and t will also change by exactly the same factor κ (cf. 3, 10). Shown in Fig. 3 are the PDFs of $\hat{x}_0$, $\hat{x}_1$ and $\bar{x}$ under $\mathcal{H}_0$ given $\sigma_t = \kappa \times \sqrt{2}\,\text{m}$, $\sigma_{\hat{x}_0} = \kappa \times \sqrt{0.2}\,\text{m}$, $L_1 = 1$ and $\text{P}_{\text{FA}} = 0.1$. The left panel corresponds with $\kappa = 1$ while the right panel shows the results of $\kappa = 1.5$. In agreement with the normal PDFs of $\hat{x}_0$ and $\hat{x}_1$, the PDF of $\bar{x}$ gets less peaked around the true value when the data in use gets less precise (κ increases).

Evaluation Under $\mathcal{H}_1$ For our analysis under $\mathcal{H}_1$, we need to consider some value for μ_{t_1} as well. With $\mathcal{P}_0$ in (12) and the definition of correct detection probability in (9), the larger the value of b_1 (bias under $\mathcal{H}_1$) and thus μ_{t_1}, the higher is the probability of correct detection. Figure 4 shows, for $\sigma_t = \sqrt{2}\,\text{m}$, $\sigma_{\hat{x}_0} = \sqrt{0.2}\,\text{m}$, $L_1 = 1$ and $\text{P}_{\text{FA}} = 0.1$, the graphs of $f_{\hat{x}_0}(\theta|\mathcal{H}_1)$, $f_{\hat{x}_1}(\theta|\mathcal{H}_1)$ and $f_{\bar{x}}(\theta|\mathcal{H}_1)$. The panels, from left to right, correspond with $\mu_{t_1} = 3\,\text{m}$, $\mu_{t_1} = 4\,\text{m}$ and $\mu_{t_1} = 7\,\text{m}$. Given on top of each panel is the corresponding probability of correct detection. We note that the PDF of the DIA-estimator under $\mathcal{H}_1$ is no longer symmetric around the center, revealing that the DIA-estimator under $\mathcal{H}_1$ is biased, i.e. $\mathsf{E}(\bar{x}|\mathcal{H}_1) \neq 0$. The larger the probability of correct detection gets, the closer the PDF of the DIA-estimator gets towards the normal PDF of $\hat{x}_1$. And ultimately with a correct detection probability larger than 0.99, the PDF $f_{\bar{x}}(\theta|\mathcal{H}_1)$ almost coincides with the PDF $f_{\hat{x}_1}(\theta|\mathcal{H}_1)$ which indeed makes sense as more than 99% of the time, the testing procedure leads to $\mathcal{H}_1$ being selected. We remark that the probability mass of $f_{\bar{x}}(\theta|\mathcal{H}_1)$ becomes more centred around the correct value, or equivalently the DIA-estimator becomes less biased under $\mathcal{H}_1$, for higher correct detection probabilities which can be achieved as a result of larger biases (as shown in Fig. 4), larger false alarm probabilities and/or more precise data.

5.2 Multiple-Alternative Case

In Sect. 5.1, we discussed the properties of the DIA-estimator through some simple examples of binary hypothesis testing applied to a single-unknown, single-redundancy model. With the insight gained from these examples, we now consider the DIA-estimator in the context of a more practical application, i.e. the well-known GNSS single-point positioning (SPP). Assuming that a single GNSS receiver is tracking the pseudorange observations of m satellites on a single frequency, the SPP model under the null hypothesis reads

$$\mathcal{H}_0: \quad \mathsf{E}(y) = [-G \quad e_m]\begin{bmatrix} x \\ dt \end{bmatrix} \quad , \quad \mathsf{D}(y) = \sigma_y^2 I_m \tag{16}$$

with $G \in \mathbb{R}^{m\times 3}$ containing the receiver-satellite unit direction vectors as its rows, $e_m \in \mathbb{R}^m$ containing ones and $I_m \in \mathbb{R}^{m\times m}$ being the identity matrix. There are four unknown parameters to be estimated ($n = 4$); $x \in \mathbb{R}^3$ the receiver

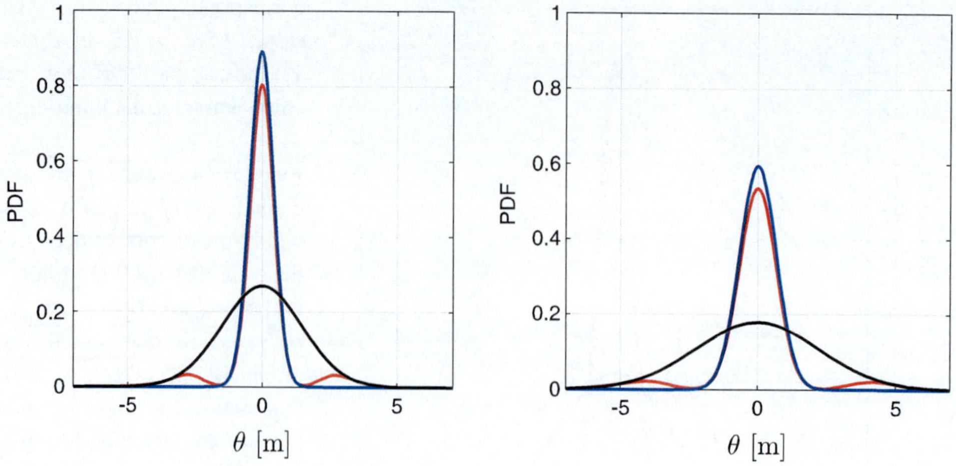

Fig. 3 PDF of the DIA-estimator $\bar{x}$ versus those of $\hat{x}_0$ and $\hat{x}_1$ under $\mathcal{H}_0$ in (14), given $\sigma_t = \kappa \times \sqrt{2}\,\mathrm{m}$, $\sigma_{\hat{x}_0} = \kappa \times \sqrt{0.2}\,\mathrm{m}$, $L_1 = 1$ and $\mathrm{P_{FA}} = 0.1$ for [left] $\kappa = 1$ and [right] $\kappa = 1.5$. *blue*: $f_{\hat{x}_0}(\theta|\mathcal{H}_0)$; *black*: $f_{\hat{x}_1}(\theta|\mathcal{H}_0)$; *red*: $f_{\bar{x}}(\theta|\mathcal{H}_0)$

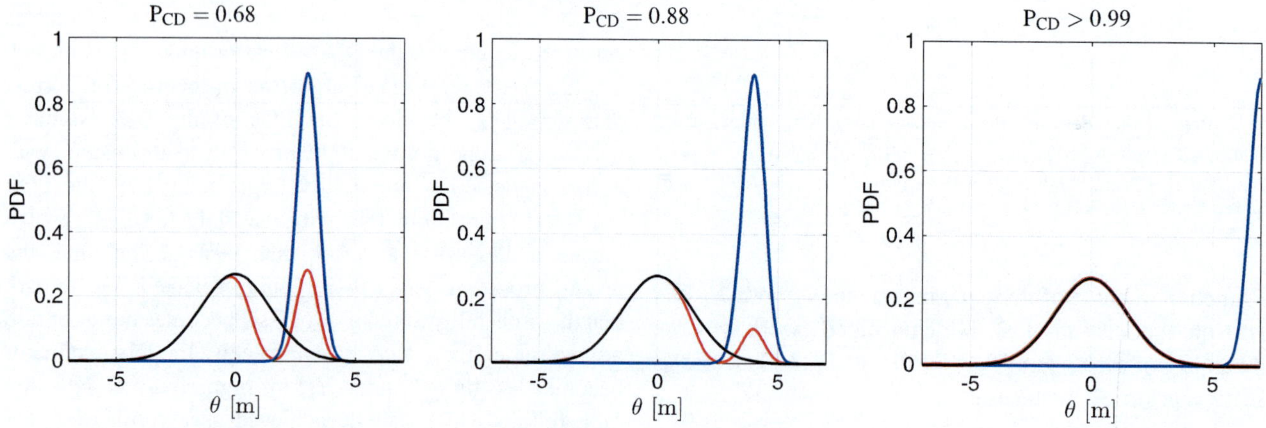

Fig. 4 PDF of the DIA-estimator $\bar{x}$ versus those of $\hat{x}_0$ and $\hat{x}_1$ under $\mathcal{H}_1$ in (14), given $\sigma_t = \sqrt{2}\,\mathrm{m}$, $\sigma_{\hat{x}_0} = \sqrt{0.2}\,\mathrm{m}$, $L_1 = 1$ and $\mathrm{P_{FA}} = 0.1$ for, [left] $\mu_{t_1} = 3\,\mathrm{m}$, [middle] $\mu_{t_1} = 4\,\mathrm{m}$ and [right] $\mu_{t_1} = 7\,\mathrm{m}$. *blue*: $f_{\hat{x}_0}(\theta|\mathcal{H}_1)$; *black*: $f_{\hat{x}_1}(\theta|\mathcal{H}_1)$; *red*: $f_{\bar{x}}(\theta|\mathcal{H}_1)$. The corresponding correct detection probabilities are given on top of each panel

coordinate components increments and $dt \in \mathbb{R}$ the receiver clock error increment. The redundancy of $\mathcal{H}_0$ is then $r = m - 4$. The dispersion of the observables is characterized through the standard deviation σ_y. At this stage, for the sake of simplicity, we do not consider a satellite elevation-dependent variance matrix.

It is assumed that the alternative hypotheses capture the outliers in individual observations. Thus, with m satellites being available, there are m alternatives $\mathcal{H}_i$ $(i = 1, \ldots, m)$ of the following form

$$\mathcal{H}_i: \quad \mathsf{E}(y) = [-G \quad e_m] \begin{bmatrix} x \\ dt \end{bmatrix} + c_i\, b_i\,, \; \mathsf{D}(y) = \sigma_y^2 I_m \tag{17}$$

where $c_i \in \mathbb{R}^m$ is a canonical unit vector having one as its ith entry and zero elsewhere, and $b_i \in \mathbb{R}$ is the scalar outlier. Note that the alternative hypotheses in (17) are identifiable provided that $c_{t_i} \nparallel c_{t_j}$ for any $i \neq j$ (Zaminpardaz 2018). For our analysis, we consider the satellite geometry illustrated in Fig. 5, comprising six satellites ($m = 6$).

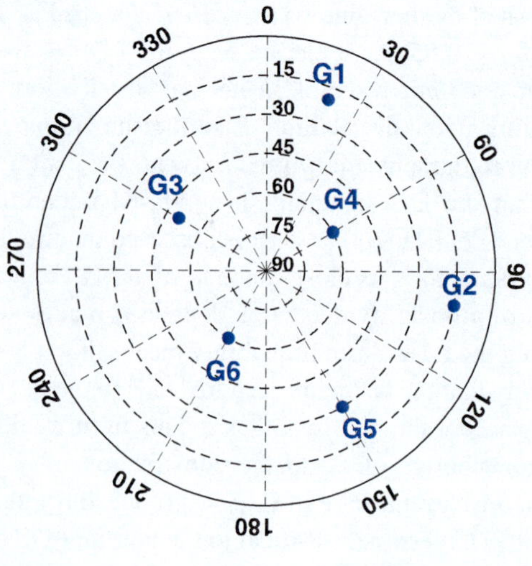

Fig. 5 Skyplot view of satellites. The six blue circles denote the skyplot position of the satellites

Therefore, six alternative hypotheses ($k = m = 6$) of the form of (17) are considered in the DIA procedure, and the

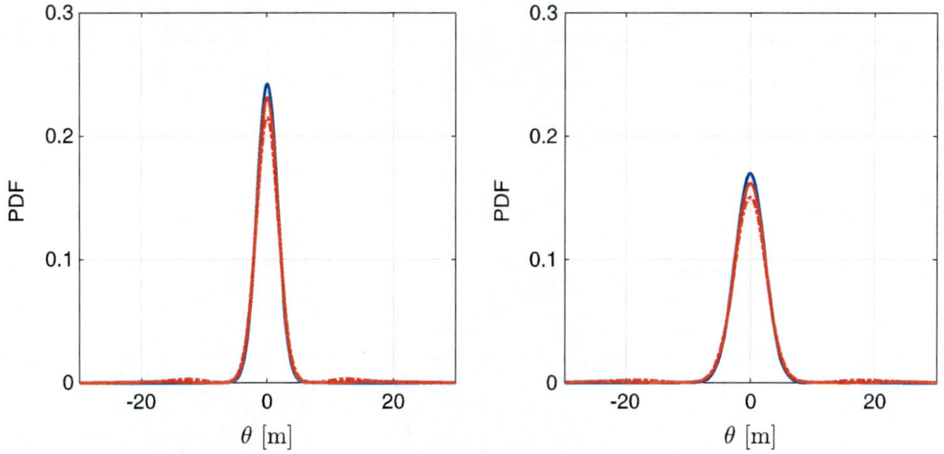

Fig. 6 PDF of the DIA-estimator $\bar{u}$ versus that of $\hat{u}_0$ under $\mathcal{H}_0$. The illustrations are given for the receiver coordinate up component in SPP model in (16) corresponding with the satellite geometry in Fig. 5 for [left] $\sigma_y = 0.7$ m and [right] $\sigma_y = 1$ m. *blue*: $f_{\hat{u}_0}(\theta|\mathcal{H}_0)$; *red dashed-dotted*: $f_{\bar{u}}(\theta|\mathcal{H}_0)$ for $\mathrm{P_{FA}} = 0.3$; *red solid*: $f_{\bar{u}}(\theta|\mathcal{H}_0)$ for $\mathrm{P_{FA}} = 0.1$

redundancy under the null hypothesis is $r = 2$ ($t \in \mathbb{R}^2$). We also remark that for this satellite geometry, all the six alternatives are identifiable. Our illustrations will be shown for receiver coordinate up component, denoted by u, under $\mathcal{H}_0$, $\mathcal{H}_1$ (outlier in G1 observation) and $\mathcal{H}_4$ (outlier in G4 observation). However, we note that our conclusions will be valid for any unknown parameter in (16). Without loss of generality, we also assume that the true value of the up component is zero.

Evaluation Under $\mathcal{H}_0$ In Fig. 6, the PDFs of $\hat{u}_0$ and $\bar{u}$ are depicted in, respectively, blue and red color. The left panel shows the results corresponding with $\sigma_y = 0.7$ m while the right one shows the results corresponding with $\sigma_y = 1$ m. In each panel two red graphs are illustrated; *dashed-dotted*: $\mathrm{P_{FA}} = 0.3$, *solid*: $\mathrm{P_{FA}} = 0.1$. Again the symmetry of the DIA-estimator PDF around the true value indicates the unbiasedness of the DIA-estimator under the null hypothesis. In addition, the peakedness of the DIA-estimator PDF, like the PDF of $\hat{u}_0$, around the true value decreases when the data precision gets poorer (σ_y increases). Similar to the single-alternative example, we expect that the difference between the red and blue graphs will diminish if the false alarm probability decreases. This is indeed corroborated by comparing the red dashed-dotted graphs with the red solid ones in Fig. 6.

Evaluation Under $\mathcal{H}_1$ *and* $\mathcal{H}_4$ For our analysis under alternative hypotheses, we, as example, take the two alternatives $\mathcal{H}_1$ and $\mathcal{H}_4$. Assuming $\sigma_y = 1$ m and $\mathrm{P_{FA}} = 0.1$, Fig. 7 shows the PDFs of $\hat{u}_0$, $\hat{u}_i$ and $\bar{u}$ under $\mathcal{H}_i$. The top panels are given for $i = 1$ while the bottom panels are obtained for $i = 4$. The values for b_i under the mentioned two alternatives are, from left to right, set to $b_i = 3$, 7 and 15m. Here, because of having more than one single alternative, in addition to the correct detection probability, we also compute the correct identification probability, both of which are shown on top of each panel.

The significant departure between the red graph and the other two normal curves in each panel is an indicator of how misleading the post-DIA quality assessments would be if one neglects the conditioning on testing outcome. For example, let us assume that, for the case of lower-right panel, $\mathcal{H}_4$ is selected through the DIA procedure. As shown in Fig. 7, the PDF of $\hat{u}_4$ has larger probability mass around the true value than that of $\bar{u}$. Therefore, assessments on the basis of $f_{\hat{u}_4}(\theta|\mathcal{H}_4)$, rather than $f_{\bar{u}}(\theta|\mathcal{H}_4)$, would lead to optimistic/misleading quality descriptions (precision, accuracy, integrity, etc.). As the bias value b_i increases, both the correct detection and identification probabilities increase as well, resulting in less discrepancies between $f_{\hat{u}_i}(\theta|\mathcal{H}_i)$ and $f_{\bar{u}}(\theta|\mathcal{H}_i)$. Note, however, that the difference between the red graphs and the corresponding black ones becomes small only for large *correct identification* probabilities, and not necessarily for large *correct detection* probabilities. For example, for the case where bias under $\mathcal{H}_1$ is $b_1 = 7$ m (upper-middle panel), despite having a large correct detection probability of $\mathrm{P_{CD_1}} = 0.94$, there is a big difference between the red and black curve as the correct identification probability is only $\mathrm{P_{CI_1}} = 0.60$.

6 Summary and Concluding Remarks

There is a close link between estimation and testing in any quality control procedure. By highlighting this link and its consequences, we revealed its impact on the quality evaluations usually performed and elaborated on the circumstances under which negligence of this link may still be considered acceptable. In doing so, we provided a comparative study of the probabilistic properties of the actual DIA outcome $\bar{x}$ derived from the characteristics of both estimation and testing, and the individual estimators $\hat{x}_i$ corresponding with the hypotheses at hand $\mathcal{H}_i$ ($i = 0, 1, \ldots, k$) neglecting the uncertainty of the testing decision process. Our analyses were conducted assuming that the observations are normally distributed and that the underlying models are linear.

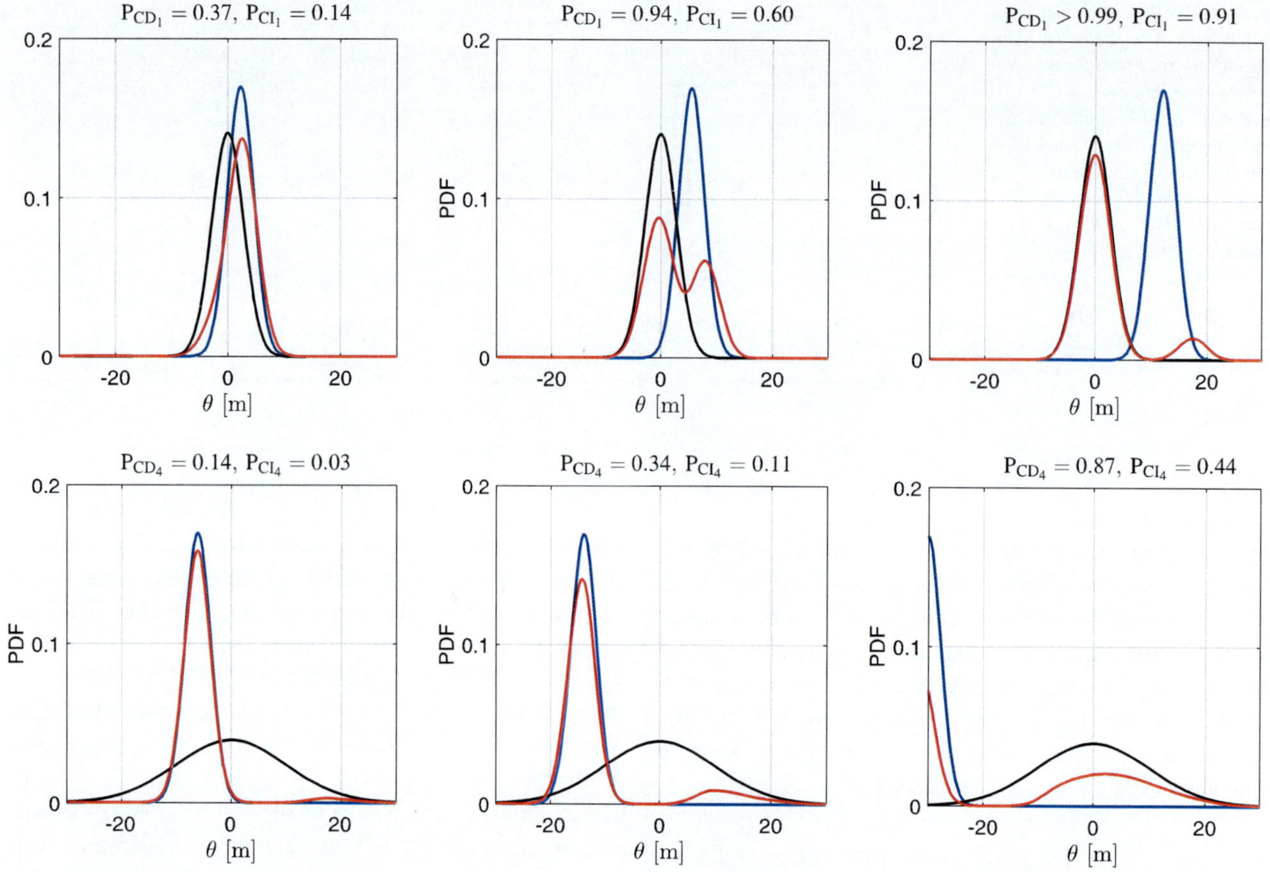

Fig. 7 PDF of the DIA-estimator $\bar{u}$ versus those of $\hat{u}_0$ and $\hat{u}_i$ under $\mathcal{H}_i$ in (17) for [top] $i = 1$ and [bottom] $i = 4$. The illustrations are given for $\sigma_y = 1$ m and for [left] $b_i = 3$ m, [middle] $b_i = 7$ m and [right] $b_i = 15$ m. *blue*: $f_{\hat{u}_0}(\theta|\mathcal{H}_i)$; *black*: $f_{\hat{u}_i}(\theta|\mathcal{H}_i)$; *red*: $f_{\bar{u}}(\theta|\mathcal{H}_i)$. The corresponding correct detection and identification probabilities are given on top of each panel

We started with simple examples of single alternative hypothesis where a single-unknown, single-redundancy model was considered. The DIA-estimator was then constructed by $\hat{x}_0$, $\hat{x}_1$ and t. It was demonstrated that the distribution of the DIA-estimator, unlike its individual constructing components, is not normal, but multi modal. However, the non-normal PDF of $\bar{x}|\mathcal{H}_0$ ($\bar{x}|\mathcal{H}_1$) will approach the normal distribution of $\hat{x}_0|\mathcal{H}_0$ ($\hat{x}_1|\mathcal{H}_1$) if P_{FA} ($P_{MD} = 1 - P_{CD}$) decreases. The impact of the data precision on the DIA-estimator PDF was also illustrated. For example, under $\mathcal{H}_0$, the more precise the observations are, the more peacked the DIA-estimator PDF gets around the true value. It was also shown that while $\bar{x}$ is unbiased under $\mathcal{H}_0$, it is biased under $\mathcal{H}_1$. The bias of $\bar{x}|\mathcal{H}_1$ gets, however, smaller when the correct detection probability gets larger as more probability mass of $f_{\bar{x}}(\theta|\mathcal{H}_1)$ becomes centred around the true value.

Having investigated the single-alternative case, we then applied the DIA-method to the satellite-based single point positioning model where multiple alternative hypotheses, describing outliers in individual observations, were considered. For our illustrations, we showed the results corresponding with the receiver coordinate up component u. We however remark that the following conclusions are valid for any other unknown parameter and linear model. Similar to the single-alternative example, it was shown that the PDF of the DIA-estimator $\bar{u}$ cannot be characterized by a normal distribution. Depending on the underlying settings, there could be significant departures between the PDF of the DIA-estimator and that of the estimator associated with the identified hypothesis. It was highlighted that if the uncertainty of the statistical testing is not taken into account, then one may end up with a too optimistic quality description of the final estimator. Nevertheless, depending on the requirements of the application at hand, the DIA-estimator PDF may be well approximated by $\hat{u}_0|\mathcal{H}_0$ under $\mathcal{H}_0$ and by $\hat{u}_i|\mathcal{H}_i$ under $\mathcal{H}_i$ for, respectively, small P_{FA} and large P_{CI_i}. It is therefore important that one always properly evaluates identification probabilities, as a large probability of correct detection not necessarily implies a large correct identification probability (Teunissen 2017b).

References

Baarda W (1967) Statistical concepts in geodesy. Netherlands Geodetic Commission, Publ. on geodesy, New series 2(4)

Baarda W (1968) A testing procedure for use in geodetic networks. Netherlands Geodetic Commission, Publ. on geodesy, New Series 2(5)

DGCC (1982) The Delft Approach for the Design and Computation of Geodetic Networks. In: "Forty Years of Thought" Anniversary edition on the occasion of the 65th birthday of Prof W Baarda, vol 2, pp 202–274. By staff of the Delft Geodetic Computing Centre (DGCC)

Khodabandeh A, Teunissen PJG (2016) Single-epoch GNSS array integrity: an analytical study. In: Sneeuw N, Novák P, Crespi M, Sansò F (eds) VIII Hotine-Marussi symposium on mathematical geodesy: proceedings of the symposium in Rome, 17–21 June, 2013. Springer International Publishing, Cham, pp 263–272

Klein I, MatsuokaMatheus MT, Guzatto MP, Nievinski FG, Veronez MR, Rofatto VF (2018) A new relationship between the quality criteria for geodetic networks. J Geodesy 1–16. https://doi.org/10.1007/s00190-018-1181-8

Perfetti N (2006) Detection of station coordinate discontinuities within the Italian GPS fiducial network. J Geodesy 80(7):381–396

Salzmann M (1995) Real-time adaptation for model errors in dynamic systems. Bull Geodesique 69:81–91 (1995)

Teunissen PJG (1990) An integrity and quality control procedure for use in multi-sensor integration. In: Proc of ION GPS-1990. ION, pp 513–522

Teunissen PJG (2000) Testing theory: an introduction, Series on mathematical geodesy and positioning. Delft University Press, Delft

Teunissen PJG (2017a) Batch and recursive model validation. In: Teunissen PJG, Montenbruck O (eds) Springer handbook of global navigation satellite systems, chapter 24, pp. 727–757

Teunissen PJG (2017b) Distributional theory for the DIA method. J Geodesy 1–22. https://doi.org/10.1007/s00190-017-1045-7

Teunissen PJG, Imparato D, Tiberius CCJM (2017) Does RAIM with correct exclusion produce unbiased positions? Sensors 17(7):1508

Tiberius CCJM (1998) Recursive data processing for kinematic GPS surveying. Netherlands Geodetic Commission, Publ. on Geodesy, New series (45)

Tienstra J (1956) Theory of the adjustment of normally distributed observation. Argus, Amsterdam

Zaminpardaz S, Teunissen PJG (2018) DIA-datasnooping and identifiability. J Geodesy 1–17 (2018). https://doi.org/10.1007/s00190-018-1141-3

Zaminpardaz S, Teunissen PJG, Nadarajah N, Khodabandeh A (2015) GNSS array-based ionospheric spatial gradient monitoring: precision and integrity analysis. In: Proceedings of the ION Pacific PNT meeting. ION, pp 799–814

Check for updates

Controlling the Bias Within Free Geodetic Networks

Burkhard Schaffrin and Kyle Snow

Abstract

It is well known that the MInimum NOrm LEast-Squares Solution (MINOLESS) minimizes the bias uniformly since it coincides with the BLUMBE (Best Linear Uniformly Minimum Biased Estimate) in a rank-deficient Gauss-Markov Model as typically employed for free geodetic network analyses. Nonetheless, more often than not, the partial-MINOLESS is preferred where a selection matrix $S_k := \text{Diag}(1, \ldots, 1, 0, \ldots, 0)$ is used to only minimize the first k components of the solution vector, thus resulting in larger biases than frequently desired. As an alternative, the Best LInear Minimum Partially Biased Estimate (BLIMPBE) may be considered, which coincides with the partial-MINOLESS as long as the rank condition $\text{rk}(S_k N) = \text{rk}(N) = \text{rk}(A) =: q$ holds true, where N and A are the normal equation and observation equation matrices, respectively. Here, we are interested in studying the bias divergence when this rank condition is violated, due to $q > k \geq m - q$, with m as the number of all parameters. To the best of our knowledge, this case has not been studied before.

Keywords

Bias control · Datum deficiency · Free geodetic networks

1 Introduction

It has long been recognized that the adjustment of free geodetic networks will lead to biased point coordinate estimates when using a rank-deficient Gauss-Markov Model (GMM). Only quantities that can be expressed as functions of the observables may result in unbiased estimates after a (weighted) least-squares adjustment. For more details, we refer to Grafarend and Schaffrin (1974).

A far wider scope is reflected in the Springer book *Optimization and Design of Geodetic Networks*, edited by Grafarend and Sansó (1985), which represents the status of research at the time. For the discussion in the following contribution, most relevant should be the paper "Network Design" by Schaffrin (1985) that can be found in that volume. In particular, estimates of type MINOLESS (MInimum NOrm LEeast-Squares Solution), BLESS (Best Least-Squares Solution), and BLUMBE (Best Linear Uniformly Minimum Biased Estimate) are derived therein and given various representations, including proofs that show their identity under certain conditions.

On the other hand, for many practical applications (e.g., Caspary 2000), the general MINOLESS is being replaced by *partial* MINOLESS for which only the S_k-weighted norm of the estimated parameter vector is minimized, where $S_k := \text{Diag}(1, \ldots, 1, 0, \ldots, 0)$ is called a "*selection matrix.*" Unfortunately, Snow and Schaffrin (2007) would then show that the (full) MINOLESS is the only Least-Squares Solution that minimizes the bias *uniformly* so that the bias of the partial MINOLESS must be monitored separately. Moreover, when Schaffrin and Iz (2002) began studying estimators that minimize the bias at least partially, it turned out that the best among them (BLIMPBE: Best LInear Minimum

B. Schaffrin
Geodetic Science Program, School of Earth Sciences, The Ohio State University, Columbus, OH, USA
e-mail: aschaffrin@earthlink.net

K. Snow (✉)
Topcon Positioning Systems, Inc., Columbus, OH, USA

P. Novák et al. (eds.), *IX Hotine-Marussi Symposium on Mathematical Geodesy*,
International Association of Geodesy Symposia 151, https://doi.org/10.1007/1345_2019_69

Partially Biased Estimate) will not belong to the LEast-Squares Solutions (LESS) unless a certain rank criterion is fulfilled that involves the particular choice of the selection matrix.

This phenomenon ultimately led to the study in the following where it is tried to describe the growth of bias in dependence of the selection matrix, in particular when the number k of selected parameters approaches the minimum $m - q$. After a review of the rank-deficient Gauss-Markov Model in Sect. 2, the various relevant estimators will be presented and characterized by their essential properties. Then, in Sect. 3, the rank of a certain matrix product is shown to be a very good indicator of the expected bias growth for $k \to m - q$. Finally, in Sect. 4, a simple 1-D network from leveling data will be analyzed, and the results will be summarized in a brief section on conclusions.

2 A Review of the Rank-Deficient Gauss-Markov Model and Some of Its Most Relevant Parameter Estimates

Let the Gauss-Markov Model (GMM) in linearized form be defined by

$$y = \underset{n\times m}{A}\,\xi + e, \quad e \sim (\mathbf{0}, \sigma_0^2 P^{-1}), \tag{1}$$

where

- y denotes the $n \times 1$ vector of observational increments,
- A denotes the $n \times m$ coefficient matrix with $q := \operatorname{rk} A < m \le n$,
- ξ denotes the (unknown) $m \times 1$ vector of parameter increments,
- e denotes the (unknown) $n \times 1$ random error vector.

Also, $E\{e\} = \mathbf{0}$, where E denotes the *expectation* operator, and $\Sigma = E\{ee^T\} = \sigma_0^2 P^{-1}$ with

- σ_0^2 as (unknown) variance component. Here,
- P^{-1} represents the symmetric positive-definite $n \times n$ cofactor matrix, resp.
- P the corresponding $n \times n$ weight matrix.

Note that the dispersion matrix Σ is assumed to be *nonsingular*; for the more general case of a singular matrix Σ, see Schaffrin (2013/2014), for instance.

2.1 General Representation of the Weighted LESS

A weighted LEast-Squares Solution (LESS) would come from the variational principle

$$e^T P e = \min_{e,\xi} \text{ subject to } y - A\xi - e = \mathbf{0}. \tag{2}$$

The corresponding *Lagrange target function* reads

$$\Phi(e, \xi, \lambda) := e^T P e + 2\lambda^T (y - A\xi - e), \tag{3}$$

and, after making it stationary, leads to the *Euler-Lagrange necessary conditions*, which finally provide the "orthogonality condition"

$$A^T P \tilde{e} = \mathbf{0} \tag{4a}$$

for the *residual vector* $\tilde{e} := y - A\hat{\xi}$, and then the "normal equations"

$$\underset{m\times m}{N}\,\hat{\xi} = c, \tag{4b}$$

$$\operatorname{rk} N = \operatorname{rk} A = q < m, \quad \text{for } \left[N \middle| c\right] := A^T P \left[A \middle| y\right].$$

Equation (4) allows for a multitude of LESS's that can be characterized in various ways, for instance by

$$\hat{\xi}_{\text{LESS}} \in \{N^- c \mid N N^- N = N\} = \tag{5}$$
$$= \{N_{\text{rs}}^- c \mid N N_{\text{rs}}^- N = N,\ N_{\text{rs}}^- = N_{\text{rs}}^- N N_{\text{rs}}^- = (N_{\text{rs}}^-)^T\}$$

using generalized inverses or – more efficiently – reflexive symmetric g-inverses; alternatively, any LESS can be described in the form

$$\hat{\xi}_{\text{LESS}} \in \{(N + K^T K)^{-1} c \mid K \text{ an } (m - q) \times m \text{ matrix, } KE^T \text{ nonsingular}\}, \tag{6a}$$

where E is any $(m - q) \times m$ matrix with

$$\operatorname{rk} E = m - q \text{ and } AE^T = 0. \tag{6b}$$

Obviously, this particular LESS would fulfill the so-called *minimum constraints*

$$K\hat{\xi}_{\text{LESS}} = \mathbf{0}, \tag{6c}$$

due to the identity

$$A(N + K^T K)^{-1} K^T = A \cdot E^T (KE^T)^{-1} = 0. \tag{7}$$

2.2 Conventional Choices of Particular LESS's

Let the MINOLESS be defined by

$$\hat{\xi}^T \hat{\xi} = \min_{\hat{\xi}} \text{ subject to } c - N\hat{\xi} = \mathbf{0}; \tag{8a}$$

then, it can be characterized by

$$\bar{\xi} = N(NN)^- c \quad \text{for any g-inverse } (NN)^-. \tag{8b}$$

Alternatively, the BLESS may be defined through

$$\operatorname{tr} D\{\bar{\xi}\} = \min\{\operatorname{tr} D\{\hat{\xi}_{\text{LESS}}\} = \sigma_0^2 \cdot \operatorname{tr} N_{\text{rs}}^- N (N_{\text{rs}}^-)^T\}, \tag{9a}$$

leading to the representation

$$\bar{\xi} = N^+ c \quad \text{with } N^+ \text{ as "pseudo-inverse" of } N. \tag{9b}$$

As a third alternative, the BLUMBE may be considered by establishing its three properties

(i) "linear" $\bar{\xi} = Ly$ for some $m \times n$ matrix L,
(ii) "uniformly minimum biased"

$$E\{\bar{\xi}\} - \xi = L \cdot E\{y\} - \xi = -(I_m - LA)\xi =: \underset{m\times m}{B} \cdot \xi \tag{10a}$$

with B as "bias matrix." Then:

$$\operatorname{tr}(BB^T) = \min_{L^T} \Leftrightarrow AA^T L^T = A, \tag{10b}$$

(iii) "best"

$$\operatorname{tr} D\{\bar{\xi}\} = \min_{L^T}\{\sigma_0^2 \cdot \operatorname{tr} LP^{-1}L^T \mid AA^T L^T = A\}. \tag{10c}$$

These three properties of BLUMBE lead to the representation

$$\bar{\xi} = (N + E^T E)^{-1} c \tag{10d}$$

for any matrix E satisfying (6b).

Theorem 1 (cf. Schaffrin 2013/2014)

$$\bar{\xi}_{\text{MINOLESS}} = \bar{\xi}_{\text{BLESS}} = \bar{\xi}_{\text{BLUMBE}}$$

In addition, the partial MINOLESS is frequently introduced by using the "selection matrix" $S_k :=$ Diag$(1, \ldots, 1, 0, \ldots, 0)$ *with* $m \geq k \geq m - q$ *for*

$$\operatorname{tr} \hat{\xi}^T S_k \hat{\xi} = \min_{\hat{\xi}} \ \textit{subject to} \ c - N\hat{\xi} = 0, \tag{11a}$$

yielding the "extended normal equations"

$$\begin{bmatrix} S_k & I_m \\ N & 0 \end{bmatrix} \begin{bmatrix} \bar{\bar{\xi}} \\ N\bar{\bar{\lambda}} \end{bmatrix} = \begin{bmatrix} 0 \\ c \end{bmatrix}, \ \textit{resp.} \tag{11b}$$

$$\begin{bmatrix} S_k + N & I_m \\ N & 0 \end{bmatrix} \begin{bmatrix} \bar{\bar{\xi}} \\ N\bar{\bar{\lambda}} \end{bmatrix} = \begin{bmatrix} c \\ c \end{bmatrix}, \tag{11c}$$

and, hence, the representations

$$\bar{\bar{\xi}} = (N + S_k E^T E S_k)^{-1} c = \tag{12a}$$

$$= (S_k + N)^{-1} N [N (S_k + N)^{-1} N]^- c, \tag{12b}$$

provided that $E S_k E^T$ *is nonsingular, resp.* $\operatorname{rk}(S_k + N) = m$. *In this case, the partial BLESS turns out to coincide with the partial MINOLESS, whereas the BLUMBE will be the partial BLUMBE already for all choices of* S_k. *In particular, it can be shown (Snow and Schaffrin 2007) that the (full) MINOLESS is the only LESS that does minimize the bias uniformly by fulfilling (10b). So, by using the partial MINOLESS instead, more bias than necessary is introduced. This led to the interesting question whether the bias can be kept under control, at least partially, while minimizing the trace of the dispersion matrix. This approach led to the BLIMPBE.*

2.3 The Best Estimate Under Partial Bias Control

Let the BLIMPBE (Best LInear Minimum Partially Biased Estimate) be defined by its three properties

(i) "linear" $\breve{\xi} = Ly$ for some $m \times n$ matrix L,
(ii) "minimum partial bias"

$$B := -(I_m - LA) \ \text{as "bias matrix;" then:}$$

$$\operatorname{tr}(B S_k B^T) = \min_{L^T} \Leftrightarrow A S_k A^T L^T = A S_k, \tag{13a}$$

(iii) "best"

$$\operatorname{tr} D\{\breve{\xi}\} = \min_{L^T}\{\sigma_0^2 \cdot \operatorname{tr} LP^{-1}L^T \mid A S_k A^T L^T = A S_k\}. \tag{13b}$$

It may then be represented by

$$\breve{\xi} = S_k N (N S_k N S_k N)^- N S_k \cdot c \notin \{\hat{\xi}_{\text{LESS}}\}, \tag{13c}$$

in general.

Theorem 2 (cf. Schaffrin and Iz 2002) *The BLIMPBE will be a LESS if, and only if,*

$$\operatorname{rk}(S_k N) = \operatorname{rk} N = q. \tag{14a}$$

In this case,

$$\breve{\xi} = \mathring{\xi} := S_k N (N S_k N)^- \cdot c \in \{\hat{\xi}_{\text{LESS}}\}, \tag{14b}$$

the latter being introduced as "twin to BLIMPBE," which looks like an S_k*-weighted form of MINOLESS (8b), but does* *not necessarily coincide with the partial MINOLESS either as long as*

$$ES_k \cdot \breve{\bar{\boldsymbol{\xi}}} \neq \mathbf{0} = ES_k \cdot \bar{\bar{\boldsymbol{\xi}}}. \tag{15}$$

3 The Various Choices of the Selection Matrix

In view of the rank condition (14a), it is important to distinguish a number of cases within the interval for $k \in [m-q, m]$, plus a few sub-cases; note the following implications:

$$\text{rk}(S_k + N) = m = \text{rk}\left(\left[N \middle| S_k\right]\begin{bmatrix} I_m \\ I_m \end{bmatrix}\right) \Rightarrow \text{rk}\left[N \middle| S_k\right] = m,$$

$$\text{rk}\left[N \middle| S_k\right] < m \Rightarrow \text{rk}(S_k + N) < m.$$

Case 1: $m - q = k$ such that $\text{rk}(S_k + N) = m = \text{rk}\left[N \mid S_k\right]$,
Case 1′: $m - q = k$ such that $\text{rk}\left[N \mid S_k\right] < m$,
Case 1″: $m-q = k$ such that $\text{rk}(S_k+N) < \text{rk}\left[N \mid S_k\right] = m$;
Case 2: $m - q < k < q$ such that $\text{rk}(S_k + N) = m = \text{rk}\left[N \mid S_k\right]$,
Case 2′: $m - q < k < q$ such that $\text{rk}\left[N \mid S_k\right] < m$,
Case 2″: $m - q < k < q$ such that $\text{rk}(S_k + N) < \text{rk}\left[N \mid S_k\right] = m$;
Case 3: $q \le k < m$ such that $\text{rk}(S_k N) = \text{rk}\, N = q$,
Case 3′: $q \le k < m$ such that $\text{rk}(S_k N) < \text{rk}\, N = q$;
Case 4: $k = m \Rightarrow S_k = I_m, \ \ \bar{\boldsymbol{\xi}} = \bar{\bar{\boldsymbol{\xi}}} = \breve{\boldsymbol{\xi}} = \breve{\bar{\boldsymbol{\xi}}}$.

Some of these cases may turn out *empty*, which is obvious for the *Cases 2, 2′, and 2″* when $2q \le m$. Also, so far *Cases 1″ and 2″* were never encountered in any practical example, in spite of the fact that $\text{rk}(S_k + N)$ and $\text{rk}[N \mid S_k]$ may be different:

$$\text{rk}(S_1 + N) = 1 < 2 = \text{rk}[N \mid S_1] \ \text{ for } \ N := \begin{bmatrix} 0 & 1 \\ 1 & 1 \end{bmatrix}.$$

Apparently, N is *not positive-semidefinite* here, and this may be decisive for this counter-example (as we could not find one for a positive-semidefinite N).

4 Numerical Experiments: A Leveling Network

Here, a simple 2-loop leveling network is considered as in Fig. 1, where heights of four stations (P_1, P_2, P_3, P_4) must be estimated from observed height differences, giving rise to a datum deficiency of one.

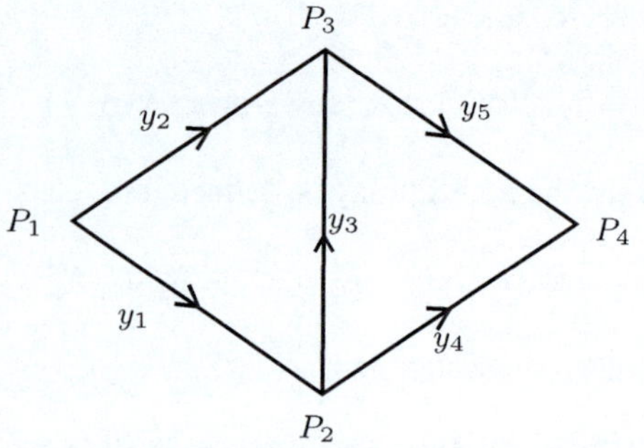

Fig. 1 A 2-loop leveling network with equal path lengths between points. Arrows point in the direction of the level runs

Table 1 Station heights in meters: true, biased, additive bias

Station	True $\boldsymbol{\Xi}_t$	Biased $\boldsymbol{\Xi}_0$	Bias $\boldsymbol{\beta}_t$
P_1	281.130	281.090	−0.040
P_2	269.131	269.109	−0.022
P_3	290.128	290.133	0.005
P_4	258.209	258.255	0.046

Then, the GMM reads

$$\underset{5\times1}{\boldsymbol{y}} = \underset{5\times4}{A}\ \underset{4\times1}{\boldsymbol{\xi}} + \underset{5\times1}{\boldsymbol{e}}, \quad \boldsymbol{e} \sim (\mathbf{0}, \Sigma = \sigma_0^2 I_5),$$

$$\text{with } A := \begin{bmatrix} -1 & +1 & 0 & 0 \\ -1 & 0 & +1 & 0 \\ 0 & -1 & +1 & 0 \\ 0 & -1 & 0 & +1 \\ 0 & 0 & -1 & +1 \end{bmatrix}, \quad P = I_5,$$

$$n = 5 > m = 4 > q = 3 > m - q = 1,$$

$$\text{nullspace: } \mathcal{N}(A) = \{\boldsymbol{\tau} \cdot \alpha \mid \alpha \in \mathbb{R}\} = \mathcal{R}(\boldsymbol{\tau}) \ \text{ for}$$

$$\underset{4\times1}{\boldsymbol{\tau}} := \left[1, 1, 1, 1\right]^T \ \text{ as “summation vector.”}$$

Simulation of Observations and Parameter (Height) Biases
The observations were simulated by adding random noise in the interval $[-0.02, +0.02]$ m to differences of *“true”* heights. Such a level of simulated observational noise might be representative of leveling work across lines of length 10 km, for example. A vector of true heights $\boldsymbol{\Xi}_t$ was given, and random noise on the interval of $[-0.05, +0.05]$ m was sampled to determine a “known bias vector” $\boldsymbol{\beta}_t$, which was added to the true heights to obtain biased heights $\boldsymbol{\Xi}_0$, i.e., $\boldsymbol{\Xi}_0 = \boldsymbol{\Xi}_t + \boldsymbol{\beta}_t$. Thus, the simulated bias level added to the true heights is about two and a half times larger than the simulated measurement noise. The biased heights $\boldsymbol{\Xi}_0$ were used as initial approximations in the formulas for the four estimators. Their numerical values, along with those of the simulated observations, are shown exactly in Tables 1 and 2, respectively. It is noted that the norm of the simulated height bias is $\|\boldsymbol{\beta}_t\| = 0.0650$ m.

Table 2 Observations in meters: true, simulated, additive noise

Obs.	From → to	True	Simulated	Noise
y_1	$P_1 \to P_2$	−11.999	−11.9864	0.0126
y_2	$P_1 \to P_3$	8.998	9.0142	0.0162
y_3	$P_2 \to P_3$	20.997	20.9821	−0.0149
y_4	$P_2 \to P_4$	−10.922	−10.9055	0.0165
y_5	$P_4 \to P_3$	−31.919	−31.9243	−0.0053

Table 3 Number of selection matrices S_k for each paring of case number and number of selected parameters k

Case/k	1	2	3	4	Totals
1	4	0	0	0	4
2	0	6	0	0	6
3	0	0	4	0	4
4	0	0	0	1	1
Totals	4	6	4	1	15

Table 4 Cases 1–4 for partial MINOLESS

$k/\dagger$	$10^3 \Omega$	$\text{tr}\, Q_{\bar{\bar{\xi}}}$	$\|\boldsymbol{\beta}\|$	$\|S_k\boldsymbol{\beta}\|$	$\|B\|$	$\|S_kB\|$	$\|BS_k\|$	$[S_k(i,i)]$
1/1	0.464	2.250	0.058	0.040	2.000	1	2.000	[1, 0, 0, 0]
1/1	0.464	1.750	0.052	0.022	2.000	1	2.000	[0, 1, 0, 0]
1/1	0.464	1.750	0.016	0.005	2.000	1	2.000	[0, 0, 1, 0]
1/1	0.464	2.250	0.078	0.046	2.000	1	2.000	[0, 0, 0, 1]
2/2	0.464	1.375	0.055	0.045	1.414	1	1.414	[1, 1, 0, 0]
2/2	0.464	1.375	0.029	0.027	1.414	1	1.414	[1, 0, 1, 0]
2/2	0.464	1.250	0.018	0.015	1.414	1	1.414	[1, 0, 0, 1]
2/2	0.464	1.250	0.026	0.012	1.414	1	1.414	[0, 1, 1, 0]
2/2	0.464	1.375	0.020	0.017	1.414	1	1.414	[0, 1, 0, 1]
2/2	0.464	1.375	0.044	0.036	1.414	1	1.414	[0, 0, 1, 1]
3/3	0.464	1.139	0.036	0.035	1.155	1	1.155	[1, 1, 1, 0]
3/3	0.464	1.083	0.018	0.017	1.155	1	1.155	[1, 1, 0, 1]
3/3	0.464	1.083	0.017	0.016	1.155	1	1.155	[1, 0, 1, 1]
3/3	0.464	1.139	0.019	0.017	1.155	1	1.155	[0, 1, 1, 1]
4/3	0.464	1.000	0.016	0.016	1.000	1	1.000	[1, 1, 1, 1]

In column 1, † denotes $\text{rk}(S_k N)$. Note that $\|B\| = \|BS_k\| = \sqrt{m/k}$ for this estimator. The MINOLESS appears for $k = 4$, $\dagger = 3$

Table 5 Cases 1–4 for BLIMPBE

$k/\dagger$	$10^3 \cdot \Omega$	$\text{tr}\, Q_{\breve{\xi}}$	$\|\boldsymbol{\beta}\|$	$\|S_k\boldsymbol{\beta}\|$	$\|B\|$	$\|S_kB\|$	$\|BS_k\|$	$[S_k(i,i)]$
1/1	6.825	0.500	0.056	0.023	1.871	0.707	0	[1, 0, 0, 0]
1/1	4.829	0.333	0.062	0.007	1.826	0.577	0	[0, 1, 0, 0]
1/1	7.212	0.333	0.065	0.003	1.826	0.577	0	[0, 0, 1, 0]
1/1	2.628	0.500	0.046	0.003	1.871	0.707	0	[0, 0, 0, 1]
2/2	2.407	1.000	0.051	0.020	1.789	1.095	0	[1, 1, 0, 0]
2/2	6.804	1.000	0.057	0.024	1.789	1.095	0	[1, 0, 1, 0]
2/2	2.043	1.000	0.032	0.023	1.732	1.000	0	[1, 0, 0, 1]
2/2	4.820	0.750	0.062	0.011	1.732	1.000	0	[0, 1, 1, 0]
2/2	2.016	1.000	0.041	0.008	1.789	1.095	0	[0, 1, 0, 1]
2/2	0.479	1.000	0.055	0.030	1.789	1.095	0	[0, 0, 1, 1]
3/3	0.464	2.250	0.078	0.063	2.000	1.732	0	[1, 1, 1, 0]
3/3	0.464	1.750	0.016	0.016	2.000	1.732	0	[1, 1, 0, 1]
3/3	0.464	1.750	0.052	0.047	2.000	1.732	0	[1, 0, 1, 1]
3/3	0.464	2.250	0.058	0.041	2.000	1.732	0	[0, 1, 1, 1]
4/3	0.464	1.000	0.016	0.016	1.000	1.000	1	[1, 1, 1, 1]

In column 1, † denotes $\text{rk}(S_k N)$. Note that Ω reveals that BLIMPBE is a LESS when $k \geq 3 = q$, and hence identical to its twin

Comments on Numerical Computations The case numbers are listed in Table 3. Key to our work is the computation of norms of various bias quantities. We define *computed bias* as $\boldsymbol{\beta} := \boldsymbol{\Xi}^* - \boldsymbol{\Xi}_t$, where $\boldsymbol{\Xi}^*$ stands for the estimated parameters provided by either partial MINOLESS ($\bar{\bar{\boldsymbol{\xi}}}$), BLIMPBE ($\breve{\boldsymbol{\xi}}$), BLIMPBE's twin ($\breve{\breve{\boldsymbol{\xi}}}$), or MINOLESS ($\bar{\boldsymbol{\xi}}$), and $\boldsymbol{\Xi}_t$ stands for the true parameters as noted above. Note that uppercase $\boldsymbol{\Xi}$ is used to denote full parameter quantities, while lowercase $\boldsymbol{\xi}$ denotes incremental quantities in the case of linearization.

It is important to recall that the biased heights $\boldsymbol{\Xi}_0$ were used as initial approximations in the various formulas for the four estimators. This is relevant because (partial) MINOLESS minimizes the norm of the vector of differences between the initial, approximate and the final, estimated parameters, according to (8a) and (11a), whereas BLIMPBE will reproduce the parameters (i.e., the initial approximations) that are *not* selected by matrix S_k.

The following list of norms were computed and tabulated below:

1. $\Omega := \tilde{e}^T P \tilde{e}$ being the (P-weighted) sum of squared residuals, minimized for the LESS by (2);
2. $\|\boldsymbol{\beta}\| = \sqrt{\boldsymbol{\beta}^T \boldsymbol{\beta}}$ being the norm of the computed bias vector;
3. $\|S_k \boldsymbol{\beta}\| = \sqrt{\boldsymbol{\beta}^T S_k S_k \boldsymbol{\beta}} = \sqrt{\boldsymbol{\beta}^T S_k \boldsymbol{\beta}}$ being the S_k-weighted norm of the computed bias vector;
4. $\|B\| = \sqrt{\text{tr}(B^T B)} = \sqrt{\text{tr}(BB^T)}$ being the norm of the bias matrix $B := -(I_m - LA)$, minimized for MINOLESS = BLUMBE by (10b);
5. $\|S_k B\| = \sqrt{\text{tr}(B^T S_k B)}$ being a type of S_k-weighted norm of the bias matrix $B := -(I_m - LA)$;
6. $\|BS_k\| = \sqrt{\text{tr}(BS_k B^T)}$ being another type of S_k-weighted norm of the bias matrix $B := -(I_m - LA)$, minimized for BLIMBPE by (13).

While we list numerical values of all of these norms for all the computed estimates and for all cases of selection matrices S_k below, we note again that only the norms listed in items 1, 4, and 6 were actually minimized for particular estimators.

Comments on Numerical Results Tables 4, 5, and 6 list relevant quantities computed using the four estimators of interest for each case listed in Table 3, where these abbreviations are used in the column labels: $Q_{\xi^*} := \sigma_0^{-2} \cdot D\{\xi^*\}$ for the trace of the cofactor matrix of the estimated parameters, and † stands in for $\text{rk}(S_k N)$.

There are a number of noteworthy observations to be made from Tables 4, 5, and 6. A few are enumerated here, with some associated entries highlighted in the tables.

1. It was confirmed that the matrix norm $\|BS_k\| = \sqrt{\text{tr}(BS_k B^T)}$, with bias matrix B, turned out to be smaller for BLIMPBE and its twin than for partial MINOLESS for all cases and instances of S_k, $k < m$;

Table 6 Cases 1–4 for BLIMPBE's twin

$k/\dagger$	$10^3 \cdot \Omega$	$\operatorname{tr} Q_{\check{\xi}}$	$\|\boldsymbol{\beta}\|$	$\|S_k\boldsymbol{\beta}\|$	$\|B\|$	$\|S_k B\|$	$\|BS_k\|$	$[S_k(i,i)]$
1/1	7.357	0.556	0.065	0.039	2.000	1.000	0	[1, 0, 0, 0]
1/1	4.829	0.333	0.062	0.007	1.826	0.577	0	[0, 1, 0, 0]
1/1	7.212	0.333	0.065	0.003	1.826	0.577	0	[0, 0, 1, 0]
1/1	2.693	0.556	0.046	0.003	2.000	1.000	0	[0, 0, 0, 1]
2/2	2.645	1.082	0.051	0.021	1.980	1.385	0	[1, 1, 0, 0]
2/2	7.580	1.082	0.069	0.046	1.980	1.385	0	[1, 0, 1, 0]
2/2	2.043	1.000	0.032	0.023	1.732	1.000	0	[1, 0, 0, 1]
2/2	4.820	0.750	0.062	0.011	1.732	1.000	0	[0, 1, 1, 0]
2/2	2.206	1.082	0.043	0.016	1.980	1.385	0	[0, 1, 0, 1]
2/2	0.480	1.082	0.054	0.029	1.980	1.385	0	[0, 0, 1, 1]
3/3	0.464	2.250	0.078	0.063	2.000	1.732	0	[1, 1, 1, 0]
3/3	0.464	1.750	0.016	0.016	2.000	1.732	0	[1, 1, 0, 1]
3/3	0.464	1.750	0.052	0.047	2.000	1.732	0	[1, 0, 1, 1]
3/3	0.464	2.250	0.058	0.041	2.000	1.732	0	[0, 1, 1, 1]
4/3	0.464	1.000	0.016	0.016	1.000	1.000	1	[1, 1, 1, 1]

In column 1, † denotes $\operatorname{rk}(S_k N)$. Note that Ω reveals that BLIMPBE's twin is a LESS when $k \geq 3 = q$, and hence identical to BLIMPBE itself

however, the same cannot be said about the corresponding bias vectors $\boldsymbol{\beta} = B\boldsymbol{\xi}$.

2. Interestingly, the square of the norm, $\|BS_k\|^2$, turns out to be an integer that depends on the number of selected parameters k and, in the case of BLIMPBE and its twin, the rank of matrix $S_k N$, but in the case of partial MINOLESS, the dimension of N. The following relations hold:
   (a) For BLIMPBE (and its twin):

   $$\|BS_k\|^2 = \begin{cases} k - \operatorname{rk}(S_k N) & \text{for } k \geq \operatorname{rk}(S_k N) \\ 0 & \text{otherwise} \end{cases}$$

   This property is even more apparent in investigations by the authors involving a 2D-network, where cases $1'$, $2'$, and $3'$ were not empty sets and where a larger datum deficiency allowed for a greater range of k beyond $\operatorname{rk}(S_k N)$.
   (b) For partial MINOLESS: $k \cdot \|BS_k\|^2 = m$, with $\dim(N) = m$. However, we did not find this to be the case in the 2D-network mentioned in the preceding item. Note also that $\|B\| = \|BS_k\|$, but this may be an artifact of our example problem, rather than a general rule. Also note that $\|S_k B\| = 1$ everywhere, independent of the number of selected parameters k.
3. The values for the SSR, Ω, appearing in the last five rows of all three tables reveal that both BLIMPBE and BLIMPBE's twin yield least-squares solutions when the rank condition (14a) is satisfied, which is certainly as expected. Moreover, the tables reveal that partial MINOLESS has the smallest trace of cofactor matrix, resp. dispersion matrix for $\boldsymbol{\xi}^*$ (partial BLESS) and the smallest norm of bias matrix $\|B\|$ in these cases, showing that both BLUMBE criteria (10b) and (10c) were fulfilled whenever $\operatorname{rk}(S_k N) = q = 3$.
4. The following relationships can be seen from careful inspection of the tables:
   (a) $\operatorname{tr} Q_{\text{BLIMPBE}} \leq \operatorname{tr} Q_{\text{twin}} \leq \operatorname{tr} Q_{\text{part.MINOLESS}}$ if $k < q = 3$.
   (b) $\hat{\boldsymbol{\xi}}_{\text{BLIMPBE}} = \hat{\boldsymbol{\xi}}_{\text{twin}}$, $\boldsymbol{\beta}_{\text{BLIMPBE}} = \boldsymbol{\beta}_{\text{twin}}$, etc. if $k \geq q = 3$.
   (c) $\|\boldsymbol{\beta}_{\text{BLIMPBE}}\| < \|\boldsymbol{\beta}_{\text{twin}}\|$, $\|B_{\text{BLIMPBE}}\| < \|B_{\text{twin}}\|$, etc. if $k < q = 3$.
   (d) $\|S_k B\|_{\text{BLIMPBE}}$ smaller whenever k gets smaller ($k \leq q = 3$).

5 Conclusions and Outlook

We have summarized the properties of some commonly used estimators for estimating parameters in a rank-deficient Gauss-Markov Model, with particular focus on their usefulness in the presence of bias induced by treatment of the rank (datum) deficiency via a selection matrix S_k. Some known properties of both partial MINOLESS and BLIMPBE were confirmed numerically, and a host of perhaps not so well known characteristics where enumerated above. Further studies should reveal, which, if any, of these characteristics can be expected for network adjustments in general.

It was also made obvious through the numerical example, that the minimization of the bias matrix $B := -(I_m - LA)$ does not imply that the corresponding bias vector $\boldsymbol{\beta} = B\boldsymbol{\xi}$ will be a minimum, an important truth to keep in mind when considering which estimator to use for a particular problem.

There are still open questions, as our study does not yet allow us to make definitive statements regarding preference of estimator for any particular cases. We do believe the subject warrants further research, and hope that similar studies involving classical 2D- and 3D-networks might shed more light on the matter. In fact, they are already underway.

References

Caspary W (2000) Concepts of network and deformation analysis. Monograph 11, School of Geomatic Engineering, University of New South Wales

Grafarend E, Sansò F (eds) (1985) Optimization and design of geodetic networks. Springer, Berlin

Grafarend E, Schaffrin B (1974) Unbiased free net adjustment. Surv Rev 22:200–218

Schaffrin B (1985) Network design. In: Grafarend EW, Sansó F (eds) Optimization and design of geodetic networks. Springer, Berlin, pp 548–597

Schaffrin B (2013/2014) Modern adjustment computations: a model-based approach, Slides prepared for the IAG Summer School in Taiwan and a Summer Course at Tech. University, Graz/Austria

Schaffrin B, Iz HB (2002) BLIMPBE and its geodetic applications. In: Adam J, Schwarz K (eds) Vistas for geodesy in the new millenium. In: Springer IAG-symposium, vol 125. Springer, Berlin, pp 377–381

Snow K, Schaffrin B (2007) GPS network analysis with BLIMPBE: an alternative to least-squares adjustment for better bias control. J Surv Eng 133(3):114–122

Regularized Solutions of the Two Layers Inverse Gravimetric Problem in the Space of Bounded Variation Functions

Martina Capponi, Daniele Sampietro, and Fernando Sansò

Abstract

The two layers inverse gravimetric problem is to determine the shape of the two layers in a body B, generating a given gravitational potential in the exterior of B. If the constant density of the two layers is given, the problem is reduced to the determination of the geometry of the interface between the two. The problem is known to be ill posed and therefore it needs a regularization, that for instance could have the form of the optimization of a Tikhonov functional. In this paper it is discussed why neither L^2 nor $H^{1,2}$ are acceptable choices, the former giving too rough solutions, the latter too smooth. The intermediate Banach space of functions of Bounded Variation is proposed as a good solution space to allow for discontinuities, but not too wild oscillations of the interface. The problem is analyzed by standard variational techniques and existence of the optimal solution is proved.

Keywords

Bounded variation functions · Inverse gravimetric problem · Regularization methods

1 Introduction to the Problem

The inverse gravimetric problem is to determine the mass distribution in a body B from the known gravity potential in the complementary region $\Omega = \overline{B}^C$. As it is well known the problem is undetermined (Sansò 2014; Ballani and Stromeyer 1982) and it is generally solved by restricting the space where a solution can be sought. One typical model that, under suitable hypotheses on the regularity of the surfaces involved (Barzaghi and Sansò 1986; Isakov 1990), implies the uniqueness of the solution, is that of a body B consisting by two layers, each of known constant density. This can be considered as a perturbation of a larger known gravity field that acts as reference and imposes the main direction of the vector $\mathbf{g}$. Here we will work on the problem when the reference field is considered as parallel and opposed to the verse of the z axis of the Cartesian system in which we frame the problem.

The two layers (see Fig. 1) are the upper layer, $-(H + \delta H) \leq Z \leq -H_+$ and the lower layer $-H_- \leq Z \leq -(H + \delta H)$ with densities respectively ρ_+ and ρ_-; we will call $\delta\rho = \rho_+ - \rho_-$ the density contrast and $\mu = G\delta\rho$, with G the universal gravitational constant. Always referring to Fig. 1, we will assume that the interface S between the two layers, $\{z = -(H + \delta H(\mathbf{x}))\}$, is agreeing with the plane $\{z = -H\}$ outside the support C, that is a bounded set on the horizontal plane. Moreover, we will set the hypothesis that S has to be constrained, to stay in depth within the layer $-H_- \leq z \leq -H_+$, for geophysical/geological reasons. We will simplify this constraint by assuming that

$$|\delta H(\boldsymbol{\xi})| \leq L\,, \quad \boldsymbol{\xi} \in C\,, \quad L = \max(-H_+ + H, -H + H_-)$$

without any loss of generality. Finally we assume that the modulus of gravity $g_o(\mathbf{x})$, is observed on the $z = 0$ plane. Such a hypothesis is just comfortable for the

M. Capponi · F. Sansò
Politecnico di Milano, DICA, Milano, Italy
e-mail: martina.capponi@polimi.it; fernando.sanso@polimi.it

D. Sampietro (✉)
Geomatics Research & Development srl, Lomazzo, Italy
e-mail: daniele.sampietro@g-red.eu

P. Novák et al. (eds.), *IX Hotine-Marussi Symposium on Mathematical Geodesy*,
International Association of Geodesy Symposia 151, https://doi.org/10.1007/1345_2019_77

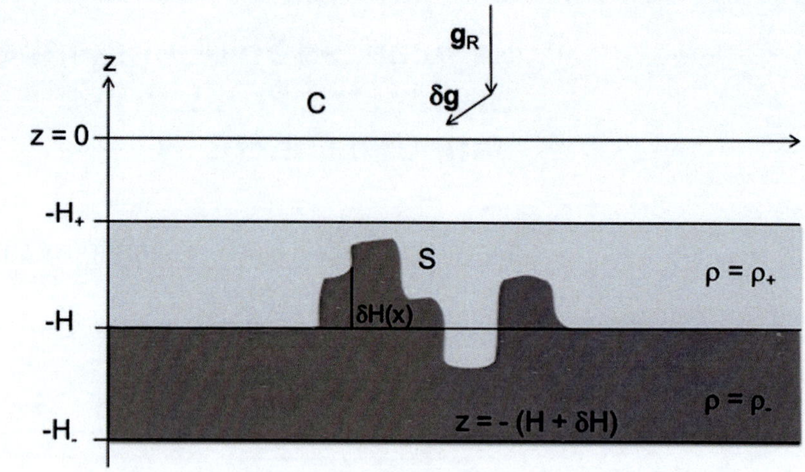

Fig. 1 The two layers problem in a Cartesian reference; ρ_+ is the density of the upper layer, ρ_- that of the lower layer; the surface S, that can have sharp discontinuities, in the plane $z = -H$ but for the set C on which $z = -(H + \delta H)$; S is constrained to stay in the layer $\{-H_- < z < -H_+\}$

analysis though not essential. Note that, above $-H_+$ and below $-H_-$, we suppose no masses at all, meaning that their gravitational effect has already be removed from $g_o(\mathbf{x})$.

Just to make an example where such hypothesis are meaningful, one can consider $z = 0$ as the surface of (or a piece of) the sea, the plane $\{z = -H_+\}$ as the sea floor, the upper layer as constituted by sediments while the lower layer by the bedrock. Of course we could complicate this model by assuming that $\{z = -H_+\}$ and $\{z = -H_-\}$ are surfaces with a different (known) shape, but since this would not alter the mathematical properties of the problem we analyse, we prefer to stay with the present simple hypothesis.

Starting from g_o, by subtracting to it a reference value g_{ref} (i.e. the normal gravity field evaluated in the same observation point) and the two Bouguer plates g_{Boug} (with densities ρ_+ between $-H_+$ and $-H$ and ρ_- between $-H$ and $-H_-$), we are left with the following linearized observation equation

$$\delta g = g_o - g_{ref} - g_{Boug} \equiv -\boldsymbol{\delta g} \cdot \mathbf{e}_z. \tag{1}$$

We call δg the reduced gravity disturbance generated by the body enclosed between S and $z = -H$, with density $\delta\rho$ where $\delta H < 0$ and $-\delta\rho$ where $\delta H > 0$. Let us notice that the choice of the functional δg as observable is by no means obliged: we could choose in fact any other functional of the anomalous field that uniquely determines itself in the harmonicity domain $R^3_+ = z \geq 0$ (Sansò et al. 2018).

One advantage of the above model is that the forward operator computing δg from δH, namely

$$\delta g = F(\delta H), \tag{2}$$

can be explicitly written in its exact non linear form, as

$$\delta g(\mathbf{x}) = -\mu \int_C \mathrm{d}_2\xi \int_{-H}^{-(H+\delta H)} \frac{\zeta}{[|\mathbf{x} - \boldsymbol{\xi}|^2 + \zeta^2]^{3/2}} \mathrm{d}\zeta \equiv F(\delta H) \tag{3}$$

$$F(\delta H) \equiv \mu \int \mathrm{d}_2\xi \left\{ \frac{1}{[|\mathbf{x} - \boldsymbol{\xi}|^2 + (H + \delta H)^2]^{1/2}} - \frac{1}{[|\mathbf{x} - \boldsymbol{\xi}|^2 + H^2]^{1/2}} \right\}.$$

Summarizing we have a first provisory formulation of the problem: to find δH satisfying the constraints

$$\begin{aligned} &\delta H(\boldsymbol{\xi}) = 0 \quad \boldsymbol{\xi} \in \overline{C}^c, \\ &|\delta H(\boldsymbol{\xi})| \leq L \quad \boldsymbol{\xi} \in C \end{aligned} \tag{4}$$

and the observation equation

$$\delta g = F(\delta H). \tag{5}$$

An important remark should be added about the domain C: as a matter of fact no real gravity map can suggest, or even more prove, that a certain gravity datum is generated be an isolated anomaly in the layers interface. Nevertheless, it is costumacy in applied geophysics to look for bounded anomalies, supposing that data outside the examined region do depends from other causes. So the extent of the region C is somehow dictated by the targets and experience of the analyst.

It is perfectly known that the problem formulated as written above is improperly posed according to Tikhonov (Sansò et al. 2018). This means that for any reasonable space X on which δH can vary and Y, to which we expect δg to belong, the operator $F(\delta H)$ has not a continuous inverse because, to invert it, we need first to downward continue δg to S or to $\{z = -H\}$ and this is a (linear) strongly discontinuous operation (Sansò et al. 2018).

Rather than solving the inverse of Eq. (5), the problem is reformulated in order to find its approximate (regularized) solution by minimizing a Tikhonov functional (Tikhonov et al. 2013; Tikhonov and Arsenin 1977)

$$T(\delta H) = \|\delta g - F(\delta H)\|_Y + \lambda J(\delta H); \tag{6}$$

the choice of the space Y and of the stabilizing functional $J(\delta H)$ (actually of the space X to which δH has to belong) is the object of next section.

2 Choice of the Tikhonov Functional

We start the discussion by the choice of the regularizing functional $J(\delta H)$. Often, though not always, $J(\delta H)$ is chosen as the norm of the space X of the unknown function δH. In our case, we could figure to set the space X equal to L^p or $H^{1,p}$. However in our opinion $X = H^{1,p}$, for whatever $p \geq 1$, is not satisfactory because $X = H^{1,1}$, which is the largest of these spaces, can not accommodate sharply discontinuous functions and produces too smooth surfaces (see Fig. 2 case c). The reason why we would like to admit surfaces S that have jumps is because this is in general a salient characteristic of true geological mass distributions. On the other hand, the hypothesis $X = L^p$ is instead too week to get a proper regularization of the problem (see Fig. 2 case a). As a matter of fact, in literature even stronger regularizations are used, for instance $X = H^{2,2}$ (see Richter 2016); however in our opinion this implies a quite unnatural smoothing of S. So, we are looking for a space, possibly a Banach space, to apply standard variational techniques, that is intermediate between $H^{1,1}(R^2)$ and $L^1(R^2)$ (see Fig. 2 case b).

A space of this kind can be the space of the so called Bounded Variation (BV) functions, as proposed by Ennio De Giorgi (see Giusti and Williams 1984), and already used in geophysical literature (see Acar and Vogel 1994).

Let us here recall the definition of BV and its relevant properties, useful for this work.

Definition 1 We say that $f(\mathbf{x}) \in BV(R^2)$, if

$$f \in L^1(R^2), \quad \|f\|_{L^1} = \int_{R^2} |f(\mathbf{x})| \mathrm{d}_2 x < +\infty, \tag{7}$$

and, after defining a set $\mathscr{T}$ of test functions as

$$\mathscr{T} = \{\varphi(\mathbf{x}) \in C^1(R^2), \quad \varphi = 0 \text{ in } \overline{C}^c, \quad |\varphi| \leq 1\}; \tag{8}$$

we have

$$J_0(f) = \sup_{\varphi \in \mathscr{T}} \int \nabla \cdot \varphi f(\mathbf{x})\, \mathrm{d}_2 x < \infty. \tag{9}$$

We denote as Df the Radon measure such that

$$\int \nabla \cdot \varphi f \, \mathrm{d}_2 x = \int \varphi \cdot Df \, \mathrm{d}_2 x \equiv < Df, \varphi > \tag{10}$$

and we shall further put

$$J_0(f) = \int_{R^2} |Df| \, \mathrm{d}_2 x. \tag{11}$$

Let us observe that since φ is identically zero in the open set $\overline{C}^c$ the relation in (11) could be written also in this other form

$$J_0(f) \equiv \int_{C_+} |Df| \, \mathrm{d}_2 x \tag{12}$$

where C_+ is any open bounded set containing $\overline{C}$; in fact $\overline{C}$ is the support of the vector measure Df and so also of $|Df|$.

Finally we define the normed space BV as

$$f \in BV \quad \Longleftrightarrow \quad \|f\|_{BV} = \|f\|_{L^1} + J_0(f). \tag{13}$$

The following propositions are essential in the present work and we report them here without any proof, that the interested reader can find in the book (Giusti and Williams 1984).

Proposition 1 *BV is a Banach space, i.e. Cauchy sequences $\{f_n\} \in BV$ have always a limit $f \in BV$ (see Giusti and Williams 1984, 1.12).*

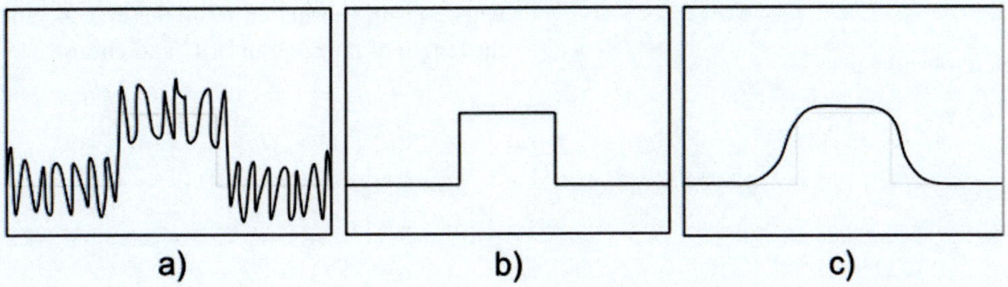

Fig. 2 Qualitative example in planar approximation of solution surfaces obtained from the choice of different $J(\delta H)$: **(a)** case of $X = L^p$, **(b)** case of $X = BV$, **(c)** case of $X = H^{1,1}$

Proposition 2 *The norm $\|f\|_{BV}$ is a lower semi-continuous functional for sequences $\{f_n\}$ bounded in BV and converging to f in L^1. Since $\|f\|_{L^1}$ is obviously continuous with respect to such a sequence, this means that given $\{f_n\,;\ \|f\|_{BV} \leq c\,;\ \|f_n - f\|_{L^1} \longrightarrow 0\}$ one has*

$$\underline{\lim}\ J_0(f_n) \geq J_0(f)\,, \tag{14}$$

or $f \in BV$ too and

$$\underline{\lim}\ \|f_n\|_{BV} \geq \|f\|_{BV}. \tag{15}$$

(see Giusti and Williams 1984, 1.9).

Proposition 3 *The embedding of BV in L^1 is (sequentially) compact, namely given any bounded $\{f_n\} \in BV$, there is a sub-sequence $\{f_{n_k}\}$ which is Cauchy in L^1 and therefore there is an $f \in L^1$ such that $f_{n_k} \longrightarrow f$; according to Proposition 2 then, $f \in BV$ too and*

$$\|f\|_{BV} \leq \underline{\lim}\ \|f_{n_k}\|_{BV}. \tag{16}$$

(Giusti and Williams 1984, 1.19).

Proposition 4 *Let us consider an $f \in BV$ such that $f(\mathbf{x}) = |f(\mathbf{x})| \geq 0$, $f(\mathbf{x}) = 0$ in $\overline{C}^c$, and define the Epigraph of f as usual as*

$$\text{Epi}(f) \equiv \{(\mathbf{x}\,, z)\,;\quad z \leq f(\mathbf{x})\} \cap R^3\,;$$

let us further define

$$\begin{cases} S \equiv \partial\text{Epi}(f) \cap \{z > 0\} = \partial\text{Epi}(f) \cap R^3_+ \\ B = \text{Epi}(f) \cap R^3_+ \end{cases}$$

then, denoting by $V(B)$ the Lebesgue volume of B, by $H(S)$ the Hausdorff measure of S normalized to coincide with the Lebesgue measure when f is smooth (see Falconer 1986, theorem 1.12), $|C|$ the measure of the base set C and by

$$J(f) = \|f\|_{L^1} + J_0(f) = \|f\|_{BV}\,,$$

one has

$$\frac{1}{\sqrt{2}}(J(f) + |C|) \leq V(B) + H(S) \leq J(f) + |C|\,. \tag{17}$$

Proof It is enough to observe that

$$\|f\|_{L^1} = V(B)$$

so that Eq. (17) is the same as

$$\frac{1}{\sqrt{2}}(J_0(f) + |C|) \leq H(S) \leq J_0(f) + |C|. \tag{18}$$

On the other hand, when f is sufficiently smooth (implying that $f = 0$ on ∂C)

$$\frac{1}{\sqrt{2}}(1 + |\nabla f|) \leq \sqrt{(1 + |\nabla f|^2)} \leq 1 + |\nabla f|\,. \tag{19}$$

Since $|\nabla f| = \tan(I)$, with I the inclination of the area element of S on the plane (x, y), one has, in this case,

$$H(S) = \int_C \sqrt{1 + |\nabla f|^2} \mathrm{d}_2 x.$$

Then, an integration of Eq. (19) over C proves Eq. (18) when f is smooth. Finally an approximation of f by a sequence $\{f_n\} \in \mathscr{D}$ and the use of the density of $\mathscr{D}$ in BV (cfr. Giusti and Williams 1984, Theorem 1.17) completes the proof. □

The above proposition, provides a geometrical interpretation of our variational principle in that, since $|C|$ is a constant, when we try to keep small $J(f)$ we equivalently impose to $V(B)$ and $H(S)$ to be small. In other words we control the extent and the smoothness of the body B. The next Example 1, though elementary, is designed to show how the definition of $|Df|$ and of $J_0(f)$ are capable of accounting for the lateral surface of S, that is generated when $f(\mathbf{x})$ is discontinuous through ∂C.

Example 1 Let us take

$$f(\mathbf{x}) = h\ \chi_{\overline{C}}(\mathbf{x}) = \begin{cases} h & \mathbf{x} \in \overline{C} \\ 0 & \mathbf{x} \in \overline{C}^c \end{cases}$$

then S is in this case as in Fig. 3. It is clear that $|Df| \equiv 0$ in C open i.e. the support of the measure $|Df|$ is on the contour line $\mathscr{L}$. We can write for any $\boldsymbol{\varphi} \in \mathscr{T}$,

$$\int_{R^2} f \nabla \cdot \boldsymbol{\varphi}\ \mathrm{d}_2 x \equiv h \int_C \nabla \cdot \boldsymbol{\varphi}\ \mathrm{d}_2 x \equiv h \int_{\mathscr{L}} \boldsymbol{\varphi} \cdot \mathbf{n}\ dl\,,$$

where $\mathbf{n}$ is the normal in R^2 to the line $\mathscr{L}$. Then, call $\mu_1(\mathscr{L})$ the length of the contour of C and choose

$$\boldsymbol{\varphi}(\mathbf{x}) \equiv \mathbf{n}\psi(\mathbf{x})$$

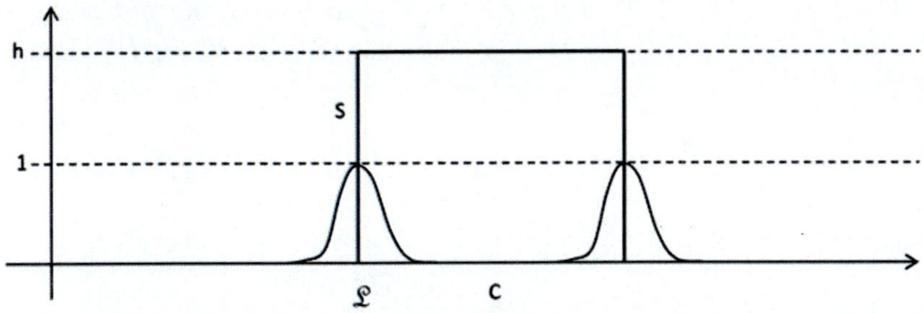

Fig. 3 The surface S for f as in Example 1; $\mathscr{L}$ is the trace of the contour line of C; the two bells represent a function $\psi(\mathbf{x})$ which is $0 \leq \psi(\mathbf{x}) \leq 1$ and goes up to 1 on $\mathscr{L}$

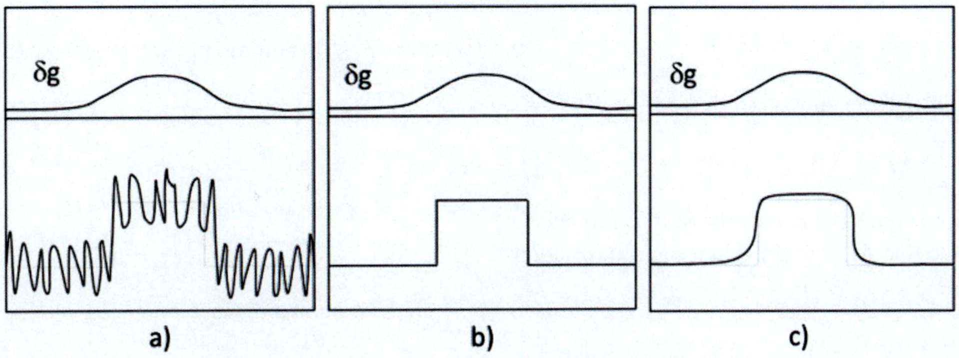

Fig. 4 Three interfaces: (**a**) in L^1, (**b**) in BV, (**c**) in $H^{1,1}$ that generate a similar gravity signal on $z = 0$

with ψ as in Fig. 3. Assuming that $\mathscr{L}$ is regular so that $\mathbf{n}$ is continuous, the above choice is such that $\varphi \in \mathscr{T}$. We find

$$J_0(f) = \sup_{\varphi \in \mathscr{T}} \int \nabla \cdot \varphi \ f \ d_2x = h\mu_1(\mathscr{L}),$$

which is in fact the measure of the lateral surface of S, we see that in this case

$$h\mu_1(\mathscr{L}) + \mu_2(C) = \|f\|_{BV}$$

is proved.

As help to the intuition of the discussion of the regularity of S, we provide in Fig. 4 three qualitative examples of surfaces, one in L^1, one in BV and one in $H^{1,1}$ that generate quite similar gravity signals on the plane $z = 0$. As a first conclusion of this discussion we fix the choice

$$J(\delta H) = \|\delta H\|_{BV} \tag{20}$$

with the understanding that $\delta H \in BV$ implies that the interface S between the two layers can be discontinuous, however its area is bounded and, thanks to the Tikhonov principle, it should not be too large.

We come now to the choice of Y. Let us observe that a classical choice for Tikhonov theory would be $Y = L^2(R^2)$. However we have to consider that we will look for the minimum of $T(\delta H)$, not on the whole Y but just in a subset K (see (4))

$$K \equiv \{\delta H \quad \text{measurable} ; \quad |\delta H| \leq L\chi_C(\mathbf{x})\} ; \tag{21}$$

now it is clear that the same K is a subset of all L^p, $p \geq 1$. Therefore the choice of p in this case is arbitrary. However in order to be more homogeneous with the regularizing functional, we make the choice $p = 1$, namely we put

$$T(\delta H) = \|\delta g - F(\delta H)\|_{L^1} + \lambda \|\delta H\|_{BV}. \tag{22}$$

A last remark is that (22) gives a quite classical form of the Tikhonov functional, which however does not cover the case of continuous Wiener noise. This in fact is well known to have no measurable realizations.

3 Existence of the Minimum

We want to propose the following theorem, which is a classical result in regularization theory (see for example Freeden and Nashed 2018).

Theorem 1 *Let us put*

$$M = \inf_{\delta H \in BV \cap K} T(\delta H) \tag{23}$$

with $T(\delta H)$ as in (22), then there is a $\delta H \in BV$ such that

$$M = T(\delta\overline{H}) \tag{24}$$

namely M is the minimum of T in $BV \cap K$.

Proof The proof comes in four steps:

(i) let us call B_R the sphere of radius R in BV, i.e.

$$B_R \equiv \{\delta H \in BV\,;\quad \|\delta H\|_{BV} \le R\}; \tag{25}$$

of course $B_R \cap K$ is a closed, bounded and convex set in BV. For any minimizing sequence $\delta H_n \in K$,

$$M = \lim\ T(\delta H_n)\,, \tag{26}$$

we can find an R large enough so that $\delta H \in B_R \cap K$. In fact

$$M \le T(0) = \|\delta g\|_{L^1}. \tag{27}$$

Would M be equal to $T(0)$ the theorem is proved, so we can assume that $M < T(0)$. Therefore, for n larger than a suitable N,

$$T(\delta H_n) = \|\delta g - F(\delta H_n)\|_{L^1} + \lambda\|\delta H_n\|_{BV} \le \|\delta g\|_{L^1}$$

and so,

$$n > N\,,\quad \|\delta H_n\| \le \frac{\|\delta g\|_{L^1}}{\lambda} \tag{28}$$

and we can choose $R = \frac{\|\delta g\|_{L^1}}{\lambda}$;

(ii) according to Proposition 3, $B_R \cap K$ is L^1 compact, so if δH_n is a minimizing sequence there is a sub-sequence that we call again δH_n such that

$$\delta H_n \underset{L^1}{\longrightarrow} \delta\overline{H} \in BV. \tag{29}$$

Then, according to (16),

$$\|\delta\overline{H}\|_{BV} \le \underline{\lim}\ \|\delta H_n\|_{BV} \le R \tag{30}$$

i.e. $\delta\overline{H} \in B_R$. Furthermore, from (29), knowing that a sub-sequence of δH_n converges almost everywhere to $\delta\overline{H}$, we see that $\delta\overline{H} \in K$ (i.e. $|\delta H| \le L\chi_c(\mathbf{x})$). Therefore one has $\delta\overline{H} \in B_R \cap K$;

(iii) let us notice that $F(\delta H)$ is continuous $L^1 \longrightarrow L^1$. In fact, by using Taylor's theorem, (3) can be written

$$F(\delta H) = \mu \int_C \mathrm{d}_2\xi \frac{\delta H(\boldsymbol{\xi})}{[|\mathbf{x} - \boldsymbol{\xi}|^2 + (H + \theta\delta H)^2]^{3/2}}; \tag{31}$$

where θ is function of $\boldsymbol{\xi}$ and $\boldsymbol{x}$, such that $0 \le \theta \le 1$, so that $(H + \theta\delta H)^2 \ge H_+^2$. Then,

$$\begin{aligned}\|F(\delta H)\|_{L^1} &\le \mu \textstyle\int_C \mathrm{d}_2\xi\ |\delta H(\boldsymbol{\xi})| \int_{R^2} \mathrm{d}_2 x\ \frac{1}{[|\mathbf{x}-\boldsymbol{\xi}|^2 + H_+^2]^{3/2}}\\ &= \mu\frac{2\pi}{H_+}\|\delta H\|_{L^1}\end{aligned} \tag{32}$$

namely F is bounded $L^1 \longrightarrow L^1$. In a similar way one proves that

$$\|F(\delta H_1) - F(\delta H_2)\|_{L^1} \le \mu\frac{2\pi}{H_+}\|\delta H_1 - \delta H_2\|_{L^1}, \tag{33}$$

i.e. F is continuous $L^1 \longrightarrow L^1$;

(iv) since δH_n is minimizing one has, from (iii) and (30),

$$M = \lim\ T(\delta H_n) = \|\delta g - F(\delta\overline{H})\|_{L^1} + \underline{\lim}\ J(\delta H_n) \tag{34}$$

$$\begin{aligned}&\|\delta g - F(\delta\overline{H})\|_{L^1} + \underline{\lim}\ J(\delta H_n)\\ &\ge \|\delta g - F(\delta\overline{H})\|_{L^1} + \lambda J(\delta\overline{H}) = T(\delta\overline{H})\,;\end{aligned}$$

on the other hand, since $\delta\overline{H} \in B_R \cap K$, one has too

$$T(\delta\overline{H}) \ge M; \tag{35}$$

with (34) and (35) proving (24).

□

Remark 1 Let us notice that for every λ we have a regularized solution $\delta\overline{H}_\lambda$; of course this says not so much about the right choice of λ. Yet in this respect it is classical a theorem in Tikhonov theory claiming that if η_n are decreasing errors in the "observation" equation $\delta g_{o,n} = \delta g + \eta_n = F(\delta H) + \eta_n$,

$$\delta_n = \|\eta_n\|_{L^1} \longrightarrow 0, \tag{36}$$

there is a strategy of $\lambda_n = \lambda(\delta_n)$, namely

$$\lambda_n > c\ \delta_n\,,\ (c = \text{constant}),$$

such that the corresponding solutions $\delta\overline{H}_{\lambda_n}$ admit a sub-sequence L^1 converging to the correct solution δH (Bertero 1982; Freeden and Nashed 2018).

4 A Numerical Example

This example has been developed to show how the theory presented above can lead to a practical solution where data are, by necessity, discrete and finite in number and the solution too has to be restricted to a finite dimensional subspace of BV.

It is our purpose: (1) to show how to discretize the Tikhonov principle with the use of prisms; (2) to see how the regularized solution depends on the observational noise, σ_ν^2; (3) to compare the solution foreseen in the paper with the one obtained with another Tikhonov functional, e.g. by using

$$J_2(\delta H) = \|\delta H\|^2_{H^{1,2}}$$

$$T_2(\delta H) = \|\delta g - F(\delta H)\|^2_{L^2} + \lambda J_2(\delta H), \quad (37)$$

and with the one resulting from a simple least squares solution (which is the discretized version of (37)). In order to keep the example as simple as possible, we decided to reduce it to a so called $2\frac{1}{2}$ dimensional case, where the "true" body is constituted by three prisms (see Fig. 5) of length 1 km in x, each of them, width 5 km in the y-direction and depth respectively of 1.0 km, 1.4 km and 0.6 km.

Therefore if we take measurements only along the central line in the x-direction and on the same time we impose to our model that the width in y is constant and equal to the correct value, we have all the information we need, almost as if the whole picture would be exactly bidimensional. To generate the exact values of δg we have chosen $\delta\rho = 500\,\text{kg/m}^3$ which is quite a value considering that $\delta\rho$ has the meaning of a density contrast.

According to the above picture, we will model our unknown interface S by 12 prisms (see Fig. 5), with bases of length 0.25 km in x, width 5 km in y as we said and unknown depth $z = -H_k$ ($k = 1, 2 \ldots 12$). The 36 observations simulated in this example, are evenly distributed on a grid on the x-axis with an inter-distance of 0.1 km, so as to have three more observation points at each side of the body, along x.

If we call, B_k, $k = 1, 2 \ldots 12$, the bases of the model prisms we can write

$$\delta g(\mathbf{x}) = \sum_{k=1}^{12} F_k(\delta H_k; \mathbf{x})$$

where

$$F_k(\mathbf{x}) = \mu \int_{B_k} \mathrm{d}_2\xi \left\{ \frac{1}{[|\mathbf{x} - \boldsymbol{\xi}|^2 + H_k^2]^{1/2}} - \frac{1}{|\mathbf{x} - \boldsymbol{\xi}|} \right\}. \quad (38)$$

Please note that in this context instead of using $\delta H_k = H_k - H$ as unknown, we are using the whole H_k. For this reason, when coming to discretization of $\int |\delta H|\, d_2\xi$, to be close to the theory developed, we discretize the regularizing functionals by the formulas

$$\sum_{k=1}^{12} |H_k - m_H| \quad \text{or} \quad \sum_{k=1}^{12} |H_k - m_H|^2$$

where

$$m_H = \frac{1}{2} \sum_{k=1}^{12} H_k.$$

So we start now by first generating 36 exact values $\delta g(x_i)$, $i = 1, 2 \ldots 36$, and then we add to them Gaussian independent noises ν_i once drown from $N(0; 1\ \text{mGal}^2)$ and in a second case from $N(0; 4\ \text{mGal}^2)$.

The optimization is then done with the discretized Tikhonov functional

$$T_1(H_k) \equiv \sum_{i=1}^{36} |\delta g_o(x_i) - \sum_{k=1}^{12} F_k(H_k; x_i)| + \lambda J_1(H_k). \quad (39)$$

Here a comment is required: as we can see, the discretization of the BV norm is done with the L^1 term already discretized

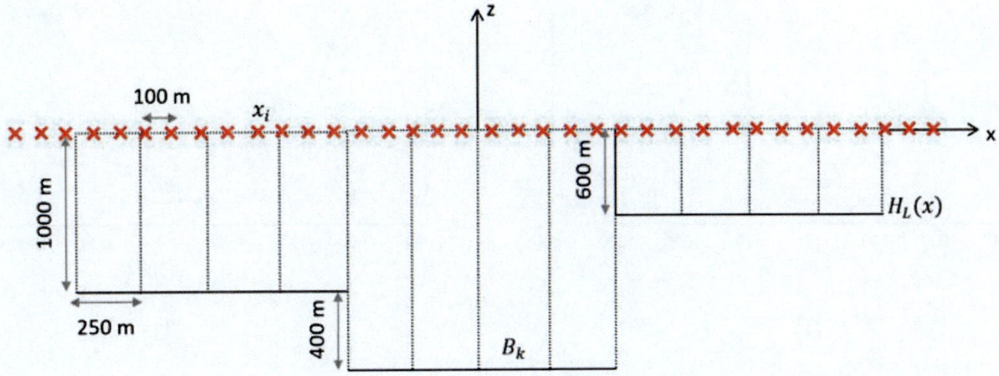

Fig. 5 The three prisms section constituting the "true" body of the numerical example and the simulated observation points on a regular grid set at $z = 0$ (red crosses)

and with the incremental term

$$\sum_{k=1}^{11} |H_{k+1} - H_k|.$$

According to our Example 1, this term, proportional to the lateral surfaces of the whole set of prisms, is also proportional to the measure of the lateral surfaces of S, which is (so to say) the discretized version of $\int |DH|\, d_2\xi$. Now we might observe that (39) could also be considered as a discretization of a Tikhonov principle regularized by the functional $\|\delta H\|_{H^{1,1}}$. This is in fact the case and it means that when we come to a discretized numerical computation, the two cases $\delta H \in BV$ and $\delta H \in H^{1,1}$ cannot be distinguished. This indeed does not cancel the theoretical characteristics of the two, discussed in the paper. As for the value of λ to be used in (39) we have chosen, in analogy with the L^2 theory, and its stochastic interpretation (Sansò 1986),

$$\lambda_1 = \frac{\sigma_1}{J_1(\delta H_{\text{theo}})}. \tag{40}$$

The results are displayed in terms of observations $\delta g_o(x_i)$ and interpolated function $\delta\hat{g}(\mathbf{x})$ as well as theoretical $H(\mathbf{x})$ and estimated $\hat{H}(\mathbf{x})$, once with a noise of $\sigma_\nu = 1\,\text{mGal}$, in Fig. 6a, and then with a noise of $\sigma_\nu = 2\,\text{mGal}$, in Fig. 6b. As it can be observed, the proposed regularization is stable, having a similar behavior with the two noise realizations.

A second numerical check, was aimed to the comparison of $\hat{H}_1(\mathbf{x})$, derived by minimizing $T_1(H_k)$, with $\hat{H}_2(\mathbf{x})$ derived by minimizing

$$T_2(H_k) = \sum_{i=1}^{36} |\delta g_o(x_i) - \sum_{k=1}^{12} F_k(H_k; x_i)|^2 + \lambda J_2(H_k), \tag{41}$$

$$J_2(H_k) = \sum_{k=1}^{12}(H_k - m_H)^2 + \sum_{k=1}^{11}(H_{k+1} - H_k)^2. \tag{42}$$

In this case λ has been chosen to be

$$\lambda_2 = \frac{\sigma_\nu^2}{J_2(H_{true})}$$

Furthermore a second comparison is done with $\hat{H}_0(\mathbf{x})$ derived by a simple least squares criterion, namely by minimizing $T_2(H_k)$ after we have put $\lambda = 0$. The results are displayed in Fig. 7. The conclusions that can be drawn from this small numerical experiment are:

1. the proposed method seems to give in all cases reasonable results, fitting quite well the real shape of the interface (standard deviations of the differences between real and

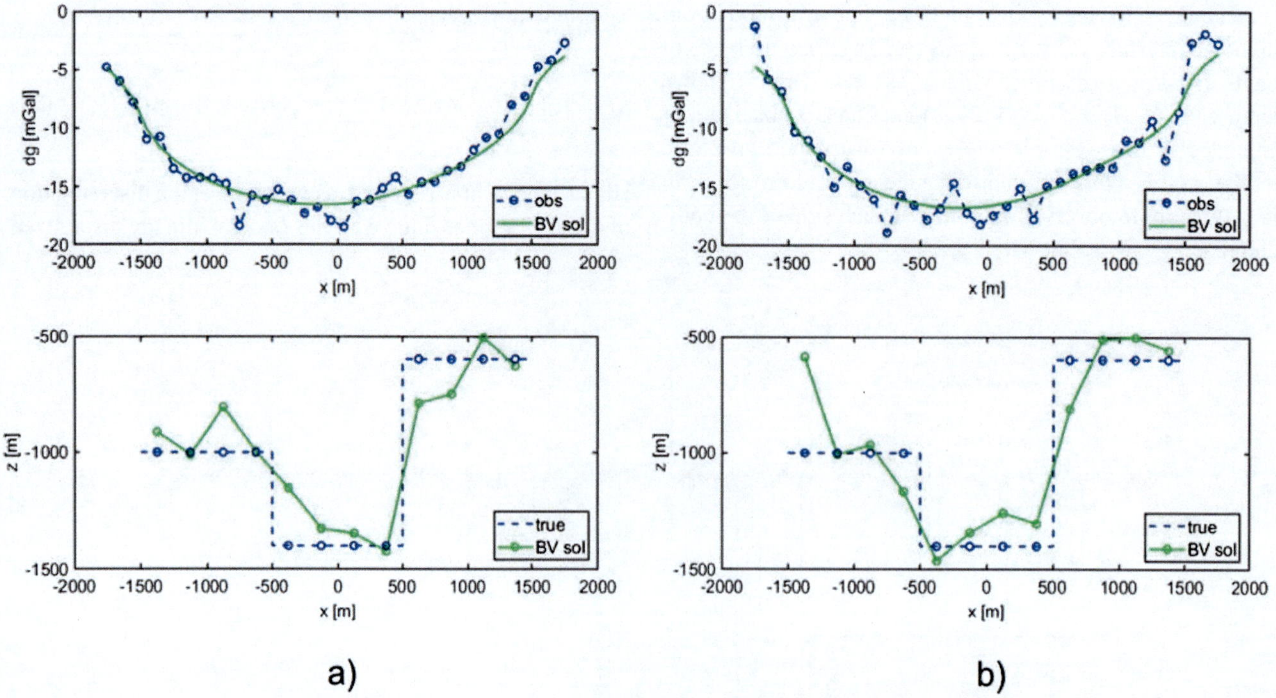

Fig. 6 Observed values and interpolating function $\delta\hat{g}(\mathbf{x})$ and below the true interface $\delta H(\mathbf{x})$ and estimated $\delta\hat{H}(\mathbf{x})$: (**a**) case with $\sigma_\nu = 1\,\text{mGal}$; (**b**) case with $\sigma_\nu = 2\,\text{mGal}$

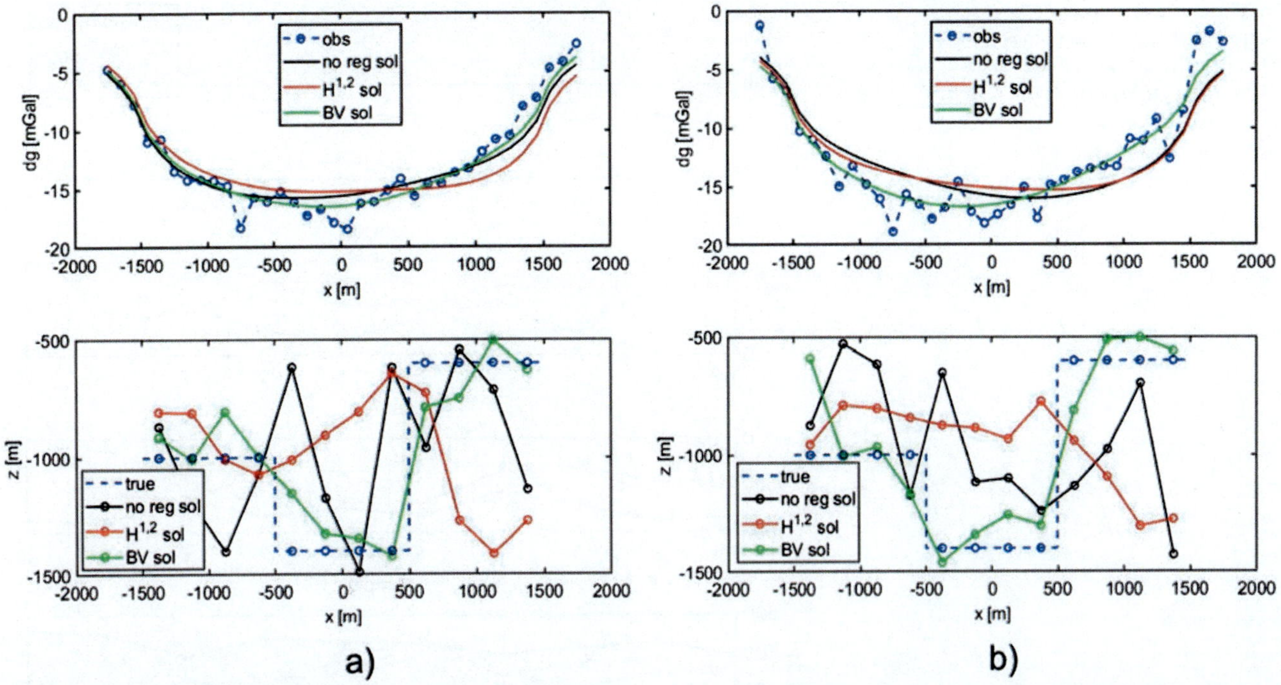

Fig. 7 Comparison between different regularizations: $\hat{H}_0$ the l.s. solution (no regularization), $\hat{H}_1$, the BV (or $H^{1,1}$) solution, $\hat{H}_2$ the $H^{1,2}$ solution

estimated depths are 125 m and 158 m respectively for Fig. 6a, b);

2. the comparison with the least squares solution $\hat{H}_o$ shows that this last one is typically less stable than $\hat{H}_1$ (standard deviations for least squares solutions are 419 m and 454 m respectively for Fig. 7a, b);
3. the comparison between $\hat{H}_1$ and $\hat{H}_2$ shows that, at least for the presented simple experiment, the first one performs better than the second both in terms of gravity signal fitting and depths estimation (standard deviations for $\hat{H}_2$ solutions are 519 m and 481 m respectively for Fig. 7a, b).

To measure quantitatively the quality of our results, the differences between the three solutions ($\hat{H}_0$, $\hat{H}_1$, $\hat{H}_2$) with respect to the true surface have been analyzed. In details the previous experiment has been repeated ten times with different noise realizations (with $\sigma_\nu = 1$ mGal). Results are reported in Fig. 8 where it can be seen that the behaviour (and the consequent conclusions drawn for the single realization presented in Figs. 6 and 7) are systematic. In particular, it can be noticed that the rmse of the differences of depths estimates by means of BV regularization results to be always the smallest one as well as the std. The same consideration can be done also looking at the minimum and maximum values, which means that the solution obtained by using BV regularization is the closest to the reality. All these statistics, confirm the goodness of the method and the effective improvement with respect to the two other solutions.

5 Conclusions

The inverse gravimetric problem, in a certain frame with a two layers configuration has been discussed. In particular, the meaning of a Tikhonov regularization has been analyzed in terms of prior requirements on the regularity of the interface S. A reasonable choice has been to put the condition that $\delta H \in BV$, namely that S has a finite and possibly small area. A theorem of existence of the solution of the corresponding variational principle has been proved. A numerical example, though very simple, seem to indicate that the foresought properties of the BV solution are in fact found and they seem to give better results than a no-regularization solution, or a stronger regularization solution, at least in an averaged sense.

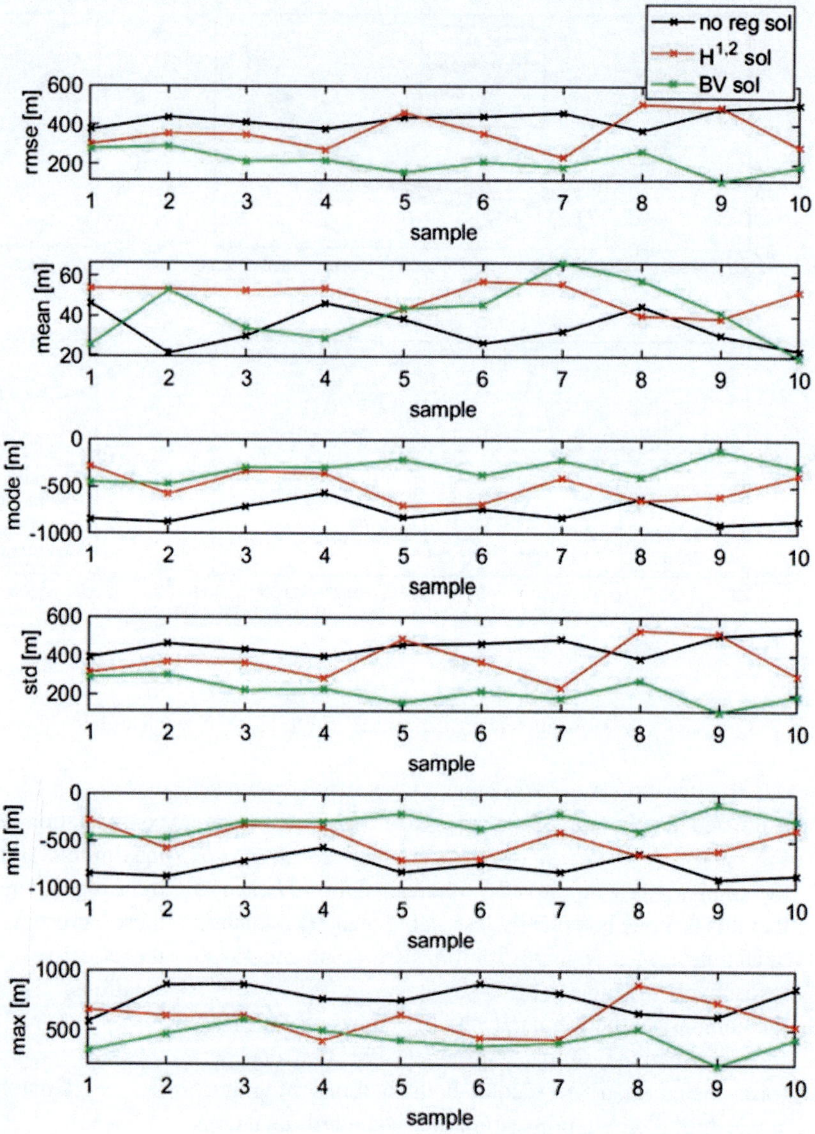

Fig. 8 Statistical analysis of the differences of the three solutions with respect to the true surface: $(H_k - \hat{H}_{0,k})$ the l.s. solution in black, $(H_k - \hat{H}_{1,k})$ the BV (or $H^{1,1}$) solution in green, $(H_k - \hat{H}_{2,k})$ the $H^{1,2}$ solution in red

References

Acar R, Vogel CR (1994) Analysis of bounded variation penalty methods for ill-posed problems. Inv Probl 10(6):1217

Ballani L, Stromeyer D (1982) The inverse gravimetric problem: a Hilbert space approach. In: Proceedings of the international symposium figure of the earth, the moon, and other planets, pp 359–373

Barzaghi R, Sansò F (1986) Bollettino di geodesia e scienze affini. Geol Lithos 45(2):203–216

Bertero M (1982) Problemi lineari non ben posti e metodi di regolarizzazione, Consiglio Nazionale delle Ricerche. Istituto di Analisi Globale e Applicazion

Falconer KJ (1986) The geometry of fractal sets, vol 85. Cambridge University Press, Cambridge

Freeden W, Nashed MZ (2018) Ill-posed problems: operator methodologies of resolution and regularization. Handbook of mathematical geodesy, pp 201–314. Springer, Basel

Giusti E, Williams GH (1984) Minimal surfaces and functions of bounded variation, vol 2(3). Springer, Basel

Isakov V (1990) Inverse source problems, vol 34. American Mathematical Society, Providence

Richter M (2016) Discretization of inverse problems. In: Inverse problems, pp 29–75. Springer, Basel

Sansò F (1986) Statistical methods in physical geodesy. In: Mathematical and numerical techniques in physical geodesy, pp 49–155. Springer, Vienna

Sansò F (2014) On the regular decomposition of the inverse gravimetric problem in non-L^2 and L^2 spaces. GEM-Int J Geomath 5(1):33–61

Sansò F, Capponi M, Sampietro D (2018) Up and down through the gravity field. In: Freeden W, Rummel R (eds) Handbuch der Geodäsie. Springer Reference Naturwissenschaften. Springer Spektrum, Berlin. https://doi.org/10.1007/978-3-662-46900-2_93-1

Tikhonov AN, Arsenin VY (1977) Solutions of ill-posed problems, vol 330. WH Winston, Washington

Tikhonov AN, Goncharsky A, Stepanov VV, Yagola AG (2013) Numerical methods for the solution of ill-posed problems, vol 328. Springer, Berlin

Converted Total Least Squares Method and Gauss-Helmert Model with Applications to Coordinate Transformations

Jianqing Cai, Dalu Dong, Nico Sneeuw, and Yibin Yao

Abstract

In this paper, the three kind of solutions of TLS problem, the common solution by singular value decomposition (SVD), the iteration solution and Partial-EIV model are firstly reviewed with respect to their advantages and disadvantages. Then a newly developed Converted Total Least Squares (CTLS) dealing with the errors-in-variables (EIV) model is introduced. The basic idea of CTLS has been proposed by the authors in 2010, which is to take the stochastic design matrix elements as virtual observations, and to transform the TLS problem into a traditional Least Squares problem. This new method has the advantages that it cannot only easily consider the weight of observations and the weight of stochastic design matrix, but also deal with TLS problem without complicated iteration processing, if the suitable approximates of parameters are available, which enriches the TLS algorithm and solves the bottleneck restricting the application of TLS solutions. CTLS method, together with all the three TLS models reviewed here has been successfully integrated in our coordinate transformation programs and verified with the real case study of 6-parameters Affine coordinate transformation. Furthermore, the comparison and connection of this notable CLTS method and estimation of Gauss-Helmert model are also discussed in detail with applications of coordinate transformations.

Keywords

Converted TLS · Errors-In-Variables (EIV) · Gauss-Helmert model · Total Least Squares (TLS) · Virtual observation

1 Introduction

Total Least Squares (TLS) is a method of fitting that is appropriate when there are errors in both the observation vector and in the design matrix in computational mathematics and engineering, which is also referred as Errors-In-Variables (EIV) modelling or orthogonal regression in the statistical community. The TLS/EIV principle studied by Adcock (1878) and Pearson (1901) already more than one century ago. Kendall and Stuart (1969) described this problem as structural relationship model models. In geodetic application this method was discussed by Koch (2002) and studied recently by Schaffrin (2005). How to obtain the best parameter estimation values and give the statistical information of parameters in the EIV model is not 'perfectly' solved. Nevertheless, the EIV model is still becoming increasingly widespread in remote sensing (Felus and Schaffrin 2005) and geodetic datum transformation (Schaffrin and Felus 2005, 2008; Schaffrin and Wieser 2008; Akyilmaz 2007; Cai and Grafarend 2009; Shen et al. 2011;Amiri-Simkooeil and Jazaeri 2012).

In 1980, the mathematical structure of TLS was completed by Golub and Van Loan (1980), who gave the first numerically stable algorithm based on matrix singular value decomposition. With the rapid development of the

J. Cai (✉) · D. Dong · N. Sneeuw
Institute of Geodesy, University of Stuttgart, Stuttgart, Germany
e-mail: cai@gis.uni-stuttgart.de

Y. Yao
School of Geodesy and Geomatics, Wuhan University, Wuhan, China

P. Novák et al. (eds.), *IX Hotine-Marussi Symposium on Mathematical Geodesy*,
International Association of Geodesy Symposia 151, https://doi.org/10.1007/1345_2020_116

numerical method over the last decade, various approach for TLS emerged. These include singular value decomposition (SVD), the completely orthogonal approach, the Cholesky decomposition approach, the iterative approach, and so on (Van Huffel and Zha 1993; Van Huffel and Vandewalle 1991; Van Huffel and Lemmerling 2002; Schaffrin 2003),the most representative of which are the SVD and iterative solutions. However, there are some problems in the both methods. In the SVD method, some elements of design matrix may be non-stochastic, or some elements containing errors could appear more than once. To perform the minimum norm constraint without this consideration is inappropriate and may result in large deviations. By the Iteration method, since the iteration is the gradual approximation of the true value of parameter, iteration solutions can be a problem if there is a high degree of nonlinearity. In addition, this method has also the problem by the repetition of parameters in design matrix.

Recent years, a further method reformed from EIV model called Partial-EIV has a relative good solution to this kind of problems (Xu et al. 2012; Wang et al. 2016). However, almost all the Partial-EIV models focus on the calculation with iterations, which makes the mathematical algorithm very complicate. According to the research results by Yao et al. (2010), one method called Converted Total Least Squares (CTLS) was developed since 2010. This method can perform the processes with just one step iteration (i.e., proper approximates estimated by LS) and at the same time solve the problem by the repetition of elements and the non-stochastic elements containing errors in design matrix. In the bachelor thesis by Dong (2017), the CTLS method was also systematically introduced and applied to the coordinate transformation in Baden-Württemberg together with other three estimators.

In this paper, based on a short review of TLS and EIV model, CTLS method will be described in detail in Sect. 2. In the following Sect. 3, CTLS, together with all the three TLS models reviewed here has been successfully integrated in 6-parameters Affine coordinate transformation and verified with the real case study of 131 BWREF points in Baden-Württemberg, Germany. As a comparison in Sect. 4, the transformation parameters estimated by LS, TLS (SVD), Partial-EIV model and CTLS will be represented and discussed. Furthermore, in Sect. 5 the comparison and connection of this notable CLTS method and estimation of Gauss-Helmert model are also discussed in detail with applications of coordinate transformations. The conclusions and further studies are presented in the last section.

2 Converted Total Least Squares

2.1 The Derivation of Converted Total Least Squares

The Total Least Squares Estimator (TLS) or Errors-In-Variables (EIV) modelling

$$\begin{aligned} &(\mathbf{y}-\mathbf{e}) = (\mathbf{A}-\mathbf{E_A})\ \boldsymbol{\xi} \\ &E\{[(\mathrm{vec}\mathbf{E_A}),\mathbf{e}]\} = 0, C\{\mathrm{vec}\mathbf{E_A},\mathbf{e}\} = 0, \\ &D\{\mathbf{e}\} = \boldsymbol{\Sigma}_0 \otimes \mathbf{Q_y}, D\{\mathrm{vec}\mathbf{E_A}\} = \boldsymbol{\Sigma}_0 \otimes \mathbf{Q_a} \end{aligned} \tag{1}$$

The Total Least Squares Euler-Lagrange Approach for $\mathbf{Q_y} = \mathbf{Q_a} = \mathbf{I}_{2n}$ is according to the criteria

$$\mathbf{e}^T\mathbf{e} + (\mathrm{vec}\mathbf{E_A})^T\ (\mathrm{vec}\mathbf{E_A}) = \min\left(\mathbf{e},\mathbf{E_A},\boldsymbol{\xi}\right). \tag{2}$$

The Converted Total Least Squares (CTLS) is proposed to deal with the errors-in-variables (EIV) model. Firstly, we take classic Gauss-Markov model as basis observation equation.

$$\mathbf{y} = \mathbf{A}\boldsymbol{\xi} + \mathbf{e_y}. \tag{3}$$

Taking into account the design matrix's stochastic errors in model (1) will lead to difficulties for the parameter estimation and accuracy assessment. Particularly, one cannot apply the traditional error propagation law directly, since the law is established based on linear relations. The basic idea of CTLS is to take the stochastic design matrix elements as virtual observations. Based on the original observation Eq. (1), the number of observation equation is augmented by taking the design matrix elements as new observation vector, and some of the design matrix elements are estimated as parameters in the new approach. The advantage of such strategy is the ability to estimate these required parameters, where the design matrix is constructed by the initial value of design matrix parameters, which has no longer random properties. The estimated parameters are the linear functions of the observation vector. After this treatment, Eq. (3) can be solved with the classical LS adjustment theory.

Augmenting the observation equations that take functionally independent random elements of design matrix elements as virtual observation based on the original error equation.

$$\mathbf{y_a} = \boldsymbol{\xi}_\mathbf{a} + \mathbf{e_a}, \tag{4}$$

where $\mathbf{y_a}$ is comprised of the design matrix elements that contain errors, and $\boldsymbol{\xi}_\mathbf{a}$ is comprised of the new parameters.

With the combination of observation Eqs. (1) and (2), a new mathematical observation model can be obtained:

$$\begin{cases} \mathbf{y} = \mathbf{A}\boldsymbol{\xi} + \mathbf{e_y} \\ \mathbf{y_a} = \ \boldsymbol{\xi}_\mathbf{a} + \mathbf{e_a} \end{cases}. \tag{5}$$

Note that $\mathbf{y_a}$ contains only the observations of design matrix. To distinguish the design matrix in the original model the symbol $\mathbf{A}_\xi$ is used to denote the design matrix in (1), which consists of the initial value of parameters $\boldsymbol{\xi}_\mathbf{a}$ and other elements of design matrix without errors.

Based on the above model and according to our previous research results (Yao et al. 2010) we can get the following derivations:

$$\begin{aligned} -\mathbf{e_y} &= \left(\mathbf{A}_\xi^0 + \mathbf{E_A}\right)\left(\boldsymbol{\xi}^0 + \Delta\boldsymbol{\xi}\right) - \mathbf{y} \\ &= \mathbf{A}_\xi^0 \Delta\boldsymbol{\xi} + \mathbf{E_A}\boldsymbol{\xi}^0 + \mathbf{A}_\xi^0\boldsymbol{\xi}^0 - \mathbf{y} + \mathbf{E_A}\Delta\boldsymbol{\xi} \quad \to \mathbf{E_A}\Delta\boldsymbol{\xi} \approx 0 \\ &= \mathbf{A}_\xi^0 \Delta\boldsymbol{\xi} + \mathbf{B}\Delta\mathbf{a} + \mathbf{A}_\xi^0\boldsymbol{\xi}^0 - \mathbf{y} \\ -\mathbf{e_a} &= \boldsymbol{\xi}_\mathbf{a} - \mathbf{y_a} \\ &= \left(\mathbf{a}^0 + \Delta\mathbf{a}\right) - \mathbf{y_a} \\ &= \Delta\mathbf{a} + \left(\mathbf{a}^0 - \mathbf{y_a}\right) \end{aligned} \tag{6}$$

Where $\mathbf{E_A}$ is composed of $\Delta\mathbf{a}$, the corrections to the new parameters, and $\mathbf{B}\Delta\mathbf{a}$ is the rewritten form of $\mathbf{E_A}\boldsymbol{\xi}^0$. The conversion of $\mathbf{E_A}\boldsymbol{\xi}^0$ to $\mathbf{B}\Delta\mathbf{a}$ is the key step for this approach, where the vectorization of matrix product equation

$$\underset{n\cdot n\cdot 1}{\mathbf{B}\,\mathbf{x}} = \underset{n\otimes np}{\left(\mathbf{x}^T \otimes \mathbf{I}\right)}\underset{np\cdot 1}{\mathrm{vec}\,(\mathbf{B})}$$

is applied. is composed of non-stochastic elements in the design matrix and the initial value $\mathbf{a}^0$. In addition, the first equation of (6) ignores second-order item $\mathbf{E_A}\Delta\boldsymbol{\xi}$ under the condition that the approximate values $\mathbf{A}_\xi^0$ and $\boldsymbol{\xi}^0$ are sufficiently close to $\mathbf{A}_\xi$ and $\boldsymbol{\xi}$, respectively. This leads to avoid the further iteration. Actually this condition can be satisfied in the case of estimation of the similar or affine coordinate transformation parameters, where the suitable approximates are provider by normal LS, more detail refer to Sect. 3.

Define

$$\begin{aligned} \mathbf{z} &= \begin{bmatrix} \mathbf{y} - \mathbf{A}_\xi^0\boldsymbol{\xi}^0 \\ \mathbf{y_a} - \mathbf{a} \end{bmatrix}, \mathbf{A}_\eta = \begin{bmatrix} \mathbf{A}_\xi^0 & \mathbf{B} \\ \mathbf{0} & \mathbf{I} \end{bmatrix}, \\ \Delta\boldsymbol{\eta} &= \begin{bmatrix} \Delta\boldsymbol{\xi} \\ \Delta\mathbf{a} \end{bmatrix}, \mathbf{e_z} = \begin{bmatrix} \mathbf{e_y} \\ \mathbf{e_a} \end{bmatrix} \end{aligned} \tag{7}$$

The new observations (6) can be represented as

$$\mathbf{z} = \mathbf{A}_\eta \Delta\boldsymbol{\eta} + \mathbf{e_z} \tag{8}$$

with the new weight matrix or the variance-covariance matrix

$$\mathbf{P_z} = \begin{bmatrix} \mathbf{P_y} & \mathbf{0} \\ \mathbf{0} & \mathbf{P_a} \end{bmatrix}, \text{ or } \boldsymbol{\Sigma}_\mathbf{z} = \sigma_{\mathbf{z}0}^2 \begin{bmatrix} \mathbf{P_y}^{-1} & \mathbf{0} \\ \mathbf{0} & \mathbf{P_a}^{-1} \end{bmatrix} \tag{9}$$

where $\mathbf{e_z}$is the residual vector of all observations, $\mathbf{A}_\eta$is the new design matrix, formed by the initial values of the parameters, and $\Delta\boldsymbol{\eta}$ is comprised of the corrections to all parameters.

The estimation criterion is still to get the minimum of the residual sums of the squares:

$$\mathbf{e_z}^T\mathbf{P_z}\mathbf{e_z} = \mathbf{e_y}^T\mathbf{P_y}\mathbf{e_y} + \mathbf{e_a}^T\mathbf{P_a}\mathbf{e_a} \to \min \tag{10}$$

With these derivations and condition, the TLS problem is successfully converted into the classical LS problem. The estimation can be completed by following the classical LS principle (10) and with considering the weights or variance-covariances of observations and virtual observations by:

$$\Delta\hat{\boldsymbol{\eta}} = \left(\mathbf{A}_\eta^T\mathbf{P_z}\mathbf{A}_\eta\right)^{-1}\mathbf{A}_\eta^T\mathbf{P_z}\mathbf{z} \tag{11}$$

subject to the related dispersion matrix

$$D\left\{\Delta\hat{\boldsymbol{\eta}}\right\} := \boldsymbol{\Sigma}_{\Delta\hat{\eta}} = \sigma_{\mathbf{z}0}^2\left(\mathbf{A}_\eta^T\mathbf{P_z}\mathbf{A}_\eta\right)^{-1}. \tag{12}$$

2.2 Estimation Formula of Unit Weight Variance

Normally, the estimation formula of unit weight variance of the TLS is difficult to determine. In considering the design matrix errors, the question of whether or not the degree of freedom for adjustment model changes arises. As we have described in Sect. 2.1, the TLS problem is converted into a classical LS problem. With the estimator (11) the correction of observations and virtual observations can be also derived:

$$\hat{\mathbf{e}}_\mathbf{z} = \mathbf{A}_\eta\Delta\hat{\boldsymbol{\eta}} - \mathbf{z}. \tag{13}$$

And the statistics for accuracy assessment, such as unit weight variance for CTLS can be straightforward estimated:

$$\hat{\sigma}_{\mathbf{z}0}^2 = \frac{\hat{\mathbf{e}}_\mathbf{z}^T\mathbf{P_z}\hat{\mathbf{e}}_\mathbf{z}}{(n+u)-(m+u)} = \frac{\hat{\mathbf{e}}_\mathbf{z}^T\mathbf{P_z}\hat{\mathbf{e}}_\mathbf{z}}{n-m}. \tag{14}$$

where n is the number of original observations that the observation vector $\mathbf{y}$ contains, u is the number of functionally independent random elements in the design matrix $\mathbf{A}$. m is the number of unknown parameters, here only the number of original parameters in model (3). The estimation of the variance-covariance matrix can further achieved by

$$\hat{\boldsymbol{\Sigma}}_{\Delta\hat{\eta}} = \hat{\sigma}^2_{\mathbf{z}_0}\left(\mathbf{A}_{\eta}^T\mathbf{P}_{\mathbf{z}}\mathbf{A}_{\eta}\right)^{-1}. \tag{15}$$

This approach of CTLS provides not only one practical solution of TLS problems, but also the estimation of unit weight and variance-covariance matrix of original and virtual observation, which achieves one successful integration of the TLS theory with the classical LS approach.

3 Implementation of CTLS Method to 2-D 6-Parameter Affine Transformation

In the most applications of the plane transformation the 6-parameter affine models is applied and are also recommended by the Surveying Authorities of the States of the Federal Republic of Germany (AdV). Here the new CTLS method will be implemented to the 6-parameter affine transformation model in order to estimate the parameters of the plane transformation parameters based on one real case study with 131 collocated points in Baden-Württemberg.

With the planar affine transformation, where six parameters are to be determined, both coordinate directions are rotated with respect to two different angles α and β, so that not only the distances and the angles are distorted, but usually also the original orthogonality of the axes of coordinates is lost. An affine transformation preserves collinearity and ratios of distances. While an affine transformation preserves proportions on lines, it does not necessarily preserve angles or lengths.

The 6-parameter affine transformation model between any two plane coordinates systems, e.g. from Gauß-Krueger coordinate (H, R) in DHDN (G) directly to the UTM-Coordinate (N, E) in ETRS89 (Cai and Grafarend 2009) can be written as

$$\begin{bmatrix} N \\ E \end{bmatrix} = \begin{bmatrix} m_H \cos\alpha & -m_R \sin\beta \\ m_H \sin\alpha & m_R \cos\beta \end{bmatrix} \begin{bmatrix} H \\ R \end{bmatrix} + \begin{bmatrix} t_N \\ t_E \end{bmatrix} \tag{16}$$

where t_N and t_E are translation parameters; α and β are rotation parameters; m_H and m_H are scale corrections.

We apply here TLS solutions to the centralized 6-parameter affine transformation model where the translation parameters are vanished. Let rewrite the 6-parameter affine transformation in 6 parameters as

$$\begin{aligned} \begin{bmatrix} N \\ E \end{bmatrix} &= \begin{bmatrix} m_H \cos\alpha & -m_R \sin\beta \\ m_H \sin\alpha & m_R \cos\beta \end{bmatrix} \begin{bmatrix} H \\ R \end{bmatrix} + \begin{bmatrix} t_N \\ t_E \end{bmatrix} \\ &=: \begin{bmatrix} \xi_{11} & \xi_{21} \\ \xi_{12} & \xi_{22} \end{bmatrix} \begin{bmatrix} H \\ R \end{bmatrix} + \begin{bmatrix} \xi_{31} \\ \xi_{32} \end{bmatrix} \end{aligned} \tag{17}$$

Then the observations and old coordinates are centered around their average values in the form:

$$\begin{bmatrix} \underline{N} \\ \underline{E} \end{bmatrix} =: \begin{bmatrix} \xi_{11} & \xi_{21} \\ \xi_{12} & \xi_{22} \end{bmatrix} \begin{bmatrix} \underline{H} \\ \underline{R} \end{bmatrix}, \tag{18}$$

with

$$\underline{N} = N - mean(N),\ \underline{E} = E - mean(E)$$
$$\underline{H} = H - mean(H),\ \underline{R} = R - mean(R)$$

For the n couple of coordinates we have an empirically suited representation of affine transformation model constructed in the form of observation model (3)

$$E\left\{\begin{bmatrix} \underline{N}_1 \\ \cdots \\ \underline{N}_n \\ \underline{E}_1 \\ \cdots \\ \underline{E}_n \end{bmatrix}\right\} = E\left\{\begin{bmatrix} \underline{H}_1 & \underline{R}_1 & 0 & 0 \\ \cdots & \cdots & \cdots & \cdots \\ \underline{H}_n & \underline{R}_n & 0 & 0 \\ 0 & 0 & \underline{H}_1 & \underline{R}_1 \\ \cdots & \cdots & \cdots & \cdots \\ 0 & 0 & \underline{H}_n & \underline{R}_n \end{bmatrix}\right\} \begin{bmatrix} \xi_{11} \\ \xi_{21} \\ \xi_{12} \\ \xi_{22} \end{bmatrix}. \tag{19}$$

According to the CTLS model (6) we have the correspond matrixes for 6-parameter affine transformation model

$$\underset{(2n\times 2n)}{\mathbf{E}_{\mathbf{A}}} = \begin{bmatrix} \Delta\underline{H}_1 & \Delta\underline{R}_1 & 0 & 0 \\ \vdots & \vdots & \vdots & \vdots \\ \Delta\underline{H}_n & \Delta\underline{R}_n & 0 & 0 \\ 0 & 0 & \Delta\underline{H}_1 & \Delta\underline{R}_1 \\ \vdots & \vdots & \vdots & \vdots \\ 0 & 0 & \Delta\underline{H}_n & \Delta\underline{R}_n \end{bmatrix},$$

$$\underset{(4\times 1)}{\boldsymbol{\xi}^0} = \begin{bmatrix} \xi_{11}^0 \\ \xi_{21}^0 \\ \xi_{12}^0 \\ \xi_{22}^0 \end{bmatrix},\quad \underset{(2n\times 1)}{\Delta\mathbf{a}} = \begin{bmatrix} \Delta\underline{H}_1 \\ \vdots \\ \Delta\underline{H}_n \\ \Delta\underline{R}_1 \\ \vdots \\ \Delta\underline{R}_n \end{bmatrix} \tag{20}$$

The key step here is to convert $\mathbf{E}_{\mathbf{A}}\boldsymbol{\xi}^{0}$ to $\mathbf{B}\Delta\mathbf{a}$

$$\mathbf{E}_{\mathbf{A}}\boldsymbol{\xi}^{0}=\mathbf{B}\Delta\mathbf{a}=\left(\begin{bmatrix}\xi_{11}^{0} & \xi_{21}^{0}\\ \xi_{12}^{0} & \xi_{22}^{0}\end{bmatrix}\otimes\mathbf{I}_{n}\right)\Delta\mathbf{a} \tag{21}$$

where

$$\underset{2n\times 2n}{\mathbf{B}}=\begin{bmatrix}\xi_{11}^{0} & \xi_{21}^{0}\\ \xi_{12}^{0} & \xi_{22}^{0}\end{bmatrix}\otimes\mathbf{I}_{n}=$$

$$=\begin{bmatrix}\xi_{11}^{0} & 0 & 0 & \xi_{21}^{0} & 0 & 0\\ 0 & \ddots & 0 & 0 & \ddots & 0\\ 0 & 0 & \xi_{11}^{0} & 0 & 0 & \xi_{21}^{0}\\ \xi_{12}^{0} & 0 & 0 & \xi_{22}^{0} & 0 & 0\\ 0 & \ddots & 0 & 0 & \ddots & 0\\ 0 & 0 & \xi_{12}^{0} & 0 & 0 & \xi_{22}^{0}\end{bmatrix} \tag{22}$$

The estimate of CTLS is (11)

$$\Delta\hat{\boldsymbol{\eta}}=\left(\mathbf{A}_{\eta}^{T}\mathbf{P}_{\mathbf{z}}\mathbf{A}_{\eta}\right)^{-1}\mathbf{A}_{\eta}^{T}\mathbf{P}_{\mathbf{z}}\mathbf{z}. \tag{23}$$

The solution $\Delta\hat{\boldsymbol{\eta}}$ is a $(2n+4)\times 1$ vector. The first four elements of $\Delta\hat{\boldsymbol{\eta}}$ are the corrections of $\boldsymbol{\xi}$ and the following $2n$ elements are the corrections of $\mathbf{a}$, which are the independent corrections for the initial design matrix $\mathbf{A}$. The final estimate of transformation parameters are $\hat{\boldsymbol{\xi}}=\boldsymbol{\xi}^{0}+\Delta\hat{\boldsymbol{\xi}}$ and $\hat{\boldsymbol{\xi}}_{\mathbf{a}}=\hat{\mathbf{a}}=\mathbf{a}^{0}+\Delta\hat{\mathbf{a}}$, where $\boldsymbol{\xi}^{0}$ are calculated from the classic LS solution of (3).

4 Comparison and Analysis of the Results with CTLS and Other TLS Methods

Through the TLS solution where the errors in the design matrix $\mathbf{A}$ are considered the remaining transformation coordinate residuals of collocated DHDN points in B-W are reduced from 12 cm to 4 cm (Fig. 1), which are illustrated in Fig. 2 in comparison with Fig. 1 in detail. The statistics of these residual in comparison with LS methods are listed in Table 1. The following statistical terms shows us the difference between the quadratics sums of the residuals $\hat{\mathbf{e}}_{LS}^{T}\hat{\mathbf{e}}_{LS}$ related LS and $\hat{\mathbf{e}}_{TLS}^{T}\hat{\mathbf{e}}_{TLS}$ related TLS, together with

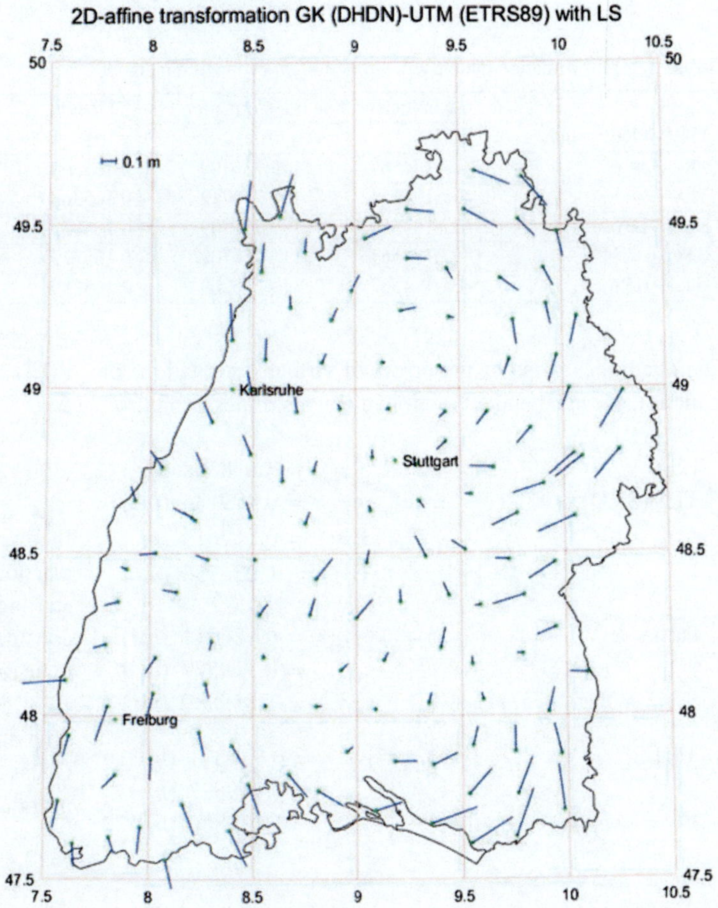

Fig. 1 Horizontal residuals after 6-Parameter Affine transformation in Baden-Württemberg Network with LS method

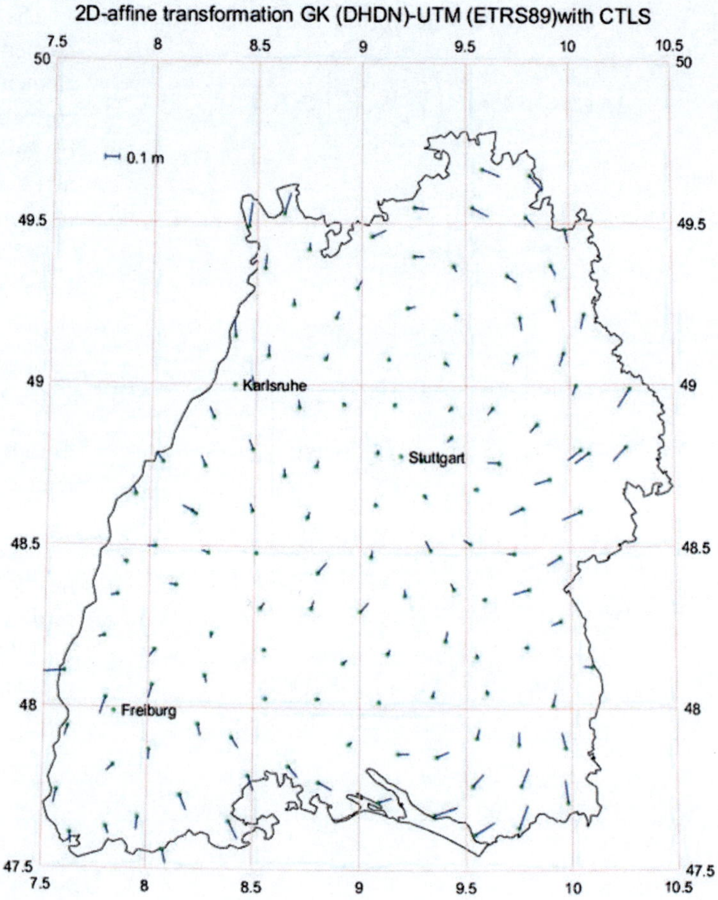

Fig. 2 Horizontal residuals after 6-Parameter Affine transformation in Baden-Württemberg Network with CTLS method

Table 1 Transformation parameters with different estimation methods

	6-parameter affine transformation GK (DHDN) – UTM (ETRS89)					
131 BWREF points	t_N (m)	t_E (m)	α (″)	β (″)	dm_1 ($\times 10^{-4}$)	dm_2 ($\times 10^{-4}$)
LS	437.194567	119.756709	0.165368	−0.196455	−3.996797	−3.988430
TLS	437.194554	119.756712	0.165368	−0.196455	−3.996797	−3.988430
Partial-EIV	437.194556	199.756709	0.165375	−0.196445	−3.996797	−3.988430
CTLS	437.194567	119.756709	0.165375	−0.196445	−3.996797	−3.988430
Gauss-Helmert	437.194567	119.756709	0.165289	−0.196397	−3.996797	−3.988430

the quadratics sums of the errors of virtual observations of functionally independent random elements of design matrix:

$$\text{LS}: \quad \hat{\mathbf{e}}_{LS}^T\hat{\mathbf{e}}_{LS} = 3.678308\ \left(m^2\right)$$

$$\begin{aligned} \text{TLS (SVD)}: \quad \hat{\mathbf{e}}_{TLS}^T\hat{\mathbf{e}}_{TLS} &= 0.409136\ \left(m^2\right) \\ \hat{\mathbf{E}}_{TLS}^T\hat{\mathbf{E}}_{TLS} &= 0.817619\ \left(m^2\right) \\ \hat{\mathbf{e}}_{TLS}^T\hat{\mathbf{e}}_{TLS} + \hat{\mathbf{E}}_{TLS}^T\hat{\mathbf{E}}_{TLS} &= 1.226756\ \left(m^2\right) \end{aligned}$$

$$\begin{aligned} \text{Partial-EIV}: \quad \hat{\mathbf{e}}_{TLSP}^T\hat{\mathbf{e}}_{TLSP} &= 0.920311\ \left(m^2\right) \\ \hat{\mathbf{e}}_{\mathbf{a}_{TLSP}}^T\hat{\mathbf{e}}_{\mathbf{a}_{TLSP}} &= 0.919577\ \left(m^2\right) \\ \hat{\mathbf{e}}_{TLSP}^T\hat{\mathbf{e}}_{TLSP} + \hat{\mathbf{e}}_{\mathbf{a}_{TLSP}}^T\hat{\mathbf{e}}_{\mathbf{a}_{TLSP}} &= 1.839889\ \left(m^2\right) \end{aligned}$$

$$\begin{aligned} \text{CTLS}: \quad \hat{\mathbf{e}}_{CTLS}^T\hat{\mathbf{e}}_{CTLS} &= 0.920311\ \left(m^2\right) \\ \hat{\mathbf{e}}_{\mathbf{a}_{CTLS}}^T\hat{\mathbf{e}}_{\mathbf{a}_{CTLS}} &= 0.919577\ \left(m^2\right) \\ \hat{\mathbf{e}}_{CTLS}^T\hat{\mathbf{e}}_{CTLS} + \hat{\mathbf{e}}_{\mathbf{a}_{CTLS}}^T\hat{\mathbf{e}}_{\mathbf{a}_{CTLS}} &= 1.839889\ \left(m^2\right) \end{aligned}$$

5 Connection of CTLS Estimator and the Estimator of Gauss-Helmert Model

In one independent study about the parameter estimation of coordinate transformations, where the coordinates of the starting system and the coordinates of the final system are considered as random variables with identical covariance matrices, Koch (2002) has proven that the estimated parameters are identical, if the Gauss-Helmert model (G-H-M), after introducing additional unknown parameters with the Gauss-Markov model (G-M-M). Since the approach with introducing additional unknown parameters is just the same

Table 2 Statistical comparison the results of the coordinate transformation with different estimation methods

		Absolute mean residuals		Max. of absolute residuals			
Transformation models	Collocated sites	|VN| (m)	|VE| (m)	|VN| (m)	|VE| (m)	RMS (m)	Standard deviation of unit weight (m)
LS	B-W 131	0.1048	0.0804	0.3288	0.3226	0.1187	0.119868
TLS	B-W 131	0.0350	0.0268	0.1097	0.1076	0.0396	0.0400
Partial-EIV	B-W 131	0.0525	0.0402	0.1645	0.1614	0.0594	0.0848
CTLS	B-W 131	0.0525	0.0402	0.1645	0.1614	0.0594	0.0848
Gauss-Helmert	B-W 131	0.0525	0.0402	0.1644	0.1613	0.0594	0.0848

as the ideas of CTLS, which brings the connection of G-H-M with TLS model.

Koch (2002) established Taylor linearized Gauss-Helmert Model (Wolf 1978) with respect to the transformation parameters and coordinates for the case of coordinate transformation (17) directly

$$\mathbf{x}_{zi} + \mathbf{e}_{zi} = \mathbf{t} + \mathbf{RM}\left(\mathbf{x}_{si} + \mathbf{e}_{si}\right), \quad \text{for } i \in \{1, \ldots, p\}, \tag{24}$$

With new observation vector—$\mathbf{y}_{zi} = \mathbf{t}_0 + \mathbf{R}_0\mathbf{M}_0\mathbf{x}_{si} - \mathbf{x}_{zi}$, new notations of $\boldsymbol{\beta} = [\Delta t, \Delta\alpha, \Delta m]'$ and the related derivative term $\mathbf{A}$ of Taylor linearization (Pope 1972) we have the Gauss-Helmert model

$$\begin{gathered}\mathbf{y}_{zi} = \mathbf{A}\boldsymbol{\beta} + \mathbf{R}_0\mathbf{M}_0\mathbf{e}_{si} - \mathbf{e}_{zi} = \mathbf{A}\boldsymbol{\beta} + \mathbf{Ze} \\ \text{with } E\{\mathbf{e}\} = \mathbf{0} \text{ and } D\{\mathbf{e}\} = \sigma^2\boldsymbol{\Sigma}_{\mathbf{ee}}\end{gathered} \tag{25}$$

He converted Gauss-Helmert model into Gauss-Markov model for the same case of transformation with additional new parameters/observations, in which the error vector coordinates of starting system $\mathbf{e}_{si}$ is substituted by vector $\Delta\mathbf{x}_{si}$

$$\begin{aligned}\mathbf{y}_{zi} + \mathbf{e}_{zi} &= \mathbf{A}\boldsymbol{\beta} + \mathbf{R}_0\mathbf{M}_0\Delta\mathbf{x}_{si} \\ \mathbf{y}_{si} + \mathbf{e}_{si} &= \mathbf{I}\Delta\mathbf{x}_{si}\end{aligned} \tag{26}$$

With new notation of $\mathbf{B} = \mathbf{R}_0\mathbf{M}_0$ and $\boldsymbol{\gamma} = \Delta\mathbf{x}_{si}$ we can reform (26) as

$$\begin{cases}\mathbf{y} = \mathbf{A}\boldsymbol{\beta} + \mathbf{B}\boldsymbol{\gamma} + \mathbf{e}_{\mathbf{y}} \\ \mathbf{y}_{\boldsymbol{\gamma}} = \qquad \boldsymbol{\gamma} + \mathbf{e}_{\boldsymbol{\gamma}}\end{cases} \tag{27}$$

He has also proved that both the G-H-M (25) and converted G-M-M (27) produce identical estimated parameters.

Note that this new converted G-M-M (27) has the same structure with (6) or (8). After the convert of the random elements in design matrix into new random parameters together with the argumentation of virtual observations to Gauss-Markov model we arrives the same models:

$$\begin{cases}\mathbf{y} - \mathbf{A}_{\xi}^{0}\boldsymbol{\xi}^{0} = \mathbf{A}_{\xi}^{0}\Delta\boldsymbol{\xi} + \mathbf{B}\Delta\mathbf{a} + \mathbf{e}_{\mathbf{y}} \\ \mathbf{y}_{\mathbf{a}} - \mathbf{a}^{0} = \qquad\qquad + \Delta\mathbf{a} + \mathbf{e}_{\mathbf{a}}\end{cases} \tag{28}$$

The numerical transformation results with converted G-M-M are listed in last line of Tables 1 and 2, and also shown in Fig. 3.

In addition, the following statistical terms shows that there are no difference between the quadratics sums of the residuals $\hat{\mathbf{e}}_{CTLS}^{T}\hat{\mathbf{e}}_{CTLS}$ related CTLS and $\hat{\mathbf{e}}_{GHM}^{T}\hat{\mathbf{e}}_{GHM}$ related G-H-M, together with the quadratics sums of the errors of virtual observations of functionally independent random elements of design matrix:

$$\text{CTLS}: \quad \begin{aligned}\hat{\mathbf{e}}_{CTLS}^{T}\hat{\mathbf{e}}_{CTLS} &= 0.920311\ \left(m^2\right) \\ \hat{\mathbf{e}}_{\mathbf{a}_{CTLS}}^{T}\hat{\mathbf{e}}_{\mathbf{a}_{CTLS}} &= 0.919577\ \left(m^2\right) \\ \hat{\mathbf{e}}_{CTLS}^{T}\hat{\mathbf{e}}_{CTLS} + \hat{\mathbf{e}}_{\mathbf{a}_{CTLS}}^{T}\hat{\mathbf{e}}_{\mathbf{a}_{CTLS}} &= 1.839889\ \left(m^2\right)\end{aligned}$$

$$\text{G-H-M} \quad \begin{aligned}\hat{\mathbf{e}}_{z}^{T}\hat{\mathbf{e}}_{z} &= 0.920311\ \left(m^2\right) \\ \hat{\mathbf{e}}_{s}^{T}\hat{\mathbf{e}}_{s} &= 0.919577\ \left(m^2\right) \\ \hat{\mathbf{e}}_{z}^{T}\hat{\mathbf{e}}_{z} + \hat{\mathbf{e}}_{s}^{T}\hat{\mathbf{e}}_{s} &= 1.839888\ \left(m^2\right)\end{aligned}$$

It can be concluded that this notable development of the CLTS has also revealed that the connection/relationship or identity of estimator of the CTLS and the converted G-M-M estimator of Gauss-Helmert model in dealing with EIV models, especially in the case of similar or affine coordinate transformations.

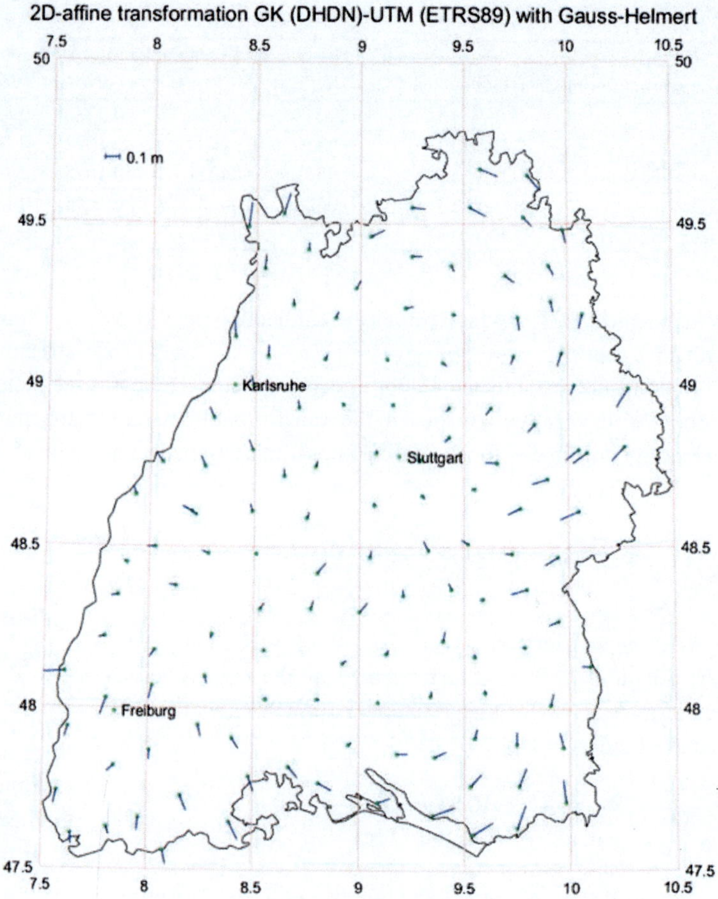

Fig. 3 Horizontal residuals after 6-Parameter Affine transformation in Baden-Württemberg Network with Gauss-Helmert Model

6 Conclusions and Further Studies

The traditional techniques used for solving the linear estimation problems are based on classical LS. However, only the errors of observation vector are considered, and the design matrix is assumed to be accurate without any errors. This makes LS not valid for most cases. Further study based on Errors-in-Variables (EIV), Total Least Squares method considers the errors in design matrix as well. The problem of which is, the repetition of parameters in design matrix has a deviation influence on the minimum norm constraint. Reform from EIV-model to Partial-EIV model and the Converted Total Least Squares could solve the Problem. Compared with Partial-EIV model, the solution of Converted Total Least Squares does not need the iteration. Based on these analyses and comparisons with different estimation methods the following points can be concluded:

- The traditional SVD method of TLS has a theoretical weakness in that it cannot be applied directly when only part of the design matrix contains errors.
- The Converted Total Least Squares (CTLS) can be used to deal with stochastic design matrix in TLS problem, where the TLS problem has been successfully converted into a LS problem.
- CTLS can be easily applied with considering the weight of observations and the weight of stochastic elements of design matrix. (Completely!)
- Although the estimated transformation parameters of Partial-EIV model and CTLS are almost identical, our CTLS has its advantage without complicated iteration processing. (Efficiently!)
- This study develops one converted approach for TLS problem, which provides statistical information of parameters and stochastic design matrix, enriches the TLS algorithm, and solves the bottleneck restricting the application of TLS.
- This notable development of the CLTS reveals that CTLS estimator is identical to Gauss-Helmert model estimator in dealing with EIV models, especially in the case of coordinate transformation.

Further studies should focus on the study of a general connection and even identical estimates of CTLS and Gauss-Helmert model. Further applications can be performed to 7- or 14-parameters similarity coordinate transformations among IERS ITRF realizations.

Acknowledgements The authors thank two anonymous reviewers for his many constructive comments, which helped to clarify a number of points in the revision.

References

Adcock RJ (1878) A problem in least squares. Analyst 5:53–54

Akyilmaz O (2007) Total least-squares solution of coordinate transformation. Surv Rev 39(303):68–80

Amiri-Simkooeil A, Jazaeri S (2012) Weighted total least squares formulated by standard least squares theory. J Geodetic Sci 2(2):113–124

Cai J, Grafarend E (2009) Systematical analysis of the transformation between Gauss-Krüger-Coordinate/DHDN and UTM-Coordinate/ETRS89 in Baden-Württemberg with different estimation methods. In: Hermann Drewes (ed) IAG Symposium, vol 134. Geodetic Reference Frame 2009

Dong D (2017) Study on the converted Total Least Squares method and its application in coordinate transformation. Bachelor Thesis, University of Stuttgart

Felus Y, Schaffrin B (2005) Performing similarity transformations using errors-in-variables-model. Proceedings of the ASPRS meeting, Baltimore, MD

Golub GH, Van Loan CF (1980) An analysis of the total least squares problem. SIAM J Numer Anal 17(6):883–893, December 1980

Kendall MG, Stuart A (1969) The advanced theory of statistics - Volume 2. Charles Griffin & Co Ltd, London

Koch R (2002) Räumliche Helmert-Transformation variabler Koordinaten im Gauß-Helmert und im Gauß-Markov-Modell. ZfV, Zeitschrift für Geodäsie, Geoinformation und Landmanagement 127:147–152

Pearson K (1901) On lines and planes of closest fit to points in space. Philos Mag 2:559–572

Pope A (1972) Some pitfalls to be avoided in the iterative adjustment of nonlinear problems. In: Proceeding of the 38th Annual Meeting American Society of Photogrammetry

Schaffrin B (2003) A note on constrained Total Least-Squares estimation. In: Computational statistics and data analysis, eingereicht Oct 2003

Schaffrin B (2005) Generalizing the Total Least-Squares estimator for empirical coordinate transformation, lecture in Geodetic Institute. University of Stuttgart, Stuttgart

Schaffrin B, Felus Y (2005) On total least-squares adjustment with constraints, In: Sansò F. (ed) A window on the future of geodesy. International association of geodesy symposia, vol 128. Springer, Berlin, Heidelberg

Schaffrin B, Felus Y (2008) Multivariate total least squares adjustment for empirical affine transformations. In: Xu P, Liu J, Dermanis A (eds) International association of geodesy symposia: VI Hotine-Marussi symposium on the theoretical and computational geodesy, 29 May-2 June 2006,Wuhan, China, vol 132. Springer, Berlin, pp 238–242

Schaffrin B, Wieser A (2008) On weighted Total Least Squares for linear regression. J Geod 82:415–421

Shen Y, Li B, Chen Y (2011) An iterative solution of weighted total least squares adjustment. J Geod 85:229–238

Van Huffel S, Lemmerling P (eds) (2002) Total least squares and errors-in-variables modeling. Kluwer Academic Publishers, Dordrecht

Van Huffel S, Vandewalle J (1991) The Total Least Squares problem computational aspects and analysis. Society for Industrial and Applied Mathematics, Philadelphia

Van Huffel S, Zha H (1993) The Total Least-Squares problem. In: Rao CR (ed) Handbook of statistics: computational statistics, vol 9. North Holland Publishing Company, Amsterdam, pp 377–408

Wang B, Li J, Liu C (2016) A robust weighted total least squares algorithm and its geodetic applications. Stud Geophys Geod 60(2016):180–194

Wolf H (1978) Das geodätische Gauß-Helmert-Modell und seine Eigenschaften. zfv 103:41–43

Xu P, Liu J, Shi C (2012) Total least squares adjustment in partial errors-in-variables models: algorithm and statistical analysis. J Geod 86:661–675

Yao Y, Kong J, Cai J, Sneeuw N (2010) New method for Total Least Squares: the combination LS solution considering the random errors of design matrix, internal research report. University of Stuttgart, Stuttgart

Check for updates

A Bayesian Nonlinear Regression Model Based on t-Distributed Errors

Alexander Dorndorf, Boris Kargoll, Jens-André Paffenholz, and Hamza Alkhatib

Abstract

In this contribution, a robust Bayesian approach to adjusting a nonlinear regression model with t-distributed errors is presented. In this approach the calculation of the posterior model parameters is feasible without linearisation of the functional model. Furthermore, the integration of prior model parameters in the form of any family of prior distributions is demonstrated. Since the posterior density is then generally non-conjugated, Monte Carlo methods are used to solve for the posterior numerically. The desired parameters are approximated by means of Markov chain Monte Carlo using Gibbs samplers and Metropolis-Hastings algorithms. The result of the presented approach is analysed by means of a closed-loop simulation and a real world application involving GNSS observations with synthetic outliers.

Keywords

Bayesian nonlinear regression model · Gibbs sampler · Markov Chain Monte Carlo · Metropolis-Hastings algorithm · Scaled t-distribution

1 Introduction

The estimation of model parameters is a fundamental task in geodetic applications. One possibility for accomplishing this task is provided by Bayesian inference, which is based on Bayes' theorem and utilizes probability density functions of observations and parameters. Bayesian inference also enables hypothesis testing and the determination of confidence regions. In comparison to classical non-Bayesian statistics, Bayesian inference is more intuitive and "methods become apparent which in traditional statistics give the impression of arbitrary computational rules" (according to Koch 2007, p. 1). The fields of application of Bayesian statistics are diverse and include disciplines such as biological, social and economic sciences (see Gelman et al. 2014, for the fundamental basics of Bayesian inference and some application examples from these disciplines). Bayesian statistics has also been used for different geodetic applications for decades (Bossler 1972; Koch 1988, 2007, 2018; Riesmeier 1984; Schaffrin 1987; Yang 1991; Zhu et al. 2005). However, in all of these studies simple linear functions are used, simple conjugate prior density functions are assumed or outlier-affected observations are not considered. A particular problem that arises with nonlinear models or non-conjugate priors in connection with popular classes of algorithms, such as Markov chain Monte Carlo (MCMC) methods (cf. Gamerman and Lopes 2006; Gelman et al. 2014), one generally cannot sample directly from the posterior density function. The class of MCMC methods includes the well known Metropolis-Hastings algorithm and Gibbs sampler. The latter has been originally developed for the Bayesian restoration of digital images and later used for a variety of problems of Bayesian inference. Such problems include nonlinear inverse problems (see, for instance, Haario et al. 2006; Johnathan et al. 2014). In a geodetic context Gibbs sampler methods have been used for the purpose of error propagation and inversion of large matrices (cf. Koch 2017; Alkhatib and Schuh 2006; Gundlich et al. 2003).

A. Dorndorf (✉) · B. Kargoll · J.-A. Paffenholz · H. Alkhatib
Geodetic Institute, Leibniz University Hannover, Hannover, Germany
e-mail: dorndorf@gih.uni-hannover.de; kargoll@gih.uni-hannover.de; paffenholz@gih.uni-hannover.de; alkhatib@gih.uni-hannover.de

P. Novák et al. (eds.), *IX Hotine-Marussi Symposium on Mathematical Geodesy*,
International Association of Geodesy Symposia 151, https://doi.org/10.1007/1345_2019_76

The likelihood function and prior distribution are generally assumed to be Gaussian in this context, so that these approaches are not robust against outliers. To obtain a robust approach, the family of normal distributions is replaced by a family of heavy-tailed distributions. Koch and Kargoll (2013) introduced the scaled t-distribution as a heavy-tailed error law in a geodetic application involving a linear model. An extension of this approach to nonlinear models was developed in Alkhatib et al. (2017). In both works, the solution was obtained by means of an expectation maximization algorithm, which is only based on the likelihood function and thereby does not allow for the integration of prior knowledge about the parameters. In a Bayesian context involving linear models with t-distributed errors, a variety of solution approaches utilizing MCMC methods, in particular a Gibbs sampler, exist (e.g., Geweke 1993; Gelman et al. 2014). In our current contribution, we extend the previous approaches to a Bayesian approach to solving a nonlinear model based on the t-distribution error law. The suggested approach allows for the integration of prior model parameters by assuming any family of prior distributions.

2 Bayesian Inference

2.1 Fundamentals of Bayesian Inference

In many geodetic applications one typically works with models which depend upon parameters to be estimated. We limit ourselves in this paper to the linear or nonlinear regression models. Let $\boldsymbol{l}$ be a vector of data and $\boldsymbol{\Theta}$ be a vector which contains the parameters for a model which seeks to explain $\boldsymbol{l}$. The relationship between observations and unknown parameters is described by means of a functional model $\boldsymbol{f}(\boldsymbol{\Theta})$:

$$\boldsymbol{l} + \boldsymbol{\epsilon} = \boldsymbol{f}(\boldsymbol{\Theta}). \tag{1}$$

The residuals ("errors") $\boldsymbol{\epsilon}$ arise in an overdetermined system and describe the precision of the observations. The usual assumption is that the residuals are normally distributed with zero mean and variance-covariance matrix $\boldsymbol{\Sigma}$. The vector-valued function $\boldsymbol{f}(\boldsymbol{\Theta})$ can be linear (e.g., distance estimation from repeated measurements) or nonlinear (e.g., estimation of geometric shape parameters from 2D or 3D points), but the focus is on the nonlinear case in this paper. The previous model defines a Gauss-Markov model (GMM), which is usually adjusted by means of the method of least squares (cf. Koch 1999).

As an extension of that approach, Bayesian inference uses probability distributions to determine the unknown parameters of a model and is based on Bayes' theorem:

$$p(\overline{\boldsymbol{\Theta}}|\boldsymbol{l}) \propto p(\underline{\boldsymbol{\Theta}}) \cdot p(\boldsymbol{l}|\boldsymbol{\Theta}). \tag{2}$$

Here, $p(\overline{\boldsymbol{\Theta}}|\boldsymbol{l})$ is the so-called posterior density, from which the unknown model parameters can be derived for given observations $\boldsymbol{l}$. The prior density $p(\underline{\boldsymbol{\Theta}})$ expresses all additional information about the unknown parameters and may be obtained from, e.g., results of a previous adjustment or a manufacturer's data sheet. $p(\boldsymbol{l}|\boldsymbol{\Theta})$ is the likelihood function, which represents the information of the observations conditional on the unknown parameters. In case of Bayesian inference the solution approach is based on marginal and conditional densities, where one distinguishes between conjugated and non-conjugated prior distributions. To evaluate the mean of the posterior density as a point estimate, one can solve the integral $E(\boldsymbol{\Theta}|\boldsymbol{l}) = \int \boldsymbol{\Theta} p(\overline{\boldsymbol{\Theta}}|\boldsymbol{l}) d\boldsymbol{\Theta}$. However, except for special cases involving linear functional models and normally distributed residuals, it is impossible to evaluate this integral analytically. To overcome this limitation, Monte Carlo (MC) techniques can be employed to approximate desired statistical measures such as expectations, variances, covariances, skewness, and kurtosis (cf. Koch 2017; Gelman et al. 2014).

2.2 A Robust Bayesian Model

The general assumption of normally distributed residuals in Eq. (1) does not account for outliers, which may therefore deteriorate the inference about the model parameters. To deal with outliers, hypothesis tests for outlier detection or a robust adjustment should be carried out. The focus of this contribution is on the latter. For this purpose, the normal distribution may be replaced by a longer-tailed family of distributions (cf. Gelman et al. 2014), for instance, by the family of scaled (Student's) t-distributions, frequently used in Bayesian and likelihood inference. Accordingly, the stochastic model for each residual is assumed to be

$$\epsilon_i \sim \mathrm{t}_\nu(0, s_t^2). \tag{3}$$

The degree of freedom ν controls the thickness of the tails. Outliers, being located in the tails, are more abundant for small ν, whereas the t-distribution approaches a normal distribution with increasing ν. Inference about ν is possible, so that estimators based on the t-distribution model have been called *adaptive* or *self-tuning* robust estimators (cf. Koch and Kargoll 2013; Parzen 1979). The residuals are assumed to have expectation 0, and the scale factor s_t is related to their variances through

$$\sigma_\epsilon^2 = \frac{\nu}{\nu - 2} s_t^2, \tag{4}$$

defined for $\nu > 2$. The t-distribution model Eq. (3) can be re-formulated conveniently and equivalently as the rescaled normal distribution (cf. Gelman et al. 2014)

$$\begin{aligned} \epsilon_i &\sim \mathrm{N}(0, \alpha^2 W_i), \\ W_i &\sim \text{Inv-}\chi^2(\nu, \tau^2), \\ \text{with } s_t^2 &= \alpha^2\tau^2. \end{aligned} \tag{5}$$

This equivalence is enabled by the introduction of additional, scaled inverse-chi-square distributed weights W_i. Whereas ν itself is also a degree of freedom of that Inv-χ^2-distribution, the scale factor s_t is factorized into the scale factor α with respect to the normal distribution and the parameter τ of the Inv-χ^2-distribution.

As this t-distribution results in a non-conjugated prior and as the functional model is nonlinear, MCMC methods are required to calculate the posterior density. The fundamentals and further discussion of the solution of non-conjugated prior densities by means of MCMC can be found, e.g., in Kroese et al. (2011) and Gelman et al. (2014). In this contribution, the Gibbs sampler is used for the generation of Markov chains, by sequentially drawing the unknown posterior parameters from their conditional densities. The purpose of the next section is to provide the required fundamentals of MCMC as well as a calculation procedure based on the Gibbs sampler.

3 A Bayesian Approach to Parameter Estimation

3.1 Specification of the Bayesian Model

Without loss of generality, the calculation procedure is developed for the typical task of adjusting n 3D points. More specifically, each observation

$$\boldsymbol{l}_i = [x_i, y_i, z_i] \quad (i = 1, \ldots, n), \tag{6}$$

is defined by an x-, a y- and a z-coordinate. To demonstrate the flexibility of the Bayesian model, outliers are assumed to occur only in the z-coordinate, so that a t-distribution is associated with the corresponding residuals, whereas the residuals of the other two coordinate components are assumed to be normally distributed at the outset. The residuals for all three coordinate components may have different levels of variance, and stochastic dependencies between them are currently neglected. Thus, the stochastic model reads

$$\begin{aligned} \epsilon_{x_i} &\sim \mathrm{N}(0, \sigma_x^2), \\ \epsilon_{y_i} &\sim \mathrm{N}(0, \sigma_y^2), \\ \epsilon_{z_i} &\sim \mathrm{t}_\nu(0, s_z^2). \end{aligned} \tag{7}$$

The functional model f generally involves a parameter vector $\boldsymbol{\Theta} = [\theta_1, \theta_2, \ldots, \theta_u]^{\mathrm{T}}$ consisting of u unknowns. As part of a Bayesian model, prior distributions are assumed for all unknown parameters, including the three scale factors and the degree of freedom ν of the t-distribution. An informative prior is assigned to the functional model parameters $\boldsymbol{\Theta}$ and non-informative priors to all other parameters. As non-informative priors may be defined by constant probability, only the prior of $\boldsymbol{\Theta}$ remains to be specified, which choice depends on the inferential procedure about the parameters $\underline{\boldsymbol{\Theta}}$. If the redundancy of the prior adjustment problem is large, it is frequently adequate to assume a multivariate normal distribution

$$p(\boldsymbol{\Theta}) = \frac{1}{\sqrt{(2\pi)^u \det \boldsymbol{\Sigma}_{\underline{\boldsymbol{\Theta}}}}} \exp\left(-\frac{(\boldsymbol{\Theta}-\underline{\boldsymbol{\Theta}})^{\mathrm{T}}\, \boldsymbol{\Sigma}_{\underline{\boldsymbol{\Theta}}}^{-1}\, (\boldsymbol{\Theta}-\underline{\boldsymbol{\Theta}})}{2}\right). \tag{8}$$

The prior knowledge of the model parameters $\underline{\boldsymbol{\Theta}}$ is the expected value of the multivariate normal distribution, and $\boldsymbol{\Sigma}_{\underline{\boldsymbol{\Theta}}}$ controls the spread of the distribution. $\boldsymbol{\Theta}$ is an arbitrary realization of the vector of functional model parameters. If the redundancy of the prior adjustment is small, a multivariate t-distribution would be an adequate choice for the prior. The likelihood function is determined by the observation equations $\boldsymbol{l}_i + \boldsymbol{\epsilon}_i = \boldsymbol{f}_i(\boldsymbol{\Theta})$ and the stochastic model (7) for the residuals. Expressing the t-distribution for the residuals of the z-component as the rescaled normal distribution (5), the likelihood function is defined by the (factorized) multivariate normal distribution

$$\begin{aligned} p\left(\boldsymbol{l}|\boldsymbol{\Theta}, \sigma_x, \sigma_y, \alpha, \boldsymbol{W}, \nu, \tau\right) &= \prod_{i=1}^{n} \frac{1}{\sqrt{(2\pi)^k \det \boldsymbol{\Sigma}_{ll_i}}} \\ &\times \exp\left(-\frac{(\boldsymbol{l}_i - \boldsymbol{f}_i(\boldsymbol{\Theta}))^T\, \boldsymbol{\Sigma}_{ll_i}^{-1}\, (\boldsymbol{l}_i - \boldsymbol{f}_i(\boldsymbol{\Theta}))}{2}\right), \\ \text{with } \boldsymbol{\Sigma}_{ll_i} &= \begin{bmatrix} \sigma_x^2 & 0 & 0 \\ 0 & \sigma_y^2 & 0 \\ 0 & 0 & \alpha^2 W_i \end{bmatrix} \text{ and } k = 3, \end{aligned} \tag{9}$$

where the variance-covariance matrix $\boldsymbol{\Sigma}_{ll_i}$ of one observed point fulfills the assumption of uncorrelated coordinate components. The following section demonstrates the calculation of the posterior density.

3.2 Calculation of Posterior Parameters

The posterior density $p\left(\overline{\boldsymbol{\Theta}}, \overline{\sigma_x}, \overline{\sigma_y}, \overline{\alpha}, \overline{\boldsymbol{W}}, \overline{\nu}, \overline{\tau}|\boldsymbol{l}\right)$ summarizes the information of the prior (8) and the likelihood function (9) via Bayes's theorem (2). As indicated in Sect. 2,

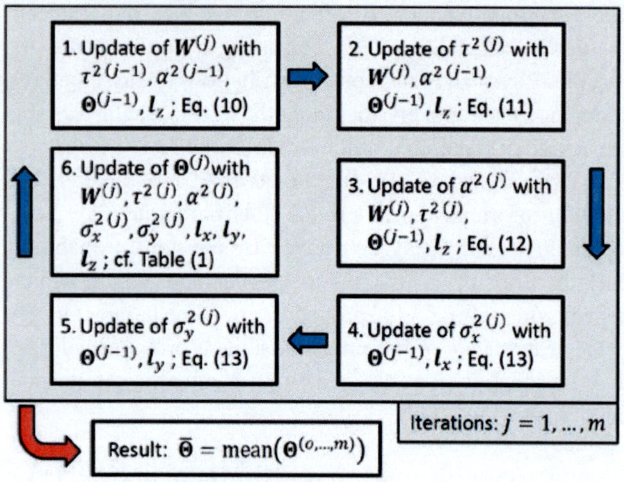

Fig. 1 Computational steps of the Gibbs sampler for adjusting 3D points based on the Bayesian model described in Sect. 3.1 with fixed degree of freedom ν

the prior is non-conjugated for the current model, so that the posterior is solved for numerically. The calculation of the degree of freedom is an intricate step, for which various approaches have been proposed in the case of a linear regression model (see Geweke 1993; Gelman et al. 2014). Due to limited space, the current model is simplified by assuming that the degree of freedom is known with $\nu = 4$, which value has been recommended for the purpose of a robust estimation (cf. Gelman et al. 2014). For the calculation of the posterior unknowns a Markov chain is generated by means of a Gibbs sampler (see Fig. 1). The convergence of Gibbs sampler depends strongly on the choice of the initial parameter values $\Theta^{(0)}$. Less critically, initial values for the parameters τ and α must also be specified, e.g., setting $\tau^{2(0)} = 1$ and $\alpha^{2(0)}$ equal to the variance of the z-coordinate (thus assuming that the z-coordinate is initially normally distributed).

For the generation of Markov chains using the Gibbs sampler, the conditional distributions of the unknowns are required, which are generally well known from Bayesian literature in the context of linear models (e.g. Gelman et al. 2014). These are now adapted to the present nonlinear model. The Gibbs sampler starts with step 1. in any iteration step j, where the weights for the z-coordinate are drawn depending on τ, α, Θ from the preceding iteration $j-1$. According to

$$W_i|\tau^2, \alpha^2, \Theta, l_{z_i} \sim \text{Inv-}\chi^2\left(\nu+1, \frac{\nu\tau^2 + (l_{z_i} - f_{z_i}(\Theta))^2/\alpha^2}{\nu+1}\right), \quad (10)$$

the weights are Inv-χ^2-distributed, where the parameter α scales the residual square of a measured z_i; ν and τ control the robustness. In step 2. of the Gibbs sampler, the distribution of τ is updated depending on the new weights W as well as parameter values Θ and α of the iteration before:

$$\tau^2|W, \alpha^2, \Theta, l_z \sim \text{Gamma}\left(\frac{n\nu}{2}, \frac{\nu}{2}\sum_{i=1}^{n}\frac{1}{W_i}\right). \quad (11)$$

For the generation of the gamma-distributed random variable τ, the parameter α and the observations l_z are not used directly. These values are included indirectly in the weights W and the number of observations n. In step 3., the scale factor α is generated from a Inv-χ^2-distribution as follows:

$$\alpha^2|W, \tau^2, \Theta, l_z \sim \text{Inv-}\chi^2\left(n, \frac{1}{n}\sum_{i=1}^{n}\frac{(l_{z_i} - f_{z_i}(\Theta))^2}{W_i}\right). \quad (12)$$

Now, the generation of the variance σ_x^2 within step 4. differs from (12) only by the replacement of the z_i by the x_i-components while omitting the weight W_i, that is,

$$\sigma_x^2|\Theta, l_x \sim \text{Inv-}\chi^2\left(n, \frac{1}{n}\sum_{i=1}^{n}(l_{x_i} - f_{x_i}(\Theta))^2\right). \quad (13)$$

The generation of the variance σ_y^2 in step 5. is then carried out in the same way as σ_x^2, using y_i- instead of x_i-components.

In the remaining step 6., the generation of $\Theta^{(j)}$ is not possible by means of a conditional distribution as the latter is unknown. This, however, is not a problem as random numbers can be directly generated from the combination of the prior density and the likelihood function by means of a Metropolis-Hastings (MH) algorithm (Hastings 1970), as shown in Table 1. Firstly, a random number θ_i^{new} is generated using the values $\theta_i^{(j-1)}$ from preceding iteration step and using a scale factor λ_{θ_i}. The selection of the distribution family for this random number generation influences the efficiency of the MH algorithm. In view of the application given

Table 1 Generation of posterior $\Theta^{(j)}$ by means of the MH algorithm

1. Generate	$\theta_i^{\text{new}} \sim \text{N}\left(\theta_i^{(j-1)}, \lambda_{\theta_i}\right)$
2. Set	$\Theta^{\text{new}} = \left[\theta_1^{(j)}, \theta_2^{(j)}, \cdots, \theta_i^{\text{new}}, \cdots, \theta_u^{(j-1)}\right]^{\text{T}}$
	$\Theta^{\text{old}} = \left[\theta_1^{(j)}, \theta_2^{(j)}, \cdots, \theta_i^{(j-1)}, \cdots, \theta_u^{(j-1)}\right]^{\text{T}}$
3. Calculate	$\Psi = \min\left[1, \frac{p(\Theta^{\text{new}}) * p(\Theta^{\text{new}}\|l, \sigma_x, \sigma_y, \alpha, W, \nu, \tau)}{p(\Theta^{\text{old}}) * p(\Theta^{\text{old}}\|l, \sigma_x, \sigma_y, \alpha, W, \nu, \tau)}\right]$
4. Accept or reject	
Generate:	$\rho \sim \text{U}(0, 1)$
If $\rho \leq \Psi$:	$\theta_i^{(j)} = \theta_i^{\text{new}}$
If $\rho > \Psi$:	$\theta_i^{(j)} = \theta_i^{(j-1)}$

in Sect. 4, a normal distribution is employed for this purpose since the posterior of $\boldsymbol{\Theta}$ will approximately be gaussian. The efficiency of the MH algorithm also depends on the scale factor λ_{θ_i}. An approach to choosing adequate value for λ_{θ_i} in application is presented in Sect. 4.3. In step 2. the parameter vectors $\boldsymbol{\Theta}$ are set up with θ_i^{new} and $\theta_i^{(j-1)}$. The parameters $\theta_1^{(j)}, \cdots, \theta_{i-1}^{(j)}$ are from the current iteration j, which means that these values have been updated before θ_i by the MH algorithm. The model parameter values $\theta_{i+1}^{(j-1)}, \cdots, \theta_u^{(j-1)}$ stem from the iteration before. These values are gradually regenerated randomly by means of the MH algorithm after updating θ_i. In step 3. the probability ratio of $\boldsymbol{\Theta}^{\text{new}}$ and $\boldsymbol{\Theta}^{\text{old}}$ is calculated according to Eqs. (8) and (9). The value Ψ is used for the decision in step 4., whether θ_i^{new} or $\theta_i^{(j-1)}$ is the new generated $\theta_i^{(j)}$. A convenient feature of this procedure is the direct usage of the functional model $\boldsymbol{f}(\boldsymbol{\Theta})$, without the need for derivatives.

With the conclusion of the MH algorithm, the Gibbs sampler complete one iteration. This procedure is carried out in total m times and thereby yields the Markov chain results for the unknown posterior parameters. By means of resulting Markov chains, the posterior results for the key parameters (functional model parameters, variances σ_x^2 and σ_y^2, the unknown weights and scale parameters of t-distribution) can be approximated. The first half of the drawn chains are considered as *burn-in replications* and discarded. The remaining samples may serve for the estimation of the parameters from $\boldsymbol{\Theta}^{(o,\ldots,m)}$, as their mean value (see Fig. 1). By choosing m sufficiently large, the approximation error implicit in the estimate can be reduced (see, e.g, Kroese et al. 2011; Gelman et al. 2014).

4 Application and Results

4.1 Application

In our real-world application we use a multi-sensor-system (MSS) composed by a laser scanner and two firmly attached GNSS equipment proposed by Paffenholz (2012). The aim of the MSS is to efficiently geo-reference 3D point clouds by means of 3D coordinates in a superior coordinate frame. For ease of understanding, we consider only the obtained 3D coordinates by one GNSS equipment according Eq. (6) for n observed 3D points. These points describe a circle in 3D due to the rotation of the laser scanner around its vertical z-axis. For the geo-referencing of the laser scanner the unknown circle parameters must be estimated. The parameterisation of a circle in 3D is given by: $\boldsymbol{\Theta} = \left[c_x, c_y, c_z, r, \omega, \varphi\right]^{\text{T}}$.

The parameter $\boldsymbol{c}$ is the centre point of the circle for the x-, y- and z-coordinate and r is the radius. The angles ω and φ describe the orientation of the circle in 3D by means of the rotations around the x- and y-axis. By the parameter $\boldsymbol{\Theta}$ the functional model can be set up for the different coordinate components as follows:

$$f_{x_i}(\boldsymbol{\Theta}) = r\cos(t_i)\cos(\varphi) + c_x, \tag{14}$$

$$f_{y_i}(\boldsymbol{\Theta}) = r\cos(t_i)\sin(\varphi)\sin(\omega) + r\sin(t_i)\cos(\omega) + c_y,$$

$$f_{z_i}(\boldsymbol{\Theta}) = -r\cos(t_i)\sin(\varphi)\cos(\omega) + r\sin(t_i)\sin(\omega) + c_z,$$

$$\text{with } \boldsymbol{f}_i(\boldsymbol{\Theta}) = \left[f_{x_i}(\boldsymbol{\Theta}), f_{y_i}(\boldsymbol{\Theta}), f_{z_i}(\boldsymbol{\Theta})\right].$$

The splitting of the functional model in the three coordinate components corresponds to a nonlinear regression model, which allows for the estimation of the unknown parameters by means of a GMM. The equations arise from the combination of the polar coordinate equation of the circle and a 3D rotation matrix. The variable $\boldsymbol{t}$ is the rotation angle of the laser scanner around its z-axis. To simplify the model we assume that $\boldsymbol{t}$ is known and error free. This assumption is possible, because the horizontal angle measurement of the laser scanner is significantly preciser than the 3D points obtained by the GNSS equipment. In addition to the GNSS observations, prior knowledge is available from calibration measurements by means of a laser tracker. Therefore, the value $\underline{r}$ and the corresponding variance $\underline{\sigma}_r^2$ are known. We assume a non-informative prior for the other model parameters $\boldsymbol{c}$, ω and φ. Hence, we use a normal distribution as prior instead of the general presented multiple prior distribution in Eq. (8). For the stochastic model the assumption of Eq. (7) is used. For further information about the MSS such as the specific sensors, the calibration and the geo-referencing approach employed, see Paffenholz (2012).

4.2 Closed Loop Simulation

The investigation of the presented Bayesian estimation approach in Sect. 3 is based on a closed loop (CL) simulation and on a real data example. In this section we describe the generation of the CL simulation and the results. Firstly, we define the true parameter values $\boldsymbol{\Theta}_{\text{true}} = [1716.00\,\text{cm}, 3012.00\,\text{cm}, 1054.00\,\text{cm}, 30.00\,\text{cm}, 0.40°, 0.08°]$ for the 3D circle. With the functional model in Eq. (14) and $\boldsymbol{\Theta}_{\text{true}}$ we calculated 50 uniformly distributed observations on the 3D circle. After that, random normal distributed values are generated and added to these observations. For the noise generation we use the following parameters derived from the real data: $\sigma_x = 0.2$ cm, $\sigma_y = 0.1$ cm and $\sigma_z = 0.4$ cm. In addition to the normally distributed errors, we create 15% of the z-observations as outliers by means of the uniform distribution U (0.93, 1.71) with minimum value 0.93 and maximum value 1.71. The outliers are ill-conditioned distributed and are spread out over the true circle. In each iteration j_{CL} of the CL simulation the noise and outliers are generated randomly. The simulation is repeated 10,000 times, where three different approaches are used for the estimation of $\hat{\boldsymbol{\Theta}}^{(j_{\text{CL}})}$. j_{CL} denotes the counter of the CL simulations. The approaches are based on:

(1) a linearised non-robust GMM (cf. Koch 1999),
(2) the presented robust Bayesian approach with non-informative prior, and
(3) the presented robust Bayesian approach with prior information. The prior knowledge about the radius is defined here to be the true value $\underline{r} = r_{\text{true}} = 30\,\text{cm}$ as well as the standard deviation $\underline{\sigma}_r = 0.05\,\text{cm}$. Basics and further information about CL simulation are presented, e.g., in Saltelli et al. (2008).

In each iteration of the CL simulation initial values for the unknown parameters $\boldsymbol{\Theta}$ are required. For $\boldsymbol{c}^{(0)}$ we use the mean of the observations $\boldsymbol{l}$. $r^{(0)}$ is derived by means of the euclidean distance between $\boldsymbol{c}^{(0)}$ and an arbitrarily observed point. The initial angles $\omega^{(0)}$ and $\varphi^{(0)}$ can be set to zero since the MMS is levelled. These initial values are used for all three estimation approaches. Furthermore the Gibbs sampler requires starting values for α^2, τ^2 (cf. Sect. 3.2), $\boldsymbol{\lambda}_{\boldsymbol{\Theta}}$ and m. The last two parameter values are essential for the convergence of the Gibbs sampler, and they depend on the quality of the initial values of the unknown parameters. For example, if the initial values $\boldsymbol{\Theta}^{(0)}$ deviate significantly from the true parameter values $\boldsymbol{\Theta}_{\text{true}}$ then the number of generated samples in each Gibbs sampler chain m should be increased. In each iteration of the CL simulation a Markov chain with $m = 7{,}000$ and a warm-up period of $o = 3{,}500$ is generated. The convergence of the Gibbs sampler and the definition of $\boldsymbol{\lambda}_{\boldsymbol{\Theta}}$ are discussed in Sect. 4.3.

The results of the CL simulation are the estimates $\hat{\boldsymbol{\Theta}}$ for the three approaches. The means of the 10,000 vectors $\left[\hat{c}_x, \hat{c}_y, \hat{r}, \hat{\omega}, \hat{\varphi}\right]$ differ from $\boldsymbol{\Theta}_{\text{true}}$ by less than 10^{-4} [cm] resp. [deg]. Only the mean of $\hat{c}_z$ deviates significantly from $c_{z_{\text{true}}}$ (see Table 2) due to the generated outliers in the z-coordinate.

Table 2 Estimated results of the CL simulation

	GMM	Bayes non-informative	Bayes informative
Mean (RMSE) [cm]	0.25	0.19	0.19
σ_{RMSE} [cm]	0.05	0.06	0.06
Mean $(\hat{c}_z) - c_{z_{\text{true}}}$ [cm]	0.21	0.14	0.14
$\hat{\sigma}_{c_z}$ [cm]	0.053	0.063	0.063
Mean $(\hat{r})$ [cm]	30.00	30.00	30.00
$\hat{\sigma}_r$ [cm]	0.020	0.020	0.018

The outliers have a smaller influence on the estimate $\hat{c}_z$ by the robust Bayesian approach than on this estimate by the non robust GMM. For a comparison between the three approaches in terms of all estimates $\hat{\boldsymbol{\Theta}}$ simultaneously, the root mean square error (RMSE) is calculated. The RMSE represents the average euclidean distance between predicted and true 3D points.

$$\text{RMSE}^{(j_{\text{CL}})} = \sqrt{\frac{1}{n}\sum_{i=1}^{n}\left(d_{xi}^2 + d_{yi}^2 + d_{zi}^2\right)},$$

$$\text{with } d_{xi}^2 = \left[f_{x_i}\left(\boldsymbol{\Theta}_{\text{true}}\right) - f_{x_i}\left(\hat{\boldsymbol{\Theta}}^{(j_{\text{CL}})}\right)\right]^2, \tag{15}$$

$$d_{yi}^2 = \left[f_{y_i}\left(\boldsymbol{\Theta}_{\text{true}}\right) - f_{y_i}\left(\hat{\boldsymbol{\Theta}}^{(j_{\text{CL}})}\right)\right]^2,$$

$$d_{zi}^2 = \left[f_{z_i}\left(\boldsymbol{\Theta}_{\text{true}}\right) - f_{z_i}\left(\hat{\boldsymbol{\Theta}}^{(j_{\text{CL}})}\right)\right]^2.$$

The corresponding distributions of the 10,000 RMSE results are shown in Fig. 2. The average RMSE for the GMM (blue) is larger than the values for the Bayesian non-informative (green) and informative (red) approaches, which are identical (see Table 2). The only difference between the Bayesian results can be observed for the estimated radius (see Table 2): In the informative case, the prior knowledge about the radius

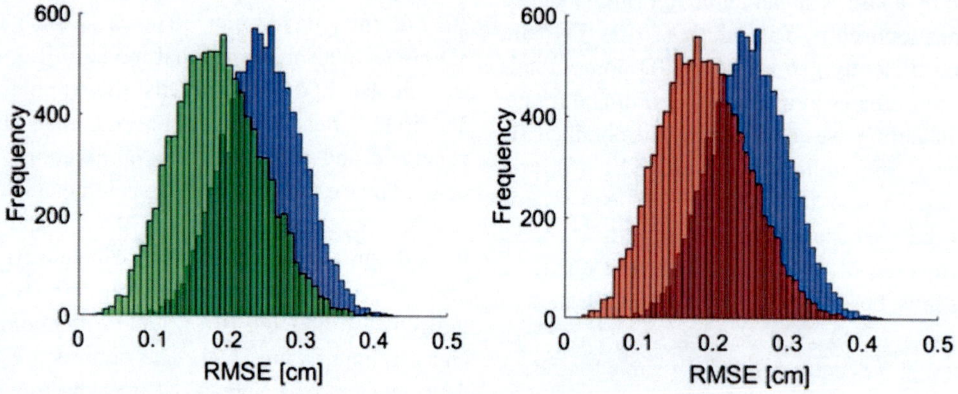

Fig. 2 Distribution of the 10,000 RMSE results for the three approaches: GMM (blue), Bayesian non-informative (green), and Bayesian informative (red)

reduces the standard deviation of the estimated radius but the difference is very small.

4.3 Convergence Analysis of the Markov Chains

Any posterior simulation approach such as the Gibbs sampler presented in Sect. 3.2 provides us with an $\hat{\mathbf{\Theta}}$ which is an estimate of $\mathrm{E}(\boldsymbol{f}(\mathbf{\Theta})|\boldsymbol{l})$. By choosing m sufficiently large and the starting values for the parameters close enough to the true values, the convergence of the Markov chain is very probable. To analyse the convergence behaviour of the Gibbs sampler, the CL simulation is repeated. Contrary to the calculated values for $\mathbf{\Theta}^{(0)}$ in Sect. 4.2 we generate these values randomly. The random generated values are now not close to the values $\mathbf{\Theta}_{\text{true}}$ as the calculated initial values given in Sect. 4.2. We will demonstrated here that the Markov chain for the random values $\mathbf{\Theta}^{(0)}$ will independently converge to the true values $\mathbf{\Theta}_{\text{true}}$. The random initial values for $\mathbf{\Theta}^{(0)}$ are generated by means of uniform distributions with the intervals:

$$\boldsymbol{c}^{(0)} \in [\boldsymbol{c}_{\text{true}} - 50, \boldsymbol{c}_{\text{true}} + 50]\ [\text{cm}],$$

$$r^{(0)} \in [r_{\text{true}} - 10, r_{\text{true}} + 10]\ [\text{cm}],$$

$$\omega^{(0)} \in [\omega_{\text{true}} - 10, \omega_{\text{true}} + 10]\ [\text{deg}],$$

$$\varphi^{(0)} \in [\varphi_{\text{true}} - 10, \varphi_{\text{true}} + 10]\ [\text{deg}].$$

In case of random initial values $\mathbf{\Theta}^{(0)}$ the choice of an adequate value for λ_{θ_i} is a challenging task. The use of a fixed λ_{θ_i} has the following difficult problem. If the value is too large, then only a few new random numbers for θ_i will be accepted in the MH algorithm. Conversely, a very small λ_{θ_i} results in a high acceptance rate for θ_i. In both cases a convergence of the Gibbs sampler cannot be guaranteed. Alternatively, an iterative adaptation can be used to determine λ_{θ_i}. The used approach is based on the assumption that an adequate λ_{θ_i} is chosen if the acceptance rate is about 50% (cf. Gelman et al. 2014). After every 250 iterations of the Gibbs sampler the acceptance rate of the currently created chain θ_i is calculated. If the acceptance ratio is less than 40%, $\lambda_{\theta_i}^{(j)}$ will be reduced by the factor γ. In contrast, an acceptance ratio greater than 60% will be increased the value $\lambda_{\theta_i}^{(j)}$ by the factor γ. For the presented application the values $\gamma = 2$ and $\gamma = 5$ have proven to be suitable. Both values are used in alternating order (cf. Fig. 3). In a future work we will investigate the dependency of the value of the alternating factor γ on the number of iterations m.

In the CL simulation the initial values $\boldsymbol{\lambda}_{\mathbf{\Theta}}^{(0)} = [5, 5, 5, 1, 1, 1]$ are used. The results of the adaptively estimated $\boldsymbol{\lambda}_{\mathbf{\Theta}}$ for one CL simulation are shown in Fig. 3. It can be seen that these estimates become constant around the warm up period. Figure 4 shows the results of the Markov chain for $\mathbf{\Theta}$ for the same simulation. The chains spread constantly around their mean values close to $\mathbf{\Theta}_{\text{true}}$. These results are similar to the results of the other 9,999 iterations of the CL simulation. Consequently, it can be assumed that the Markov chain for $\mathbf{\Theta}$ converges for the used start values m, o, $\mathbf{\Theta}^{(0)}$ and $\boldsymbol{\lambda}_{\mathbf{\Theta}}^{(0)}$. For a final statement of convergence a hypotheses test would be required. Due to the limited space in this contribution a general examination of the determination of λ_{θ_i} and convergence analysis is not feasible and will be discussed in future studies.

4.4 Real Data Example

For the real data example we use a data set with 1,580 GNSS observations. The noise of the x- and y-coordinates are approximately equal to the variances in the CL simulation. Contrary to the simulated data, the residuals of the measured z-coordinates have additional error effects,

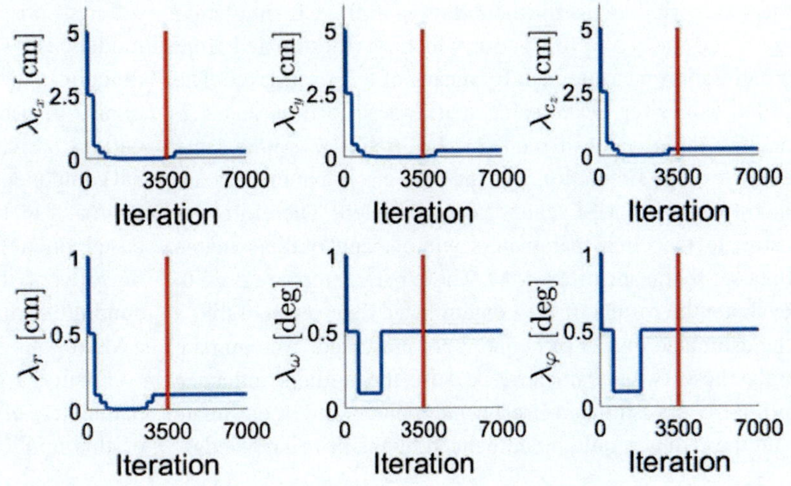

Fig. 3 Results of $\boldsymbol{\lambda}_{\mathbf{\Theta}}$ for one CL simulation: iterative adaptation of λ_{θ_i} (blue), and limit of warm-up period (red)

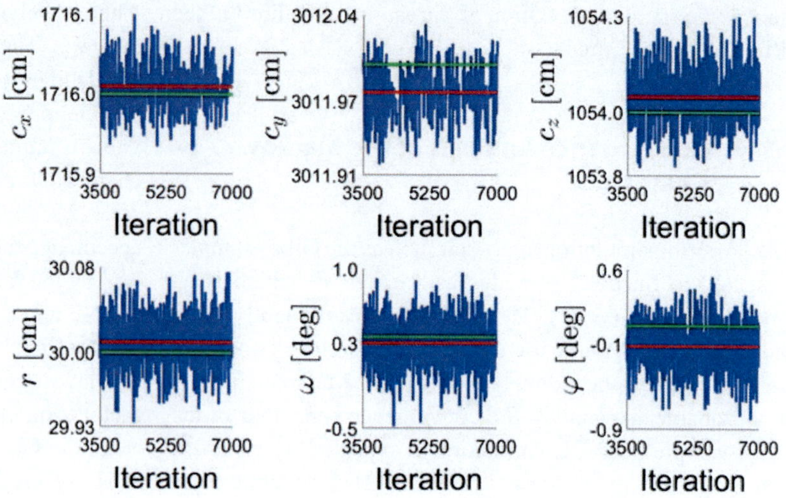

Fig. 4 Results of estimating parameters $\boldsymbol{\Theta}$ for one CL simulation: Markov chains of the Gibbs sampler after warm-up period (blue), mean values of these chains (red), and $\boldsymbol{\Theta}_{\text{true}}$ (green)

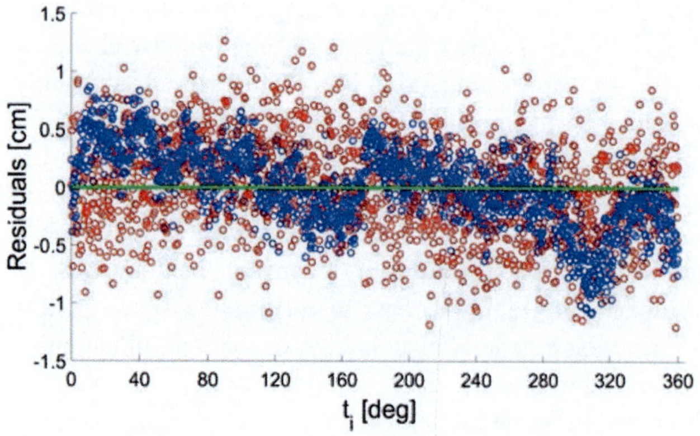

Fig. 5 Residuals of z-coordinate. $[z_i - \text{Mean}(z)]$ of the real data (blue), and noise generated from the normal distribution N $(0, 0.4\ [\text{cm}])$ as in the CL simulation (red)

Table 3 Estimation results for the real data example

Estimation approach	$\hat{c}_z$ [cm]	$\hat{\sigma}_{c_z}$ [cm]	$\hat{r}$ [cm]	$\hat{\sigma}_r$ [cm]
GMM without outliers	1.70	0.008	29.75	0.004
GMM	1.81	0.009	29.75	0.004
Bayes non-informative	1.73	0.010	29.73	0.004
Bayes informative	1.73	0.010	29.73	0.004

which are not represented by a normal distributed noise (see Fig. 5). However, these are no significant outliers but probably time-dependent effects which are not considered further in this paper. Hence, we generate randomly outliers for 15% of the z-coordinate by means of the uniform distribution U $(0.5, 1.0)$. The informative Bayesian approach uses the same prior values for the radius as in the CL simulation ($\underline{r} = 30\,\text{cm}$, $\underline{\sigma}_r = 0.05\,\text{cm}$), which were estimated from a calibration measurement by means of a laser tracker. The initial values for $\boldsymbol{\Theta}$ are calculated as described in Sect. 4.2, and the initial values for α^2, τ^2, $\boldsymbol{\lambda}_{\Theta}$, o and m are the same as in the CL simulation. As the true circle parameters are unknown, the RMSE cannot be calculated. Therefore, we estimate the circle parameters without any outliers in the data set by means of a GMM. These parameters are used to evaluate the results of the contaminated date set. In Table 3 the estimated results of $\hat{c}_z$ and $\hat{r}$ are presented. We compare only these two parameters, because the outliers influence primarily the estimated circle center point of the z-coordinate and the radius should be influenced by the prior knowledge.

The centre points of the x- and y-coordinate and the angles do not differ significantly between the different estimations. The results of $\hat{c}_z$ show that the robust Bayesian results are closer to the result of the GMM without outliers as the result of the GMM. In case of $\hat{r}$ the results of all estimations are close together without significant differences.

5 Conclusions and Outlook

In this contribution, a robust Bayesian approach to adjusting a nonlinear functional model based on normally or t-distributed residuals was presented. In this approach one observation was introduced as a 3D point, to demonstrate that different observation groups may have different stochastic models. The selection of a prior distribution and a likelihood function was described with regard to geodetic applications. t-distributions, having longer tails than normal distributions, were used within the likelihood function for individual components to account for expected outliers. A numerical approach to calculating the unknown posterior parameters based on a Gibbs sampler was suggested. As the nonlinearity of the functional model excludes the usage of a conditional distribution for the functional model parameters, a Metropolis-Hastings algorithm was outlined.

Finally, a geodetic application was presented where the parameters of a 3D circle have been of interest. The results of this application show that the introduced t-distribution in the Bayesian model reduces the influence of outliers on the estimated parameters. The introduced prior information for the radius affects only the precision $\hat{\sigma}_r$ in the range of 0.02 mm in the closed loop simulation. This is to be expected since r_{true} and $\underline{r}$ are identical. In case of the real world data no significant differences between non-informative and informative Bayesian approach is detectable for $\hat{r}$ and $\hat{\sigma}_r$. The reason for this is the larger number of observations in the real world data. The likelihood function thus dominates the posterior density, which can be interpreted as a down weighting of the prior density in the estimation.

In future studies the robust Bayesian approach should be improved to deal with fully correlated observations. Furthermore, the approach should be extended by the estimation of the degree of freedom of the t-distribution.

References

Alkhatib H, Schuh W-D (2006) Integration of the Monte Carlo covariance estimation strategy into tailored solution procedures for large-scale least squares problems. J Geod 81:53–66

Alkhatib H, Kargoll B, Paffenholz J-A (2017) Further results on robust multivariate time series analysis in nonlinear models with autoregressive and t-distributed errors. In: Rojas I, Pomares H, Valenzuela O (eds) Time series analysis and forecasting. ITISE 2017. Contributions to statistics. Springer, Cham, pp 25–38

Bossler JD (1972) Bayesian inference in geodesy. Dissertation, The Ohio State University, Columbus, Ohio, USA

Gamerman D, Lopes HF (2006) Markov chain Monte Carlo – stochastic simulation for bayesian inference, 2nd edn. Chapman and Hall/CRC, Boca Raton

Gelman A, Carlin JB, Stern HS, Dunson DB, Vehtari A, Rubin DB (2014) Bayesian data analysis, 3rd edn. CRC Press, Taylor & Francis Group, Boca Raton

Geweke J (1993) Bayesian treatment of the independent student-t linear model. J Appl Economet 8:19–40

Gundlich B, Koch KR, Kusche J (2003) Gibbs sampler for computing and propagating large covariance matrices. J Geod 77:514–528

Haario H, Laine M, Mira A, Saksman E (2006) DRAM: efficient adaptive MCMC. Stat Comput 16:339–354

Hastings WK (1970) Monte Carlo sampling methods using Markov chains and their applications. Biometrika 57:97–109

Johnathan MB, Solonen A, Haario H, Laine M (2014) Randomize-then-optimize: a method for sampling from posterior distributions in nonlinear inverse problems. SIAM J Sci Comput 36:A1895–A1910

Koch KR (1988) Bayesian statistics for variance components with informative and noninformative priors. Manuscr Geodaet 13:370–373

Koch KR (1999) Parameter estimation and hypothesis testing in linear models, 2nd edn. Springer, Berlin

Koch KR (2007) Introduction to Bayesian statistics, 2nd edn. Springer, Berlin

Koch KR (2017) Monte Carlo methods. Int J Geomath 9:117–143

Koch KR (2018) Bayesian statistics and Monte Carlo methods. J Geod Sci 8:18–29

Koch KR, Kargoll B (2013) Expectation-maximization algorithm for the variance-inflation model by applying the t-distribution. J Appl Geod 7:217–225

Kroese DP, Taimre T, Botev ZI (2011) Handbook of Monte Carlo methods. Wiley, Hoboken

Paffenholz J-A (2012) Direct geo-referencing of 3D point clouds with 3D positioning sensors. Deutsche Geodätische Kommission, Series C (Dissertation), No. 689, Munich

Parzen E (1979) A density-quantile function perspective on robust estimation. In: Launer L, Wilkinson GN (eds) Robustness in statistics. Academic Press, New York, pp 237–258

Riesmeier K (1984) Test von Ungleichungshypothesen in linearen Modellen mit Bayes-Verfahren. Deutsche Geodätische Kommission, Series C (Dissertation), No. 292, Munich

Saltelli A, Ratto M, Andres T, Campolongo F, Cariboni J, Gatelli D, Saisana M, Tarantola S (2008) Global sensitivity analysis: the primer. Wiley, Chichester

Schaffrin B (1987) Approximating the Bayesian estimate of the standard deviation in a linear model. Bull Geod 61(3):276–280

Yang Y (1991) Robust Bayesian estimation. Bull Geod 65(3):145–150

Zhu J, Santerre R, Chang X-W (2005) A Bayesian method for linear, inequality-constrained adjustment and its application to GPS positioning. J Geod 78:528–534

The GNSS for Meteorology (G4M) Procedure and Its Application to Four Significant Weather Events

Lorenzo Benvenuto, Ilaria Ferrando, Bianca Federici, and Domenico Sguerso

Abstract

The authors conceived the GNSS for Meteorology (G4M) procedure to remote-sense the Precipitable Water Vapor (PWV) content in atmosphere with the aim to detect severe meteorological phenomena. It can be applied over an orographically complex area, exploiting existing networks of Global Navigation Satellite System (GNSS) Permanent Stations (PSs) and spread meteorological sensors, not necessarily co-located. The results of a posteriori analysis of four significant meteorological events are here presented, also in comparison with rain gauge data, to show the effectiveness of the method. The potentiality of G4M to detect and locate in space and time intense rainfall events is highlighted. The upcoming application of G4M in near-real time could provide a valuable support to existing Decision Support System for meteorological alerts.

Keywords

GNSS for Meteorology (G4M) procedure · Precipitable Water Vapor (PWV) · Severe meteorological events monitoring

1 Introduction

The use of Global Navigation Satellite System (GNSS) allows the monitoring of meteorological phenomena, including severe ones, at detailed temporal and spatial scales. It is independent from the other observation techniques (e.g. rain gauges, meteorological radars, satellite images), thus it can represent an innovative source of data, which could help in improving the existing observational systems (Barindelli et al. 2018; Inoue and Inoue 2007; Oigawa et al. 2015). GNSS-derived products, Zenith Total Delay (ZTD) and Precipitable Water Vapor (PWV) in particular, are already routinely assimilated into some forecasting Numerical Weather Prediction (NWP) models (Guerova et al. 2016; Oigawa et al. 2018) and estimated within now-casting NWP models (Douša and Vaclavovic 2014).

In this context, the authors have conceived an automatic procedure, termed GNSS for Meteorology (G4M) (Ferrando et al. 2018). G4M is intended to produce 2D PWV maps and describe its spatial and temporal evolution by means of ΔPWV maps obtained by time differentiation with respect to a reference epoch. The input data are ZTD estimates from GNSS observations, Pressure (P) and Temperature (T) measurements, both derived from existing sensors spread over the considered domain and not necessarily co-located. Starting from Bevis' formulation (Bevis et al. 1992), that describes how to obtain 1D PWV values from co-located ZTD, P and T data, the G4M procedure adds a simplified mathematical model to describe P and T fields, besides data interpolation and map algebra (performed in a GIS environment), to create PWV maps (Ferrando et al. 2018). Despite the low density

L. Benvenuto
DICCA - Department of Civil, Chemical and Environmental Engineering, University of Genoa, Genoa, Italy

Gter s.r.l. Innovazione in Geomatica, GNSS e GIS, Genoa, Italy
e-mail: lorenzo.benvenuto@gter.it

I. Ferrando (✉) · B. Federici · D. Sguerso
DICCA - Department of Civil, Chemical and Environmental Engineering, University of Genoa, Genoa, Italy
e-mail: ilaria.ferrando@edu.unige.it; bianca.federici@unige.it; domenico.sguerso@unige.it

P. Novák et al. (eds.), *IX Hotine-Marussi Symposium on Mathematical Geodesy*,
International Association of Geodesy Symposia 151, https://doi.org/10.1007/1345_2020_111

and the different spatial configurations of the GNSS PSs and meteorological sensors, the procedure seems to be able to detect meteorological events with reliable results, also thanks to an index (termed Heterogeneity Index, HI), conceived by the authors, based on ΔPWV spatial variability.

In the present work, the G4M workflow is briefly introduced (Sect. 2), and the procedure is applied to four significant meteorological case studies whose results are compared with rain gauge data (Sect. 3). Final considerations conclude the paper.

2 The G4M Workflow

The G4M workflow starts from ZTD, P and T data deriving from variously distributed sensors over the analyzed domain, uploaded on a PostgreSQL+PostGIS geoDataBase (Benvenuto et al. 2018). Hence, it employs a procedure which is implemented in a GRASS GIS environment.[1] This consists in 2D data interpolation and a simplified mathematical model of the atmospheric behaviour (conceived by the authors and under evaluation for a patent; Ferrando et al. 2016, 2017). It allows to produce PWV, ΔPWV and HI 2D maps, as depicted in Fig. 1.

So far, G4M has been applied to the orographically complex territory on the French-Italian border region. A hint on input and output data is here reported.

2.1 Input Data

The GNSS data of 181 Permanent Stations (PSs), from global, trans-national, national and regional networks on the French-Italian border region, were processed with the GAMIT/GLOBK software[2] in network mode, to estimate a set of homogeneous tropospheric parameters. The mean spacing of PSs is about 40 km. ZTDs are estimated for each station, simultaneously with a daily positioning solution. The details on processing settings for ZTD estimation can be found in Sguerso et al. (2013, 2016). The ZTD estimates, from January 1998 to December 2015 with a temporal step of 2 h, have been included in the RENAG DataBase.[3] To derive PWV from ZTD through Bevis' relations, several P and T stations, with a mean spacing of 150 km and a temporal step of 1 h or higher, have been selected. P and T data are available in NCEI[4] (National Centers for Environmental Information) archive of global historical weather and climate data.

The GNSS PSs and P/T networks are depicted in Fig. 2 with black and red dots, respectively.

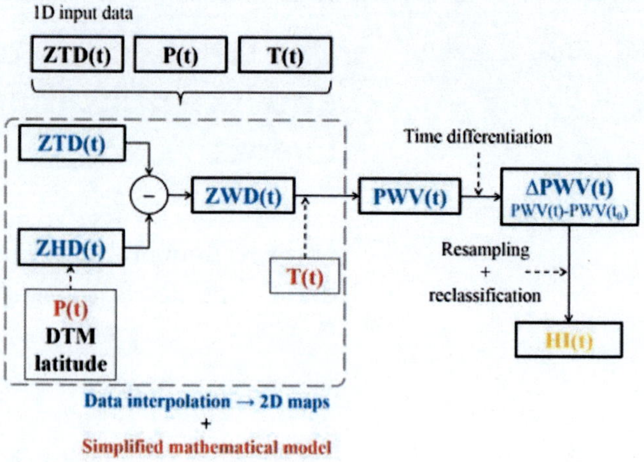

Fig. 1 G4M workflow

2.2 Output Maps

ΔPWV and HI maps are the outputs of the G4M procedure.

ΔPWV consists in a differentiation in time with respect to a reference epoch. This operation removes the orographic effect which affects PWV maps (Ferrando et al. 2018) and results in the observation of PWV variations with respect to the reference epoch. Figure 3 depicts the evolution of ΔPWV over GENO (Italy) for 9th and 10th October 2014. The blue, red, yellow and green dots represent PWV values of 9th and 10th October 2014 differenced with respect to 00:00 UTC of 6th (PWV = 26 mm), 7th (PWV = 26 mm), 8th (PWV = 35 mm) and 9th (PWV = 31 mm) October 2014, respectively. The substantial independence from the reference epoch is evident, due to the nearly identical behaviour of the plots, which seem to only contain a bias shift. Such reference epochs are quite close the severe rain event (94, 70, 46 and 22 h before, respectively), but their PWV "absolute" values are different. In particular, 6th and 7th October, 00:00 UTC have a lower PWV value, corresponding to "calm" moment with limited PWV content. Note that differentiating with respect to a "calm" epoch helps in the interpretation of the meteorological phenomenon, because positive ΔPWV values correspond to an increase in PWV with respect to the reference epoch.

The color scale for ΔPWV maps has been chosen by computing the maximum and the minimum values (respectively

[1]Geographic Resources Analysis Support System (GRASS GIS) Software, Version 7.4, GRASS Development Team, Open Source Geospatial Foundation, 2018, http://grass.osgeo.org.

[2]http://geoweb.mit.edu/gg/.

[3]ftp://renag.unice.fr/products/GPS_climatology_Sguerso_Labbouz_Walpersdorf.

[4]https://www.ncei.noaa.gov/.

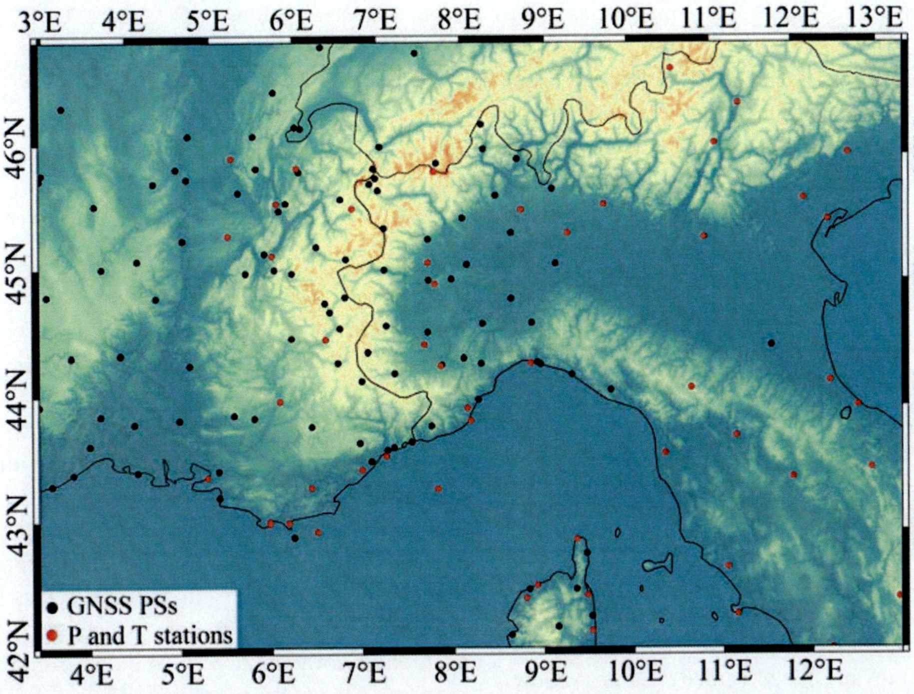

Fig. 2 GNSS PSs and P/T networks

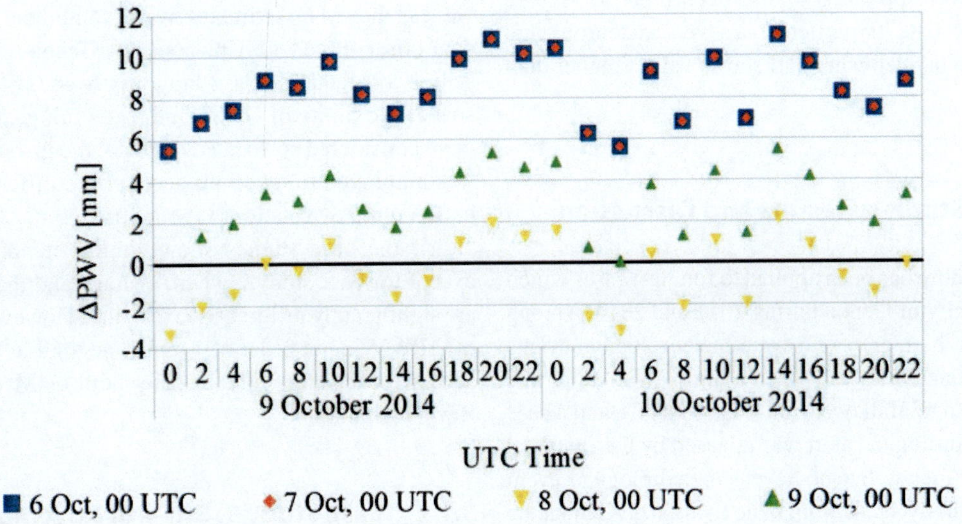

Fig. 3 ΔPWV values over GENO for 9th–10th October 2014 differenced with respect to 00:00 UTC of 6th (blue), 7th (red), 8th (yellow) and 9th (green) October

intense red and intense blue) in the area of interest, as shown in Sect. 3.

HI has been conceived to relate ΔPWV with the occurrence of rainfall. Several studies point out that local fluctuations of PWV, which are associated with water vapor increase or decrease, are responsible of convection, that can result in heavy rainfalls (Seko et al. 2004; Shoji 2013; Oigawa et al. 2015). Those researches revealed a correspondence between PWV spatial inhomogeneity and severe rain events, and observed that the inhomogeneity increases before the rain occurrence, thus it can be used as a predictive parameter. In light of this, HI is computed performing a "block standard deviation" by re-sampling the ΔPWV maps to a four times coarser grid (in both directions) and computing the value of the standard deviation. Thus, each pixel of the HI map represents the empirical standard deviation of the 16 ΔPWV map pixels inside it (Ferrando et al. 2018).

The color scale for 2D HI maps has been chosen on the basis of the statistical parameters computed in the area of interest. In particular, six steps have been selected and defined as follows: 0 (cyan), μ+2σ (green), μ+3σ (yellow), μ+4σ (orange), μ+5σ (brown), > μ+5σ (red), where μ and σ are HI mean value and standard deviation, respectively. This color scale is considered suitable for the identification of HI values corresponding to the occurrence of severe meteorological events. In particular, a severe meteorological event is likely to happen where HI shows values higher than μ+4σ.

3 Case Studies: Results and Discussion

The G4M procedure has been applied to four test cases which occurred in the city of Genoa during 2011 and 2014. Among the case studies there are two severe meteorological events (case studies 1 and 2), a case when a high PWV value was detected and little rainfall occurred (case study 3), and a case in which a meteorological alert was released by the Regional Environmental Agency but no severe meteorological event happened (case study 4). Although the focus is on Genoa, the study area is a wide region surrounding the French-Italian border, between 2°E and 15°E in longitude, and between 41°N and 47°N in latitude. The analysis resolution has been set to 32″ in both north-south and east-west directions, which corresponds to a regular grid of about (1 × 1) km. Two days have been considered for each case study. The chosen temporal step is 2 h. The starting time of the analyses, i.e. the reference epoch, is 00:00 UTC of a day before the investigated event. The most critical time for each case study was identified by observing the rainfall data registered in Genoa by University of Genoa DICCA's (Department of Civil, Chemical and Environmental Engineering) rain gauge.[5] The same rainfall data were also used in combination with ΔPWV and HI maps to interpret the meteorological events.

Table 1 Differences between G4M and 1D Bevis' ΔPWV values over GENO PS for the four case studies

CS	μ (mm)	σ (mm)	Maximum (mm)
1	0.2	0.4	0.7
2	1.2	0.3	1.7
3	0.1	0.3	0.8
4	0.1	0.5	−0.8

The following dedicated sections report the ΔPWV and the HI maps for the individuated most significant epoch for each case study, together with a plot of the ΔPWV, HI and observed rainfall obtained querying the G4M-derived maps over the DICCA meteorological station for the considered time span of two days. In the 2D maps, the GNSS PSs are depicted with black dots, except for GENO (reported as a white dot) which was excluded from the interpolation process of the G4M procedure. Thus, a validation has been performed on GENO comparing the ΔPWV values extracted from the G4M 2D maps and the ones computed applying Bevis' equations (Table 1). For the Bevis-derived ΔPWV values, the ZTD estimates were combined with P and T data observed at one University of Genoa's meteorological station located less than 1 km away from GENO. Due to the spatial proximity of GNSS and meteorological sensors, they were considered co-located. Instead, the ipsometric equation was employed to take into account the difference in height between the two stations (about 40 m).

All the case studies show similar standard deviations, except for case study 2 where average and maximum values are significantly different due to a bias. However, a difference of ΔPWV in the order of 1 mm is negligible with respect to several millimeters, obtained by both G4M model and 1D Bevis formula.

3.1 Case Study 1: 3rd–4th November 2011

The results of G4M procedure applied to 3rd and 4th November 2011 showed an increase of PWV in the entire considered 2D domain for the analyzed 48 h with respect to 3rd November, 00:00 UTC. Focusing on Genoa (Fig. 4), a gradual increase of ΔPWV is highlighted from 8:00 to 10:00 UTC of 4th November, where a local maximum occurs (ΔPWV = 12 mm). At 12:00 UTC, a decrease of ΔPWV is evident, which may be associated to the occurrence of

[5] http://www.dicca.unige.it/meteo.

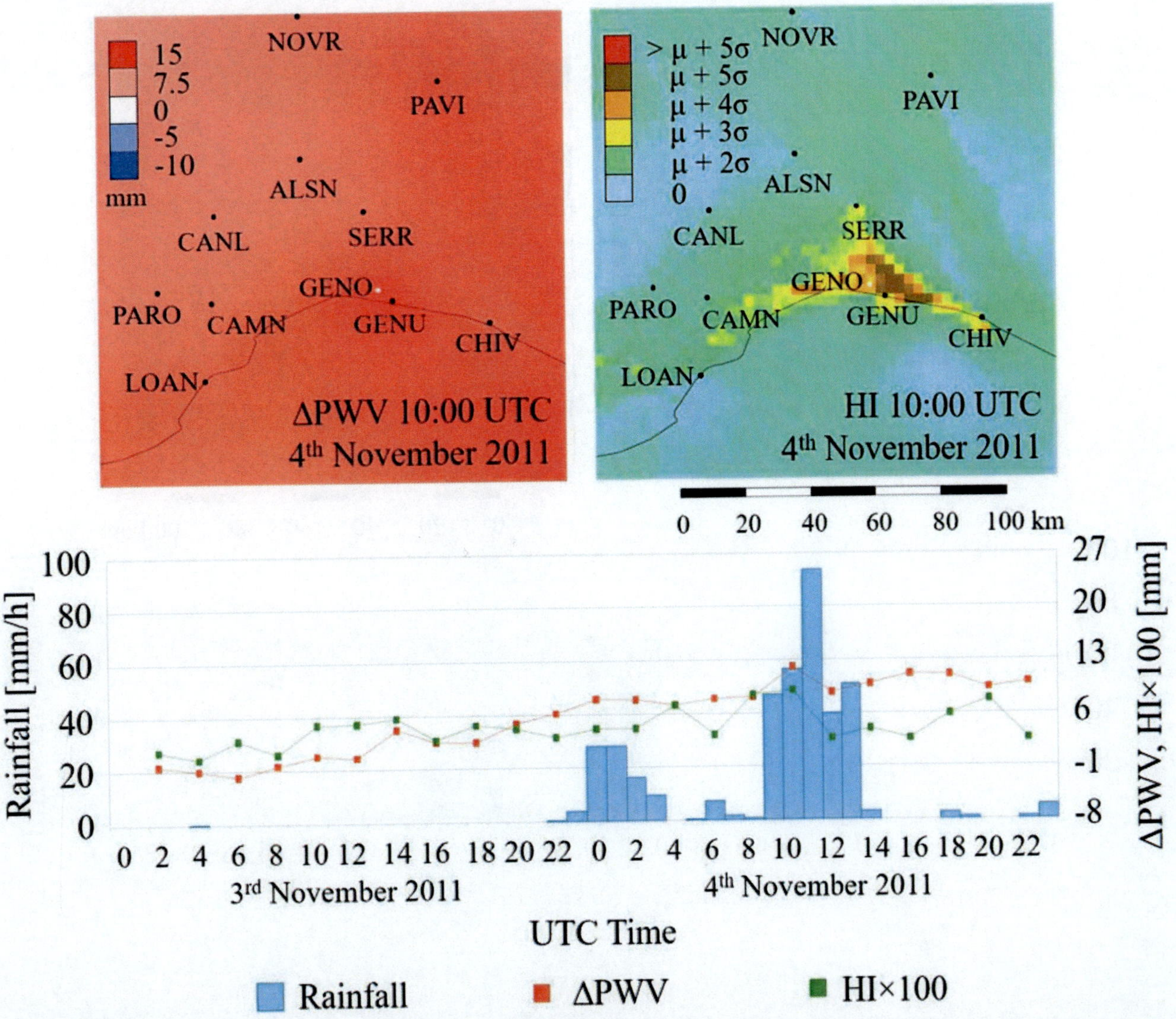

Fig. 4 Case study 1: ΔPWV and HI 2D maps for the most critical moment; ΔPWV, HI and observed rainfall for 48 h over DICCA meteorological station

a severe meteorological event, causing the reduction of the water vapor content in atmosphere (Sapucci et al. 2016).

Focusing on HI map, a strong concentration of high HI values over the city of Genoa is highlighted from 6:00 to 10:00 UTC, meaning the possibility of an intense meteorological event. Observing the temporal evolution of ΔPWV and HI over the DICCA meteorological station (Fig. 4), the most critical moment seems at 10:00 UTC, when a peak of both ΔPWV and HI is present. Looking at the hourly rainfall data of the 4th November 2011, represented in Fig. 4, it is clear that the most intense precipitation occurred between 9:00 and 13:00 UTC, with a maximum peak of 93.8 mm/h between 10:00 and 11:00 UTC, coherently with the previously highlighted results.

3.2 Case Study 2: 9th–10th October 2014

The second case study concerns the intense meteorological event that took place over Genoa in the night between 9th and 10th October 2014. The 2D ΔPWV maps did not show significant variations of PWV with respect to the reference epoch (6th October, 00:00 UTC); on the contrary, HI maps showed high values concentrated only in the area between Genoa and Chiavari (CHIV), a nearby city along the East coast, while HI values were low elsewhere (Fig. 5).

Observing the temporal evolution of HI, it is possible to deduce that the critical moment for this case study was between 20:00 UTC of 9th and 02:00 UTC of 10th October, when HI has a sharp increase. Looking at the rainfall data of

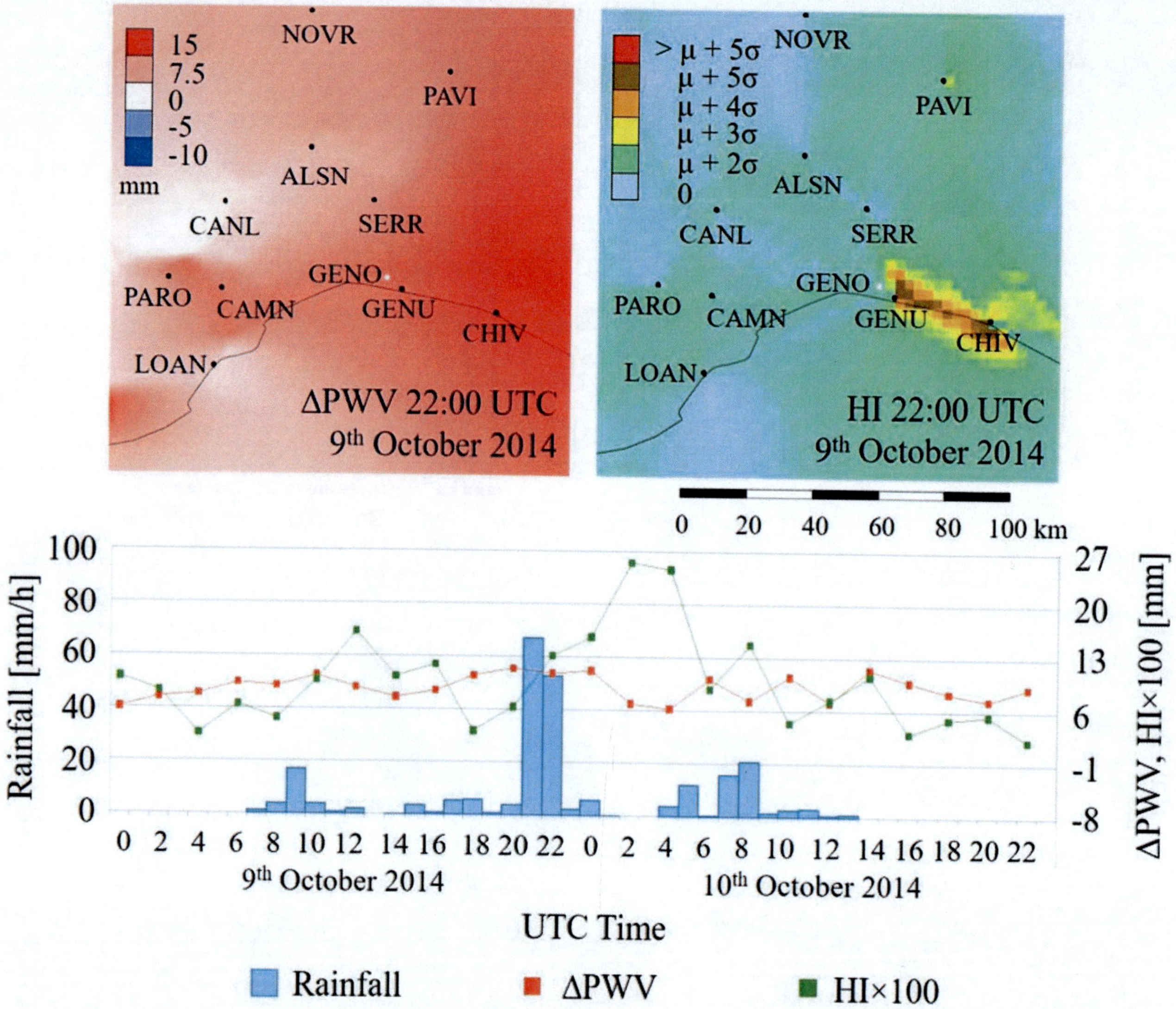

Fig. 5 Case study 2: ΔPWV and HI 2D maps for the most critical moment; ΔPWV, HI and observed rainfall for 48 h over DICCA meteorological station

9th October 2014 (Fig. 5), the highest rainfall peak occurred between 20:00 and 21:00 UTC, having a value of 67.0 mm/h. Since the temporal resolution of the G4M maps is 2 h, only the congruence between rainfall and high values of HI can be appreciated. Note that the highest peak of HI, between 02:00 and 04:00 UTC of 10th October, is associated to a decrease of ΔPWV due to the rainfall. In fact, high values of HI highlight both positive and negative variations of ΔPWV.

3.3 Case Study 3: 3rd–4th September 2011

The present case study was considered interesting because it represents one of the highest peaks of the 2011 PWV time series with respect to the PWV climatological average, i.e. the average of corresponding 2-hourly estimates of each year over all available years (1998–2015). The PWV climatological average has been computed exploiting the RENAG DB's ZTD estimates long time series (Sguerso et al. 2013), and P and T observations from University of Genoa's meteorological station. As a first rough approximation, the overcoming of a threshold defined by the PWV climatological average confidence interval could mean a possible critical situation. In several episodes of 2011 PWV overcomes this value: among them, the Genoa storm analysed in case study 1 (see Sect. 3.1) and the present case study. In both these cases, PWV values were significantly high; despite this, the total rainfall amounts were substantially different: very intense in the first case and almost absent in the present case. Thus, this unexpected behaviour has been taken into consideration.

PWV is quite evenly distributed with rather high values with respect to 00:00 UTC of 3rd September 2011, especially

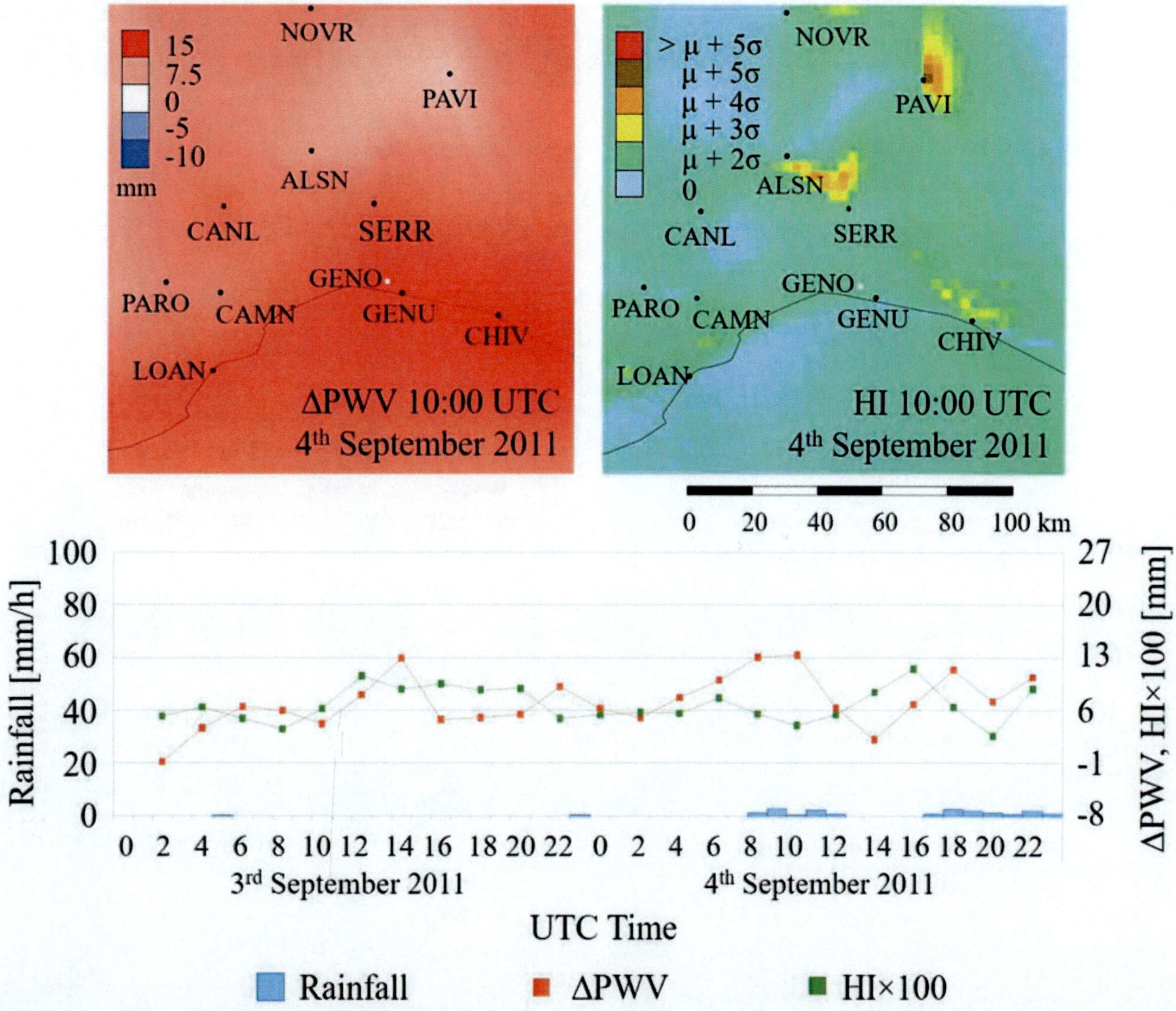

Fig. 6 Case study 3: ΔPWV and HI 2D maps for the chosen moment; ΔPWV; HI and observed rainfall for 48 h over DICCA meteorological station

between 08:00 and 10:00 UTC of 4th September (ΔPWV = 13.3 mm) in Genoa (Fig. 6). HI maps do not show any peak over the city of Genoa for the entire analyzed period. This could indicate that ΔPWV, although it has quite high values, is evenly distributed in space, hence an intense meteorological event is not likely to happen. This is confirmed by the analysis of the rainfall data observed at DICCA meteorological station, which has an hourly maximum of 2.8 mm/h. On the contrary, a peak is visible in HI maps for Pavia (PAVI) and Alessandria (ALSN) in Fig. 6. For both it corresponds to low values of ΔPWV, which are typically not related to severe meteorological events.

3.4 Case Study 4: 30th November–1st December 2014

On 1st December 2014, a meteorological alert was released over the whole Ligurian region by the Regional Environmental Agency, but the rainfall event was not severe at all. For all the analyzed period, the ΔPWV maps highlight very low and negative values, which means a decrease of PWV with respect to the starting time of the analysis (00:00 UTC of 30th November 2014). The analysis started from a not-rainy moment (30th November 2014, 00:00 UTC) quite close to the examined event, even if the PWV value

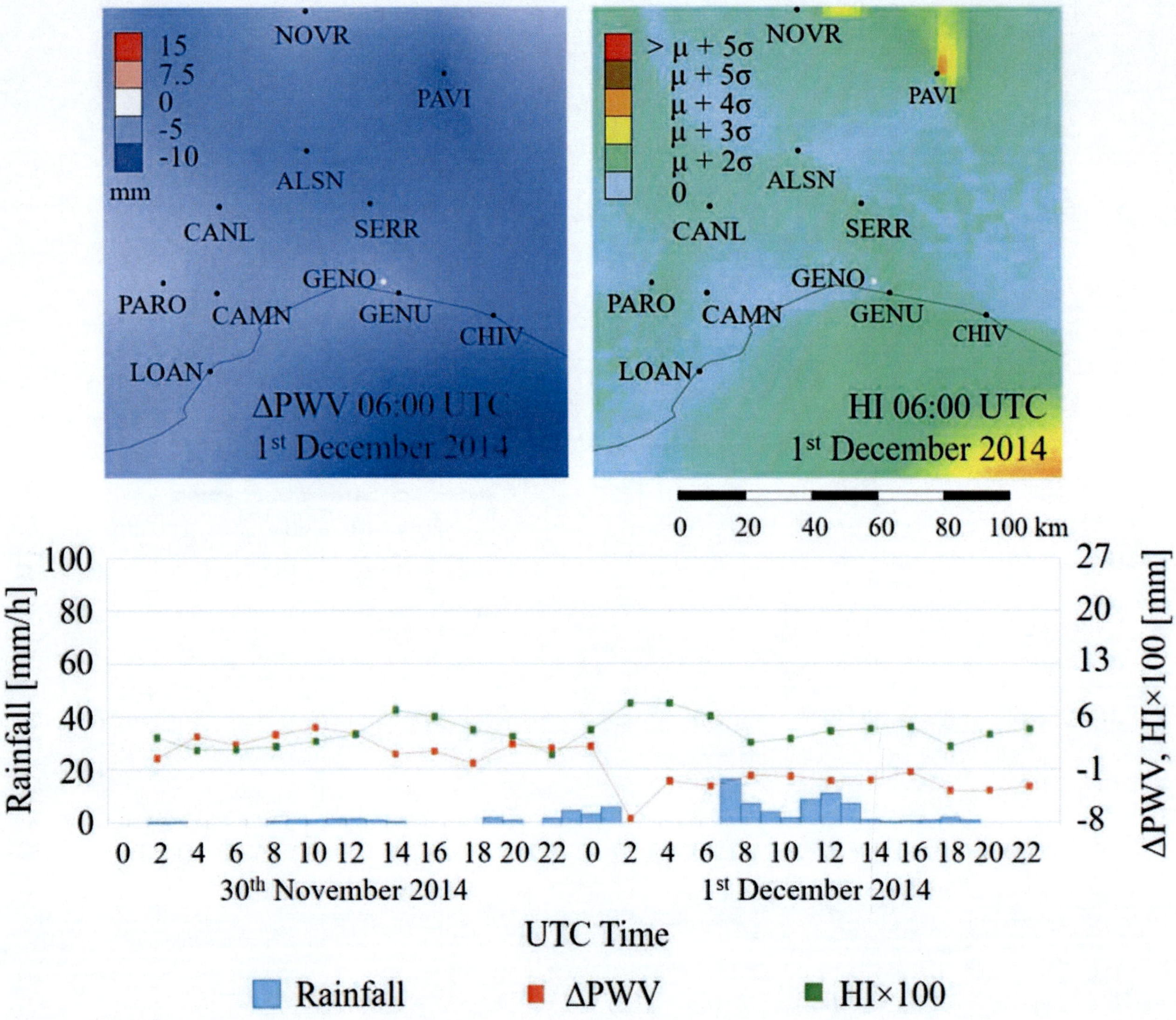

Fig. 7 Case study 4: *Δ*PWV and HI 2D maps for the chosen moment; *Δ*PWV, HI and observed rainfall for 48 h over DICCA meteorological station

(PWV = 25 mm) was rather high over Genoa. The HI maps do not show particularly high values over the city of Genoa, indicating that an intense meteorological event is not likely to happen. In fact, rainfall had a modest intensity (Fig. 7), with a maximum peak of 10.8 mm/h. The interpretation of HI maps and the analysis of its evolution leads to exclude that a severe meteorological event could occur, despite the observation of negative values of *Δ*PWV and its decreasing trend indicates a reduction of the water vapor content in the atmosphere, which can be due to rainfall.

4 Conclusions and Future Perspectives

The present work focuses on the possibility to remote-sense the water vapor content in atmosphere starting from GNSS observations, over an orographically complex area, using existing infrastructures, with a procedure called G4M. The application of G4M to four case studies focused on the city of Genoa in the years 2011 and 2014 showed its potential in analyzing different meteorological conditions. Among the case studies there are two severe meteorological events, a case characterized by a high value of PWV and little rainfall occurred, and a case in which a meteorological alert was released by the Regional Environmental Agency, but no severe meteorological event happened. The G4M outputs were compared with the rainfall data from the DICCA rain gauge, confirming that the G4M procedure is suitable for detecting and localizing intense meteorological events in space and time. In fact, a correlation was noticed between high values of *Δ*PWV associated with its high variability, represented by the Heterogeneity Index (HI), and intense rainfall. Note that the 1D analysis of HI for a specific site is not sufficient to localize a severe meteorological event in time. 2D HI and ΔPWV maps analysis is needed to detect severe meteorological events in time and space.

A future perspective regards the realization of a web application for the G4M procedure to make its application and results accessible to potential users. The original idea is to implement the procedure so that a potential user can obtain ΔPWV and HI maps as output of an a posteriori study of the desired event. Moreover, other tests on different case studies and different study areas will be undertaken to statistically evaluate the G4M performances. The applicability of G4M in near-real time will be studied in depth too, mainly concerning computational effort, needed data and criteria to raise a meteorological alert, as a contribute to Decision Support Systems.

Acknowledgements The authors wish to thank the Regional Environmental Agency (ARPAL), in particular Dr. Federico Cassola and Dr. Elisabetta Trovatore, for their cooperation in individuating significant meteorological scenarios.

The present work is a contribute to the IGS Troposphere Working Group from one of the authors, as a member.

References

Barindelli S, Realini E, Venuti G, Fermi A, Gatti A (2018) Detection of water vapor time variations associated with heavy rain in northern Italy by geodetic and low-cost GNSS receivers. Earth Planets Space 70(1):28

Benvenuto L, Marzocchi R, Ferrando I, Federici B, Sguerso D (2018) A procedure to manage open access data for post-processing in GIS environment. Technical report, PeerJ Preprints

Bevis M, Businger S, Herring TA, Rocken C, Anthes RA, Ware RH (1992) GPS meteorology: remote sensing of atmospheric water vapor using the Global Positioning System. J Geophys Res Atmos 97(D14):15787–15801

Douša J, Vaclavovic P (2014) Real-time zenith tropospheric delays in support of numerical weather prediction applications. Adv Space Res 53:1347–1358

Ferrando I, De Rosa P, Federici B, Sguerso D (2016) Spatial interpolation techniques for a near real-time mapping of pressure and temperature data. Technical report, PeerJ Preprints

Ferrando I, Federici B, Sguerso D (2017) Zenith total delay interpolation to support GNSS monitoring of potential precipitations. Geoing Ambient Mineraria 151(2):85–90

Ferrando I, Federici B, Sguerso D (2018) 2D PWV monitoring of a wide and orographically complex area with a low dense GNSS network. Earth Planets Space 70:54

Guerova G, Jones J, Douša J, Dick G, Haan Sd, Pottiaux E, Bock O, Pacione R, Elgered G, Vedel H, et al (2016) Review of the state of the art and future prospects of the ground-based GNSS meteorology in Europe. Atmos Meas Tech 9(11):5385–5406

Inoue HY, Inoue T (2007) Characteristics of the water-vapor field over the Kanto District associated with summer thunderstorm activities. SOLA 3:101–104

Oigawa M, Realini E, Tsuda T (2015) Study of water vapor variations associated with meso-γ scale convection: comparison between GNSS and non-hydrostatic model data. Scientific online letters on the atmosphere: SOLA 11:27–30

Oigawa M, Tsuda T, Seko H, Shoji Y, Realini E (2018) Data assimilation experiment of precipitable water vapor observed by a hyper-dense GNSS receiver network using a nested NHM-LETKF system. Earth Planets Space 70(1):74

Sapucci LF, Machado LAT, Menezes de Souza E, Campos TB (2016) GPS-PWV jumps before intense rain events. Atmos Meas Tech Discuss 2016:1–27

Seko H, Nakamura H, Shoji Y, Iwabuchi T (2004) The meso-γ scale water vapor distribution associated with a thunderstorm calculated from a dense network of GPS receivers. J Meteorol Soc Jpn Ser II 82(1B):569–586

Sguerso D, Labbouz L, Walpersdorf A (2013) 14 years of GPS tropospheric delays in the French-Italian border region: a data base for meteorological and climatological analyses. The International Archives of the Photogrammetry, Remote Sensing and Spatial Information Sciences XL-5/W3:7–14

Sguerso D, Labbouz L, Walpersdorf A (2016) 14 years of GPS tropospheric delays in the French–Italian border region: comparisons and first application in a case study. Appl Geomat 8(1):1–13

Shoji Y (2013) Retrieval of water vapor inhomogeneity using the Japanese nationwide GPS array and its potential for prediction of convective precipitation. J Meteorol Soc Jpn Ser II 91(1):43–62

Part IV

Advanced Numerical Methods in Geodesy

Check for updates

Modeling the Gravitational Field by Using CFD Techniques

Zhi Yin and Nico Sneeuw

Abstract

The Laplace equation represents harmonic (i.e., both source-free and curl-free) fields. Despite the good performance of spherical harmonic series on modeling the gravitational field generated by spheroidal bodies (e.g., the Earth), the series may diverge inside the Brillouin sphere enclosing all field-generating mass. Divergence may realistically occur when determining the gravitational fields of asteroids or comets that have complex shapes, known as the Complex-boundary Value Problem (CBVP). To overcome this weakness, we propose a new spatial-domain numerical method based on the equivalence transformation which is well known in the fluid dynamics community: a potential-flow velocity field and a gravitational force vector field are equivalent in a mathematical sense, both referring to a harmonic vector field. The new method abandons the perturbation theory based on the Laplace equation, and, instead, derives the governing equation and the boundary condition of the potential flow from the conservation laws of mass, momentum and energy. Correspondingly, computational fluid dynamics (CFD) techniques are introduced as a numerical solving scheme. We apply this novel approach to the gravitational field of comet 67P/Churyumov-Gerasimenko with a complex shape. The method is validated in a closed-loop simulation by comparing the result with a direct integration of Newton's formula. It shows a good consistency between them, with a relative magnitude discrepancy at percentage level and with a maximum directional difference of 5°. Moreover, the numerical scheme adopted in our method is able to overcome the divergence problem and hence has a good potential for solving the CBVPs.

Keywords

CFD techniques · Comet 67P/Churyumov-Gerasimenko · Finite Volume Method · Gravitational field modeling · Potential flow

1 Introduction

A gravitational field is the influence that a mass body extends into the space around itself, producing a force on another mass body. The gravitational field outside the masses has two important properties: source free and curl free (i.e., harmonicity). The main problem of physical geodesy is to establish a gravitational model, on the Earth's surface and in the outer space, to the extent made possible by existing data, mathematical analytical tools and numerical computational tools (Sansò et al. 2012). The analytical solution of the Laplace equation (e.g., spherical harmonics) plays an important role in building advanced geopotential models. The convergence of spherical harmonics is guaranteed outside a mass-enclosing reference sphere, also referred to as the Brillouin sphere (Hobson 2012). However, when it comes to the gravitational field modeling of complex shaped asteroids

Z. Yin (✉) · N. Sneeuw
Institute of Geodesy, University of Stuttgart, Stuttgart, Germany
e-mail: zhi.yin@gis.uni-stuttgart.de; yinzhi1221@sina.com

P. Novák et al. (eds.), *IX Hotine-Marussi Symposium on Mathematical Geodesy*,
International Association of Geodesy Symposia 151, https://doi.org/10.1007/1345_2019_72

or comets, the series at points near the surface of the body might diverge due to the large discrepancy between the Brillouin sphere and the actual shape of the body, indicating that spherical harmonics are adequate for representing the gravitational field of sphere-like bodies rather than those of complex shaped bodies. In this study, the gravitational field determination problem for complex shaped bodies is referred to as complex-boundary value problem (CBVP), which is still not well solved so far compared to other boundary value problems, e.g., those solvable by Newton's formula, Stokes's integral and Molodensky's series.

In the field of computational physics, benefiting from increasing computer power, Computational Fluid Dynamics (CFD) techniques developed rapidly in the past decades. CFD is the analysis of systems involving fluid flow, heat transfer and associated phenomena by means of computer-based simulation (Versteeg and Malalasekera 1995). The fundamental basis of almost all CFD problems are the Navier-Stokes equations defining single-phase (gas or liquid, but not both) fluid flows. By removing terms or by setting specific parameters, the equations for the general flow can be simplified to formulate various fluid flows, among which the potential flow is an idealized one that occurs in the case of incompressible, inviscid and irrotational fluid particles. In particular, the velocity field of potential flow is harmonic. It is known that the gravitational vector field and the potential flow velocity field are equivalent in the sense that both are harmonic fields. For this reason, we propose to utilize the governing equations of potential flow as well as CFD techniques to model gravitational fields, especially the CBVPs. The governing equation of potential flow is adopted as an alternative to the Laplace equation when modeling the gravitational field.

This paper mainly has two objectives: (1) to illustrate the principle and workflow of the new gravitational field modeling method; (2) to apply the new method to Comet 67P/Churyumov-Gerasimenko (hereafter referred to as Comet 67P). The two parts are arranged in Sects. 2 and 3, respectively. In Sect. 2, first, the principle of the equivalence transformation is outlined (Sect. 2.1), involving the issues of the three dimensional potential flow, the CFD techniques and the pipe transformation; then, the three issues are addressed in the Sects. 2.2–2.4, respectively. In Sect. 3, a computational workflow is devised to model the gravitational field of Comet 67P and the result is validated in a closed-loop simulation.

2 Methodology

2.1 Equivalence Transformation

The main idea of our method is based on the equivalence transformation between the gravitational vector field and the potential flow velocity field. Table 1 compares the two physical problems regarding their common and different points. Although belonging to different research fields, both vector fields are harmonic, thereby mutually equivalent. The gravitational vector $\boldsymbol{g}$ and the potential flow velocity vector $\boldsymbol{v}$ are interchangeable when formulating either vector field. In the following sections, Sects. 2.2–2.4, we first derive the mathematical expressions (i.e., the governing equation and boundary condition) of the potential flow, then introduce the principle of the CFD techniques exemplified with a pipe flow, and after that transform the pipe to make it suitable for the gravitational field modeling. Finally, after substituting all the velocity terms $\boldsymbol{v}$ with the gravitational terms $\boldsymbol{g}$ in the derived equations, the framework of our new method is established.

2.2 Three Dimensional Potential Flow

In this section, we first give the mathematical description of a general three dimensional fluid flow, and then impose on them specific constraints (listed in Table 2) to make the fundamental equations meet the potential flow's properties.

Table 1 Equivalence and difference between the gravitational field and the potential flow velocity field

		Gravitational field	Potential flow velocity field
Equivalence	Property	Source-free and curl-free	Source-free and curl-free
	Conceptual mapping	Gravitational potential	Velocity potential
		Gravitational force $\boldsymbol{g}$	Velocity $\boldsymbol{v}$
		Plumb line	Stream line
Difference	Fundamental theory	Perturbation theory	Potential flow theory
	Governing equation	Laplace's equation	• Continuity equation (mass conservation) • Momentum equation (momentum conservation)
	Boundary condition	• Fundamental equation of physical geodesy • Regularity condition at infinity	Bernoulli equation (energy conservation)
	Solving method	Spherical harmonics inversion (analytical method)	CFD techniques (numerical method)
	Processing domain	Spectral domain	Spatial domain

Table 2 Constraints imposed on a general fluid flow and the corresponding functions

Constraint	Function
$\rho = 1\ kg/m^3$	Incompressible (constant density)
$\frac{\partial}{\partial t}(\cdot) = 0$	Steady state
$\mu = 0$	Frictionless
$B_i = 0\ (i = 1, 2, 3)$	Body-force free (Ψ = const.)

2.2.1 Governing Equation

The governing equations of a general fluid flow can be found in the textbook written by Versteeg and Malalasekera (1995), including:

- Continuity equation:

$$\frac{\partial \rho}{\partial t} + \nabla \cdot (\rho \boldsymbol{v}) = 0 \tag{1}$$

- Momentum equations (Navier-Stokes equations):

$$\frac{\partial (\rho \boldsymbol{v})}{\partial t} + \nabla \cdot (\rho \boldsymbol{v} \otimes \boldsymbol{v}) = \nabla \cdot (\mu \nabla \boldsymbol{v}) + \boldsymbol{B} - \nabla p \tag{2}$$

The above two equations are under Eulerian description and the variables are functions of space and time: $\boldsymbol{v}$ denotes the velocity vector comprising the three components v_1, v_2 and v_3; ρ is density; p is pressure; μ is dynamic viscosity that controls the fluid flow behavior due to friction; $\boldsymbol{B}$ represents body force (e.g., gravitation, coriolis force or electromagnetic force). Among these variables, the three velocity components v_1, v_2 and v_3 and the pressure p are unknowns to solve for (see the SIMPLE algorithm in Sect. 2.3.2). In Eq. (2), the operator $\otimes$ denotes outer product, and the outer product of two vectors $\boldsymbol{u}$ and $\boldsymbol{v}$ is a matrix $\boldsymbol{w}$ given by $w_{ij} = u_i u_j$. With this definition, the second term on the left hand side of Eq. (2) could be expanded as follows:

$$\nabla \cdot (\rho \boldsymbol{v} \otimes \boldsymbol{v}) = \rho (\nabla \cdot \boldsymbol{v}) \boldsymbol{v} + \rho \boldsymbol{v} \cdot \nabla \boldsymbol{v} \tag{3}$$

Applying the constraints listed in Table 2 to Eqs. (1)–(3), one can obtain the governing equations of the three dimensional potential flow as follows:

$$\begin{cases} \nabla \cdot \boldsymbol{v} = 0 \\ \boldsymbol{v} \cdot \nabla \boldsymbol{v} = -\nabla p \end{cases} \tag{4}$$

2.2.2 Boundary Condition

The Bernoulli equation is a statement of conservation of energy for an inviscid flow, which implies an increase in the speed of a fluid simultaneously with a decrease in pressure. The Bernoulli equation for unsteady inviscid flows has the following form

$$-\frac{\partial \phi}{\partial t} + \frac{v^2}{2} + \frac{p}{\rho} + \psi = C \tag{5}$$

where ϕ is the velocity potential, Ψ is the conservative body force potential defined as $\nabla \Psi = \boldsymbol{B}$, and C is a constant number. Applied with the constraints of Table 2, Eq. (5) transforms into the following formula:

$$p = -\frac{1}{2} v^2 \tag{6}$$

which holds throughout the flow and is particularly used for calculating the boundary condition in our method. Note that the constants involved in the equation reduction process have been set to zero, and such treatment has no influence on the final solution.

2.3 CFD Techniques

CFD techniques mainly include four modules: problem identification, preprocessing, solving and post processing. In CFD the Finite Volume Method (FVM) is a method for representing and evaluating partial differential equations in the form of algebraic equations. Following the textbook written by Versteeg and Malalasekera (1995), this section mainly focuses on the basic FVM steps and illustrate with a three dimensional pipe flow (see Fig. 1a). For clarity, the illustration uses structured control volumes (CVs) characterized by hexahedrons.

2.3.1 Step 1: Grid Generation

The first step in the FVM is to divide the computation domain into discrete CVs (aka cells, elements) where the variable of interest (i.e., velocities and pressure) is located at the nodal point (i.e., the centroid of the CV). Assume there are a number of nodal points located inside the domain, the faces of CVs are positioned mid-way between adjacent nodes, and each node is surrounded by a CV, as demonstrated in Fig. 1.

A system of notation is established as follows. Figure 1b shows a general nodal point identified by P with six neighboring nodes in a three-dimensional geometry identified as west, east, south, north, bottom and top nodes (W, E, S, N, B, T). The lowercase notation, w, e, s, n, b and t are used to refer to cell faces in the corresponding directions. The x_1-, x_2- and x_3-components of the coordinate system are aligned with the directions of WE, SN and BT, respectively. The distances between two geometrical elements are denoted

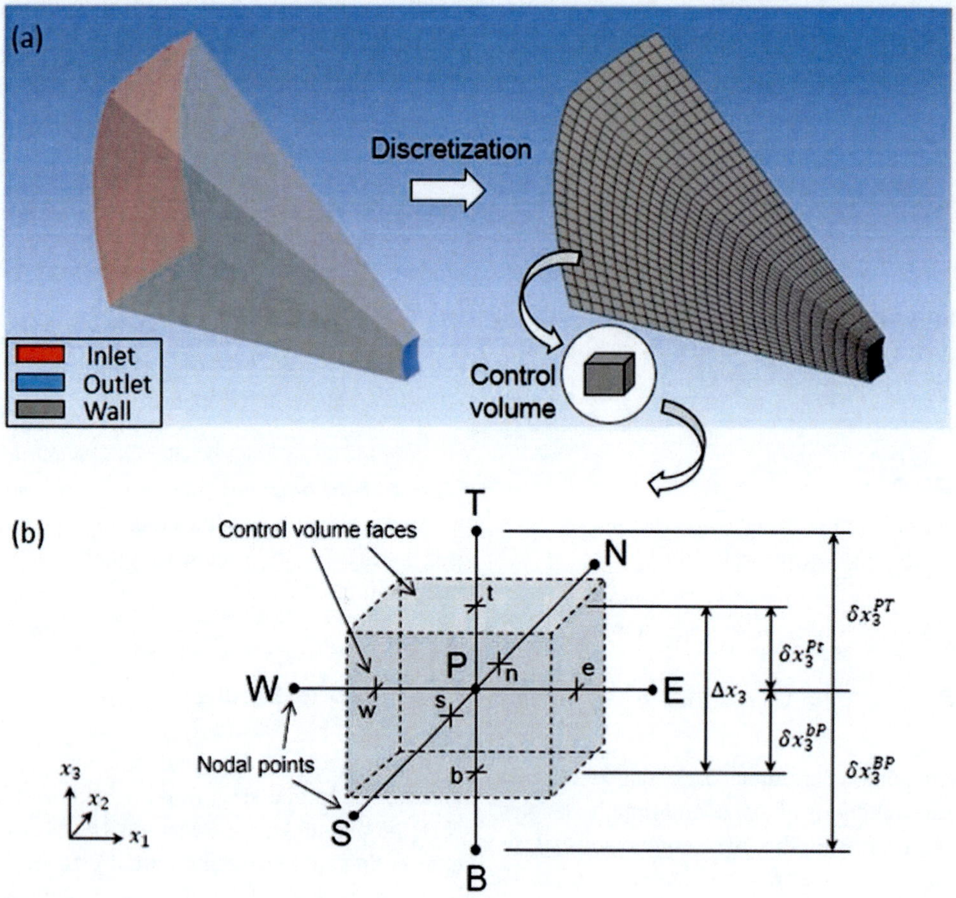

Fig. 1 (**a**) Discretization of a pipe flow domain. (**b**) Notations of a finite control volume

as $\delta x_1^{\cdot}$, $\delta x_2^{\cdot}$ and $\delta x_3^{\cdot}$, with the superscripts occupied by the node or face notations. For example, the distances between the nodes P and T, between the node P and the face t and between the faces b and t, are identified by δx_3^{PT}, δx_3^{Pt} and δx_3^{bt}, respectively. By contrast, the area A of one face t could be denoted as A^t.

2.3.2 Step 2: Discretization

A critical operation of the FVM is the integration of the governing equation (Eq. 4) over a CV to yield a discretized equation at its nodal point P. For the CV defined in Fig. 1b, this gives

$$\begin{cases} \int_{\Delta V} \nabla \cdot \boldsymbol{v} dV = 0 \\ \int_{\Delta V} \nabla \cdot (\boldsymbol{v} v_i) dV = \int_{\Delta V} \left(-\frac{\partial p}{\partial x_i}\right) dV, \ (i = 1, 2, 3) \end{cases} \tag{7}$$

where ΔV is the integral volume of each cell. The above equation set represents the flux balance of the mass and the momentum of the potential flow in a CV. Applying the Gauss theorem to the left hand side terms of equation set (Eq. 7), one can obtain

$$\begin{cases} \int_{A} \boldsymbol{n} \cdot \boldsymbol{v} dA = 0 \\ \int_{A} \boldsymbol{n} \cdot (\boldsymbol{v} v_i) dA = \int_{\Delta V} \left(-\frac{\partial p}{\partial x_i}\right) dx_1 dx_2 dx_3 \ , (i = 1, 2, 3) \end{cases} \tag{8}$$

where $\boldsymbol{n}$ is the unit normal vector of the CV face element dA. Expanding equation set (Eq. 8) yields

$$\left(A^e v_1^e - A^w v_1^w\right) + \left(A^n v_2^n - A^s v_2^s\right) + \left(A^t v_3^t - A^b v_3^b\right) = 0 \tag{9}$$

and

$$\begin{aligned} &\left(A^e v_1^e v_i^e - A^w v_1^w v_i^w\right) + \left(A^n v_2^n v_i^n - A^s v_2^s v_i^s\right) \\ &+ \left(A^t v_3^t v_i^t - A^b v_3^b v_i^b\right) = -A_i^P \Delta p_i^P, \quad (i = 1, 2, 3) \end{aligned} \tag{10}$$

Note that in Eq. (10) the unknowns of velocity components with lowercase superscripts are defined at the CV faces, rather than at the nodal points where the final solution should be defined. Therefore, the velocity components at the CV faces must be expressed by those at the nodal points with a specific interpolation scheme, to transform Eq. (10) into a linear form solvable by matrix inversion

$$a^P v_i^P = a^W v_i^W + a^E v_i^E + a^S v_i^S + a^N v_i^N + a^B v_i^B + a^T v_i^T - A_i^P \Delta p_i^P, \quad (i = 1, 2, 3) \tag{11}$$

where the coefficients of the unknowns still contain unknown variables (i.e., the velocities at faces). For this reason, an iterative algorithm is needed to solve this equation system, which is the main topic of Sect. 2.3.3.

2.3.3 Step 3: Solution of Equations

The discretized momentum equations (Eq. 11) must be set up at each nodal point in the flow domain; for the nodal points adjacent to the domain boundaries, boundary conditions can be incorporated. After constituting the discretized momentum equations for all the CVs, one can solve for the overall velocities v_i^P $(i = 1, 2, 3)$ and pressure p. At the same time, the continuity equation (Eq. 9) needs to be satisfied as well.

In CFD, the Semi-Implicit Method for Pressure Linked Equations (SIMPLE) algorithm developed by Patankar and Spalding (1972) is a widely used numerical procedure to solve the Navier-Stokes equations. It is essentially based on a guess-and-correct procedure. For the details of this algorithm, please refer to the textbook written by Versteeg and Malalasekera (1995). The SIMPLE algorithm is an iterative algorithm and does not stop until the balances of mass and the momentums of the three velocity components are satisfied.

2.4 Pipe Transformation

Up to this point, the harmonic velocity field of a three dimensional pipe flow can be solved with the CFD techniques. To model the gravitational field of one mass body, the pipe needs to be transformed following the procedure shown in Fig. 2, changing into a special pipe that only contains one inlet and one outlet (Fig. 2g). The inlet is a sphere with its center located at the center of mass, and the outlet is the surface of mass. Due to the interchangeability of the gravitational vector $\boldsymbol{g}$ and the potential flow velocity $\boldsymbol{v}$, the fundamental equations of the gravitational field determination problem are given from Eqs. (4) and (6) as follows.

- Governing equations:

$$\begin{cases} \nabla \cdot \boldsymbol{g} = 0 \\ \boldsymbol{g} \cdot \nabla \boldsymbol{g} = -\nabla p \end{cases} \tag{12}$$

- Boundary condition:

$$p = -\frac{1}{2}\boldsymbol{g}^2 \tag{13}$$

Note that only the scalar gravitation is needed to calculate the boundary value, p. For the inlet (i.e., the outer sphere) with a radius large enough (denoted as r, usually ten times the average size of object), the scalar gravitation thereon could be calculated with the Newton's formula of gravitation, $\frac{GM}{r^2}$, by approximating the body as a mass point with the total mass amount of M. For the outlet (i.e., the mass surface),

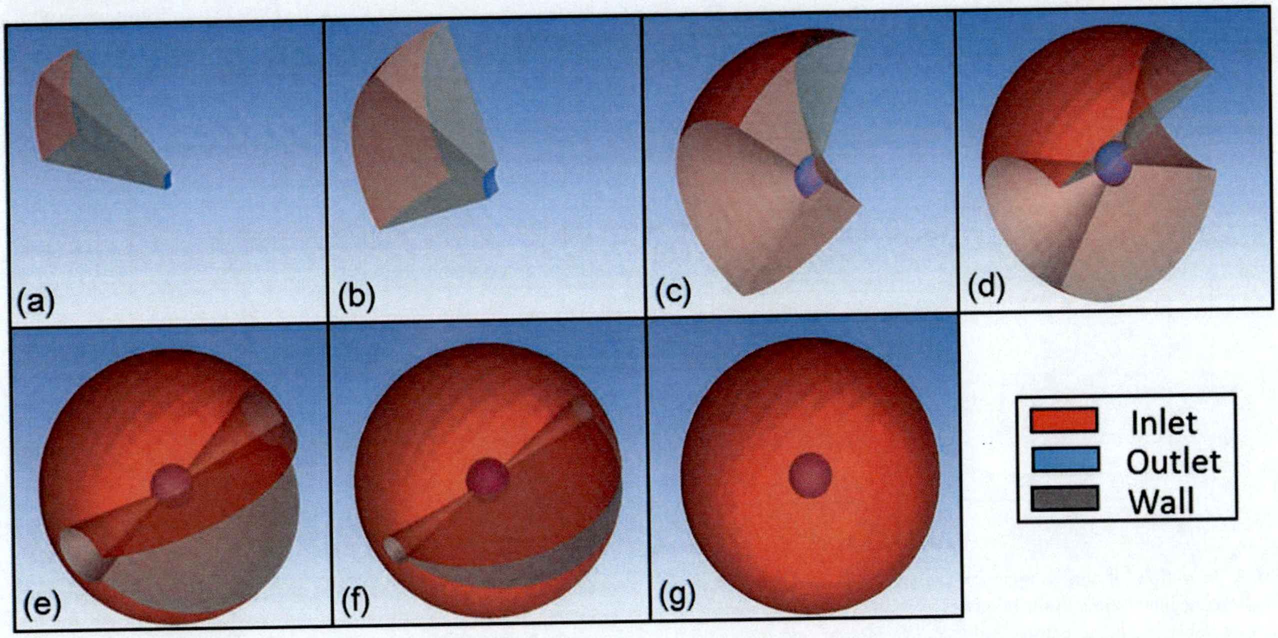

Fig. 2 Geometrical transformation of a pipe. (a) A normal pipe with an inlet, an outlet and four walls. (g) A pipe only with an inlet and an outlet. Panels (b–f) demonstrate the intermediate transformation process

the scalar gravitation could be obtained either by direct measurement or by integration from an assumed density distribution.

3 Example: Comet 67P

To validate the new method, we model the gravitational field of Comet 67P, the target of ESA's comet-chasing Rosetta mission. The shape model is developed by ESA's Rosetta archive team, consisting of 104,207 faces, matching NVC image data gathered up to October 2016. The basic information of the two-lobe comet is given in Table 3 and the shape model can be downloaded from https://imagearchives.esac.esa.int/index.php?/page/navcam_3d_models. We conduct the case study in a closed-loop simulation. Figure 3 shows the workflow. A CFD solution is derived following our new method, and, taken as a benchmark, the direct Newton integration sums up gravitational components generated by point masses (condensed by CV masses onto their centroids), forming a rigorously harmonic gravitational field. After about 600 iterations the SIMPLE algorithm converges, and we obtain the gravitational vector field of Comet 67P (Fig. 4).

Table 3 Basic information of Comet 67P/Churyumov-Gerasimenko (ESA 2016)

Size of nucleus:	
Overall dimensions	4.34 km × 2.60 km × 2.12 km
Small lobe	2.50 km × 2.14 km × 1.64 km
Large lobe	4.10 km × 3.52 km × 1.63 km
Mass	1.0×10^{13} kg
Volume	18.0 km^3
Density	533.0 kg/m^3

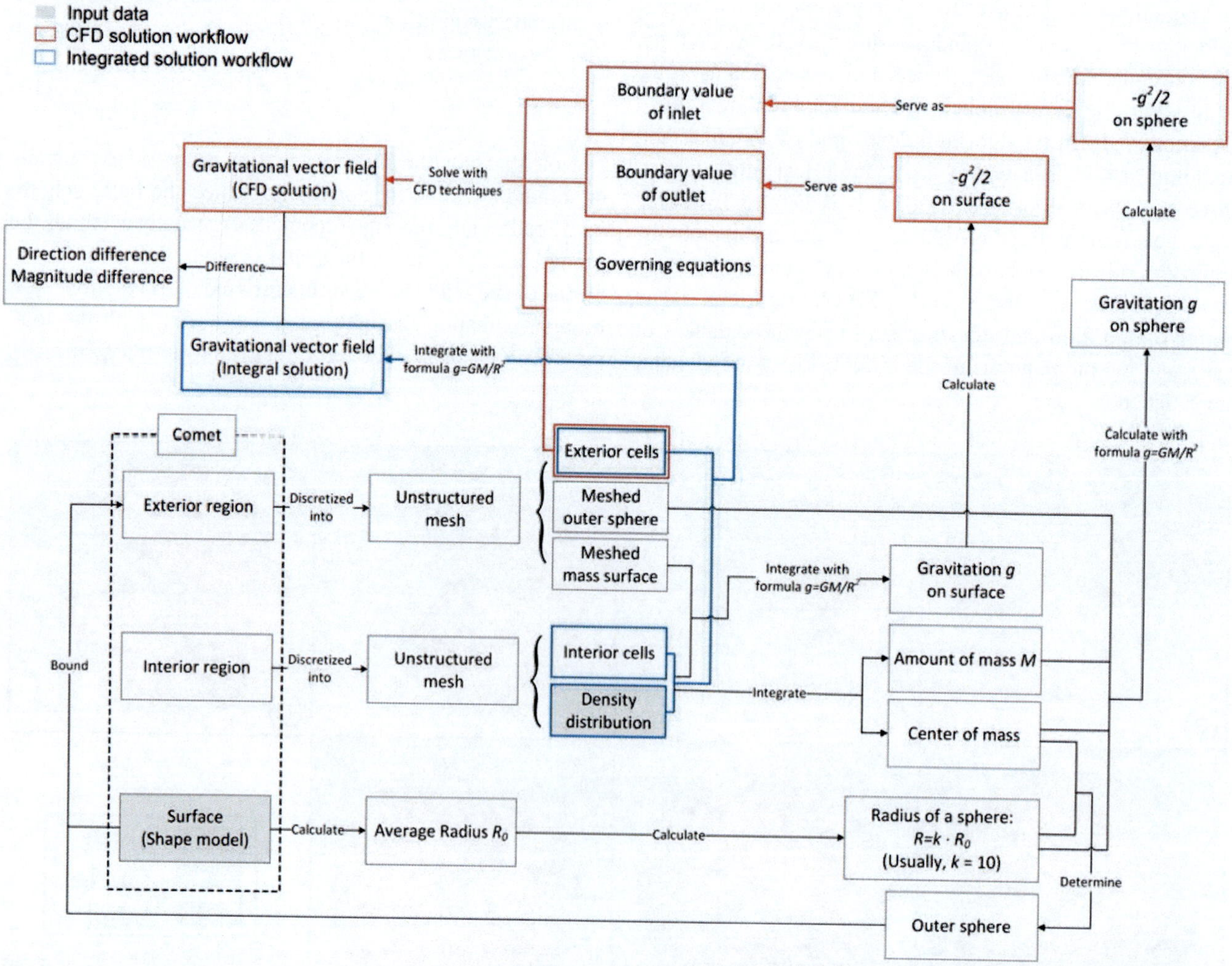

Fig. 3 Workflow of the comparison between the CFD solution and the Newton integration. Both branches are simulated from two input data: the shape and the density models. The interior and the exterior meshes are generated separately. The mass of each interior tetrahedron is assumed to be condensed as a point mass at the centroid for the subsequent Newton integration

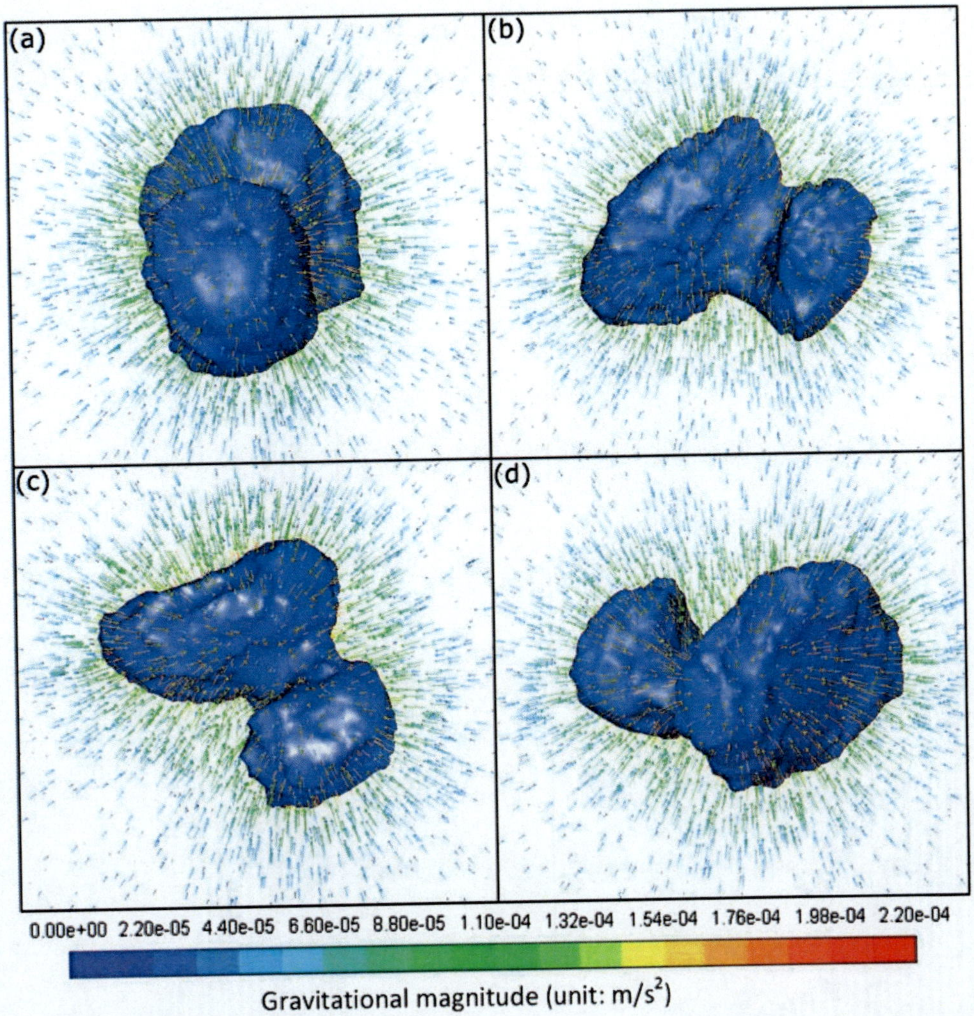

Fig. 4 The gravitational vector field of Comet 67P viewed from (a) the top, (b) and (c) the side and (d) the oblique perspective

In order to compare the CFD solution (velocity vectors denoted as $\boldsymbol{v}$) and the direct Newton integration (gravitational vectors denoted as $\boldsymbol{g}$), we adopt the following two indicators:

- Relative error

$$\Delta e = \left| \frac{|\boldsymbol{v}| - |\boldsymbol{g}|}{|\boldsymbol{g}|} \right| \tag{14}$$

- Angular difference

$$\Delta \theta = \arccos \frac{\boldsymbol{v} \cdot \boldsymbol{g}}{|\boldsymbol{v}|\,|\boldsymbol{g}|} \tag{15}$$

The above two indicators evaluate the magnitude and the directional differences, respectively, with smaller values indicating a better consistence of the two solutions. Figure 5 shows the histograms for a total of 457,793 validation points. The overall magnitude order of the relative errors is smaller than 10% and the angular difference is within 5°. Their mean values are 2% and 1.21°, respectively, marked with red lines in Fig. 5. The comparisons indicate a good consistence between the CFD solution and the benchmark solution, thereby proving the good performance of the new method on solving the CBVPs. However, the numerical error performance of the new method needs to be investigated as a further study in the future.

4 Conclusion

A new gravitational field modeling method is proposed in this research, elaborated from both the theoretical and the application aspects:

- A new equation set, including the governing equation (Eq. 12) and the boundary condition (Eq. 13), is derived from the potential flow theory and is equivalent to the Laplace equation when expressing the gravitational field.

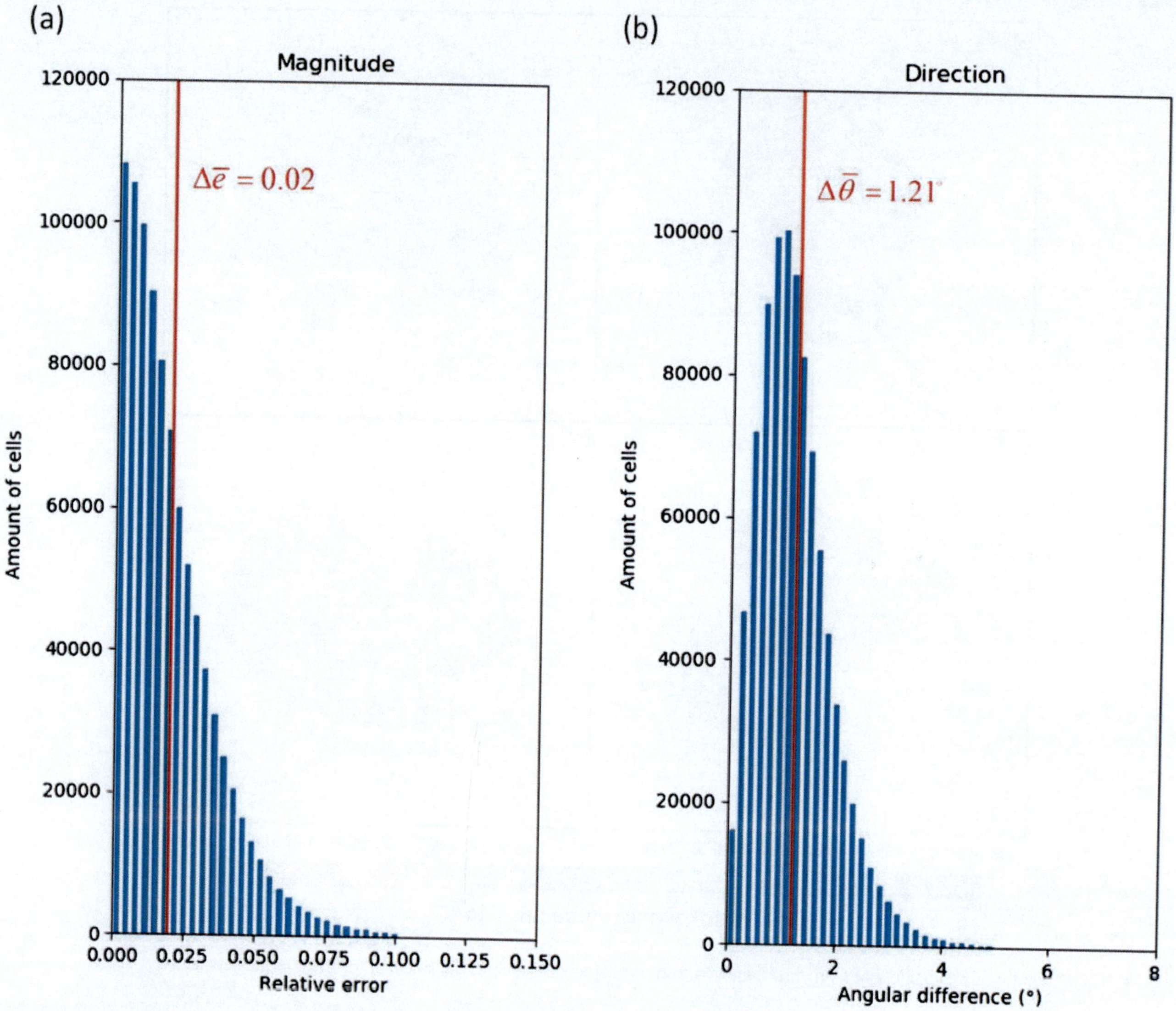

Fig. 5 Histograms of (**a**) the relative error and (**b**) the directional difference between the CFD solution and the integrated solution. Red lines mark the average values of the two indicators

- The CFD techniques are introduced as numerical tools to solve the new fundamental equation, and a gravitational field modeling workflow is devised. The gravitational field generated by Comet 67P is exemplified to validate our method.

The method is dedicated to solving gravitational fields numerically, with the advantage of circumventing the perturbation theory, the basis of the methods using the Laplace equation. From both theoretical and practical point of view, this new method has a good performance on solving Complex Boundary Value Problems, overcoming the divergence problem of conventional approaches.

Acknowledgements This work is supported by Sino-German (CSC-DAAD) Postdoc Scholarship Program. Zhi Yin gratefully acknowledges Prof. Caijun Xu for the PhD supervision allowing him to acquire necessary research skills for the realization of this work.

References

ESA (2016) Rosetta's target: Comet 67p/churyumov gerasimenko. http://sci.esa.int/rosetta/14615-comet-67p/

Hobson EW (2012) The theory of spherical and ellipsoidal harmonics. Cambridge University Press, Cambridge

Patankar S, Spalding D (1972) A calculation procedure for heat, mass and momentum transfer in three-dimensional parabolic flows. Int J Heat Mass Transf 15(10):1787–1806. https://doi.org/10.1016/0017-9310(72)90054-3

Sansò F, Barzaghi R, Carrion D (2012) The geoid today: still a problem of theory and practice. In: Sneeuw N, Novák P, Crespi M, Sansò F (eds) VII Hotine-Marussi symposium on mathematical geodesy. Springer, Berlin, Heidelberg, pp 173–180

Versteeg HK, Malalasekera W (1995) An introduction to computational fluid dynamics: the finite volume approach. Longman Scientific & Technical, Harlow

Surface Loading of a Self-Gravitating, Laterally Heterogeneous Elastic Sphere: Preliminary Result for the 2D Case

Yoshiyuki Tanaka, Volker Klemann, and Zdeněk Martinec

Abstract

Advancements in the Global Geodetic Observing System (GGOS) have enabled us to investigate the effects of lateral heterogeneities in the internal Earth structure on long-term surface deformations caused by the Glacial Isostatic Adjustment (GIA). Many theories have been developed so far to consider such effects based on analytical and numerical approaches, and 3D viscosity distributions have been inferred. On the other hand, fewer studies have been conducted to assess the effects of lateral heterogeneities on short-term, elastic deformations excited by surface fluids, with 1D laterally homogeneous theories being frequently used. In this paper, we show that a spectral finite-element method is applicable to calculate the elastic deformation of an axisymmetric spherical Earth. We demonstrate the effects of laterally heterogeneous moduli with horizontal scales of several hundred kilometers in the upper mantle on the vertical response to a relatively large-scale surface load. We found that errors due to adopting a 1D Green's function based on a local structure could amount to 2–3% when estimating the displacement outside the heterogeneity. Moreover, we confirmed that the mode coupling between higher-degree spherical harmonics needs to be considered for simulating smaller-scale heterogeneities, which agreed with results of previous studies.

Keywords

Finite element method · GGOS · Lateral heterogeneity · Mass redistribution · Surface loading

1 Introduction

A surface-loading response is a geophysical process that describes deformation of the solid Earth due to variations in surface fluids including the atmosphere, ocean, continental water and ice sheets. Short-term responses to these loads are usually modeled by elastic deformation. Farrell (1972)'s theoretical framework based on Green's function (GF) method is well known. Mathematically, it enables us to estimate local as well as global scale elastic deformations of a self-gravitating, layered sphere to a load applied at the surface of the Earth.

On the other hand, long-term responses of the Earth caused by the GIA have been modeled by viscoelastic

Y. Tanaka (✉)
Department of Earth and Planetary Science, The University of Tokyo, Tokyo, Japan
e-mail: y-tanaka@eps.s.u-tokyo.ac.jp

V. Klemann
Department 1: Geodesy, Helmholtz Centre Potsdam—GFZ German Research Centre for Geosciences, Potsdam, Germany

Z. Martinec
Faculty of Mathematics and Physics, Charles University in Prague, Praha 8, Czech Republic

School of Cosmic Physics, Dublin Institute for Advanced Studies, Dublin 2, Ireland

P. Novák et al. (eds.), *IX Hotine-Marussi Symposium on Mathematical Geodesy*,
International Association of Geodesy Symposia 151, https://doi.org/10.1007/1345_2019_62

relaxation. A number of authors have developed theoretical models for laterally homogeneous (e.g., Peltier 1974) and heterogeneous (e.g., D'Agostino et al. 1997; Kaufmann and Wolf 1999; Wu 2002) cases (see the review by Whitehouse (2018) for more details). The authors of this paper also developed a computational method in which 3D viscosity distributions and variations in the gravitational field are naturally treated on a global scale (Martinec 2000; Klemann et al. 2008; Tanaka et al. 2011). In particular, accompanied with the advancement in numerical computational ability, increasingly more models that consider a 3D heterogeneous internal structure of the Earth have been presented to interpret the deformations and gravity changes detected by GGOS. For example, Milne et al. (2018) estimated the effects of a horizontally varying thickness of the lithosphere (and thus large lateral viscosity contrasts) on the GIA in Greenland with a finite volume formulation.

Compared with the case of viscoelastic relaxation, relatively few studies have assessed the effects of 3D heterogeneous structures on elastic responses to a load, probably due to the fact that lateral heterogeneities in elastic constants and the density are much smaller than those in viscosity and harder to detect by observations. Ito and Simons (2011) employed GF for the 1D structure to constrain local density and elastic structure, analyzing ocean loading observed by the global navigation satellite systems. Dill et al. (2015) approximately estimated the effects of lateral heterogeneities on elastic responses to atmospheric and hydrological loads, using local GF for the 1D structure of the respective crustal structure below the load. More details about theoretical models of elastic surface loading and sensitivity studies regarding Earth structure parameters in past studies are summarized in Martens et al. (2016).

In the present paper, we report that the abovementioned method developed for considering 3D viscosity distribution (Tanaka et al. 2011) can be modified to take into account lateral heterogeneities in elastic constants. Preliminary computation results are presented that demonstrate the effects of a large-scale (400–1,600 km) heterogeneity on surface vertical displacement. Furthermore, we investigate the effects of mode coupling on the elastic response as Wu (2002) did for the viscoelastic response.

2 Method

2.1 A Theory of Elastic Response in the 1D Earth Structure

The governing equations for an initially hydrostatic, self-gravitating, layered sphere consist of a quasi-static equilibrium of stress, Poisson's equation and a strain-stress relation for the isotropic elastic material (Farrell 1972; Tanaka et al. 2011):

$$\nabla \cdot \boldsymbol{\tau} - \rho_0 \nabla \phi_1 + \nabla \cdot (\rho_0 \boldsymbol{u}) \nabla \phi_0 - \nabla (\rho_0 \boldsymbol{u} \cdot \nabla \phi_0) = \boldsymbol{0} \quad (1)$$

$$\nabla^2 \phi_1 + 4\pi G \nabla \cdot (\rho_0 \boldsymbol{u}) = 0 \quad (2)$$

$$\boldsymbol{\tau} = \lambda (\nabla \cdot \boldsymbol{u}) \boldsymbol{I} + 2\mu \boldsymbol{\varepsilon} \quad (3)$$

$$\boldsymbol{\varepsilon} = \frac{1}{2} (\nabla \boldsymbol{u} + \nabla^T \boldsymbol{u}) \quad (4)$$

where the subscripts 0 and 1 denote the reference state and the perturbation, respectively, and G, $\boldsymbol{I}$ and $\boldsymbol{\varepsilon}$ are the gravitational constant, the second-order identity tensor and strain tensor, respectively. The boundary condition comprises increments of normal stress and gravity potential due to a surface load. The density ρ_0 and elasticity constants λ and μ depend only on radial distance r. By applying a spherical harmonic expansion of stress $\boldsymbol{\tau}$, displacement $\boldsymbol{u}$ and the gravity potential ϕ, angular dependencies are completely separated by the orthogonality of spherical harmonics, and a differential equation with respect to r is obtained. This equation is numerically solved, and GF for a point-mass load dependent on r and θ (angular distance from the load) is obtained by summing spherical harmonics. A Heaviside-type source time function is often used. Since the inertial term is neglected in the governing equation, GF instantaneously jumps at $t = 0$ and becomes constant at $t > 0$:

$$[u, v]_{r=a} = \frac{a}{M} \sum_j \left[h_j, l_j\right]_{r=a} P_j (\cos\theta) \quad (5)$$

where u and v are vertical and horizontal displacements, respectively; j, a, M and $P_j(\cos\theta)$ represent spherical harmonic degree, Earth's radius and mass, and Legendre function, respectively; h_j and l_j denote the load Love numbers.

2.2 Modification for the 3D Elastic Case

Tanaka et al. (2011) solve a viscoelastic problem when the above strain-stress relation is replaced by Maxwell rheology:

$$\frac{\mathrm{d}}{\mathrm{d}t}\boldsymbol{\tau} = \frac{\mathrm{d}}{\mathrm{d}t}\boldsymbol{\tau}^E - \frac{\mu}{\eta}\left(\boldsymbol{\tau} - \frac{1}{3} tr\, \boldsymbol{\tau} \mathbf{I}\right) \quad (6)$$

$$\boldsymbol{\tau}^E \equiv \lambda (\nabla \cdot \boldsymbol{u}) \boldsymbol{I} + 2\mu \boldsymbol{\varepsilon} \quad (7)$$

where η is 3D viscosity dependent on (r, θ, φ) and the superscript E means an elastic part. The density and elasticity constants are still radially symmetric. In contrast to the elastic case, a transient deformation occurs for $t > 0$. Discretizing the above equation with respect to the time derivative, we have

$$\boldsymbol{\tau}^{i+1} = \boldsymbol{\tau}^{E,i+1} + \boldsymbol{\tau}^{V,i} \tag{8}$$

and

$$\nabla \cdot \boldsymbol{\tau}^{E,i+1} - \rho_0 \nabla {\phi_1}^{i+1} + \nabla \cdot \left(\rho_0 \boldsymbol{u}^{i+1}\right) \nabla \phi_0 - \nabla \left(\rho_0 \boldsymbol{u}^{i+1} \cdot \nabla \phi_0\right) = -\nabla \cdot \boldsymbol{\tau}^{V,i} + \mathbf{f}^{i+1} \tag{9}$$

where i is the index representing time step, and $\mathbf{f}$ denotes a force representing the surface boundary condition. The variational equality, which is constructed to apply a finite-element method, can be written as

$$\delta E^{i+1} = \delta F^{i}_{diss} + \delta F^{i+1}_{surf} \tag{10}$$

(eq. (11) of Tanaka et al. 2011). The left-hand side includes the energy functionals associated with elastic deformation, self-gravitation and uniqueness of the solution, and the right-hand side corresponds to the viscous dissipation and the surface boundary condition. To be more specific, the energy functionals related to bulk and shear deformations and the dissipation are written as

$$\delta E^{i+1}_{bulk} = \int_V \lambda (\nabla \cdot \boldsymbol{u}^{i+1})(\nabla \cdot \delta \boldsymbol{u}) dV, \tag{11}$$

$$\delta E^{i+1}_{shear} = 2\int_V \mu \left(\boldsymbol{\varepsilon}^{i+1} \cdot \delta \boldsymbol{\varepsilon}\right) dV, \tag{12}$$

$$\delta F^{i}_{diss} = -\int_V \left(\boldsymbol{\tau}^{V,i} \cdot \delta \boldsymbol{\varepsilon}\right) dV, \tag{13}$$

where V denotes the volume of a sphere. At the first time step ($i = 0$), δF^{i}_{diss} is set to zero, and the instantaneous elastic response is obtained from Eq. (10), which agrees with Farrell's result. At $i = 1$, the dissipation term is computed with help of the solution at $i = 0$. In the following time steps, the dissipation term is evaluated with the solution at the previous time step, and the contribution of the boundary condition at the present time step drives the elastic deformation at the present time step $i + 1$.

As in the elastic case, a spherical harmonic expansion of δE^{i+1} and δF^{i+1}_{surf} gives a differential equation regarding r. This equation is numerically solved by applying a 1D finite-element representation. However, the dissipation term includes 3D viscosity, and the orthogonality of spherical harmonics do not simplify this term as for a 1D case. Therefore, the integration with respect to θ and φ is evaluated numerically using the Gauss-Legendre quadrature and fast Fourier transform. This approach, which combines the 1D finite elements and tensor spherical harmonics, allows us to model viscoelastic deformations of the whole sphere with a natural treatment of self-gravitation. No constraints to suppress artificial translations are necessary.

We now consider a 3D heterogeneity in the rigidity. To explain the essence of the method, we write only the term associated with the shear deformation on the left-hand side of Eq. (10):

$$\delta E^{i+1}_{shear} = -\int_V \left(\boldsymbol{\tau}^{V,i} \cdot \delta \boldsymbol{\varepsilon}\right) dV + \delta F^{i+1}_{surf}. \tag{14}$$

In the elastic case, the viscous dissipation does not occur. Therefore, the elastic response is obtained by solving

$$\delta E^{i+1}_{shear} = \delta F^{i+1}_{surf}. \tag{15}$$

We decompose the rigidity on the left-hand side into the spherically symmetric part and residual. Then, we have

$$\begin{aligned} \delta E^{i+1}_{shear} &= \int_V 2\mu_0(r) \left(\boldsymbol{\varepsilon}^{i+1} \cdot \delta \boldsymbol{\varepsilon}\right) dV \\ &\quad + \int_V 2\Delta\mu\,(r,\theta,\varphi) \left(\boldsymbol{\varepsilon}^{i+1} \cdot \delta \boldsymbol{\varepsilon}\right) dV. \end{aligned} \tag{16}$$

Transferring the residual part into the right-hand side of Eq. (16) yields

$$\begin{aligned} &\int_V 2\mu_0(r) \left(\boldsymbol{\varepsilon}^{i+1} \cdot \delta \boldsymbol{\varepsilon}\right) dV \\ &= -\int_V 2\Delta\mu\,(r,\theta,\varphi) \left(\boldsymbol{\varepsilon}^{i+1} \cdot \delta \boldsymbol{\varepsilon}\right) dV + \delta F^{i+1}_{surf}. \end{aligned} \tag{17}$$

We note that Eq. (17) coincides with Eq. (14) for the viscoelastic case of $\Delta\mu = 0$ if we replace $\boldsymbol{\tau}^{V,i}$ in Eq. (14) by $2\Delta\mu\boldsymbol{\varepsilon}^{i+1}$. Now, we replace the time index $i+1$ in $2\Delta\mu\boldsymbol{\varepsilon}^{i+1}$ to i. However, the elastic response must be constant in time so that the solution should not depend on i. Here, we change the meaning of i from a time step to a number of iterations. It follows that, for $i = 0$, the elastic deformation caused by the spherically symmetric part is obtained. For $i = 1$, the elastic solution is corrected for the contribution from the residual part, which is estimated using the solution for $i = 0$. Because $\Delta\mu$ is assumed to be significantly smaller than μ_0, the correction term decreases with iterations. We confirmed that the computational results presented in the later section converge within 0.1% after 4–5 iterations. Heterogeneities in the bulk modulus can be considered in a similar manner:

$$\begin{aligned} &\int_V \lambda_0 \left(\nabla \cdot \boldsymbol{u}^{i+1}\right) (\nabla \cdot \delta \boldsymbol{u})\, dV \\ &= -\int_V \Delta\lambda (\nabla \cdot \boldsymbol{u}^{i})(\nabla \cdot \delta \boldsymbol{u}) dV + \delta F^{i+1}_{surf}. \end{aligned} \tag{18}$$

Table 1 Earth structure models used for the comparisons

Model	Remark	$\Delta\lambda$ or $\Delta\mu$	r (km)	θ (degree)
A	1D (PREM)	±0%	0–6,371	0–180
B	1D	+10%	5,701–6,371	0–180
C1	2D	+10%	5,701–6,371	0–4
C2	2D	+10%	5,701–6,371	4–8
C3	2D	+10%	5,701–6,371	12–20
D	2D	+10%	5,701–6,371	0–16

The numbers represent the relative differences in the elastic constants with respect to the PREM

2.3 The Load and Earth Models

We employed a disk load with relatively large-scale mass redistributions detectable with satellite gravity missions such as GRACE and GRACE-FO (https://gracefo.jpl.nasa.gov/). Cut-off degree of the spherical harmonics was set to 300. To investigate the behavior of GFs obtained for different Earth models, a unit mass (1 kg) was applied at $0^\circ \le \theta \le 8^\circ/16^\circ$.

Table 1 shows the Earth structure for models A, B, C1-C3 and D. In model A, the density and elastic constants of PREM (Dziewonski and Anderson 1981) is employed with those of the ocean being replaced by that of the crust. We consider 2D heterogeneities as a first step, where the model parameters for C1–C3 and D are axisymmetric, i.e., dependent on r and θ. It follows that computed GF is also independent of the azimuth. The scales of the heterogeneities were set to 4 – 16° in the horizontal direction and 670 km in the depth. A 10% increase in the elastic constants corresponds to a ~5% decrease in seismic wave velocities, which is possible in a plate subduction zone. Because the primary purpose of this paper is to report that our previous method is applicable to a laterally heterogeneous elastic deformation, we show results for only these limited Earth models.

2.4 Validation

The computational scheme for the laterally homogeneous part was already validated using independent methods (Spada et al. 2011; Tanaka et al. 2011). To confirm the above presented method, we computed the displacements for model B in two different ways. Model B is radially symmetric, so we can compute the elastic response using the method of Tanaka et al. (2011) directly (i.e., we solve the viscoelastic equation for $i = 0$ (Eq. 10)). We compared this result with that obtained by the proposed iteration method. The number of the 1D finite elements in the vertical direction is 516. The differences in the load Love numbers become smaller than 0.1% after 4–5 iterations, and this outcome is practically sufficient.

3 Results and Discussions

3.1 Effects Due to the Distribution of Lateral Heterogeneity

Figure 1 displays the vertical responses to the disk load of radius 8° for models A, B and C1. We see that the subsidence for model B is smaller than that for A, reflecting that the bulk/shear modulus of the upper mantle for model B is 10% larger than that for A. The difference in the displacements are a few percent where the load is distributed ($\theta < 8^\circ$). In model C1, the upper mantle is harder by the same amount but for $\theta < 4^\circ$ only. The displacement for model C1 agrees with that for model B nearer the center of the load. Outside the region, where the local heterogeneity is considered, the result for C1 coincides with that for model A. For $3^\circ < \theta < 5^\circ$, the response for C1 is located between the other two curves. This transitional behavior due to the local heterogeneity agrees with our intuition. This result also means that, when we employ a 1D GF that assumes a local structure below the center of the load to estimate the vertical displacement, a relative error up to 3% can occur in the area of $4^\circ < \theta < 8^\circ$.

The next example shows the cases where the heterogeneities are given outside the disk load. The left and right in Fig. 2 show the cases where a local heterogeneity is given as a ring structure at $8^\circ < \theta < 12^\circ$ (C2) and $12^\circ < \theta < 20^\circ$ (C3), respectively. In contrast to the cases in Fig. 1, the results for C2 and C3 are almost identical to the result for model A, indicating that the effects of the farther heterogeneities are much smaller than those in the cases of Fig. 1, where the heterogeneities are considered below the load. To see the effects in more detail, the differences between the results for C2/C3 and A are magnified (solid black curves). Comparing the displacements in the left and right figures, we note that the relative difference from model A decreases when the location of the local heterogeneity is farther from the load. This result indicates that lateral heterogeneities nearer the load affect the surface vertical deformation more strongly. This is understood from the fact that the integrand of the first term on the right-hand side of Eqs. (17) and (18) becomes largest when the location of the heterogeneity agrees with the area where the strain change is larger (i.e., near the load). Moreover, if we employ a 1D GF that assumes a local structure in a far field to estimate the vertical displacement near the load, an error of 3% can occur at maximum (dashed curves).

3.2 Some Remarks on Mode Coupling

In the presence of lateral heterogeneities, coupling between spheroidal and toroidal modes generally occurs. However,

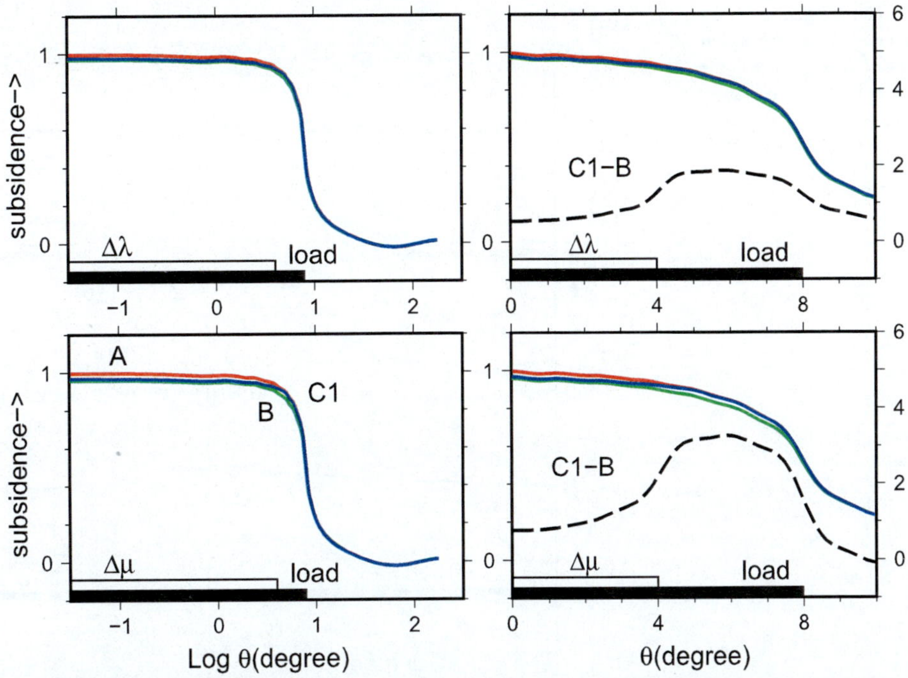

Fig. 1 Normalized vertical displacements for models A (red), B (green) and C1 (blue) when the bulk and shear moduli are considered laterally heterogeneous, respectively. Subsidence is positive and normalization is to central displacement of model A. The horizontal axis denotes the angular distance measured from disk center. The near-load region is magnified on the right, where the dashed lines denote the difference of C1 from B relative to the maximum displacement of B (the right vertical axis). The black and white thick bars denote the angular extension of the load and the elastic heterogeneity, respectively

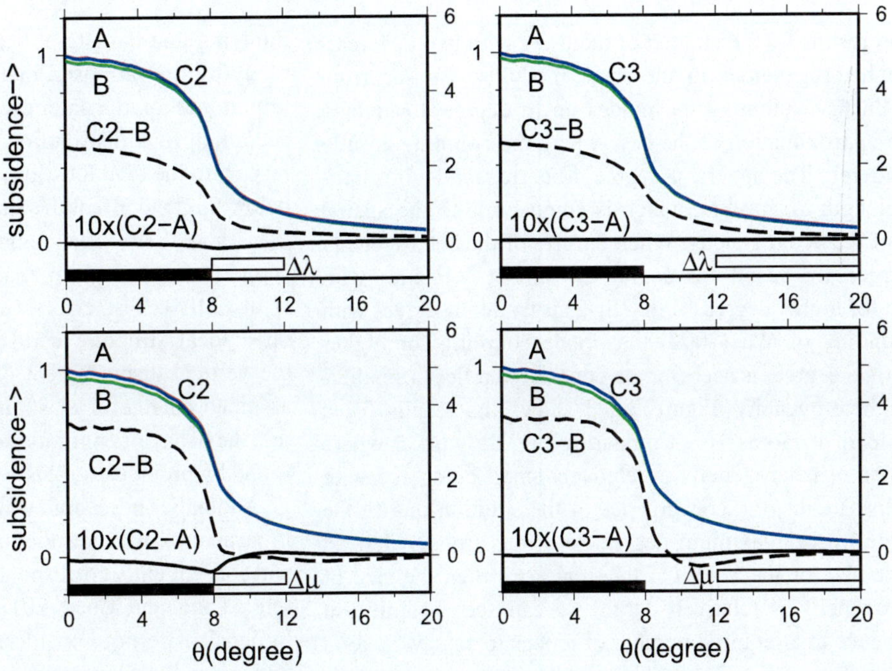

Fig. 2 The same as in Fig. 1 but for models C2/C3 (left/right). The blue curves (C2/C3) almost coincide with the red ones (A). The solid black lines denoting the difference of C2/C3 from A are amplified by a factor of 10

the toroidal mode is not excited for a case where both the load and the heterogeneity are axisymmetric (Martinec and Wolf 1999; Wu 2002). The fact that the toroidal mode becomes zero can be confirmed also in our formulation. The contribution of mode coupling is evaluated by the integral of the first term on the RHS of Eqs. (17) and (18). As a result of the first iteration, only the spheroidal mode is excited by the load, which enters into the strain tensor in the integrand on the RHS. Then, the coefficients of strain-tensor variations in the weak formulation for the toroidal mode vanish. ($W_{jm} = F_{jm} = H_{jm} = 0$ in eqs. (88-92) and $(A, B, C)_{r\varphi} = (B, C)_{\theta\varphi} = 0$ in (103) of Martinec (2000) and the coefficients of $\delta\varepsilon^3$ and $\delta\varepsilon^4$ become zero. See also the description below eq. (110) of the same paper.) Accordingly, only the coupling between spheroidal modes has to be discussed.

In Figs. 1 and 2, all the modes up to the cut-off degree of 300 were considered. By reducing the maximum degree in the abovementioned integral for evaluating the mode coupling, we investigated how the solution changed. Figure 3a, b

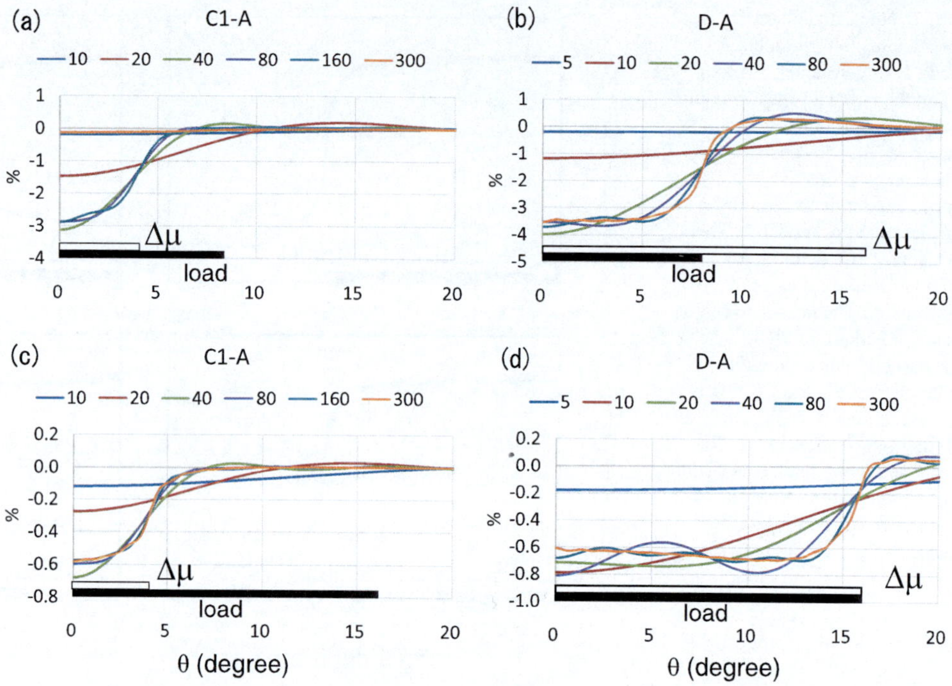

Fig. 3 The vertical displacements for various cut-off degrees in the mode coupling (numbers above the plots). Relative differences to the reference model A are shown

shows results for a disk load of radius 8° with two different-scale heterogeneities in the shear modulus. We see from Fig. 3a that summing the modes up to degree 40 gives a good approximation of the case where the coupling is fully considered. The spherical degree 40 corresponds to a half-wavelength of 500 km, which is comparable to the spatial size of the heterogeneity. When the size of the heterogeneity is larger, the result for degree 20 almost coincides with those for higher degrees (Fig. 3b). This tendency agrees with the finding of Wu (2002) that mode coupling for higher spherical degrees is necessary to consider to reflect a smaller-scale heterogeneity. Figure 3c, d shows the results for a disk load of radius 16°. Comparing Fig. 3c with a, where the size of heterogeneity is relatively small compared with the disk radius, the convergence of the solution due to the change in the maximum degree does not strongly depend on the size of the load. On the contrary, when the size of heterogeneity is relatively large, we can see a significant difference in a required number of modes to achieve a good approximation (Fig. 3b, d). This result indicates that the role of the mode coupling depends both on the sizes of loads and heterogeneities, which is also consistent with Wu (2002).

4 Conclusions

We have shown that the spectral finite-element approach is applicable to quantify the effects of laterally heterogeneous elastic constants to surface loading. Preliminary results indicate that the behavior of the elastic vertical response obtained for different structures is in agreement with our intuition, implying that the lateral heterogeneities are considered correctly by the proposed method. We found that the relative importance of the laterally heterogeneous structure varies according to a distribution of the local heterogeneity with respect to the load location. In particular, the heterogeneities below the load affect the surface deformation more strongly than those below the regions away from the load, which supports the assumption made in the local GF approach (Dill et al. 2015). The errors caused by using a 1D GF based on a local structure can amount to 3% when estimating the vertical displacement outside the heterogeneity for the assumed change of elasticity by 10%. Also, we confirmed that the behaviors of mode coupling are consistent with those found by previous studies.

Anomalies in seismic wave velocity are related to changes in structure and thermodynamic state (e.g., Karato 2008), and consequently correspond also to heterogeneity in density (e.g., Ding and Chao 2018). Since lateral heterogeneities in density are not considered in the presented method, we will extend it to more general 3D cases including lateral heterogeneities in the density so that we can investigate the effects of laterally heterogeneous structures on surface loading, using GGOS data and realistic loads in a future study.

Acknowledgements We appreciate valuable comments from two anonymous reviewers. This work was initiated when YT was visiting the GFZ German research center, and he acknowledges the hospitality of the institute, and, through VK, contributes to the Helmholtz project, "Advanced Earth system Modelling Capacity (ESM)". This study was partly supported by JSPS KAKENHI Grant Numbers JP15K17746, JP16H02219 and JP16H06474.

References

D'Agostino G, Spada G, Sabadini R (1997) Postglacial rebound and lateral viscosity variations: a semi-analytical approach based on a spherical model with Maxwell rheology. Geophys J Int 129:F9–F13

Dill R, Klemann V, Martinec Z, Tesauro M (2015) Applying local Green's functions to study the influence of the crustal structure on hydrological loading displacements. J Geodyn 88:14–22

Ding H, Chao BF (2018) A 6-year westward rotary motion in the Earth: detection and possible MICG coupling mechanism. Earth Planet Sci Lett 495:50–55

Dziewonski AM, Anderson A (1981) Preliminary reference Earth model. Phys Earth Planet Inter 25:297–356

Farrell WE (1972) Deformation of the Earth by surface loads. Rev Geophys 10:761–797

Ito T, Simons M (2011) Probing asthenospheric density, temperature, and elastic moduli below the western United States. Science 332:947–951. https://doi.org/10.1126/science.1202584

Karato S (2008) Deformation of earth materials: an introduction to the rheology of the solid earth. Cambridge University Press, Cambridge, p 463

Kaufmann G, Wolf D (1999) Effects of lateral viscosity variations on postglacial rebound: an analytical approach. Geophys J Int 137:489–500

Klemann V, Martinec Z, Ivins ER (2008) Glacial isostasy and plate motions. J Geodyn 46:95–109

Martens HR, Rivera L, Simons M, Ito T (2016) The sensitivity of surface mass loading displacement response to perturbations in the elastic structure of the crust and mantle. J Geophys Res Solid Earth 121:3911–3938. https://doi.org/10.1002/2015JB012456

Martinec Z (2000) Spectral-finite element approach to three-dimensional viscoelastic relaxation in a spherical Earth. Geophys J Int 142:117–141

Martinec Z, Wolf D (1999) Gravitational viscoelastic relaxation of eccentrically nested spheres. Geophys J Int 138:45–66

Milne GA, Latychev K, Schaeffer A, Crowley JW, Lecavalier BS, Audette A (2018) The influence of lateral Earth structure on glacial isostatic adjustment in Greenland. Geophys J Int 214:1252–1266

Peltier WR (1974) The impulse response of a Maxwell Earth. Rev Geophys Space Phys 12:649–669

Spada G, Barletta VR, Klemann V, Riva REM, Martinec Z, Gasperini P, Lund B, Wolf D, Vermeersen LLA, King MA (2011) A benchmark study for glacial isostatic adjustment codes. Geophys J Int 185:106–132

Tanaka Y, Klemann V, Martinec Z, Riva REM (2011) Spectral finite-element approach to viscoelastic relaxation in a spherical compressible Earth: application to GIA modeling. Geophys J Int 184:220–234

Whitehouse PL (2018) Glacial isostatic adjustment modelling: historical perspectives, recent advances, and future directions. Earth Surf Dyn 6:401–429

Wu P (2002) Mode coupling in a viscoelastic self-gravitating spherical Earth induced by axisymmetric loads and lateral viscosity variations. Earth Planet Sci Lett 202:49–60

Check for updates

Using Structural Risk Minimization to Determine the Optimal Complexity of B-Spline Surfaces for Modelling Correlated Point Cloud Data

Corinna Harmening and Hans Neuner

Abstract

The increased use of areal measurement techniques in engineering geodesy requires the development of adequate areal analysis strategies. Usually, such analysis strategies include a modelling of the data in order to reduce the amount of data while preserving as much information as possible. Free form surfaces like B-splines have been proven to be an appropriate tool to model point clouds. The complexity of those surfaces is among other model parameters determined by the number of control points. The selection of the appropriate number of control points constitutes a model selection task, which is typically solved under consideration of parsimony by trial-and-error procedures. In Harmening and Neuner (J Appl Geod 10(3):139–157, 2016; 11(1):43–52, 2017) a model selection approach based on structural risk minimization was developed for this specific problem. However, neither this strategy, nor standard model selection methods take correlations into account. For this reason, the performance of the developed model selection approach on correlated data sets is investigated and the respective results are compared to those provided by a standard model selection method, the Bayesian Information Criterion.

Keywords

B-spline surfaces · Correlated point clouds · Model selection · Point cloud modelling · Structural risk minimization · VC-dimension

1 Introduction

With the development of the terrestrial laser scanner (TLS) a measuring instrument which allows a fast, efficient and contactless data acquisition moved into focus of engineering geodesy (Heunecke et al. 2013). The acquired data is of high spatial and temporal resolution and, therefore, forms an excellent basis to solve engineering geodetic tasks like geometric state descriptions or spatio-temporal deformation analyses. However, despite of its many advantages, the use of laser scanners also holds new challenges (see for example Mukupa et al. 2016 or Holst and Kuhlmann 2016).

One of the major challenges is the development of appropriate deformation analysis strategies which are able to deal with the huge amount of data. When developing areal analysis strategies, the choice is between five possible ways to handle point clouds with respect to a subsequent deformation analysis (Ohlmann-Lauber and Schäfer 2011).

Among them, a frequently used strategy is geometry-based. It includes a geometric modelling of the point clouds in order to reduce the amount of data while preserving as much information as possible. Applications of the geometry-based approach using geometric primitives like planes or cylinders can be found in Erdélyi et al. (2017), Lindenbergh and Pfeifer (2005) or Vezočnik et al. (2009). However, when applying this approach to complex structures like domes, arch bridges or even natural objects, flexible mathematical functions are required. Free form surfaces like B-splines have been proven to be particularly suitable to model laser scanner point clouds (see for example Harmening and

C. Harmening (✉) · H. Neuner
Department of Geodesy and Geoinformation, TU Wien, Vienna, Austria
e-mail: corinna.harmening@tuwien.ac.at

P. Novák et al. (eds.), *IX Hotine-Marussi Symposium on Mathematical Geodesy*, International Association of Geodesy Symposia 151, https://doi.org/10.1007/1345_2019_88

Neuner 2015, Ioannidis and Valani 2006 or Paffenholz et al. 2017).

B-spline surfaces gain their flexibility due to a variety of parameter types which have to be determined in an appropriate way. The focus of this contribution lies on the determination of the optimal number of control points which influence the B-spline's complexity to a large extent. Usually, this parameter type is chosen quite arbitrarily by intuitive trial-and-error-procedures, finding a balance between the surface's parsimony and its approximation quality.

In this contribution the performance of structural risk minimization with regard to the choice of the optimal number of B-spline control points is compared to that one of the Bayesian Information Criterion. Both approaches assume identically and independently distributed (iid) data. As this requirement is not met in reality, the influence of correlations in the data sets on the respective results is furthermore investigated.

The article is structured as follows: Sect. 2 gives a short overview of the estimation of B-spline surfaces. In Sect. 3 the task of model selection is introduced using the example of B-spline curves with different numbers of control points. The general principles of model selection and two possibilities to implement them (the Bayesian Information Criterion and Structural Risk Minimization) are described. Both approaches are applied to the estimation of B-spline surfaces in order to determine the optimal number of control points in Sect. 4. The results are analysed, evaluated and compared. Finally, a conclusion is drawn and an outlook is given in Sect. 5.

2 Estimation of B-Spline Surfaces

A B-spline surface of degrees p and q is defined by Piegl and Tiller (1997):

$$\hat{\mathbf{S}}(u, v) = \sum_{i=0}^{n_\mathbf{P}} \sum_{j=0}^{m_\mathbf{P}} N_{i,p}(u) N_{j,q}(v) \mathbf{P}_{ij}. \tag{1}$$

According to Eq. (1), a surface point $\hat{\mathbf{S}}(u, v)$ is located on the surface by the surface parameters u and v and is computed as the weighted average of the $(n_\mathbf{P} + 1)$ x $(m_\mathbf{P} + 1)$ control points $\mathbf{P}_{ij}$. The corresponding weights are defined by the functional values of the B-spline basis functions $N_{i,p}(u)$ and $N_{j,q}(v)$ which can be recursively computed by means of the Cox-de-Boor-algorithm (see Cox 1972; Boor 1972). Two knot vectors $\mathbf{U} = [u_0, \ldots, u_r]$ and $\mathbf{V} = [v_0, \ldots, v_s]$ in direction of the surface parameter u and v respectively split the B-spline's domain into knot spans. This subdivision of the B-spline's domain leads to the property of locality, meaning that the shifting of one control point changes the surface only locally.

The tensor product representation of a B-spline surface given by Eq. (1) illustrates the relationship between B-spline surfaces and B-spline curves: Due to the product of two one-dimensional basis functions depending on different parameters u and v respectively, the surface consists of an infinite number of B-spline curves running in two different directions (given by u and v).

When determining an optimal B-spline surface, usually, only the location of the control points is estimated in a linear Gauß-Markov model. In order to obtain a linear relationship between the l observations $\mathbf{l} = \mathbf{S}_k(u, v)$ with $(k = 1, \ldots l)$ and the unknown control points $\mathbf{P}_{ij}$, the B-spline's knots as well as its degrees are specified a priori. In this study, the use of cubic B-splines with $p = 3$ and $q = 3$ is applied as this is a generally accepted choice due to their C^2-continuity (Piegl and Tiller 1997). Methods for determining appropriate knot vectors can be found in Piegl and Tiller (1997), Schmitt and Neuner (2015) or Bureick et al. (2016).

Another prerequisite for the estimation of the control points' location is the allocation of convenient surface parameters u and v to the observations (cf. Harmening and Neuner 2015).

3 Model Selection

3.1 Problem Definition

The remaining parameter type to be specified prior to the estimation of the control points' location is the number of control points $(n_\mathbf{P} + 1)$ x $(m_\mathbf{P} + 1)$. The importance of an appropriate choice is demonstrated in Fig. 1 for the case of B-spline curves. Due to the B-spline surface's composition of B-spline curves this motivation can be straightforwardly extended to surfaces.

Figure 1 shows a simulated noisy point cloud (blue points) which forms the basis for the estimation of three B-spline curves with different numbers of control points. The more control points are estimated, the more complex the curve becomes and the better it approximates the data in terms of a smaller sum of squared residuals.

However, although the black curve with 30 control points leads to the smallest sum of squared residuals, it is much too complex and, therefore, not able to separate the noise from the actual phenomenon; this model has a large **variance** as it **overfits** the data by encoding noise in the estimated control points (Cherkassky and Mulier 2007).

In contrast, the green curve with only four control points is too inflexible to model the complete phenomenon. Such models have a large **bias**, they **underfit** the data (Burnham and Anderson 2002).

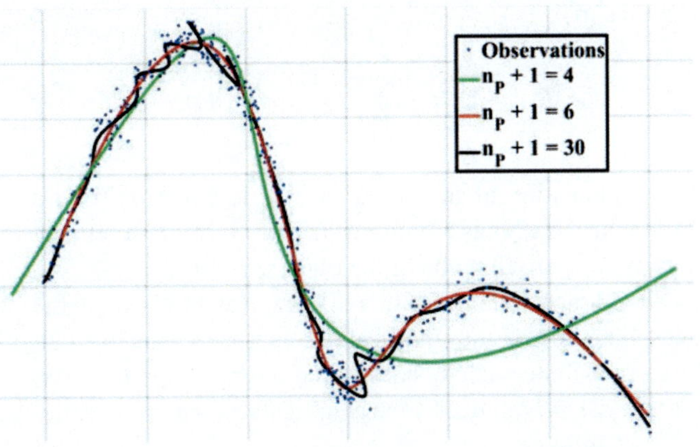

Fig. 1 B-spline curves with different numbers of control points $n_P + 1$

The choice of the optimal number of control points thus corresponds to the finding of a balance between simplicity (small variance, large bias) and complexity (small bias and large variance), which is known as the **bias-variance trade-off** (Cherkassky and Mulier 2007). Information criteria and structural risk minimization are tools which find an optimal model by implementing this trade-off. Both follow the **principle of parsimony**, which states that a good model should approximate the phenomenon as well as possible while it is as simple as possible (Burnham and Anderson 2002).

3.2 Model Selection by Means of the Bayesian Information Criterion

The common tools to find a trade-off between a model's approximation quality and its simplicity are information criteria. Based on a set of candidate models these information criteria evaluate each model according to an optimality criterion and choose that model to be optimal which achieves the best score (Burnham and Anderson 2002).

The Bayesian Information Criterion (BIC), which was introduced by Schwarz (1978), is one of the most popular information criteria and is defined to be

$$\mathrm{BIC} = -2\log(\mathcal{L}(\hat{\boldsymbol{\theta}}|\mathrm{data})) + \log(l)K. \tag{2}$$

The first part of the criterion, the log-likelihood of the estimated model parameters $\hat{\boldsymbol{\theta}}$ given the data set of size l, evaluates the approximation quality of the respective model. As the approximation quality increases with a growing number of model parameters, the second term penalizes an increase of the model's complexity in terms of the number of estimated model parameters K.

The BIC chooses that model to be optimal which is a posteriori the most likely (Cavanaugh and Neath 1999) and is based on the assumption that the true dependency is contained in the set of candidate models. Thus, it belongs to the class of asymptotic consistent criteria which identify the correct model with probability of 1 in the asymptotic case.

Although information criteria are a standard tool in model selection, they have some disadvantages: They are only applicable to linear models, their known properties are valid only for the asymptotic case and the specification of a function's complexity in terms of the number of free parameters is not adequate for every function type (Cherkassky and Mulier 2007).

3.3 Model Selection by Means of Structural Risk Minimization

These drawbacks demand for an alternative which can be found in statistical learning theory (SLT). SLT provides a theoretical framework for the estimation of functional dependencies from a given set of data (Vapnik 1999). In contrast to model selection, SLT describes function estimation in finite samples and takes explicitly the sample size into account. Furthermore, SLT can also be applied to nonlinear models (Cherkassky and Mulier 2007).

The general learning problem of SLT aims to select that function $\tilde{g}(\mathbf{x}, \boldsymbol{\theta}_0)$ from a set of functions $g(\mathbf{x}, \boldsymbol{\theta})$, which approximates the dependency between input vector $\mathbf{x}$ and output vector $\mathbf{y}$ in an optimal manner (Cherkassky and Mulier 2007). This task includes the choice of the optimal function type $\tilde{g}$ as well as the choice of the optimal parameters $\boldsymbol{\theta}_0$. The quality of this approximation is measured by a loss function $L(\mathbf{y}, g(\mathbf{x}, \boldsymbol{\theta}))$ whose expectation w.r.t. the data

$$R(\boldsymbol{\theta}) = \int L(\mathbf{y}, g(\mathbf{x}, \boldsymbol{\theta}))\, dP(\mathbf{x}, \mathbf{y}). \tag{3}$$

is minimized by the optimal function $\tilde{g}(\mathbf{x}, \boldsymbol{\theta}_0)$. Due to the unknown joint probability density of the data $dP(\mathbf{x}, \mathbf{y})$, the risk functional (3) cannot be computed and the minimization of the true risk is usually replaced by the minimization of the empirical risk

$$R_{emp}(\boldsymbol{\theta}) = \frac{1}{l}\sum_i L(\mathbf{y}_i, g(\mathbf{x}_i, \boldsymbol{\theta})). \quad (4)$$

The computation of the empirical risk is based on the iid training data $(\mathbf{x}, \mathbf{y})$ as a finite realization from $dP(\mathbf{x}, \mathbf{y})$ and, thus, is an approximation of the true risk in Eq. (3).

Structural risk minimization (SRM) gives an answer to the question how well the empirical risk approaches the true risk in case of finite sample sizes: According to Vapnik (1998) the true risk is bounded by the empirical risk with probability $1 - \eta$ as follows:

$$R(\boldsymbol{\theta}) \leq \frac{R_{emp}(\boldsymbol{\theta})}{(1-\sqrt{e})_+} = R_{str} \quad (5)$$

with

$$e = \frac{h\left(\ln\frac{l}{h}+1\right) - \ln(\eta/4)}{l}. \quad (6)$$

and

$$\eta = \min\left(\frac{4}{\sqrt{l}}, 1\right). \quad (7)$$

The bound defined by Eq. (5) is a function of the sample size l and the VC-dimension h which is an essential tool in SLT as it describes the complexity of a set of functions without explicitly taking the number of free parameters into account (Vapnik 1998). It is defined to be the maximum number of samples which can be separated in all possible ways by the set of functions (cf. Fig. 2). The VC-dimension cannot analytically be determined in case of the majority of function classes, but there exists a general procedure for estimating the VC-dimension (Vapnik et al. 1994).

Given a set of candidate models and being able to estimate the respective VC-dimensions, SRM uses the bound in Eq. (5) to choose that model to be optimal which has minimal structural risk R_{str}. As this implies a minimization of the empirical risk and a simultaneous controlling of the function classes' complexity, SRM implements the principle of parsimony (Vapnik 1999).

3.4 Choosing the Optimal Number of B-Spline Surface Control Points by means of SRM

Although the general SRM procedure to choose an optimal model is straightforward, the application to the determination of the optimal number of control points implies one difficulty: In order to estimate the VC-dimension of B-spline surfaces, they have to be used as classifiers, or, in other words, they have to be used as discrete approximators instead of continuous ones as in their original definition (cf. Vapnik et al. 1994).

A detailed derivation of the B-spline classifier used in the following can be found in Harmening and Neuner (2016, 2017). It is strongly oriented on the Support Vector Machines (SVM, Vapnik 1998) and is based on the following main ideas:

- The B-spline classifier is based on linear decision boundaries. Thus, the classification problem benefits from all advantages of linear optimization.
- However, as the functional relationship of B-splines contains non-linearities due to the necessity to estimate surface parameters, knot locations and control points, the classifier has to be able to deal with them. By implementing a mapping of the input space into a high dimensional feature space, linear decision boundaries in this feature space result in non-linear decision boundaries in the input space.
- The problem of increased dimensionality due to the feature space transformation is solved by making use of the kernel trick. This trick allows to indirectly compute the inner product of two feature space vectors $\mathbf{x}_{FS}$ by using a Kernel function K and the original and low-dimensional input vectors $\mathbf{x}$:

$$K(\mathbf{x}, \mathbf{x}^T) = (\mathbf{x}_{FS}, \mathbf{x}_{FS}^T). \quad (8)$$

The respective B-spline kernel is given by

$$K(u_s, v_s, u_t, v_t) = \sum_{i=0}^{n}\sum_{j=0}^{m} N_{i,p}(u_s)\, N_{j,q}(v_s) \cdot N_{i,p}(u_t) N_{j,q}(v_t). \quad (9)$$

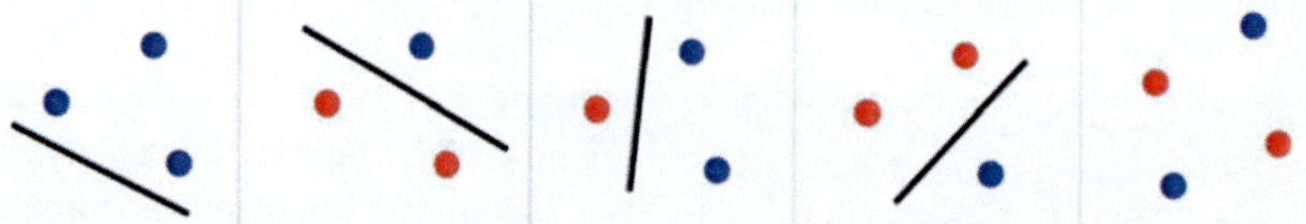

Fig. 2 A 2D-line can separate a set of three points in all possible ways, whereas there exists a formation of four points which cannot be achieved by a separating 2D-line. Thus, the VC-dimension of a 2D-line equals $h = 3$

- The application of the kernel trick requires an optimization problem in which the input vector appears solely in form of the inner product (see Eq. (8)). The loss function of the ridge regression

$$L = \sum_{i=1}^{l} \epsilon_i^2 + \gamma ||\boldsymbol{\theta}_{FS}||^2$$

is an extension of the sum of squared residuals by a regularization term, which restricts the length of the estimated parameter vector $\boldsymbol{\theta}_{FS}$.

The dual solution of this optimization problem

$$\boldsymbol{\alpha} = [(\mathbf{X}_{FS}\mathbf{X}_{FS}^T) + \boldsymbol{\lambda}\mathbf{I}]^{-1}\mathbf{y}$$

meets the requirement demanded above.

Based on these developments the optimal number of control points can be determined as sketched in Fig. 3:

- Approximation of the point cloud by means of B-spline surfaces with varying number of control points, yielding the empirical risk R_{emp} of each B-spline surface (Eq. (4)).
- Estimating the VC-dimension of each of the used B-spline surfaces according to Vapnik et al. (1994) and using the B-spline classifier developed in Harmening and Neuner (2016).
- Computing the structural risk R_{str} of each B-spline surface according to Eq. (5).
- Choosing the B-spline surface with the minimal structural risk to be the optimal one.

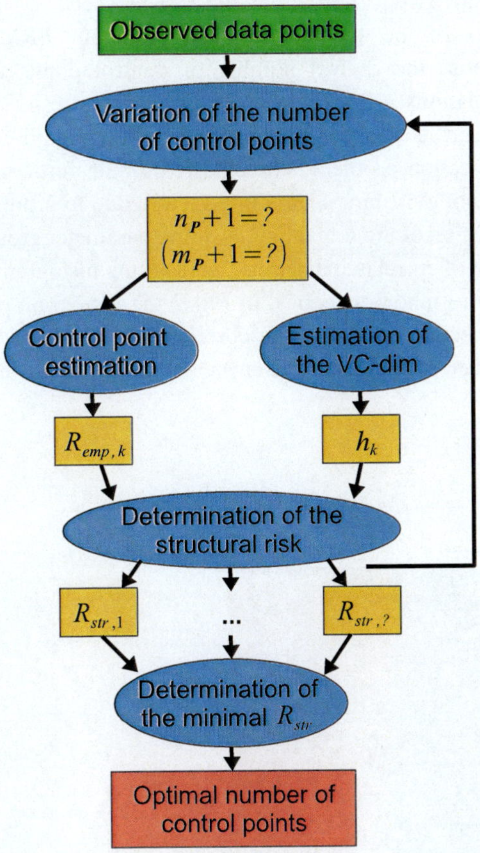

Fig. 3 Structural risk minimization to determine the optimal number of B-spline control points

4 Choosing the Optimal Number of B-Spline Control Points

Aim of this section is to study the performance of the developed SRM approach compared to that of BIC in case of point cloud data. Special focus is set on the performance in case of correlated data.

4.1 Data Sets and General Procedure

The following investigations are based on simulated data sets as they allow a comparison with nominal values as well as a controlled establishment of correlating noise.

The basis of the data simulation is formed by a B-spline surface with $n_\mathbf{P} + 1 = 9$ and $m_\mathbf{P} + 1 = 7$ control points. The scanning process of the surface is realized by a sampling and a subsequent superimposition of the sampled point cloud with noise (cf. Fig. 4). Four different types of data sets are

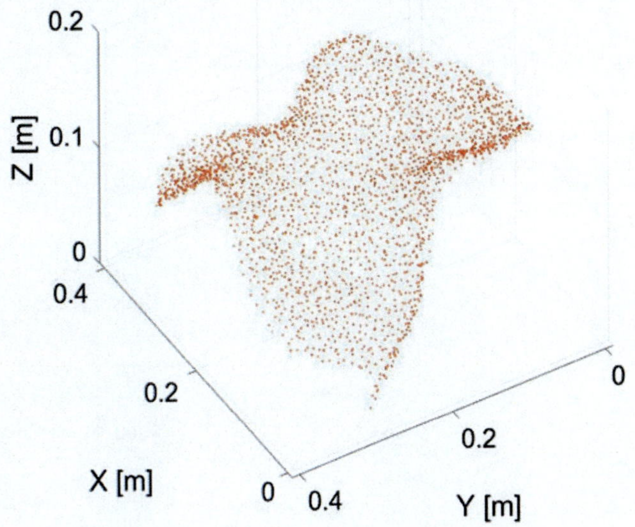

Fig. 4 Simulated B-spline surface with $n_\mathbf{P} + 1 = 9$ and $m_\mathbf{P} + 1 = 7$ control points

obtained by varying the type of noise and the used correlation function respectively:

- For the first type of data set white noise with a standard deviation of $\sigma = 2$ mm is used. This type of data set meets the model selection approaches' requirements of being iid.
- The other three types of data sets are generated by superimposing the sampled surfaces with correlated noise. The respective correlation function is of type $e^{-\alpha\Delta}$ which is an appropriate choice for representing laser scanning data of this specific test specimen (Kauker and Schwieger 2017). The correlation length is varied over the three types of data sets according to Fig. 5. These different exponents lead to correlation lengths ranging from 5 cm in case of the blue curve over 7.5 cm in case of the red curve to 10 cm in case of the yellow curve.

For each type of data 500 realization sets are generated. Each of the realizations forms the basis for the estimation of B-spline surfaces with varying number of control points and the subsequent evaluation of the estimated surfaces by means of BIC as well as by means of the SRM-approach introduced above.

As there is a strong connection between the number of control points, the degree and the number of knots, a completely isolated considerations of the number of control points is not possible. In order to reduce the effect of the remaining parameter groups, only the location of the control points is estimated, whereas the surface parameters and degree of the B-spline basis functions are set to the nominal values, which are known from the simulation process. In order to receive a knot vector which is as close as possible to the nominal one, the nominal knot positions are either complemented by further knots provided by the strategy of Piegl and Tiller (1997) or those nominal knots which are furthest from those provided by Piegl and Tiller (1997) are discarded. The variance-covariance matrix of the observations in the stochastic model of the adjustment is chosen to be the identity matrix.

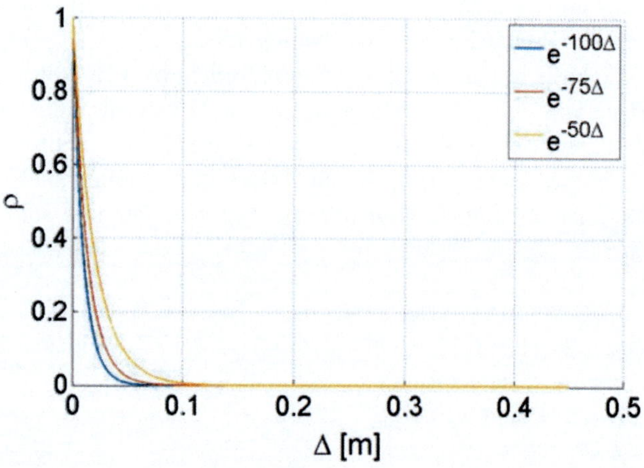

Fig. 5 Correlation functions of the simulated data sets

4.2 Results

The results of the model selection are presented in form of histograms specifying the number of control points which is chosen to be optimal according to BIC and SRM as well as the respective frequency among the 500 data sets.

In case of the uncorrelated data sets both, SRM and BIC, choose the correct number of control points in all 500 simulations as can be seen in Fig. 6. This result could be expected as this setting is optimal with regard to the model selection problem: The data is iid and, furthermore, there do not exist any uncertainty factors due to a previous determination of the remaining B-spline parameter groups.

In case of correlated data sets the resulting histograms are by far more inhomogeneous. In Fig. 7 the histogram of the model selection results of the correlated data sets with the smallest correlation length (5 cm) can be seen. A variety of

Fig. 6 Model selection results in case of uncorrelated data sets

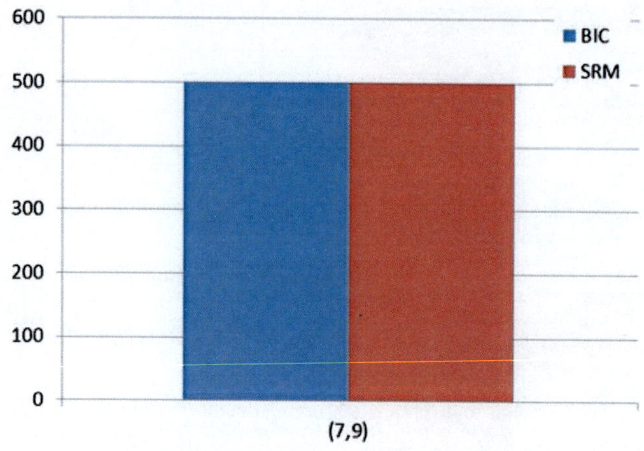

Fig. 7 Model selection results in case of correlated data sets ($\rho = e^{-100\Delta}$)

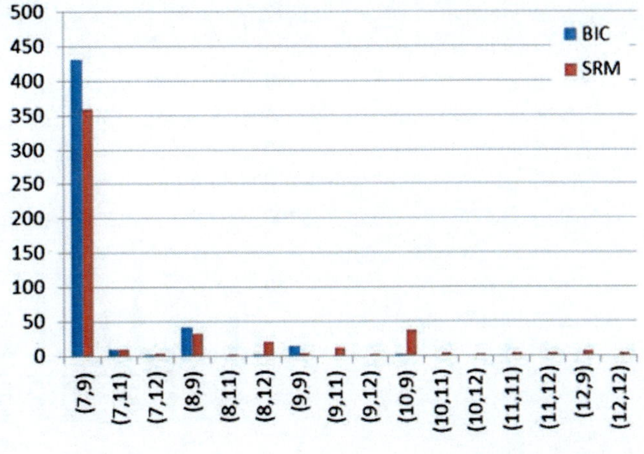

Fig. 8 Model selection results in case of correlated data sets ($\rho = e^{-75\Delta}$)

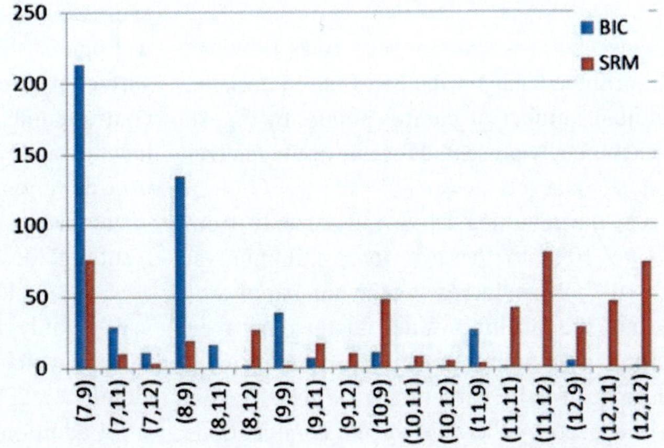

combinations of $(n_{\mathbf{P}}+1, m_{\mathbf{P}}+1)$ is chosen to be optimal over the 500 realizations, with SRM providing a broader range of results than BIC. However, both approaches identify the correct number of control points in the majority of data sets, with BIC choosing the nominal number of control points in more cases ($\approx$435) than SRM ($\approx$355). The remaining combinations of chosen control points show one (in case of BIC) or two (in case of SRM) smaller accumulations at the solutions $(8, 9)$ and $(10, 9)$ respectively. In these cases at least the correct number of control points in the v-direction is identified.

When using a correlation length of 7.5 cm during the simulation process, only BIC chooses the correct number of control points in the majority of the data sets, whereas SRM most frequently chooses the combination of $(11, 12)$ control points to be optimal (cf. Fig. 8). When comparing the results with the previous ones, it can be seen that the range of results becomes broader for both approaches and that the maxima are considerably less pronounced.

The inclusion of the results caused by the data sets with a correlation length of 10 cm reveals that the increase of the correlation length shifts the most frequent solution from the correct number of control points towards the highest number of control points which was included into the computations (cf. Fig. 9). This behaviour is more pronounced for SRM than for BIC: In case of a correlation length of 10 cm the results of BIC still cover a very broad spectrum (similar to the results of SRM in case of a correlation length of 7.5 cm), whereas the results of SRM already accumulate at the highest possible number of control points.

The general behaviour of both strategies to choose more complex models in case of correlated data sets can be explained as follows: Due to the correlations, additional systematics are inserted into the data sets. An appropriate modelling of these systematics is only possible when using more complex functions. In reality only correlated data are available. An adequate handling of these correlations, which avoids an overfitting of the data, is thus necessary.

However, the apparently weaker performance of SRM compared to BIC could not be expected.

For these two reasons further investigations are made in the following.

Fig. 9 Model selection results in case of correlated data sets ($\rho = e^{-50\Delta}$)

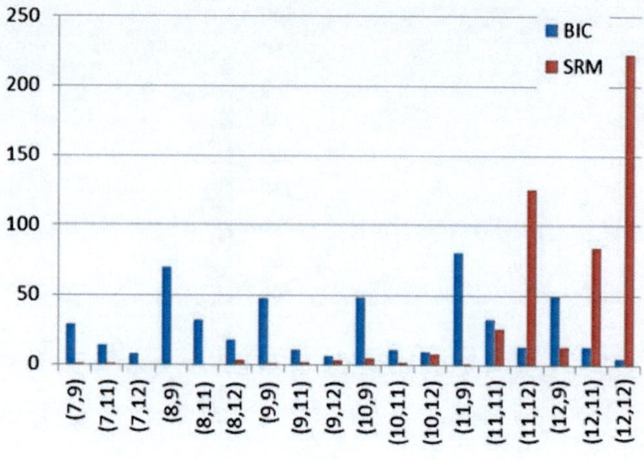

4.3 Further Investigations

In a first step a closer look is taken at the values provided by Eq. (5) which form the basis for the SRM-based decision regarding the optimal number of control points. In Fig. 10 these values are exemplarily presented for one of the realizations of the correlated data sets ($\rho = e^{-100\Delta}$).

As can be seen, the resulting curve falls steeply from $R_{str}(6,6) \approx 4.9 \cdot 10^{-5}$ to the first local minimum at $R_{str}(7,9) \approx 8.5 \cdot 10^{-6}$. When increasing the number of estimated control points, the resulting values fluctuate between four levels resulting in a multitude of local minima with almost identical values for the structural risk. The curve's periodic pattern is caused by the order of the combinations of control points on the abscissa: During the process of estimating B-spline surfaces with different numbers of control points, the number of control points in direction of the surface parameter v is fixed while increasing the number of control points in direction of the surface parameter u. When having achieved the maximum number of control points in u-direction, the number of control points in v-direction is increased by one and the procedure is repeated. The resulting local maxima represent the combinations $(8,6)$, $(9,6)$, ..., $(12,6)$ (green circles in Fig. 10). Obviously, the use of only six control points in u-direction leads to a surface which is much too simple to approximate the point cloud.

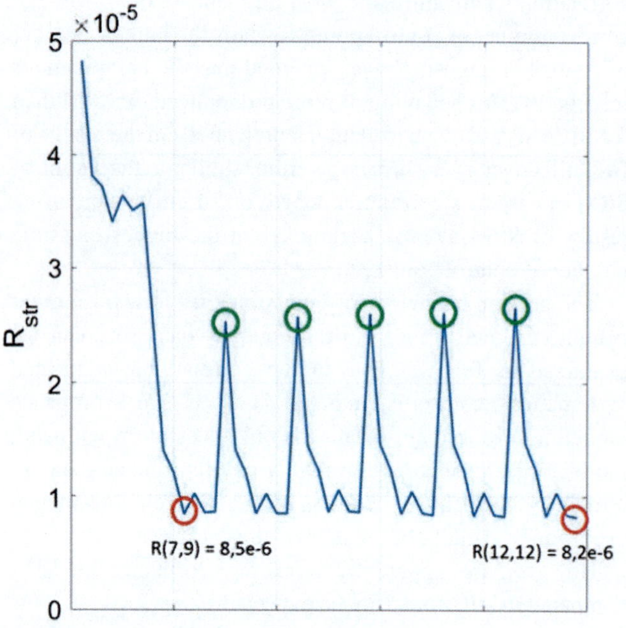

Fig. 10 Structural risk R_{str} for one realization of the correlated data set with $\rho = e^{-100\Delta}$

In addition to the periodic pattern, it is conspicuous that the overall minimum ($R_{str}(12,12) \approx 8.2 \cdot 10^{-6}$) is only slightly smaller than the structural risk provided by the correct number of control points. Thus, the course of the curve indicates that a simple numerical comparison of the computed values of the structural risk is not sufficient in this application. Rather, further investigations regarding the separability of models have to be included, allowing for a decision if two local minima differ significantly or if the principle of parsimony is even better obeyed when the most simple model leading to a local minima is chosen to be the optimal one. These further investigations will clearly improve the performance of SRM compared to that one of BIC.

In a second step the known correlation structure of the data is used to decorrelate the data. The effect of the decorrelation on the values of the structural risk can be seen in Fig. 11. The decorrelation causes a kind of distortion of the curve in Fig. 10, leading to a distinct overall minimum of the structural risk for the correct number of control points. Thus, in case the correlation structure of the data is known or can be sufficiently well estimated, a decorrelation of the data is an appropriate way to deal with stochastic relationships within the acquired data sets.

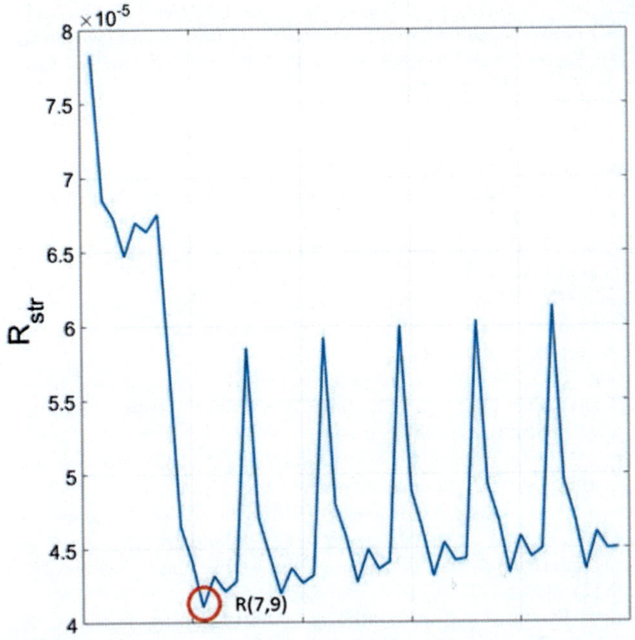

Fig. 11 Structural risk R_{str} after decorrelation of the data

5 Conclusion and Outlook

Due to the development of the terrestrial laser scanner, the extension of point-based analysis approaches to areal ones is a current research topic in engineering geodesy. One possibility to deal with the large amount of laser scanning data is the point clouds' approximation by means of continuous mathematical functions, which form the basis for further analysis steps. B-spline surfaces have been proven to be a powerful, but sophisticated tool for point cloud modelling. When estimating a best fitting B-spline surface, a variety of parameter groups has to be appropriately determined.

The focus of this contribution is on the determination of the optimal number of B-spline control points and, therefore, on the choice of the optimal complexity of a B-spline surface in case of correlated data. In this contribution structural risk minimization was used to solve this model selection problem as it overcomes some of the disadvantages of information criteria, which are only applicable to linear models and whose well known properties are only valid for the asymptotic case. The associated estimation of the VC-dimension requires the development of a B-spline classifier whose basic ideas were introduced.

In order to evaluate the approach's performance, it was applied to a variety of simulated data sets. The results were compared to the nominal values for the number of control points as well as to the results provided by BIC.

In case the data sets are iid, both, BIC and SRM chose the correct number of control points to be optimal. However, as white noise does not occur in reality, both approaches were also applied to correlated data sets. At first glance, BIC seems to outperform SRM as the latter has a stronger tendency to overfit correlated data. However, further investigations regarding the separability of models and their values for the structural risk might clearly improve the performance of SRM.

Until then and in case the data's correlation structure is known or can be reliably estimated, decorrelation of the data is an appropriate way to deal with correlations in the data.

In future, the approach's applicability to measured data sets will be investigated. Due to missing or incomplete stochastic models of terrestrial laser scanner measurements, the unknown correlations in measured data sets may cause further challenges. Therefore, the first step will be the scanning of a test specimen with known B-spline form, which also was the basis for the simulation process. An inclusion of currently developed stochastic models for laser scanning data (see for example Kauker and Schwieger 2017, Wujanz et al. 2018 or Jurek et al. 2017) is imaginable.

Additionally, the approach will be applied to non-linear models by using it for determining the optimal number of B-spline control points in a joint estimation of the control points' locations and the surface parameters.

Acknowledgements The presented paper shows results developed during the research project "Integrierte raumzeitliche Modellierung unter Nutzung korrelierter Messgrößen zur Ableitung von Aufnahmekonfiguationen und Beschreibung von Deformationsvorgängen" (IMKAD) (1706-N29), which is funded by the Austrian Science Fund (FWF).

References

de Boor C (1972) On calculating with B-splines. J Approx Theory 6(1):50–62

Bureick J, Alkhatib H, Neumann I (2016) Robust spatial approximation of laser scanner point clouds by means of free-form curve approaches in deformation analysis. J Appl Geod 10(1):27–35

Burnham KP, Anderson DR (2002) Model selection and multimodel inference: a practical information-theoretic approach, 2nd edn. Springer, New York.

Cavanaugh J, Neath A (1999) Generalizing the derivation of the Schwarz information criterion. Commun Stat Theory Methods 28(1):49–66

Cherkassky VS, Mulier F (2007) Learning from data: concepts, theory, and methods, 2nd edn. IEEE Press and Wiley-Interscience, Hoboken, NJ

Cox MG (1972) The numerical evaluation of B-splines. IMA J Appl Math 10(2):134–149

Erdélyi J, Kopáčik A, Lipták I, Kyrinovič P (2017) Automation of point cloud processing to increase the deformation monitoring accuracy. Appl Geomat 9(2):105–113

Harmening C, Neuner H (2015) A constraint-based parameterization technique for B-spline surfaces. J Appl Geod 9(3):143–161

Harmening C, Neuner H (2016) Choosing the optimal number of B-spline control points (part 1: methodology and approximation of curves). J Appl Geod 10(3):139–157

Harmening C, Neuner H (2017) Choosing the optimal number of B-spline control points (part 2: approximation of surfaces and applications). J Appl Geod 11(1):43–52

Heunecke O, Kuhlmann H, Welsch W, Eichhorn A, Neuner H (2013) Handbuch Ingenieurgeodäsie: Auswertung geodätischer Überwachungsmessungen, 2nd edn. Wichmann, H, Heidelberg, Neckar

Holst C, Kuhlmann H (2016) Challenges and present fields of action at laser scanner based deformation analyses. J Appl Geod 10(1):17–25

Ioannidis C, Valani A (2006) 3D model generation for deformation analysis using laser scanning data of a cooling tower. In: Proceedings of the 3rd IAG/12th FIG symposium

Jurek T, Kuhlmann H, Holst C (2017) Impact of spatial correlations on the surface estimation based on terrestrial laser scanning. J Appl Geod 11(3):143–155

Kauker S, Schwieger V (2017) A synthetic covariance matrix for monitoring by terrestrial laser scanning. J Appl Geod 11(2):77–87

Lindenbergh R, Pfeifer N (2005) A statistical deformation analysis of two epochs of terrestrial laser data of a lock. In: Proceedings of optimal 3D measurement techniques VII

Mukupa W, Roberts GW, Hancock CM, Al-Manasir K (2016) A review of the use of terrestrial laser scanning application for change detection and deformation monitoring of structures. Surv Rev 36(5):1–18

Ohlmann-Lauber J, Schäfer T (2011) Ansätze zur Ableitung von Deformationen aus TLS-Daten. Terrestrisches Laserscanning – TLS 2011 mit TLS-Challenge (66):147–158

Paffenholz J-A, Stenz U, Wujanz D, Neitzel F, Neumann I (2017) 3D-Punktwolken-basiertes Monitoring von Infrastrukturbauwerken am Beispiel einer historischen Gewölbebrücke. Terrestrisches Laserscanning 2017, pp 115–127

Piegl LA, Tiller W (1997) The NURBS book. Monographs in visual communications, 2nd edn. Springer, Berlin and New York

Schmitt C, Neuner H (2015) Knot estimation on B-Spline curves. Österreichische Zeitschrift für Vermessung und Geoinformation (VGI) 103(2+3):188–197

Schwarz G (1978) Estimating the dimension of a model. Ann Stat 6(2):461–464

Vapnik VN (1998) Statistical learning theory. Adaptive and learning systems for signal processing, communications, and control. Wiley, New York

Vapnik VN (1999) An overview of statistical learning theory. IEEE Trans Neural Netw (a publication of the IEEE Neural Networks Council) 10(5):988–999

Vapnik V, Levin E, Le Cun Y (1994) Measuring the VC-dimension of a learning machine. Neural Comput 6(5):851–876

Vezočnik R, Ambrožič T, Sterle O, Bilban G, Pfeifer N, Stopar B (2009) Use of terrestrial laser scanning technology for long term high precision deformation monitoring. Sensors (Basel, Switzerland) 9(12):9873–9895

Wujanz D, Burger M, Tschirschwitz F, Nietzschmann T, Neitzel F, Kersten TP (2018) Determination of intensity-based stochastic models for terrestrial laser scanners utilising 3D-point clouds. Sensors (Basel, Switzerland) 18(7):2187

On the Numerical Implementation of a Perturbation Method for Satellite Gravity Mapping

Christopher Jekeli and Nlingi Habana

Abstract

In 2008 P. Xu (Celest Mech Dyn Astron, 100:231–249) proposed a strictly kinematic perturbation method for determining the Earth's gravitational field from continuous satellite tracking. The main idea is to process orbital arcs of arbitrary length, thus minimizing superfluous parameter estimation associated with stitching together short-arc solutions, and at the same time formulating the problem in terms of standard linear parameter estimation. While the original formulation appears mathematically robust, its nested quadruple along-track integrations are computationally challenging. We reduce the formulation to double integrals and show that the method is numerically not feasible as originally envisaged. On the other hand, by abandoning the rigorous Gauss-Markov formalism, we show the numerical feasibility of processing multiple-day orbital arcs. The methodology lends itself to high-low and low-low satellite-to-satellite tracking, or combinations thereof, as for GRACE-like systems.

Keywords

GNSS satellite tracking · Gravitational field estimation · Numerical orbit integration · Satellite perturbation theory

1 Introduction

With the modern ability to track low-Earth-orbiting satellites continually and uniformly with Global Navigation Satellite Systems (GNSS), such as the Global Positioning System (GPS), the standard methods to extract estimates of the Earth's gravitational field from satellite tracking data may be re-visited. Prior to GNSS tracking, ground-based tracking created a patch-work of data as a satellite rose and set at any particular tracking station; and, accordingly, a good a priori or reference orbit was essential in stitching the observed arcs together. A perturbation theory based on Keplerian elements, for example, served to separate secular and long-period orbital variations from the more short-term resonances, which facilitated this estimation process. These techniques are still practiced today in various forms; a good review is given in the volume by Naeimi and Flury (2017). Xu (2008) proposed a radical change from this methodology specifically in view of the proven accurate tracking capabilities with GNSS. Continual and uniform high-accuracy tracking, arguably obviates piecing together short arcs and simplifies the overall problem setup as all formulations may be made with the straightforward use of Cartesian coordinates. The idea certainly has tremendous theoretical and practical appeal and in this paper we aim to elucidate this in the simplest terms, but the numerical implementation, never attempted by the originator, puts at least some limitations on the length of the orbital arc that constitutes a segment of the overall estimation process. It is this aspect of the proposed methodology that we wish to highlight and further develop in this short note.

C. Jekeli (✉) • N. Habana
Division of Geodetic Science, School of Earth Sciences, Ohio State University, Columbus, OH, USA
e-mail: jekeli.1@osu.edu

P. Novák et al. (eds.), *IX Hotine-Marussi Symposium on Mathematical Geodesy*,
International Association of Geodesy Symposia 151, https://doi.org/10.1007/1345_2019_81

As assumed by Xu (2008) we consider only the gravitational perturbations on a satellite due to a temporally fixed field, leaving the non-gravitational actions (drag, solar radiation pressure, etc.) and temporally varying effects (tidal potentials, terrestrial mass redistributions, etc.) as secondary to the main problem and outside the present scope. We may thus formulate the problem simply using Newton's equation of motion of a particle in a gravitational field,

$$\ddot{x}(t) = g\left(x(t), p\right), \quad x\left(t_0\right) = x_0, \quad \dot{x}\left(t_0\right) = \dot{x}_0 \tag{1}$$

where t is time, x is the position vector, g is the gravitational field, p is a vector of unknown parameters of the field, and x_0, $\dot{x}_0$ are initial conditions (observed). This equation holds in inertial space and one should include an explicit temporal dependence of g on the right side due to Earth's rotation. This is easy to do and it is assumed in our analyses. The differential equation (1) may be integrated to obtain a relationship between the observable position of the satellite and the parameters of the field,

$$x(t) = x_0 + \dot{x}_0 \cdot (t - t_0) + \int_{t_0}^{t}\int_{t_0}^{t'} g\left(x\left(t''\right), p\right) dt'' dt'. \tag{2}$$

That this is the solution to (1) is easily checked by back-substitution. It is the form given by Xu (2008, eq. 8). Immediately we see that its practical implementation eventually requires the numerical integration of nested integrals – not a desirable prospect. Fortunately there is an alternative, completely equivalent solution, given by

$$x(t) = x_0 + \dot{x}_0 \cdot (t - t_0) + \int_{t_0}^{t} \left(t - t'\right) g\left(x\left(t'\right), p\right) dt', \tag{3}$$

which is also easily verified by back-substitution. The following development and numerical tests are thus based on (3) rather than (2) and all conclusions derived therefrom hold equally for the development contained in (Xu 2008).

2 The Perturbation Theory

Equation (3) is a complicated, non-linear relationship between the observable positions, x, and the unknown parameters, p, since the field values, g, depend explicitly also on x. The standard way forward is to linearize the solution, which is done here independently with respect to the orbit and the field (Xu considers primarily the linearization of the orbit, but the latter is standard and straightforward practice since higher-degree reference fields are now accurately known). Firstly, with respect to the orbit, we have

$$\begin{aligned}\Delta x(t) + x_{\mathrm{ref}}(t) &= x_0 + \dot{x}_0 \cdot (t - t_0) \\ &+ \int_{t_0}^{t} \left(t - t'\right)\left(g\left(x_{\mathrm{ref}}\left(t'\right), p\right) + \left.\frac{\partial g\left(x(t), p\right)}{\partial x}\right|_{x = x_{\mathrm{ref}}(t')} \times \Delta x\left(t'\right)\right) dt'\end{aligned} \tag{4}$$

where $\Delta x(t) = x(t) - x_{\mathrm{ref}}(t)$; and, secondly, with respect to the field,

$$\begin{aligned}\Delta x(t) &= x_0 + \dot{x}_0 \cdot (t - t_0) - x_{\mathrm{ref}}(t) \\ &+ \int_{t_0}^{t} \left(t - t'\right)\left(g_{\mathrm{ref}}\left(x_{\mathrm{ref}}\left(t'\right)\right) + \left.\frac{\partial}{\partial x}\left(g_{\mathrm{ref}}\left(x(t)\right)\right)\right|_{x = x_{\mathrm{ref}}(t')} \Delta x\left(t'\right)\right) dt' \\ &+ \int_{t_0}^{t} \left(t - t'\right)\left(\Delta g\left(x_{\mathrm{ref}}\left(t'\right), p\right) + \frac{\partial}{\partial x}\left(\Delta g\left(x(t), p\right)\right)\big|_{x = x_{\mathrm{ref}}(t')} \Delta x\left(t'\right)\right) dt'\end{aligned} \tag{5}$$

The last term may be neglected, being of second order, thus yielding the model, analogous to Xu's eq. 10,

$$\begin{aligned}\Delta x(t) &= \delta x_0(t) + \int_{t_0}^{t} \left(t - t'\right) \Delta g\left(x_{\mathrm{ref}}\left(t'\right), p\right) dt' \\ &+ \int_{t_0}^{t} \left(t - t'\right) \Gamma_{\mathrm{ref}}\left(x_{\mathrm{ref}}\left(t'\right)\right) \Delta x\left(t'\right) dt',\end{aligned} \tag{6}$$

where $\Delta g(x(t),p) = g(x(t),p) - g_{\mathrm{ref}}(x(t))$, $\Gamma = \partial g/\partial x$, and

$$\begin{aligned}\delta x_0(t) &= x_0 + \dot{x}_0 \cdot (t - t_0) \\ &+ \int_{t_0}^{t} \left(t - t'\right) g_{\mathrm{ref}}\left(x_{\mathrm{ref}}\left(t'\right)\right) dt' - x_{\mathrm{ref}}(t).\end{aligned} \tag{7}$$

It is emphasized that the reference orbit is completely arbitrary and could be generated from the observed orbit by a suitable smoothing. Thus, the reference orbit is always close to the true orbit (and the linearization is legitimate for all t), but in no way must it be consistent with the reference field, i.e., $\ddot{x}_{\mathrm{ref}}(t) \neq g_{\mathrm{ref}}\left(x_{\mathrm{ref}}(t)\right)$. This is the simple beauty of Xu's proposal and is enabled by the continual tracking of satellites using GNSS. Setting the reference orbit to the observed orbit, as done, e.g., by Mayer-Gürr (2006), leads directly back to the solution (3), thus eliminating the linearization error, but also not advancing the separation of observations and parameters. We continue here with an analysis of Xu's model.

If one assumes that the field depends linearly on the parameters, as it does in a spherical harmonic series representation, then except for the last term in (6), there is a linear relationship between observables and parameters. Xu (2008) then makes the reasonable proposal to remedy that shortfall by substituting an approximation for the observable (analogous to his Eq. (11)),

$$\Delta x^{(0)}(t) = \delta x_0(t) + \int_{t_0}^{t} (t - t')\, \Delta g\left(x_{\mathrm{ref}}(t'), p\right) dt', \tag{8}$$

on the right side, yielding now a fully linear and separated relationship,

$$\begin{aligned}\Delta x(t) = \delta x_0(t) + \int_{t_0}^{t} (t - t')\, \Delta g\left(x_{\mathrm{ref}}(t'), p\right) dt' \\ + \int_{t_0}^{t} (t - t')\, \Gamma_{\mathrm{ref}}\left(x_{\mathrm{ref}}(t')\right) \Delta x^{(0)}(t')\, dt',\end{aligned} \tag{9}$$

where all quantities on the right side, except the parameters, are known exactly (it is assumed for the sake of a simplified analysis that also the initial conditions, x_0, $\dot{x}_0$, are known without error). That is, one can now write the model as

$$l = \mathbf{G}p, \tag{10}$$

where the vector of reduced observables at times, t_j, is

$$\begin{aligned}l^{\mathrm{T}} = \Bigg(\cdots\ \Delta x^{\mathrm{T}}(t_j) - \delta x_0^{\mathrm{T}}(t_j) \\ - \int_{t_0}^{t_j} (t_j - t') \left(\Gamma_{\mathrm{ref}}\left(x_{\mathrm{ref}}(t')\right) \delta x_0(t)\right)^{\mathrm{T}} dt' \ \cdots \Bigg),\end{aligned} \tag{11}$$

and $\mathbf{G}$ is a matrix of elements that are integrals of exactly known functions. The theoretical reasonableness of this proposal is based on the theory of solutions to Volterra integrals, which guarantees convergence to the true solution with such iterated substitutions (Kondo 1991, p. 327). On the other hand, it is not clear that convergence could be achieved if the reference orbit is left unchanged. Moreover, it is shown below in Fig. 3 that this theory fails to account for numerical integration error and the model thus formulated becomes unstable in short order.

3 Numerical Analysis

Before investigating the numerical feasibility of the model (10), it is important to determine the accuracy of the linearization, itself, that is, the numerical consistency of (6) under different levels of orbital perturbation. Toward that end one needs a true orbit in a given gravitational field. This is one of those rare cases where a simulation of the observable is perfect only for a trivial case – the Keplerian orbit. That is, for a non-trivial gravitational field, such as a spherical harmonic series truncated to degree, $n_{\max} > 2$, the "true" orbit is, in fact, subject to unavoidable error in the numerical integration of (1). The model error due to linearization can only be assessed to the extent that it is greater than the numerical integration errors in the model and in the reduced observable (11). Of course, with real data, no such integration error impacts the observable, $\Delta x(t)$, only the computation of integrals such as in (11) or in the elements of the matrix, $\mathbf{G}$.

We use a predictor-corrector integrator of the Adams type that was developed in the 1960s at the Jet Propulsion Laboratory (Krogh 1970). Testing this on a Keplerian orbit, which is known to arbitrary accuracy, we find that this integrator is accurate to better than 1 mm for a 1-day orbital arc of a low-Earth-orbiting satellite. This is also confirmed by comparison to an independent numerical integrator applied to a high-degree gravitational field (M. Naeimi, 2016, personal communication).

Upon thus generating the "true" orbit of a polar-orbiting satellite in a given gravitational field (EGM2008 up to a specified $n_{\max}$) and at an approximate altitude of 450 km, a small perturbation is introduced by adding a zero-mean normally distributed random variable with standard deviation, σ, to each coordinate at 1-s intervals. These perturbed positions are then smoothed by fitting a 7th-order B-spline over 100 s in each coordinate. The resulting reference orbit, $x_{ref}(t)$ then yields observables, $\Delta x(t)$, as shown, for example, in Fig. 1 for the case, $n_{\max} = 36$ and $\sigma = 0.1$ m.

Assuming a reference gravitational field up to degree, $n_{\mathrm{ref}} = 12$, the right side of (6) is computed using the numerical integrator, where because it is a variable step-size integrator, the second integrand, $(t - t')\Gamma_{\mathrm{ref}}(x_{\mathrm{ref}}(t'))\Delta x(t')$, is adequately evaluated with linear interpolation on $\Delta x(t')$. The absolute differences of the right and left sides of Eq. (6) are shown in Fig. 2, which also shows these differences if the second integral with the gradient term is omitted on the right side.

Figure 2 shows that the linearization error inherent in the model (6) is not larger than the numerical integration error in the observable, i.e., less than 1 mm (and perhaps much less). It also shows that the second integral on the right side of (6) cannot be neglected for millimeter-level accuracy

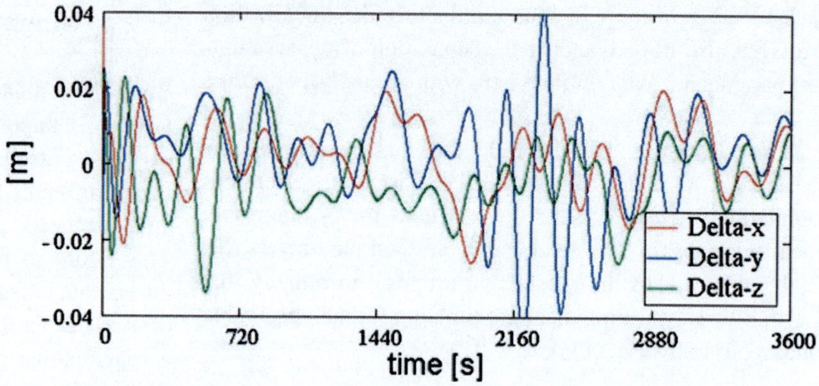

Fig. 1 Cartesian components of the simulated observables, $\Delta \mathbf{x}(t)$, for the case, $n_{max} = 36$ and $\sigma = 0.1$ m. The displayed first hour is representative of the entire 1-day orbital arc

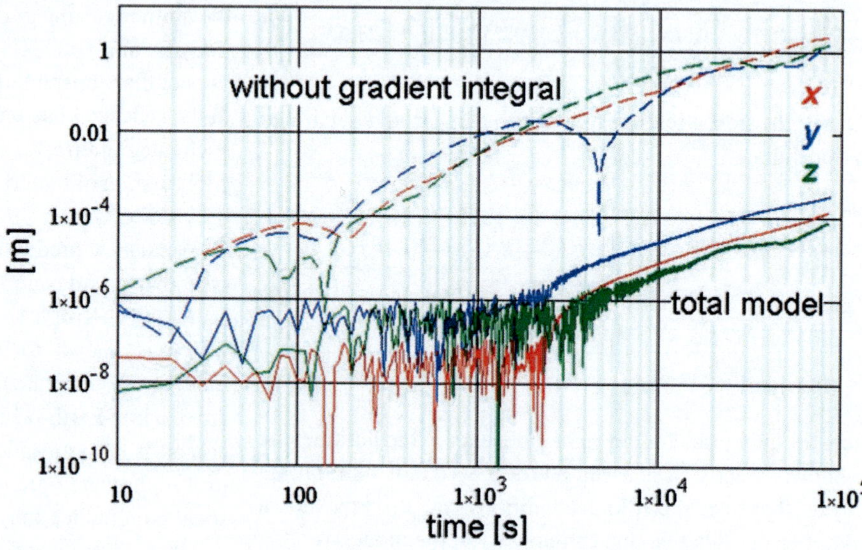

Fig. 2 Absolute differences of right and left sides of Eq. (6) with and without the gradient integral term on the right side. The components of the vector equation are color coded as indicated

Table 1 Absolute differences of right and left sides of Eq. (6) at $t = 86,400$ s

perturb. σ	n_{max}	n_{ref}	absolute error in x, y, z at $t = 86400$ s		
0.1 m	24	4	0.020,	0.145,	0.125 mm
0.1 m	36	12	0.035,	0.008,	0.013 mm
0.1 m	60	30	0.010,	0.012,	0.012 mm
0.1 m	120	30	0.020,	0.015,	0.018 mm
1 m	36	12	0.27,	1.52,	3.12 cm
1 m	60	30	2.39,	0.85,	0.46 cm
1 m	120	30	0.05,	0.002,	0.44 cm
10 m	36	12	0.35,	0.16,	0.23 m
10 m	60	30	0.29,	0.03,	0.002 m
10 m	120	30	0.09,	0.35,	0.09 m

(after a 1-day orbit) in the model. Table 1 lists linearization errors in the model at the end of a 1-day orbit for greater departures between reference and true orbits and different values of n_{max} and n_{ref}. Generally, one may conclude that the linearization is adequate if that departure is of the order of decimeter.

If one substitutes (8) as an *approximation* for $\Delta \mathbf{x}(t')$ on the right side, then the model error becomes intolerable after less than 1 h, as shown in Fig. 3, yielding errors ultimately much worse than simply neglecting the gradient integral. It is highly doubtful that iterated substitutions would yield a convergent solution, as implied by the theory of the iterated solution to Volterra's integral equation. The numerical integration errors that grow in the evaluation of the second integral on the right side of (6) simply invalidate the required premises of that theory.

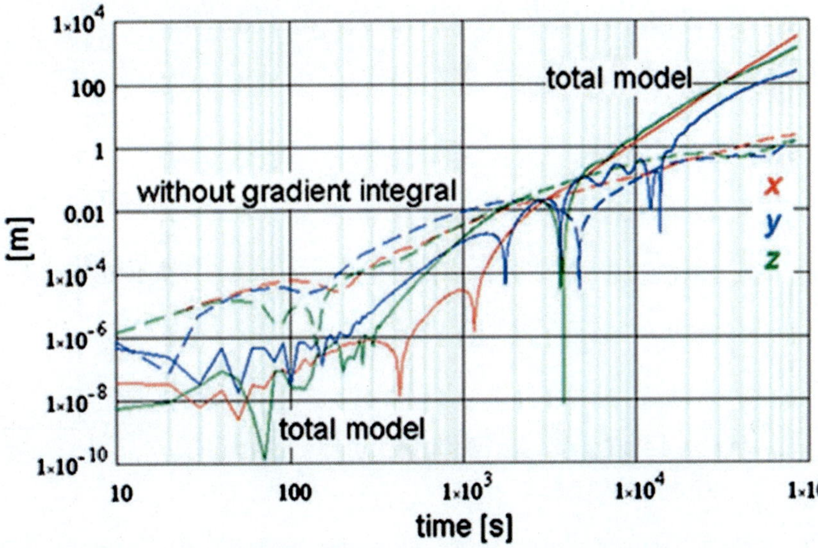

Fig. 3 Absolute differences of right and left sides of Eq. (9) with and without the gradient integral term on the right side. The components of the vector equation are color coded as indicated

Since the results presented in Fig. 2 and Table 1 for the $\sigma = 0.1$ m perturbation demonstrate millimeter accuracy, or better, in the model (10) for day-long orbital arcs, it is of interest to determine the feasibility of recovering the gravitational spectrum from such arcs. A single 24-h arc of a polar orbiting satellite (altitude = 450 km) yields approximately 16 ground tracks on the globe with roughly one repeat track each. The corresponding spatial resolution on the equator is about 22.5°, which translates to maximum harmonic degree, $n_{\max} = 8$. With $n_{\text{ref}} = 4$, the parameters to be solved are the coefficients, $\boldsymbol{p} = \left(\cdots C_{n,m} \cdots\right)^{\text{T}}$, in the series,

$$\Delta \boldsymbol{g}(t) = \sum_{n=n_{\text{ref}}+1}^{n_{\max}} \sum_{m=-n}^{n} C_{n,m} \boldsymbol{S}_{n,m}(t), \tag{12}$$

where, with mean Earth radius, R, and Newtonian gravitational constant times Earth's mass, GM,

$$\boldsymbol{S}_{n,m}(t) = \frac{GM}{R} \nabla_x \left(\left(\frac{R}{r_{\text{ref}}(t)} \right)^{n+1} \overline{Y}_{n,m}\left(\theta_{\text{ref}}(t), \alpha_{\text{ref}}(t)\right) \right), \tag{13}$$

where $r_{\text{ref}}, \theta_{\text{ref}}, \alpha_{\text{ref}}$ are spherical polar coordinates in inertial space and the $\overline{Y}_{n,m}$ are fully normalized spherical harmonic functions. Assuming no observational error (only model error in (10)), Fig. 4 shows the absolute relative errors in the coefficients estimated from reduced observations, $\boldsymbol{l}$, of a satellite tracked in a $n_{\max} = 8$ gravitational field (from EGM2008) according to

$$\hat{\boldsymbol{p}} = \left(\mathbf{G}^{\text{T}}\mathbf{G}\right)^{-1} \mathbf{G}^{\text{T}} \boldsymbol{l}. \tag{14}$$

The absolute relative accuracy and corresponding root-mean-square by degree are defined by

$$\varepsilon_{n,m} = \left| \frac{\hat{C}_{n,m} - C_{n,m}}{C_{n,m}} \right|, \quad \sigma_n = \sqrt{\frac{1}{2n+1} \sum_{m=-n}^{n} \varepsilon_{n,m}^2}. \tag{15}$$

Figure 4 includes also the case of estimating higher harmonics than might be warranted by the limited spatial resolution of a one-day ground track if the satellite is orbiting in a $n_{\max} = 24$ gravitational field. The results of the estimation show that 5-digit or better accuracy is obtained in the former case, while the latter case generally yields poorer relative accuracy at degrees $n > 8$ (consistent with the lack of high global resolution from a single 1-day orbital arc). Increasing the spatial resolution with a 3-day orbital arc reduces the errors in estimating the $n_{\max} = 24$ field by 1.5 orders of magnitude, where integration error is the likely cause for the remaining error.

4 Conclusion

A review of a new perturbation theory for estimating Earth's gravitational field from satellite tracking by GNSS shows that while the theory is elegantly simple, it falls short of its original goal to enable the analysis of arbitrarily long arcs because of unavoidable numerical integration error. One remedy is to dispense with the problematic iterative solution to the Volterra integral equation and treat the gradient term of the linearization as part of the "reduced" observable (bring the gradient integral to the left side of the equation), thus yielding a "pseudo" Gauss-Markov model for the estimation

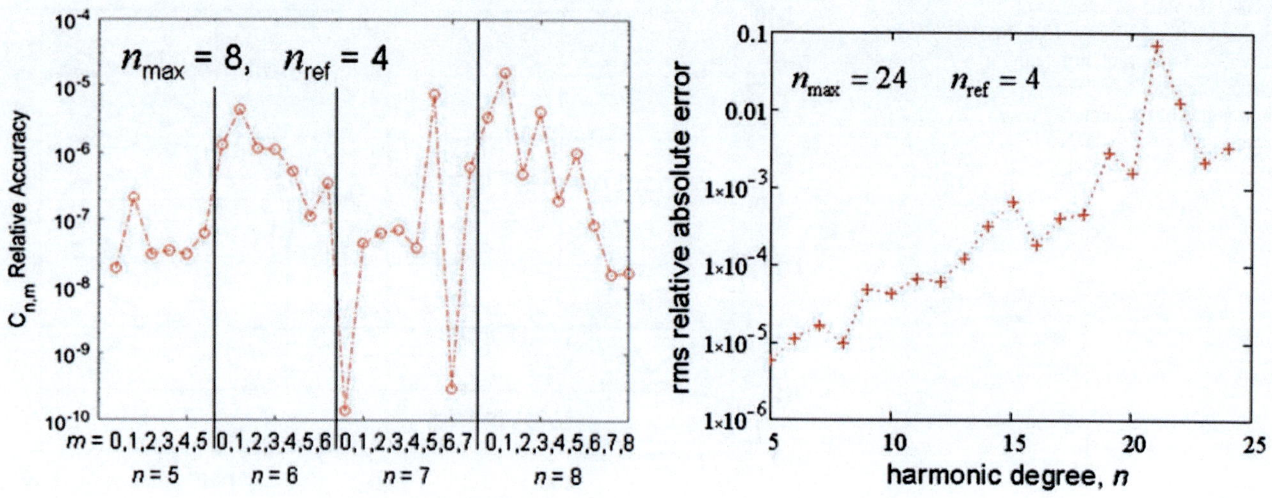

Fig. 4 On the left: absolute relative accuracy of estimated spherical harmonic coefficients obtained by tracking a satellite along a one-day orbital arc in a $n_{\max} = 8$ gravitational field ($C_{n,m}$, $m \geq 0$; similar relative accuracy is obtained for $m < 0$). On the right: the corresponding root-mean-square absolute relative error per degree for all coefficients ($5 \leq n \leq 24$) in a $n_{\max} = 24$ gravitational field. In both cases the estimation is performed using the model (10) and the least-squares solution (14)

of parameters. In addition to some other simplifications of the original proposed theory, we also introduce a reference field that presumably requires no further estimation; although, that estimation is also an option by treating corresponding parameters as stochastic. A numerical analysis shows that the linearization of the model, itself, is adequate for these purposes if the reference orbit deviates from the true orbit by less than 1 m, which is always possible since the reference orbit may be completely arbitrary and can be defined on the basis of the observed orbit. A demonstration of accurate estimation of harmonic coefficients (assuming no observation errors) also validates the accuracy of the model.

Acknowledgments The authors wish to thank Prof. Torsten Mayer-Gürr and an anonymous reviewer for valuable comments that improved the manuscript.

References

Kondo J (1991) Integral equations. Clarendon Press, Oxford

Krogh FT (1970) VODQ/SVDQ/DVDQ - variable order integrators for the numerical solution of ordinary differential equations. Tech. Util. Doc. No. CP-2308, NPO-11643, JPL, Pasadena

Mayer-Gürr T (2006) Gravitationsfeldbestimmung aus der Analyse kurzer Bahnbögen am Beispiel der Satellitenmissionen CHAMP und GRACE. Dissertation, Friedrich-Wilhelms-Universität, Bonn, Germany

Naeimi M, Flury J (eds) (2017) Global gravity field modeling from satellite-to-satellite tracking. Lecture Notes in Earth System Sciences. Springer, Berlin

Xu P (2008) Position and velocity perturbations for the determination of geopotential from space geodetic measurements. Celest Mech Dyn Astron 100:231–249

On the Numerical Implementation of a Perturbation Method for Satellite Gravity Mapping

Christopher Jekeli and Nlingi Habana

Abstract

In 2008 P. Xu (Celest Mech Dyn Astron, 100:231–249) proposed a strictly kinematic perturbation method for determining the Earth's gravitational field from continuous satellite tracking. The main idea is to process orbital arcs of arbitrary length, thus minimizing superfluous parameter estimation associated with stitching together short-arc solutions, and at the same time formulating the problem in terms of standard linear parameter estimation. While the original formulation appears mathematically robust, its nested quadruple along-track integrations are computationally challenging. We reduce the formulation to double integrals and show that the method is numerically not feasible as originally envisaged. On the other hand, by abandoning the rigorous Gauss-Markov formalism, we show the numerical feasibility of processing multiple-day orbital arcs. The methodology lends itself to high-low and low-low satellite-to-satellite tracking, or combinations thereof, as for GRACE-like systems.

Keywords

GNSS satellite tracking · Gravitational field estimation · Numerical orbit integration · Satellite perturbation theory

1 Introduction

With the modern ability to track low-Earth-orbiting satellites continually and uniformly with Global Navigation Satellite Systems (GNSS), such as the Global Positioning System (GPS), the standard methods to extract estimates of the Earth's gravitational field from satellite tracking data may be re-visited. Prior to GNSS tracking, ground-based tracking created a patch-work of data as a satellite rose and set at any particular tracking station; and, accordingly, a good a priori or reference orbit was essential in stitching the observed arcs together. A perturbation theory based on Keplerian elements, for example, served to separate secular and long-period orbital variations from the more short-term resonances, which facilitated this estimation process. These techniques are still practiced today in various forms; a good review is given in the volume by Naeimi and Flury (2017). Xu (2008) proposed a radical change from this methodology specifically in view of the proven accurate tracking capabilities with GNSS. Continual and uniform high-accuracy tracking, arguably obviates piecing together short arcs and simplifies the overall problem setup as all formulations may be made with the straightforward use of Cartesian coordinates. The idea certainly has tremendous theoretical and practical appeal and in this paper we aim to elucidate this in the simplest terms, but the numerical implementation, never attempted by the originator, puts at least some limitations on the length of the orbital arc that constitutes a segment of the overall estimation process. It is this aspect of the proposed methodology that we wish to highlight and further develop in this short note.

C. Jekeli (✉) • N. Habana
Division of Geodetic Science, School of Earth Sciences, Ohio State University, Columbus, OH, USA
e-mail: jekeli.1@osu.edu

P. Novák et al. (eds.), *IX Hotine-Marussi Symposium on Mathematical Geodesy*,
International Association of Geodesy Symposia 151, https://doi.org/10.1007/1345_2019_81

Non-Recursive Representation of an Autoregressive Process Within the Magic Square

Ina Loth, Boris Kargoll, and Wolf-Dieter Schuh

Abstract

A stochastic process can be represented and analysed by four different quantities in the time and frequency domain: (1) the process itself, (2) its autocovariance function, (3) the spectral representation of the stochastic process and (4) its spectral distribution or the spectral density function, if it exits. These quantities and their relationships can be clearly represented by the "Magic Square", where the quantities build the corners of this square and the connecting lines indicate the transformations into each other.

The real-valued, time-discrete, one-dimensional and covariance-stationary autoregressive process of order p (AR(p) process) is a frequently used stochastic process for instance to model highly correlated measurement series with constant sampling rate given by satellite missions. In this contribution, a reformulation of an AR(p) to a moving average process with infinite order is presented. The Magic Square of this reformulated process can be seen as an alternative representation of the four quantities in time and frequency, which are usually given in the literature. The results will be evaluated by discussing an AR(1) process as example.

Keywords

Autoregressive process · Moving average process · Spectral analysis · Stochastic process · Time series analysis

1 Introduction

In practice many phenomena with random characteristics exist, which cannot be represented by deterministic functions. In these cases, stochastic processes often allow for a sufficient description (see e.g. Koch and Schmidt 1994; Moritz 1980; Welsch et al. 2000). Applications of stochastic processes in geodesy tend to focus on analyses within the time domain at the level of the measurements themselves. In contrast, the usage of the relationships of a process with its spectral representation, autocovariance function and spectral distribution (or density) function is less popular, or even done incorrectly. One reason for this is that the mathematics and thus the computational aspects of these relationships and representations are rather intricate and oftentimes not readily available for a specific type of process to be used in a practical situation. To remedy this problem, Krasbutter et al. (2015) discussed a kind of Magic Square with respect to a general real-valued, one-dimensional, discrete-time, covariance-stationary stochastic process. The Magic Square has the advantage that it is a well-arranged representation of the described four quantities of a process in the time and frequency domain and their relationships. Furthermore, Krasbutter et al. (2015) expanded the Magic Square to stochastic processes, which are obtained by non-recursive filtering of an input process and evaluation of the results by

I. Loth (✉) · W.-D. Schuh
Institute of Geodesy and Geoinformation, University of Bonn, Bonn, Germany
e-mail: ina.loth@geod.uni-bonn.de

B. Kargoll
Geodetic Institute, Leibniz University Hannover, Hannover, Germany

P. Novák et al. (eds.), *IX Hotine-Marussi Symposium on Mathematical Geodesy*,
International Association of Geodesy Symposia 151, https://doi.org/10.1007/1345_2019_60

application to a q-th order moving average (MA(q)) process (see the following Sect. 2 for an overview).

In this contribution the Magic Square will be formulated for another well-known process: a discrete-time p-th order autoregressive (AR(p)) process. The use of such a process as a description of the random error term in linear observation equations seems to have been proposed first by the econometricians D. Cochrane and G. H. Orcutt (Cochrane and Orcutt 1949). In geodesy, this kind of model is becoming increasingly popular; see Koch and Schmidt (1994) and Schuh (2003) for general descriptions and Schubert et al. (2019) for a current application to the highly correlated measurement errors of the Gravity Field and Steady-State Ocean Circulation Explorer (GOCE) satellite mission. The AR(p) process may be viewed as being obtained by recursive filtering. The Magic Square of such a process is explained in Schuh et al. (2014) and further discussed in Schuh (2016).

In contrast, we study in the current contribution the Magic Square for AR(p) processes obtained through non-recursive filtering. For this purpose, we elaborate a certain reformulation of that process. In Sect. 3 the transformation is presented with a MA process of infinite order (MA(∞)) as a result. The Magic Square of this transformed stochastic process is an alternative representation of the four quantities of an AR process given in Schuh et al. (2014). The transformation can be seen as a link to the Magic Square describe in Krasbutter et al. (2015) To demonstrate the evaluation of these results, the transformation is applied to an AR(1) process and the corresponding Magic Square is compared with representations given in the literature (cf. Sect. 3). In Sect. 4 this paper is concluded with a summary and an outlook.

2 The Magic Square of a Non-Recursive Filtered Stochastic Process

A general stochastic process $\mathcal{X}_T$ is defined by a family of random variables on a probability space, symbolically

$$\mathcal{X}_T = (\Omega, \mathcal{A}, P, \{\mathcal{X}_t, t \in T\}), \tag{1}$$

where Ω denotes the sample space, $\mathcal{A}$ a σ-Algebra of events, P the probability measure and $\mathcal{X}_t$ is a random variable.

Additionally we restrict our representation to stochastic processes with the following properties:

- One-dimensional and real-valued:

$$(\Omega, \mathcal{A}) \rightarrow (\mathbb{R}, \mathcal{B}),$$

where $\mathcal{B}$ is the Borel σ-Algebra, which is generated by all real-valued, one-dimensional, left open and right closed intervals.

- Discrete in time with constant and general sampling rate Δt:

$$t = n \cdot \Delta t, n \in \mathbb{Z}, \Delta t \in \mathbb{R}.$$

On account of this, a random variable depends only on n and it will by symbolised by $\mathcal{X}_n$ in the following.

- Covariance-stationary with zero mean, variance σ^2 and autocovariance function $\gamma_k^{\mathcal{X}}$, where k is called lag (cf. Brockwell and Davis 1991, pp. 11–12).

Additionally, it is assumed, that the process $\mathcal{X}_T$ is obtained by filtering another one-dimensional, discrete-time, covariance-stationary stochastic process $\mathcal{U}_T$ by

$$\mathcal{X}_n = \sum_{j=0}^{q} \psi_j \mathcal{U}_{n-j} = \Psi(L)\mathcal{U}_n, \tag{2}$$

where $\Psi(L) = \psi_0 + \psi_1 L + \ldots + \psi_q L^q$ with lag operator notation $L^j \mathcal{U}_n = \mathcal{U}_{n-j}$ is a non-recursive, causal, absolutely summable and invertible filter. The order q of the filter can be infinite ('$q = \infty$') or finite ($q \in \mathbb{N}$).

The Magic Square of this real-valued, one-dimensional, discrete-time, covariance-stationary, non-recursive filtered stochastic process is presented in Fig. 1. In the upper left corner the stochastic process itself is given and can be seen as a collection of random and equidistant variables.

The spectral representation of the described stochastic process (upper right corner) is denoted by $d\widehat{\mathcal{Z}}_s^{\mathcal{X}}(\nu)$. In the following the hat will symbolize a quantity in the frequency domain and the superscripted factor s the fact that the corresponding representation in the time domain is discrete-valued.

However $d\widehat{\mathcal{Z}}_s^{\mathcal{X}}(\nu)$ is a stochastic orthogonal increment process with the following properties:

- One-dimensional and complex-valued,

$$\Omega \rightarrow \mathbb{C},$$

- Frequency-continuous with ν defined within the interval $[-\nu^N, \nu^N]$, where $\nu^N = \frac{1}{2\Delta t}$ is known as the Nyquist frequency.
- Orthogonal increments (cf. Brockwell and Davis 1991, pp. 138–140):

$$E\left\{d\widehat{\mathcal{Z}}_s^{\mathcal{X}}(\nu_1)(d\widehat{\mathcal{Z}}_s^{\mathcal{X}}(\nu_2))^*\right\} = 0, \quad \text{for } \nu_1 \neq \nu_2,$$

where $*$ denotes the conjugate complex.

While the proposed process can be described as a filtering of an input process $\mathcal{U}_n$ in the time domain (see (2)), the description for the frequency domain is:

$$d\widehat{\mathcal{Z}}_s^{\mathcal{X}}(\nu) = \widehat{\Psi}_s(\nu) d\widehat{\mathcal{Z}}_s^{\mathcal{U}}(\nu), \tag{3}$$

where $d\widehat{\mathcal{Z}}_s^{\mathcal{U}}(\nu)$ is the corresponding spectral representation of the input stochastic process $\mathcal{U}_n$ and

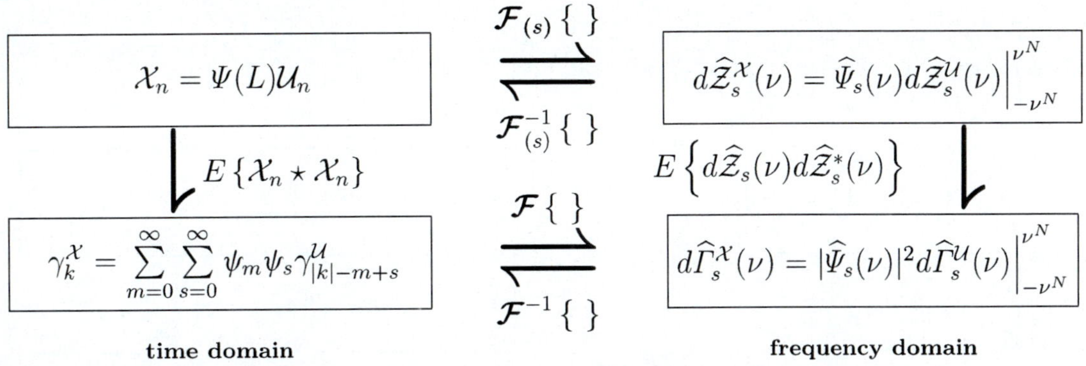

Fig. 1 Magic Square for covariance-stationary, discrete-time, non-recursive filtered stochastic process with upper left: stochastic process, lower left: autocovariance function, upper right: spectral representation of the stochastic process and lower right: spectral representation of the autocovariance function. The interrelations are symbolised by arrows with the corresponding mathematical operations, where $\mathcal{F}\{\cdot\}/\mathcal{F}^{-1}\{\cdot\}$ symbolise the Fourier transform/integral, $\mathcal{F}_{(s)}\{\cdot\}/\mathcal{F}_{(s)}^{-1}\{\cdot\}$ the stochastic Fourier transform/integral, $E\{\cdot\}$ the expectation and '$\star$' a correlation

$\widehat{\Psi}_s(\nu) = \sum_{j=0}^{q}\psi_j e^{-i2\pi\nu j\Delta t}$ is the Fourier transform of the filter $\Psi(L)$ called the transfer function (cf. Priestley 2004, pp. 263–270).

The interrelation between the stochastic process and its spectral representation, which is indicated by the harpoons, can be described by the "stochastic Fourier transform" $\mathcal{F}_{(s)}\{\cdot\}$ and the converse relationship by the "stochastic Fourier integral" $\mathcal{F}_{(s)}^{-1}\{\cdot\}$. The mathematical formulas and a detailed explanation of these relations is not the focus of this paper and are omitted. But the interested reader is referred to Krasbutter et al. (2015) and Lindquist and Picci (2015, Chapter 3).

The two quantities in the lower row of the square are the autocorrelation function $\gamma_k^{\mathcal{X}}$ (left side) and its spectral representation $d\widehat{\Gamma}_s^{\mathcal{X}}(\nu)$ (right side). This spectral representation is an increment process of the spectral distribution function $\widehat{\Gamma}_s^{\mathcal{X}}(\nu)$. If the derivative $d\widehat{\Gamma}_s^{\mathcal{X}}(\nu)/d\nu$ exits, it is called spectral density function $\widehat{\gamma}_s^{\mathcal{X}}(\nu)$.

Both quantities in the lower row of the square are deterministic functions, where the autocorrelation function is discrete and its spectral representation is frequency-continuous with ν defined in the interval $-\nu^N$ to ν^N and is continued periodically outside of this range. As described in Priestley (2004, p. 214) $\gamma_k^{\mathcal{X}}$ and $\widehat{\gamma}_s^{\mathcal{X}}(\nu)$ are even, if the related stochastic process is real-valued, which we have fixed by the above described characteristics of the process. Furthermore, the autocovariance can be formulated by the filter $\Psi(L)$ and the autocovariance of the input process $\gamma_k^{\mathcal{U}}$, while its spectral representation is given by the transfer function of the filter $\widehat{\Psi}_s(\nu)$ and the increment of the spectral distribution function of the input process $d\widehat{\Gamma}_s^{\mathcal{U}}(\nu)$. Thus, the autocorrelation of $\gamma_k^{\mathcal{X}}$ is given by

$$\gamma_k^{\mathcal{X}} = \sum_{m=0}^{\infty}\sum_{s=0}^{\infty}\psi_m\psi_s\gamma_{|k|-m+s}^{\mathcal{U}} \tag{4}$$

and the corresponding spectral representation by

$$d\widehat{\Gamma}_s^{\mathcal{X}}(\nu) = |\widehat{\Psi}_s(\nu)|^2 d\widehat{\Gamma}_s^{\mathcal{U}}(\nu). \tag{5}$$

The interrelation of these two equations can be described by the "deterministic Fourier transform" $\mathcal{F}\{\cdot\}$ and the converse relationship by the "deterministic Fourier integral" $\mathcal{F}^{-1}\{\cdot\}$. A detailed explanation of these two equations and its interrelation is given by Priestley (2004, Sect. 4.12), Brockwell and Davis (1991, Proposition 3.1.2) and Krasbutter et al. (2015).

As explained above, the upper corners are stochastic functions and in contrast to them the lower corners are deterministic functions so that the transformation from top to bottom can be interpreted as a reduction from stochastic to deterministic. This reduction is achieved by using the expectation (symbolised with $E\{\cdot\}$ in Fig. 1), which has the drawback of information loss, thus the way vice versa (from bottom to top) is not possible without additional information (indicated by the missing arrow from the lower to the upper corners). The operations from top to bottom can be seen on the left side as a stochastic correlation. The correlation is symbolised by '$\star$' in Fig. 1. The corresponding operation in the frequency is a stochastic multiplication.

3 The Magic Square of an *p*-th Order Autoregressive Process

The time-discrete, autocovariance-stationary, invertible AR(p) process with $p \in \mathbb{N}$ is defined by

$$\mathcal{X}_n := \theta_1\mathcal{X}_{n-1} + \ldots + \theta_p\mathcal{X}_{n-p} + \mathcal{E}_n,\ (n \in \mathbb{Z})$$
$$\iff \Theta(L)\mathcal{X}_n = \mathcal{E}_n, \tag{6}$$

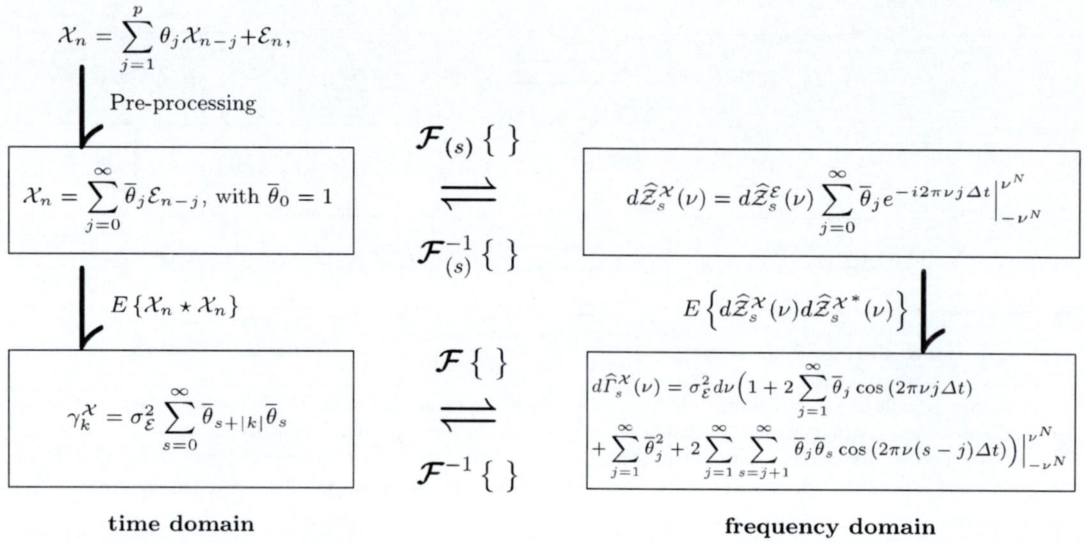

Fig. 2 Magic Square for covariance-stationary, discrete-time autoregressive process of order p (AR(p)) with general sampling rate Δt. This AR(p) process is reformulated in a pre-processing step to a moving average process of infinitive order. The Magic Square is then derived out of this reformulated process with upper left: reformulated stochastic process itself, lower left: autocovariance function, upper right: spectral representation of the reformulated stochastic process and lower right: spectral density function. The interrelations are symbolised by arrows with the corresponding mathematical operations, where $\mathcal{F}\{\cdot\}/\mathcal{F}^{-1}\{\cdot\}$ symbolises the Fourier transform/integral, $\mathcal{F}_{(s)}\{\cdot\}/\mathcal{F}^{-1}_{(s)}\{\cdot\}$ the stochastic Fourier transform/integral, $E\{\cdot\}$ the expectation and '$\star$' a correlation

where $\Theta(L) = 1 - \theta_1 L - \ldots - \theta_p L^p$ is a recursive, causal filter and $\mathcal{E}_n$ denotes white noise with $\mathcal{E}_n \sim N(0, \sigma^2_{\mathcal{E}})$ (Brockwell and Davis 1991, Definition 3.1.1). The AR process is covariance-stationary if and only if the roots of the polynomial

$$1 - \theta_1 z - \theta_2 z^2 \ldots - \theta_p z^p = 0, z \in \mathbb{C}$$

lie outside of the unit circle. Furthermore, AR processes with finite order are invertible (cf. Box and Jenkins 1976, Sect. 3.2.1).

In Schuh et al. (2014) the Magic Square of an AR(p) process is presented. In contrast, the idea of this contribution is to transform the AR(p) process given by (6) in the form of (2). The advantage of a non-recursive representation is that it is easier to evaluate concerning warm up, behaviour of the covariance and the stationarity of the process. In so doing the explained Magic Square in Sect. 2 can be applied to the reformulated process. The result is an alternative but equivalent representation of the AR(p) process within the Magic Square; the transformation can be seen as a link between these two representations of an AR(p) process within the Magic Square.

In a first step the reformulation of the AR(p) process to (2) is described and afterwards the alternative representation of the Magic Square is given. All outcomes of each step are applied to an AR(1) process and compared to the results given in the literature, where no reformulation is applied.

3.1 Reformulation of the AR(p) Process

The reformulation can be seen as a pre-processing step and is symbolised in Fig. 2, where the Magic Square of the AR(p) process is presented. This additional step starts by the multiplication of (6) with $\Theta(L)^{-1}$ on both sides:

$$\mathcal{X}_t = \Theta(L)^{-1}\mathcal{E}_n.$$

(cf. Gilgen 2006, p. 251).

To familiarize ourselves with the filter $\Theta(L)^{-1}$, whose order for instance we do not know yet, a reformulation is done. Thus, the inverse filter of $\Theta(L)$ can be formulated by

$$\Theta(L)^{-1} = \frac{1}{1 - \theta_1 L - \ldots - \theta_p L^p} \tag{7}$$

(see Hamilton 1994, Chapter 2.4). This inverse representation can be rewritten by using the infinite geometric series

$$\frac{1}{1-x} = 1 + x + x^2 + x^3 + \ldots \tag{8}$$

with $|x| < 1$ (cf. Andrews 1998, Eq. (3.3)) and results in

$$\Theta(L)^{-1} = 1 + (\theta_1 L + \ldots + \theta_p L^p) + (\theta_1 L + \ldots + \theta_p L^p)^2 + \ldots . \tag{9}$$

The next step is to expand and resort (9) to

$$\begin{aligned}\Theta(L)^{-1} &= 1 + \overline{\theta}_1 L + \overline{\theta}_2 L^2 + \ldots \\ &= \overline{\Theta}(L),\end{aligned} \quad (10)$$

where $\overline{\Theta}(L)$ is the reformulated representation of $\Theta(L)$. The determination of $\overline{\theta}_i$ for $i \in \mathbb{N}$ can be achieved by the following algorithm:

Step 1: Find all s combinations of the sorted product $\theta_1^{l_{1,j}} \cdot \theta_2^{l_{2,j}} \cdot \ldots \cdot \theta_p^{l_{p,j}}$, where $\sum_{m=1}^{p} m \cdot l_{m,j} = i$ with $l_{m,j} \in \mathbb{N}$ holds. The index $j = 1, \ldots, s$ specifies the combination number.

Step 2: Determination of

$$d_j = \frac{\left(\sum_{m=1}^{p} l_{m,j}\right)!}{l_{1,j}! \cdot l_{2,j}! \cdot \ldots \cdot l_{p,j}!}.$$

where ! denotes the factorial function. This factor indicates the number of possibilities for combining the filter coefficients $\theta_1, \theta_2, \ldots \theta_p$ of combination j.

Step 3: The last step is to determine the filter coefficient $\overline{\theta}_i$ by

$$\overline{\theta}_i = \sum_{j=1}^{s} \left(d_j \cdot \theta_1^{l_{1,j}} \cdot \theta_2^{l_{2,j}} \cdot \ldots \cdot \theta_p^{l_{p,j}}\right).$$

Hence, the alternative representation of (6) is given by

$$\mathcal{X}_n = \overline{\Theta}(L)\mathcal{E}_n = \sum_{j=0}^{\infty} \overline{\theta}_j \mathcal{E}_{n-j}, \text{ with } \overline{\theta}_0 = 1. \quad (11)$$

This process is known as moving average process of infinite order (MA(∞) process), which is a special form of the filtered stochastic process described in Sect. 2.

Example of the Transformation

The transformation is applied exemplary to an AR(1) process, which is defined by

$$\Theta(L)\mathcal{X}_n := \mathcal{E}_n, \quad (12)$$

with $\Theta(L) = 1 - \theta_1 L$ and $|\theta_1| < 1$ (cf. Hamilton 1994, p. 53). This process has only θ_1 as filter coefficient and therefore the described algorithm to determine $\overline{\theta}_i$ is simplified, because $l_{1,\cdot} = i$. Hence, the reformulated AR(1) process is given by

$$\mathcal{X}_n = \sum_{j=0}^{\infty} \theta_1^j \mathcal{E}_{n-j}.$$

The results for transformed AR processes with higher orders are much more complicated. For instance, the reformulated AR(2) process is given by

$$\begin{aligned}\overline{\Theta}(L) = &\underbrace{1}_{\overline{\theta}_0} + \underbrace{\theta_1}_{\overline{\theta}_1} L + \underbrace{(\theta_1^2 + \theta_2)}_{\overline{\theta}_2} L^2 + \underbrace{(\theta_1^3 + 2\theta_1\theta_2)}_{\overline{\theta}_3} L^3 \\ &+ \underbrace{(\theta_1^4 + 3\theta_1^2\theta_2 + \theta_2^2)}_{\overline{\theta}_4} L^4 + \underbrace{(\theta_1^5 + 4\theta_1^3\theta_2 + 3\theta_1\theta_2^2)}_{\overline{\theta}_5} L^5 \\ &+ \underbrace{(\theta_1^6 + 5\theta_1^4\theta_2 + 6\theta_1^2\theta_2^2 + \theta_2^3)}_{\overline{\theta}_6} L^6 + \ldots.\end{aligned}$$

In this contribution, due to the complexity of AR processes with higher orders, the results are applied only to AR(1) processes.

3.2 Magic Square of the Reformulated AR(p) Process

The Magic Square of the reformulated AR(p) process is described and the results are presented in Fig. 2. The derivation starts with the quantities in the time domain (corners on the left-hand side of the square), followed by the results in the frequency domain (corners on the right-hand side of the square).

3.2.1 Time Domain (Left-Hand Side)

The upper left corner of the square is given by (11), being the reformulated AR process itself. This process has an autocovariance function (lower left corner), which can be derived by using (4) and the property

$$\gamma_k^{\mathcal{E}} = \begin{cases} \sigma_{\mathcal{E}}^2 & \text{for } k = 0 \\ 0 & \text{else} \end{cases}$$

of white noise, resulting in

$$\gamma_k^{\mathcal{X}} = \sigma_{\mathcal{E}}^2 \sum_{s=0}^{\infty} \overline{\theta}_{s+|k|} \overline{\theta}_s. \quad (13)$$

Example: AR(1) Process

As described in the last section the reformulation of the AR(1) to an MA(∞) process is given by $\overline{\theta}_i = \theta_1^i$. This result is substituted into (13), leading to

$$\gamma_k^{\mathcal{X}} = \sigma_{\mathcal{E}}^2 \sum_{s=0}^{\infty} \theta_1^{s+|k|} \theta_1^s = \sigma_{\mathcal{E}}^2 \sum_{s=0}^{\infty} \theta_1^{2s+|k|}. \quad (14)$$

It can be shown, that (14) is an alternative representation of the autocovariance

$$\gamma_k^{\mathcal{X}} = \sigma_{\mathcal{E}}^2 \theta_1^{|k|} \cdot \frac{1}{1 - \theta_1^2}, \quad (15)$$

which is often mentioned in the literature (cf. Priestley 2004, Eq. (3.5.16)). To show the equivalence between these two representations of the autocovariance function, (14) is reorganised to

$$\begin{aligned}\gamma_k^{\mathcal{X}} &= \sigma_{\mathcal{E}}^2 \theta_1^{|k|} \sum_{s=0}^{\infty} \theta_1^{2s} \\ &= \sigma_{\mathcal{E}}^2 \theta_1^{|k|} (1 + \theta_1^2 + \theta_1^4 + \ldots).\end{aligned}$$

In the next step the pre-processing step is undone by using the definition (8) of the infinite geometric series. The result is (15).

3.2.2 Frequency Domain (Right-Hand Side)

The spectral representation of an AR(p) process by using (3) is defined by

$$\begin{aligned}d\widehat{\mathcal{Z}}_s^{\mathcal{X}}(\nu) &= d\widehat{\mathcal{Z}}_s^{\mathcal{E}}(\nu)\widehat{\overline{\Theta}}_s(\nu) \\ &= d\widehat{\mathcal{Z}}_s^{\mathcal{E}}(\nu) \sum_{j=0}^{\infty} \overline{\theta}_j e^{-i2\pi\nu j\Delta t}, \end{aligned} \tag{16}$$

where $\widehat{\overline{\Theta}}_s(\nu) = \sum_{j=0}^{\infty} \overline{\theta}_j e^{-i2\pi\nu j\Delta t}$ is the transfer function of the filter $\overline{\Theta}(L)$ and $d\widehat{\mathcal{Z}}_s^{\mathcal{E}}(\nu)$ the spectral representation of discrete-time white noise, also known as increment process of a Wiener process (see Lindquist and Picci 2015, Sect. 3.3.3). The spectral representation is defined for ν given in the interval $[-\nu^N, \nu^N]$ and is periodic outside of this range.

The spectral representation of the autocovariance is defined by using (5) and the property of white noise $d\widehat{\Gamma}_s^{\mathcal{E}}(\nu) = \sigma_{\mathcal{E}}^2 d\nu$:

$$d\widehat{\Gamma}_s^{\mathcal{X}}(\nu) = \sigma_{\mathcal{E}}^2 d\nu \left(\sum_{j=0}^{\infty}\sum_{s=0}^{\infty} \overline{\theta}_j \overline{\theta}_s e^{-i2\pi\nu j\Delta t} e^{i2\pi\nu s\Delta t} \right). \tag{17}$$

This sum can be reorganised to

$$\begin{aligned}d\widehat{\Gamma}_s^{\mathcal{X}}(\nu) = \sigma_{\mathcal{E}}^2 d\nu \Bigg(& \overline{\theta}_0\overline{\theta}_0 + \sum_{j=1}^{\infty} \overline{\theta}_j \overline{\theta}_0 e^{-i2\pi\nu j\Delta t} \\ & + \sum_{s=1}^{\infty} \overline{\theta}_s \overline{\theta}_0 e^{i2\pi\nu s\Delta t} + \sum_{j=1}^{\infty} \overline{\theta}_j^2 \\ & + \sum_{j=1}^{\infty} \sum_{s=1, s\neq j}^{\infty} \overline{\theta}_j \overline{\theta}_s e^{-i2\pi\nu j\Delta t} e^{i2\pi\nu s\Delta t} \Bigg).\end{aligned}$$

Now, Euler's formula and the relation $\overline{\theta}_0 = 1$ is applied and results in

$$\begin{aligned}d\widehat{\Gamma}_s^{\mathcal{X}}(\nu) = \sigma_{\mathcal{E}}^2 d\nu \Bigg(& 1 + 2\sum_{j=1}^{\infty} \overline{\theta}_j \cos(2\pi\nu j\Delta t) + \sum_{j=1}^{\infty} \overline{\theta}_j^2 \\ & + 2\sum_{j=1}^{\infty}\sum_{s=j+1}^{\infty} \overline{\theta}_j \overline{\theta}_s \cos(2\pi\nu(s-j)\Delta t) \Bigg), \end{aligned} \tag{18}$$

where the frequency ν takes values within the interval $[-\nu^N, \nu^N]$. Obviously, the derivative of (18) exits, so the spectral density function of the AR(p) process is given by

$$\begin{aligned}\widehat{\gamma}_s^{\mathcal{X}}(\nu) = \sigma_{\mathcal{E}}^2 \Bigg(& 1 + 2\sum_{j=1}^{\infty} \overline{\theta}_j \cos(2\pi\nu j\Delta t) + \sum_{j=1}^{\infty} \overline{\theta}_j^2 \\ & + 2\sum_{j=1}^{\infty}\sum_{s=j+1}^{\infty} \overline{\theta}_j \overline{\theta}_s \cos(2\pi\nu(s-j)\Delta t) \Bigg). \end{aligned} \tag{19}$$

Example: AR(1) Process

The spectral representation and the spectral density function of the reformulated AR(1) process are obtained by substituting $\overline{\theta}_j = \theta_1^j$ into (16) and (19). The spectral representation is then defined by

$$d\widehat{\mathcal{Z}}_s^{\mathcal{X}}(\nu) = d\widehat{\mathcal{Z}}_s^{\mathcal{E}}(\nu) \sum_{j=0}^{\infty} \theta_1^j e^{-i2\pi\nu j\Delta t} \tag{20}$$

and the spectral density function by

$$\begin{aligned}\widehat{\gamma}_s^{\mathcal{X}}(\nu) = \sigma_{\mathcal{E}}^2 \Bigg(& 1 + 2\sum_{j=1}^{\infty} \theta_1^j \cos(2\pi\nu j\Delta t) + \sum_{j=1}^{\infty} \theta_1^{2j} \\ & + 2\sum_{j=1}^{\infty}\sum_{s=j+1}^{\infty} \theta_1^{j+s} \cos(2\pi\nu(s-j)\Delta t) \Bigg). \end{aligned} \tag{21}$$

In (Priestley 2004, p. 238) the spectral density function of an AR(1) process is given by

$$\widehat{\gamma}_s^{\mathcal{X}}(\nu) = \frac{\sigma_{\mathcal{E}}^2}{(1 - 2\theta_1 \cos(2\pi\nu\Delta t) + \theta_1^2)}. \tag{22}$$

It can be shown that this result is equivalent to (21) by substituting Euler's formula:

$$\begin{aligned}\widehat{\gamma}_s^{\mathcal{X}}(\nu) &= \frac{\sigma_{\mathcal{E}}^2}{(1 - \theta_1 e^{i2\pi\nu\Delta t} - \theta_1 e^{-i2\pi\nu\Delta t} + \theta_1^2)} \\ &= \sigma_{\mathcal{E}}^2 \left[\left(\frac{1}{1 - \theta_1 e^{-i2\pi\nu\Delta t}} \right) \left(\frac{1}{1 - \theta_1 e^{i2\pi\nu\Delta t}} \right) \right].\end{aligned}$$

The infinite geometric series is applied and results in

$$\widehat{\gamma}_s^{\mathcal{X}}(\nu) = \sigma_{\mathcal{E}}^2 \left(\sum_{j=0}^{\infty} \theta_1^j e^{-i2\pi\nu j\Delta t} \right) \left(\sum_{s=0}^{\infty} \theta_1^s e^{i2\pi\nu s\Delta t} \right)$$
$$= \sigma_{\mathcal{E}}^2 \left(\sum_{j=0}^{\infty} \sum_{s=0}^{\infty} \theta_1^j \theta_1^s e^{-i2\pi\nu j\Delta t} e^{i2\pi\nu s\Delta t} \right).$$

As described above a rearrangement of the sums results in (21).

4 Conclusion and Outlook

Within this paper the transformation of an AR(p) into a MA(∞) process, which is in practical use easier to interpretate concerning warm-up, covariance and stationarity, is demonstrated. In so doing the graphical representation of a stochastic process in time and frequency domain given by Krasbutter et al. (2015) can be applied to determine the explicit mathematical expressions of each corner in the Magic Square for an AR(p) process. The practical application for instance to AR processes estimated by means of the data given by satellite mission GOCE and the convergence behaviour of the transformed AR(p) process is still to be examined. Due to lack of space in this contribution this investigation is omitted.

The application of the transformation to widely used stochastic processes, for instance the autoregressive moving average process (ARMA process), would be an extension of this scenario and will be considered in the future.

References

Andrews GE (1998) The geometric series in calculus. Am Math Mon 105(1):36–40. https://doi.org/10.2307/2589524

Box GEP, Jenkins GM (1976) Time series analysis forecasting and control. Holden-Day, San Francisco

Brockwell PJ, Davis RA (1991) Time series: theory and methods. 2nd edn. Springer, New York

Cochrane D, Orcutt GH (1949) Application of least squares regression to relationships containing auto-correlated error terms. J Am Stat Assoc 44(245):32–61

Gilgen H (2006) Univariate time series in geosciences-theory and examples. Springer, Berlin

Hamilton JD (1994) Time series analysis. Princeton University Press, Princeton

Koch K-R, Schmidt M (1994) Deterministische und stochastische Signale. Dümmler, Bonn

Krasbutter I, Kargoll B, Schuh W-D (2015) Magic square of real spectral and time series analysis with an application to moving average processes. In: Kutterer H et al (eds.) The 1st internationa workshop on the quality of geodetic observation and monitoring systems (QuGOMS'11), pp 9–14. Springer, New York. https://doi.org/10.1007/978-3-319-10828-5_2

Lindquist A, Picci G (2015) Linear stochastic systems. Springer, Berlin, Heidelberg

Moritz H (1980) Advanced physical geodesy. Wichmann, Karlsruhe

Priestley MB (2004) Spectral analysis and time series. Elsevier Academic, Amsterdam

Schubert T, Brockmann JM, Schuh W-D (2019) Identification of suspicious data for robust estimation of stochastic processes. In: Proceedings of the IX Hotine-Marussi symposium 2019 Springer (submitted)

Schuh W-D (2003) The processing of band-limited measurements; filtering techniques in the least squares context and in the presence of data gaps. Space Sci Rev 108(1):67–78. https://doi.org/10.1023/A:1026121814042

Schuh W-D (2016) Signalverarbeitung in der Physikalischen Geodäsie. In Freeden W, Rummel R (eds) Handbuch der Geodäsie, vol. Erdmessung und Satellitengeodäsie, pp 73–121. Springer, New York. https://doi.org/10.1007/978-3-662-47100-5_15

Schuh W-D, Krasbutter I, Kargoll B (2014) Korrelierte Messungen—was nun? In: Neuner H (ed) Zeitabhängige Messgrössen—Ihre Daten haben (Mehr-)Wert DVW-Schriftenreihe, vol 74, pp 85–101. Wißner-Verlag, Augsburg

Welsch W, Heunecke O, Kuhlmann H (2000) Handbuch Ingenieurgeodäsie, Auswertung geodätischer Überwachungsmessungen. Wichmann, Heidelberg

A Bootstrap Approach to Testing for Time-Variability of AR Process Coefficients in Regression Time Series with t-Distributed White Noise Components

Hamza Alkhatib, Mohammad Omidalizarandi, and Boris Kargoll

Abstract

In this paper, we intend to test whether the random deviations of an observed regression time series with unknown regression coefficients can be described by a covariance-stationary autoregressive (AR) process, or whether an AR process with time-variable (say, linearly changing) coefficients should be set up. To account for possibly present multiple outliers, the white noise components of the AR process are assumed to follow a scaled (Student) t-distribution with unknown scale factor and degree of freedom. As a consequence of this distributional assumption and the nonlinearity of the estimator, the distribution of the test statistic is analytically intractable. To solve this challenging testing problem, we propose a Monte Carlo (MC) bootstrap approach, in which all unknown model parameters and their joint covariance matrix are estimated by an expectation maximization algorithm. We determine and analyze the power function of this bootstrap test via a closed-loop MC simulation. We also demonstrate the application of this test to a real accelerometer dataset within a vibration experiment, where the initial measurement phase is characterized by transient oscillations and modeled by a time-variable AR process.

Keywords

Bootstrap test · EM algorithm · Monte Carlo simulation · Regression time series · Scaled t-distribution · Time-variable autoregressive process

1 Introduction

Reliable and precise estimation of geodetic time series models remains a challenging task as they frequently involve huge numbers of auto-correlated and outlier-afflicted measurements. On the one hand, a parsimonious model that allows both for the description and the estimation of auto-correlations is given by autoregressive (AR) processes (cf. Schuh 2003). On the other hand, a flexible approach to modeling multiple outliers (or more generally a heavy-tailed error law) is enabled by the assumption that the random deviations follow a scaled t-distribution (cf. Koch and Kargoll 2013).

Since adjustment techniques based on least squares are sensitive to misspecifications of the functional and stochastic observation model (cf. Kutterer 1999), as well as sensitive to outliers (cf. Baarda 1968), frequently encountered data features such as functional non-linearity, colored measurement noise and heavy-tailed error distribution should be adequately taken into account. Modern geodetic sensors often involve a data sampling at a high rate, thus producing significantly auto-correlated noise (cf. Kuhlmann 2001), in potentially huge numbers of observations. In such cases, the use of a covariance matrix easily exceeds the memory of the computer. Instead, an AR process can often be used for modeling (auto-)correlations more parsimoniously (cf. Schuh 2003). Moreover, the error law of geodetic measurements has frequently been found to be heavy-tailed, in which cases

H. Alkhatib (✉) · M. Omidalizarandi
Geodetic Institute, Leibniz University Hannover, Hannover, Germany
e-mail: alkhatib@gih.uni-hannover.de

B. Kargoll
Institut für Geoinformation und Vermessung Dessau, Anhalt University of Applied Sciences, Dessau-Roßlau, Germany

P. Novák et al. (eds.), *IX Hotine-Marussi Symposium on Mathematical Geodesy*,
International Association of Geodesy Symposia 151, https://doi.org/10.1007/1345_2019_78

robust M-estimation can be applied to the aforementioned models (cf. Wiśniewski 2014).

Kargoll et al. (2018a) recently dealt with the case where both the coefficients of the AR model of the random deviations in a linear functional model and the shape parameters of the heavy-tailed error law are unknown. As suggested by Koch and Kargoll (2013) in a geodetic context, the family of scaled t-distributions was used to model the error law. Here, the degree of freedom is a shape parameter, which controls the thickness of the tails, and which can be estimated from the given measurements jointly with the other (functional and stochastic) model parameters, in the sense of a *self-tuning* robust estimator (cf. Parzen 1979). With this kind of estimator, the unknown parameters of the functional and the AR model can conveniently be computed via iteratively reweighted least squares (IRLS).

This is not only possible when the AR process is covariance-stationary, but also when the AR coefficients are modeled as time-variable quantities through a linear regression (Kargoll et al. 2018b). Such models have been found useful in describing non-stationary effects in time series measurements that cannot be properly described as part of the deterministic model at the level of the observation equations, e.g., local distortions in Gravity Field and Steady-State Ocean Circulation Explorer (GOCE) satellite gravity gradient data (Schuh and Brockmann 2016) or transient oscillations in terrestrial accelerometer measurements (Kargoll et al. 2018b). An important issue that has not been addressed in these expositions is how such an AR error model with t-distributed white noise components can be tested for time-variability (this testing problem is defined in Sect. 2). Whereas a corresponding suitable test statistic is easily identified from standard testing theory, the associated probability distribution is not readily available due to the complexity of the model.

With difficult-too-handle problems, Monte Carlo (MC) simulation may give an adequate solution (see, e.g., Koch 2018a,b). In particular, MC bootstrap approaches (in the sense of Efron 1979) can often be applied in situations where an approximate probability distribution of a test statistic cannot be derived analytically. Besides a few geodetic fields of application (e.g., Teunissen 1998; Neuner et al. 2014; Angrisano et al. 2018; Lösler et al. 2018), bootstrap methods have been devised for rather general time series models (cf. Li and Maddala 1996; Politis 2003) and in the context of Expectation Maximization (EM) algorithms for missing-data models (cf. McLachlan and Krishnan 2008). As our self-tuning robust estimator is based on both of these, we can develop a bootstrap approach to solving the aforementioned problem of testing for time-variability of AR coefficients. In Sect. 3 we outline this procedure in the general case of testing against a linearly modeled time-variability of the coefficients of an AR(p) process, employed to model the random deviations of a possibly nonlinear regression time series. In Sect. 4, we simulate and investigate the size and power function of the test.

2 General Testing Problem

We assume that observations $\boldsymbol{\ell} = [\ell_1, \ldots, \ell_n]^T$ can be described by a nonlinear, vector-valued ('deterministic model') function $\boldsymbol{f}$ and random deviations $\boldsymbol{e} = [e_1, \ldots, e_n]^T$ through the so-called observation equations $\boldsymbol{\ell} = \boldsymbol{f}(\boldsymbol{\Xi}) + \boldsymbol{e}$, where $\boldsymbol{x} = [x_1, \ldots, x_q]^T$ constitute unknown parameters. We assume the random deviations to follow an AR(p) process

$$e_t = \sum_{j=1}^{p} \alpha_{t,j} e_{t-j} + u_t \quad (t = 1, \ldots, n) \tag{1}$$

with the time-dependent coefficients $\alpha_{t,j} = \boldsymbol{B}_t \boldsymbol{y}_j$ being described by linear combinations involving fixed vectors $\boldsymbol{B}_t$ and unknown parameters $\boldsymbol{y}_j = [y_{j,1}, \ldots, y_{j,m}]^T$. We assume the random variables $\boldsymbol{u} = [u_1, \ldots, u_n]^T$ in (1) to be independently and identically $t_\nu(0, \sigma^2)$-distributed with center 0, unknown scale parameter σ^2 and unknown degree of freedom ν. It is well known that this stochastic model is equivalent to the model $u_t \,|\, w_t \sim N(0, \sigma^2/w_t)$ involving gamma-distributed latent variables $w_t \sim G(\nu/2)$ (cf. Koch and Kargoll 2013). Common designs of the time-variability models include polynomials defined by $\boldsymbol{B}_t = [1 \;\; \tau_t^1 \;\; \cdots \;\; \tau_t^{m-1}]$. Clearly, time-dependency is eliminated either for $m = 1$ (in which case we may write $\alpha_{t,j} = y_{j,1} =: \alpha_j$), or for $y_{j,2} = \ldots = y_{j,m} = 0$ ($j = 1, \ldots, p$). If it is not known whether an AR model is time-variable or not, it makes more sense to keep these parameters $\boldsymbol{Y} = [y_{1,2}, \ldots, y_{1,m}, \ldots, y_{p,2}, \ldots, y_{p,m}]^T$ in the model and to test their joint significance. For this purpose, we define the null and the alternative hypothesis by

$$H_0 : \boldsymbol{Y} = \boldsymbol{0} \quad \text{vs.} \quad H_1 : \boldsymbol{Y} \neq \boldsymbol{0}. \tag{2}$$

A natural choice for a test statistic that measures deviations from H_0 (i.e., the degree of $\boldsymbol{Y}$ being nonzero) in a cumulative manner is given by the weighted sum $T = \hat{\boldsymbol{Y}}^T \hat{\boldsymbol{\Sigma}}_{\hat{Y}\hat{Y}}^{-1} \hat{\boldsymbol{Y}}$ of squares of the estimates of $\boldsymbol{Y}$, involving the corresponding joint (*a posteriori*) covariance matrix. In simple cases, it might be sufficient to test against a linear drift of an AR(1) process, i.e., to expand the time-variable AR(p) process up to $p = 1$ and $m = 2$, and to test $H_0 : y_2 = 0$ versus $H_1 : y_2 \neq 0$, writing more simply $\alpha_t = y_1 + \tau_t \cdot y_2$ instead of $\alpha_{t,1} = y_{1,1} + \tau_t \cdot y_{1,2}$. Then, the previous test statistic would simplify to $T = \hat{y}_2^2 / \hat{\sigma}_{\hat{y}_2}^2$. However, even in this simple setting, it is not known how well the probability distribution of this statistic can be approximated by a standard distribution (say, a t-distribution), due to the intricacy of the entire observation model. Therefore, we pursue the solution of this testing problem not via the derivation of an approximate distribution (which might even be intractable), but via the following bootstrap approach.

3 Bootstrap Approach to Solving the Testing Problem

Since the distribution of the test statistic about the time-variability parameters is unknown, we resort to simulation-based bootstrap testing. The general idea is to generate a large number of sampled values for the test statistic under the assumption that H_0 is true, and to define the critical value for the (one-sided) test as that value which is exceeded by only α % of the sampled values. This allows us then to assess whether the value of the test statistic computed from the given measurements is too large as to support H_0, which is the case when that value exceeds the previously simulated critical value. The individual steps of this approach are outlined in the following for the general problem of testing against a time-variability model with specified polynomial order m and AR model order p with given measurement results $\boldsymbol{\ell}$, deterministic model $\boldsymbol{f}(\boldsymbol{x})$, time-variability design matrix $\boldsymbol{B}$ and significance level α.

Estimation Step The modified EM algorithm described in Kargoll et al. (2018b, Section 3) was extended by the linearization procedure explained in Alkhatib et al. (2018, Section 3). This algorithm outputs the estimates $\hat{\boldsymbol{x}}$ of the functional parameters, the estimates $\hat{\boldsymbol{y}}$ of the parameters of the time-variable AR process alongside their joint covariance matrix $\hat{\boldsymbol{\Sigma}}_{\hat{y}\hat{y}}$, the estimate $\hat{\sigma}^2$ of the scale factor as well as the estimate $\hat{\nu}$ of the degree of freedom of the underlying t-distribution, and estimates $\hat{\boldsymbol{u}}$ of the white noise residuals.

Testing Step The value $T = \hat{\boldsymbol{Y}}^T \hat{\boldsymbol{\Sigma}}_{\hat{Y}\hat{Y}}^{-1} \hat{\boldsymbol{Y}}$ of the test statistic is computed.

Generation Step The generation scheme begins with the sampling of the white noise components $u_t^{(k)}$ for the time instances $t = 1, \ldots, n$ and the MC iterations $k = 1, \ldots, B$ (where B is the total number of bootstrap samples). For this purpose, we consider the following two alternatives.

(1) *Parametric bootstrapping*: random numbers with respect to the Student $t_{\hat{\nu}}(0, \hat{\sigma}^2)$-distribution are independently generated for the white noise component $u_t^{(k)}$.
(2) *Nonparametric bootstrapping*: random numbers $\lambda_t^{(k)}$ with respect to the discrete uniform distribution $\mathcal{U}(1, n)$ are drawn with replacement to define $u_t^{(k)} = \hat{u}_{\lambda_t^{(k)}}$ (thereby re-using the residuals $\hat{\boldsymbol{u}}$ from the Estimation Step).

To ensure that the B measurement series are generated under H_0, that is, under the assumption of a time-constant AR process, we take the time-wise means $\bar{\alpha}_j = \frac{1}{n}\sum_{t=1}^{n} \hat{\alpha}_{t,j}$ ($j = 1, \ldots, p$) of the estimated AR coefficients $\hat{\alpha}_{t,j} = \boldsymbol{B}_t \hat{\boldsymbol{y}}_j$. For when H_0 is true, then all of the AR coefficients are constant throughout time, and the estimated means can be expected to approximate the true constant coefficient values. Now, the previously generated white noise components are correlated recursively through evaluation of the time-constant AR model equation $e_t^{(k)} = \sum_{j=1}^{p} \bar{\alpha}_j e_{t-j}^{(k)} + u_t^{(k)}$, using the initial values $e_0^{(k)} = \ldots = e_{t-p}^{(k)} = 0$. Adding to these colored noise components the estimated deterministic model yields the sampled measurements $\ell_t^{(k)} = f_t(\hat{\boldsymbol{x}}) + e_t^{(k)}$. The resulting measurement series $\boldsymbol{\ell}^{(k)}$ is adjusted in exactly the same way as the actual measurement series $\boldsymbol{\ell}$ within the previous Estimation Step, which gives the sampled estimates $\hat{\boldsymbol{x}}^{(k)}$, $\hat{\boldsymbol{y}}^{(k)}$, $\hat{\boldsymbol{\Sigma}}_{\hat{y}\hat{y}}^{(k)}$, $(\hat{\sigma}^2)^{(k)}$ and $\hat{\nu}^{(k)}$. The sampled test statistic is obtained by $T^{(k)} = (\hat{\boldsymbol{Y}}^{(k)})^T (\hat{\boldsymbol{\Sigma}}_{\hat{Y}\hat{Y}}^{(k)})^{-1} \hat{\boldsymbol{Y}}^{(k)}$, as in the Testing Step.

Evaluation Step To determine how extreme the test value T is in comparison to the values $T^{(1)}, \ldots, T^{(B)}$ generated under H_0, we estimate the p-value by

$$\widehat{\mathrm{pv}} = \frac{1}{B} \sum_{k=1}^{B} I(T^{(k)} > T), \tag{3}$$

according to McKinnon (2007, Section 2). Here, I is the indicator function, which takes the value 1 in case the argument is true, and the value 0 if the argument is false.

Decision Step A large p-value indicates a rather large support of H_0 by the data. Thus, we reject H_0 if the estimated p-value is less than the predefined significance level α.

Note that when the random deviations of the measurements can safely be assumed to be normally distributed, then the bootstrap tests can be carried out in a similar way as described. The main differences are that

1. the degree of freedom of the t-distribution is not estimated but fixed a priori within the Estimation Steps by setting it to a large value, e.g. to 120, for which value the t-distribution closely approximates a normal distribution.
2. random numbers with respect to the normal distribution $\mathcal{N}(0, \hat{\sigma}^2)$ are independently generated for the white noise components within the parametric bootstrapping of the Generation Step.

The form of the test statistic $T = \hat{\boldsymbol{Y}}^T \hat{\boldsymbol{\Sigma}}_{\hat{Y}\hat{Y}}^{-1} \hat{\boldsymbol{Y}}$, however, remains unchanged.

4 Monte Carlo Simulation

We consider the linear regression time series model

$$\ell_t = \frac{a_0}{2} + \sum_{j=1}^{M} a_j \cos(2\pi f_j \tau_t) + b_j \sin(2\pi f_j \tau_t) + e_t \quad (4)$$

($t = 1, \ldots, n$), consisting of sine and cosine basis functions with unknown ("Fourier") coefficients $a_0, a_1, \ldots, a_M$ and $b_1, \ldots, b_M$ (collected in the parameter vector $\boldsymbol{x}$), and of random deviations. The Fourier frequencies

$$f_j = 8 \cdot j \text{ [Hz]} \qquad (j = 1, \ldots, M) \qquad (5)$$

are treated as error-free quantities. We set the number of frequencies in this simulation to $M = 12$ as a typical number encountered in vibration analysis experiments involving real accelerometer measurements (see Sect. 5 and Kargoll et al. 2018a). Furthermore, $\tau_1, \ldots, \tau_n$ are given time instances sampled with constant sampling rate $\Delta\tau = 0.00512$ s, beginning at time $\tau_1 = 67.68128$ s. The number of observations is $n = 10{,}000$. This functional model is linear, so that the design matrices $\boldsymbol{A}_{[n\times 25]}$ is immediately obtained (without linearization). Concerning the colored noise e_t, we specified a time-variable AR(1)-process using the global polynomial $\alpha_t = y_1 + y_2 \cdot \tau_t$ of degree 1. For $y_2 = 0$, the AR(1)-process is time-constant. The simulation of the p-values ($\widehat{pv}^{(i)}$) and the power function consists for every repetition $i \in \{1, \ldots, 100\}$ of the following steps:

- Generate the white noise $u_t^{(i)}$ from the true t-distribution $t_3(0, 10^{-6})$ for $t = 1, \ldots, 10{,}000$.
- Correlate the white noise by means of the AR(1) process $e_t^{(i)} = \alpha_{t,1} e_{t-1}^{(i)} + u_t^{(i)}$ with $\alpha_{t,1} = y_1 + y_2 \cdot \tau_t$. We fix here the true offset parameter $y_1 = -0.5$, and we vary the true slope parameters y_2 in steps [0 : 0.0001 : 0.003].
- Add generated colored noise to a specified oscillation model (4) to determine observations $\boldsymbol{\ell}^{(i)}$.
- Do the Estimation Step to obtain $\hat{\mathbf{x}}^{(i)}$, $\hat{\mathbf{y}}^{(i)}$, $\hat{\boldsymbol{\Sigma}}_{\hat{y}\hat{y}}^{(i)}$, $\hat{\sigma}^{(i)}$, $\hat{\nu}^{(i)}$ using exactly the same functional and time-variable AR model as described before.
- Carry out the Testing Step to compute $T^{(i)} = (\hat{y}_2^{(i)}/\hat{\sigma}_{\hat{y}_2}^{(i)})^2$.
- Carry out the Generation Step (parametric/nonparametric) to compute $T^{(i,k)}$ for $k = 1, \ldots, B$. In order to demonstrate the performance of the bootstrap test, we use exactly the same functional and time-variable AR model as in the previous Estimation Step. In particular, the model orders p and m with respect to the, respectively, AR and time-variability model are maintained.
- Carry out the Evaluation Step to compute $\widehat{pv}^{(i)}$.

This simulation was carried out for the bootstrap sample sizes $B = 19$, $B = 99$ and $B = 999$. According to Davidson and MacKinnon (2000), B should be chosen such that $\alpha \cdot (B + 1)$ is an integer. In this paper we fixed $\alpha = 0.05$, so that $B = 19$ is the least possible value. Since the loss of power of a bootstrap test is proportional to $1/B$ according to Davidson and MacKinnon (2000), we investigated the largest number $B = 999$ possible with the computer hardware available in our experiment. With this number, we can already be quite sure that the resulting p-values are not overly dependent on the particular sequence of random numbers. The results of the estimated $\widehat{pv}^{(i)}$-values for the case $y_2 = 0$ (i.e., a time-constant AR(1) process) are shown in Fig. 1. Somewhat surprisingly, varying B does not change the p-value very much within a MC run. Most importantly, the size $\alpha = 0.05$ is reproduced well on the average since the H_0-rejection rate for all 100 replications, defined by

$$\frac{1}{100} \sum_{i=1}^{100} \mathrm{I}(\widehat{pv}^{(i)} < 0.05)$$

turned out to be 0.04 for parametric and 0.05 for non-parametric bootstrapping. Figure 2 shows the empirical power function (sensitivity of the hypothesis tests) evaluated at $y_2 = [0 : 0.0001 : 0.003]$. We see that a linear

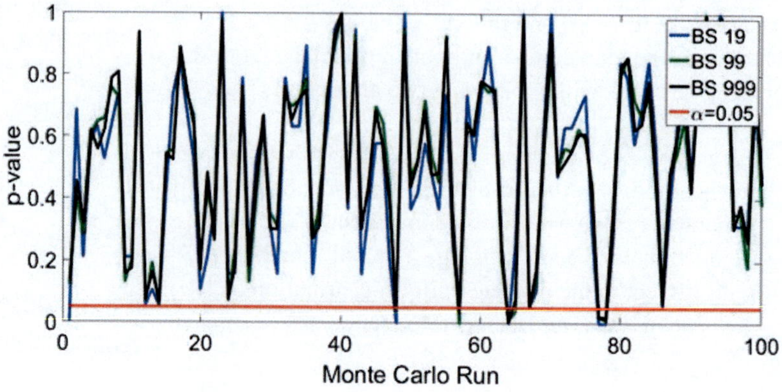

Fig. 1 Comparison of estimated $\widehat{pv}^{(i)}$-value (rejecting the null hypothesis whenever $\widehat{pv}^{(i)} < 0.05$) under the time-constant AR(1) model ($y_2 = 0$) for different bootstrap sample sizes

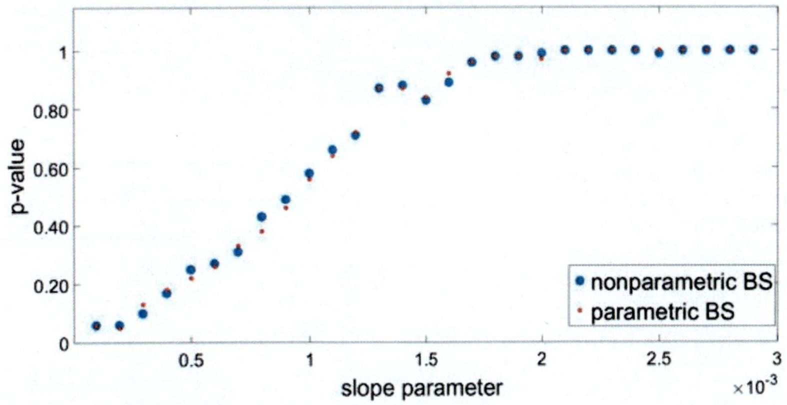

Fig. 2 Comparison of estimated $\widehat{pv}^{(i)}$-value (rejecting the null hypothesis whenever $\widehat{pv}^{(i)} < 0.05$) under the AR(1) model with time variability parameter values $y_2 = [0 : 0.0001 : 0.003]$ for parametric and nonparametric bootstrapping

time-variability with a slope of 0.0015 is detected with a relative frequency of about 0.80. This slope value may be viewed as an empirical analogue to the size of an outlier detectable with a specified probability of 0.80 within Baarda's reliability theory.

5 An Application to Vibration Analysis

We applied the bootstrap approach to testing for time-variability of an AR process to a vibration dataset measured by means of a highly accurate single-axis PicoCoulomB (PCB) Piezotronics accelerometer. As part of a vibration experiment, carried out at the Institute of Concrete Construction at the Leibniz Universität Hannover, that sensor was mounted on a shaker table, which was set to an oscillation frequency of 16 Hz. This frequency is well below Nyquist frequency of half the sampling frequency of the accelerometer (195/2 Hz) and thus detectable. The dataset, which spans about 45 min of measurements, is characterized by initial transient oscillations with irregular amplitudes but stable frequency (approximately throughout the first 1,500 data values), followed by a stationary oscillation with stable amplitudes. The dataset excluding the initial phase was previously modeled in Kargoll et al. (2018a) by the truncated Fourier series (4) with $M = 12$, with AR random deviations e_t, and with t-distributed white noise components u_t. The frequencies were treated as the fixed quantities $f_j = j \cdot 8$ Hz ($j = 1, \ldots, M$). Besides the main frequency $f_2 = 16$ Hz, 11 other frequencies at multiples of 8 Hz with associated small amplitudes were identified within the discrete Fourier transform. These were suspected to arise from the physical properties of the shaker table and thus modeled deterministically. The Fourier coefficients $a_0, a_1, \ldots, a_{12}$ and $b_1, \ldots, b_{12}$ were treated as unknowns and collected within the parameter vector $\boldsymbol{x}$. In Kargoll et al. (2018a), the initial phase was modeled by employing the observation model (4) in connection with a time-variable AR(6) model involving cubic polynomials defined by $\boldsymbol{B}_t = [1 \;\; \tau_t^1 \;\; \tau_t^2 \;\; \tau_t^3]$. This model was found by trying out different AR as well as different polynomial model orders until the periodogram excess of the estimated white noise residuals appeared to be small in comparison to theoretical white noise. The reasonable choices $m \in \{2, 3, 4, 5\}$ correspond to polynomial models defined by

- $\boldsymbol{B}_t = [1 \;\; \tau_t^1]$ for $m = 2$ (linear drift),
- $\boldsymbol{B}_t = [1 \;\; \tau_t^1 \;\; \tau_t^2]$ for $m = 3$ (quadratic polynomial),
- $\boldsymbol{B}_t = [1 \;\; \tau_t^1 \;\; \tau_t^2 \;\; \tau_t^3]$ for $m = 4$ (cubic polynomial),
- $\boldsymbol{B}_t = [1 \;\; \tau_t^1 \;\; \tau_t^2 \;\; \tau_t^3 \;\; \tau_t^4]$ for $m = 5$.

Figure 3 shows for the fixed AR model order $p = 6$ that the time-variability model with $m = 4$ results in the acceptance of the white noise test since its cumulated periodogram excess over theoretical white noise lies completely within the approximate 99% significance bounds. In contrast, the white noise hypothesis is rejected for the choices $m = 2$, $m = 3$ and $m = 5$. However, as the white noise components contain outliers when the degree of freedom of the underlying t-distribution is relatively small, the periodogram estimate might be affected and contaminated by outliers. We therefore seek to improve the model selection step with respect to the polynomial order of the time-variability model by carrying out the bootstrap test given in Sect. 3. For this purpose, we retained within the current study the AR model order of $p = 6$ and considered the testing problem (2) for $m \in \{2, 3, 4, 5\}$, Assembling the complete $\boldsymbol{B}$-matrix from one of these choices at a time, the corresponding p-value estimate was computed (as explained in Sect. 3) both under parametric and nonparametric bootstrapping, as well as for both $B = 99$ and $B = 999$ bootstrap samples (see Table 1). Since the averaging (3) yields increasingly precise p-value estimate with increasing number B of bootstrap sample, and since some of the estimated p-values change considerable when taking $B = 999$ instead of $B = 99$ bootstrap samples, we conclude that B should be at least $B = 999$. This confirms the finding of McKinnon (2007) that "it might be dangerous to use a value of B less than 999". For that number of bootstrap samples, we see that the

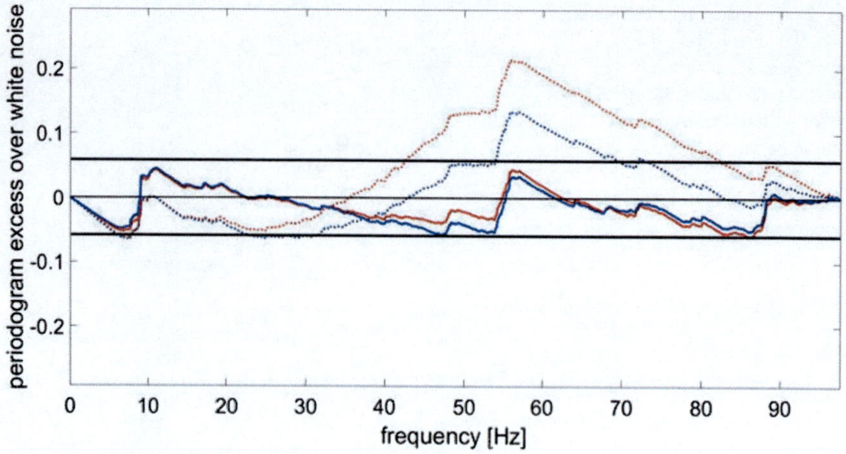

Fig. 3 Excess of the estimated periodograms of the decorrelated residuals for $m = 2$ (dotted red line), $m = 3$ (dotted blue line), $m = 4$ (solid blue line) and $m = 5$ (solid red line) with respect to the AR(6) model over the theoretical white noise periodogram (equal to 0); approximate 99% significance bounds are displayed as the two heavy black lines

Table 1 Estimated p-values for the parametric and the non-parametric version of the bootstrap test for time-variability of an AR(6) process, modeling random deviations of the observation model (4) for the initial segment of the analyzed accelerometer dataset

m	2	3	4	5
Parametric ($B = 99$)	0	0	0.03	0.17
($B = 999$)	0	0	0.02	0.20
Non-parametric ($B = 99$)	0	0	0.07	0.10
($B = 999$)	0	0	0.03	0.19

p-values obtained by non-parametric bootstrapping are very close to the values obtained parametrically by generating random numbers from the fitted t-distribution. We therefore conclude that this choice can safely be made by the user according to computational convenience. Most importantly, the p-values (for $B = 999$) are well below the standard significance level $\alpha = 0.05$ for the cubic polynomial model ($m = 4$), which previously was found to be the only adequate one. As the p-values with respect to the linear drift model ($m = 2$) and for the quadratic polynomial model ($m = 3$) are zero, the null hypothesis of 'no time-variability of the AR model' is not supported by the data, as for the test against the adequate cubic polynomial model. In a practical situation it might be desirable to formulate the more general alternative hypothesis 'The AR model is time variable.' The previous results show that any of the aforementioned specific alternative models (linear drift, quadratic, cubic polynomials) could be used within the Estimation Step and the Generation Step because each of these models implies the correct rejection of the null hypothesis. Increasing the optimal polynomial order of $m = 4$ to $m = 5$ apparently desensitizes the test since the p-values now exceed any reasonable choice for the significance level, resulting in the inadequate acceptance of H_0. In a previous study, we found that assuming an AR model order too small or too large often results in inferior model estimates, which behavior was documented by an unstable acceptance rate of a white noise test (see Kargoll et al. 2018a, Figure 9). Thus, estimated AR models have a tendency to be acceptable only within certain ranges of order. We suspect that a similar phenomenon might occur for the order of the time-variability model, but we cannot prove this finding, yet.

6 Summary, Conclusions, and Outlook

We presented a statistical procedure based on MC bootstrapping to test the null hypothesis that the random deviations of a regression time series follow a time-constant, fixed-order AR process. The alternative hypothesis may be specified by an arbitrary linear model that forces each AR coefficient to lie exactly on a time-dependent deterministic function. To take a potentially large number of outliers of unknown frequency and magnitudes into account, the input white noise to that AR process was modeled by means of a t-distribution with estimable scale factor and degree of freedom. Since the exact and even approximate test distribution are unknown, the MC bootstrapping yields an estimate of the p-value, which may be compared to a specified significance level to arrive at the test decision. A closed-loop simulation showed that the specified significance level of the bootstrap test is reproduced closely. Moreover, the application of the bootstrap test to an initial segment of an accelerometer measurement series, which was previously modeled by means of cubic polynomials with respect to a time-variable AR(6) process, showed that the null hypothesis of no time-variability should indeed by rejected in favor of linear, quadratic or cubic polynomials. We may therefore conclude that the model of linearly drifting AR coefficients may be used in a test against the general alternative that the 'AR model is time-variable'. This conjecture, however, should be further investigated in the future. The real-data study also demonstrated that at least 999 bootstrap samples should be generated to obtain adequate p-values. Both the closed-loop simulation and the real-data analysis showed that parametric bootstrapping (using the t-distribution estimated from the given dataset) and non-parametric bootstrapping (drawing at random with replacement from the estimated white noise of the real-data adjustment) in order to generate the white noise samples result in very similar p-value estimates. It appears that the presented bootstrap approach may be adapted to similar testing problems in the context of time series analysis involving intricate parametric models.

Acknowledgements Funded by the Deutsche Forschungsgemeinschaft (DFG, German Research Foundation)—386369985. The presented application of the PCB Piezotronics accelerometer within the vibration analysis experiment was performed as a part of the collaborative project "Spatio-temporal monitoring of bridge structures using low cost sensors" with ALLSAT GmbH, which is funded by the German Federal Ministry for Economic Affairs and Energy (BMWi) and the Central Innovation Programme for SMEs (ZIM Kooperationsprojekt, ZF4081803DB6). In addition, the authors acknowledge the Institute of Concrete Construction (Leibniz Universität Hannover) for providing the shaker table and the reference accelerometer used within this experiment.

References

Alkhatib H, Kargoll B, Paffenholz JA (2018) Further results on robust multivariate time series analysis in nonlinear models with autoregressive and t-distributed errors. In: Rojas I, Pomares H, Valenzuela O (eds) Time series analysis and forecasting. ITISE 2017. Contributions to statistics. Springer, Cham. pp 25–38. https://doi.org/10.1007/978-3-319-96944-2_3

Angrisano A, Maratea A, Gaglione S (2018) A resampling strategy based on bootstrap to reduce the effect of large blunders in GPS absolute positioning. J Geod 92:81–92. https://doi.org/10.1007/s00190-017-1046-6

Baarda W (1968) A testing procedure for use in geodetic networks. Publications on Geodesy (New Series), vol 2, no 5, Netherlands Geodetic Commission, Delft

Davidson R, MacKinnon JG (2000) Bootstrap tests: How many bootstraps? Economet Rev 19:55–68. https://doi.org/10.1080/07474930008800459

Efron B (1979) Bootstrap methods: another look at the jackknife. Ann Stat 7:1–26. https://doi.org/10.1214/aos/1176344552

Kargoll B, Omidalizarandi M, Loth I, Paffenholz JA, Alkhatib H (2018a) An iteratively reweighted least-squares approach to adaptive robust adjustment of parameters in linear regression models with autoregressive and t-distributed deviations. J Geod 92:271–297. https://doi.org/10.1007/s00190-017-1062-6

Kargoll B, Omidalizarandi M, Alkhatib H, Schuh WD (2018b) Further results on a modified EM algorithm for parameter estimation in linear models with time-dependent autoregressive and t-distributed errors. In: Rojas I, Pomares H, Valenzuela O (eds) Time series analysis and forecasting. ITISE 2017. Contributions to statistics. Springer, Cham, pp 323–337. https://doi.org/10.1007/978-3-319-96944-2_22

Koch KR (2018a) Bayesian statistics and Monte Carlo methods. J Geod Sci 8:18–29. https://doi.org/10.1515/jogs-2018-0003

Koch KR (2018b) Monte Carlo methods. Int J Geomath 9:177–143. https://doi.org/10.1007/s13137-017-0101-z

Koch KR, Kargoll B (2013) Expectation-maximization algorithm for the variance-inflation model by applying the t-distribution. J Appl Geod 7:217–225. https://doi.org/10.1515/jag-2013-0007

Kuhlmann H (2001) Importance of autocorrelation for parameter estimation in regression models. In: 10th FIG international symposium on deformation measurements, pp 354–361

Kutterer H (1999) On the sensitivity of the results of least-squares adjustments concerning the stochastic model. J Geod 73(7):350–361. https://doi.org/10.1007/s001900050253

Li H, Maddala GS (1996) Bootstrapping time series models. Econom Rev 15:115–158. https://doi.org/10.1080/07474939608800344

Lösler M, Eschelbach C, Haas R (2018) Bestimmung von Messunsicherheiten mittels Bootstrapping in der Formanalyse. Zeitschrift für Geodäsie, Geoinformatik und Landmanagement zfv 143:224–232

McKinnon J (2007) Bootstrap hypothesis testing. Queen's Economics Department Working Paper No. 1127, Queen's University, Kingston, Ontario, Canada

McLachlan GJ, Krishnan T (2008) The EM algorithm and extensions, 2nd edn. Wiley, Hoboken

Neuner H, Wieser A, Krähenbühl N (2014) Bootstrapping: Moderne Werkzeuge für die Erstellung von Konfidenzintervallen. In: Neuner H (ed) Zeitabhängige Messgrößen - Ihre Daten haben (Mehr-)Wert. Schriftenreihe des DVW, vol 74, pp 151–170

Parzen E (1979) A density-quantile function perspective on robust estimation. In: Launer L, Wilkinson GN (eds) Robustness in statistics. Academic, New York, pp 237–258. https://doi.org/10.1016/B978-0-12-438150-6.50019-4

Politis DN (2003) The impact of bootstrap methods on time series analysis. Statist Sci 18:219–230. https://doi.org/10.1214/ss/1063994977

Schuh WD (2003) The processing of band-limited measurements; filtering techniques in the least squares context and in the presence of data gaps. Space Sci Rev 108:67–78. https://doi.org/10.1023/A:1026121814042

Schuh WD, Brockmann JM (2016) Refinement of the stochastic model for GOCE gravity gradients by non-stationary decorrelation filters. ESA Living Planet Symposium 2016, Prag, Poster 2382

Teunissen (1998) Success probability of integer GPS ambiguity rounding and bootstrapping. J Geod 72:606–612. https://doi.org/10.1007/s001900050199

Wiśniewski Z (2014) M-estimation with probabilistic models of geodetic observations. J Geod 88(10):941–957. https://doi.org/10.1007/s00190-014-0735-7

Identification of Suspicious Data for Robust Estimation of Stochastic Processes

Till Schubert, Jan Martin Brockmann, and Wolf-Dieter Schuh

Abstract

Many geodetic measurements which are automatically gathered by sensors can be interpreted as a time series. For instance, measurements collected by a satellite platform along the satellite's track can be seen as a time series along the orbit. Special treatment is required if the time series is contaminated by outliers or non-stationarities, summarized as 'suspicious data', stemming from sensor noise variations or changes in environment. Furthermore, the collected measurements are often – for instance due to the sensor design – correlated along the track.

We propose a general estimation procedure accounting for both, correlations and the presence of suspicious data. In the estimation scheme, we adjust an autoregressive (AR) process of a given order p to model the correlations in a residual time series, which can then be used as a very flexible and general stochastic model. The AR-process estimation is iteratively refined by screening techniques based on statistical hypothesis tests and thus robustified. We incorporate different indicators to detect suspicious data or changes in the underlying process characteristics, i.e. changes in the mean value, variance and signs of the residuals.

Here, we apply the procedure to gravity gradient observations as collected by the Gravity Field and Steady-State Ocean Circulation Explorer (GOCE) satellite mission in the low orbit measurement campaign. The estimated autoregressive process is used as a stochastic model of the gravity gradients in a gradiometer-only gravity field determination following the time-wise approach. The resulting estimates are compared to the counterparts of the official EGM_TIM_RL05 processing. Additionally, with newly processed level 1B GOCE gravity gradients at hand we pursue comparison of the robust and conventional approaches for original and reprocessed data.

Keywords

AR-processes · Hypothesis tests · Outlier detection · Residual time series · Stochastic modeling · Time series

T. Schubert (✉) · J. M. Brockmann · W.-D. Schuh
Institute of Geodesy and Geoinformation, University of Bonn, Bonn, Germany
e-mail: schubert@geod.uni-bonn.de; brockmann@geod.uni-bonn.de; schuh@geod.uni-bonn.de

P. Novák et al. (eds.), *IX Hotine-Marussi Symposium on Mathematical Geodesy*, International Association of Geodesy Symposia 151, https://doi.org/10.1007/1345_2019_80

1 Introduction and Related Work

Modern sensors deliver a big treasure of measurements to determine process parameters of the Earth system. A prerequisite for a consistent model, which does not only describe the signal information but also its uncertainties in agreement with the data characteristics, is a clean and accurate modeling. Data adaptive strategies are necessary to adopt peculiarities of the measurement series.

Robust estimation techniques, introduced in 1760 by R. Boŝković, and data snooping (Baarda 1968) have a long tradition in geodesy. Already in 1905, F.R. Helmert discusses the regularity of sign chances in time series (Helmert 1905). Whereas time series at that time had a manageable number of observations present-day sensors provide a huge number of measurements. In addition, these measurements are often highly correlated and therefore the identification of erroneous data is not straightforward. Various studies cover the topic of parameterization of correlations in the time domain as well as in the spectral domain (see e.g. Kay and Marple 1981) also with special attention to time series contaminated by outliers (Fox 1972; Kleiner et al. 1979; Chang et al. 1988). Decorrelation procedures by digital filters derived from the parameterization of stochastic processes are also widely used in geodetic applications (cf. e.g. Schuh 1996, 2003; Klees et al. 2003; Siemes 2008; Krasbutter et al. 2014; Farahani et al. 2017; Schuh and Brockmann 2018). Also robust strategies with an assumed t-distribution for the errors and the data are studied (Kargoll et al. 2018a,b).

The studies carried out for this paper are part of the processing campaign of the Gravity Field and Steady-State Ocean Circulation Explorer (GOCE) satellite mission. The latest gravity field recovery with the time-wise approach has been carried out by Brockmann et al. (2014) using preprocessed level 1B GOCE gravity gradients prepared by the GOCE HPF-team. Recently, this level 1B product has been reprocessed (Siemes 2018) applying an improved calibration using quadratic factors and a new overall reprocessing campaign is established.

In this contribution, a two step approach is proposed to analyze the data and to identify suspicious data. Suspicious data refers to all data points whose characteristics do not agree with the majority. These characteristics can be small or large deviations of single or mean values, changes of the measurement noise or non-stationarity of the correlations. The analysis is performed on the residuals with respect to a prior model, e.g. an interim model or a precursor model, which is computed with a much larger amount of data and therefore has a much higher precision compared to the single measurements or a single track, which has to be analyzed. In a first step, the correlated dataset is decorrelated by a digital filter (e.g. AR-filter). The decorrelated residuals are analyzed with respect to suspicious data, which are then identified and tagged. In a second step, the stochastic characteristics of the measurement series are modeled by a robust estimation of an autoregressive process excluding the suspicious data.

The focus of this paper is on the screening methods and the processing of the new gradient data. In this study, we apply the data screening methods to estimate the AR-processes as a stochastic model for the GOCE gravity gradients. They are applied on the one hand to the 'old' L1B data, which is affected by an imperfect calibration (cf. Siemes 2018). The assumption of stationarity is strongly violated. On the other hand, it is applied to reprocessed L1B gravity gradients derived with an improved calibration (Siemes 2018).

The time-wise approach is based on a GOCE-only solution using an advanced stochastic modeling (e.g. Pail et al. 2011; Brockmann et al. 2014). Precisely, the autoregressive stochastic processes are used for decorrelation in the gravity field estimation. We intend to contribute the advance of the solutions both to the improved level 1B data and to the improved decorrelation, resulting from the improved data screening methods.

Section 2 introduces the data model and the robust estimation of the stochastic model and the identification of suspicious data. In Sect. 3, results for the stochastic model are provided and the results for the old and the new data are compared. Section 4 uses gradiometer-only gravity field solutions to validate the stochastic models and to highlight the improvements. This is followed by concluding remarks in Sect. 5.

2 Estimation of Stochastic Processes

2.1 Residual Time Series

Suppose residuals are computed with respect to some reference parameters $\hat{\mathbf{x}}$, which are either parameters from an existing reference model or results of an adjustment procedure of a previous iteration. The stochastic residuals $\overline{\mathcal{V}}$ which should be analyzed in the following are

$$\overline{\mathcal{V}} = \mathbf{A}_{\mathrm{D}}\hat{\mathbf{x}} - \mathcal{L}. \quad (1)$$

$\mathcal{L}$ are the original observations which include the deterministic part modeled by $\mathbf{A}_{\mathrm{D}}\hat{\mathbf{x}}$ and $\mathbf{A}_{\mathrm{D}}$, the deterministic design matrix. The residuals $\overline{\mathcal{V}}$ are used as purely stochastic 'observations' in the following to adjust the parameters of a stochastic process within a least squares estimation.

In the upcoming sections, it is assumed that the time series

$$\overline{\mathcal{V}} = \begin{bmatrix} \overline{\mathcal{V}}_1 \\ \overline{\mathcal{V}}_2 \\ \vdots \\ \overline{\mathcal{V}}_S \end{bmatrix} = \begin{bmatrix} \mathbf{A}_{\mathrm{D},1}\hat{\mathbf{x}} - \mathcal{L}_1 \\ \mathbf{A}_{\mathrm{D},2}\hat{\mathbf{x}} - \mathcal{L}_2 \\ \vdots \\ \mathbf{A}_{\mathrm{D},S}\hat{\mathbf{x}} - \mathcal{L}_S \end{bmatrix} \quad (2)$$

can be partitioned into S sets, such that the subsets $\overline{\mathcal{V}}_s$ are equispaced (i.e. constant Δt_s) and continuous (i.e. gap-less) sets $\overline{\mathcal{V}}_s$ of different lengths N_s.

2.2 Data Model

As the model is the same for any $\overline{\mathcal{V}}_s$, for better readability,

$$\mathcal{V} := \overline{\mathcal{V}}_s, \text{ for any } s \in \{1, \dots, S\} \tag{3}$$

is defined for the description of the model, being one of the continuous and gap-less sets with length $N := N_s$.

The data model regards each observation $\mathcal{V}_i$ for $i \in \{1, \dots, N\}$ as a composition of the stochastic signal $\mathcal{S}_i$ and noise $\mathcal{N}_i$

$$\mathcal{V}_i = \mathcal{S}_i + \mathcal{N}_i. \tag{4}$$

It is supposed that the stochastic part can be described with a stochastic process which takes the form of an AR(p) representation

$$\mathcal{V}_i = \sum_{j=1}^{p} \alpha_j \mathcal{V}_{i-j} + \mathcal{E}_i \tag{5}$$

of a fixed order $p \ll N$.

2.3 Robust Estimation of AR(p)-Processes

Special attention has to be given to the issues of outliers in autoregressive models. The definition of innovation outliers (cf. Fox 1972) is representative for the masking of contaminated noise into signal components (Eq. (5)). As a consequence, the estimation of the signal process with stationarity assumption can be significantly affected. This also raises the question whether the additive outlier can be recovered in the decorrelated residual time series.

Concepts of robust statistics include patchy outlier distributions which enables outliers to appear in patches (Kleiner et al. 1979). Furthermore, robustness is generally achieved via a bounded influence function, among which the $k\sigma$-rejection-estimator allows for a direct relation to testing for single outliers and outlier patches.

Estimating the p autoregressive parameters α_j from the data $\mathcal{V}$, least squares observation equations for the autoregressive functional model of order p read

$$\mathcal{V}_i + \mathcal{R}_i = \sum_{j=1}^{p} \alpha_j \mathcal{V}_{i-j} = \boldsymbol{a}_i^T \boldsymbol{\alpha} \tag{6}$$

with $\boldsymbol{a}_i$ being the rows of the design matrix $\boldsymbol{A}$. Note, that in contrast to a typical Gauss-Markov model the design matrix consists of observations as coefficients. The AR-process coefficients follow from the least squares estimate using a specific realization $\boldsymbol{v}$ by

$$\widetilde{\boldsymbol{\alpha}} = \left(\boldsymbol{A}^T \boldsymbol{W} \boldsymbol{A}\right)^{-1} \boldsymbol{A}^T \boldsymbol{W} \boldsymbol{v} \tag{7}$$

assuming initially a diagonal identity matrix as weight matrix, $\boldsymbol{W} = \mathbb{1}$. Therefore, the normal equations are closely related to the Yule-Walker equations (Schuh and Brockmann 2018). The filtered, i.e. decorrelated, realizations of the residuals

$$\boldsymbol{r} = \boldsymbol{A}\widetilde{\boldsymbol{\alpha}} - \boldsymbol{v} \tag{8}$$

are used in the screening and the hypothesis tests.

Robustness implies the use of a bounded influence function, among which we choose the $k\sigma$-rejection-estimator (Kleiner et al. 1979) for single outliers and outlier patches,

$$\psi_R(r_i) = \begin{cases} r_i & |r_i| \le k\sigma \\ 0 & |r_i| > k\sigma \end{cases}. \tag{9}$$

In this form one can assess a quantile k as the multiplier to the variance σ.

2.4 Residual Screening

As a first test we eliminate large single outliers. The null hypothesis $H_0 : E\{\mathcal{R}_i\} = 0$ and alternative hypothesis $H_A : E\{\mathcal{R}_i\} \neq 0$ lead to the definition of the test statistic and distribution $\mathcal{T}_{\text{outlier}} = \mathcal{R}_i \sim N(0, \sigma^2)$. From this and the realization $t_{\text{outlier},i} = |r_i|$, the decision is derived from evaluating $t_{\text{outlier},i} \le k_{1-\alpha/2}^{N(0,1)} \sigma$ which, if true, results in the acceptance or, if false, leads to rejection of the null hypothesis. The quantile value $k_{1-\alpha/2}^{N(0,1)}$ is computed from the standard normal distribution with respect to a very low level of significance α. The variance σ^2 is estimated based on the median absolute deviations (MAD) scale estimator $\tilde{s}^2_{\text{MAD}} = 1.4785\,\text{MAD}(\boldsymbol{r})$ (see e.g. Huber 1981, Sect. 5.1). For this test and all further tests we assume that $\tilde{s}^2_{\text{MAD}}$ has no uncertainty. This assumption is justified by the fact that $\tilde{s}^2_{\text{MAD}}$ is estimated from the whole data sequence with a significantly larger number of data points. Of course, all tests statistics follow their distributions only approximately, although we denote them as applying exactly. This hypothesis test and all further tests are summarized in a uniform compact representation by

$$H_0 : E\{\mathcal{R}_i\} = 0 \text{ vs. } H_A : E\{\mathcal{R}_i\} \neq 0$$

$$\mathcal{T}_{\text{outlier}} = \mathcal{R}_i \sim N(0, \tilde{s}^2_{\text{MAD}}) \tag{10}$$

$$t_{\text{outlier},i} = |r_i| \leq k^{N(0,1)}_{1-\alpha/2}\, \tilde{s}_{\text{MAD}}. \tag{11}$$

All further tests for suspicious data are applied to areas, e.g. by testing over a certain period of time T_l, which is in our case a fraction of orbit period. For the application at hand, and exemplarily for the mean test described next, we use the 1/30, 1/40 and 1/50 part of the orbit period of approximately 1.5 h, i.e. $n_{T_1} = 179$ values for $T_1 = 1/30$ orbit. These values change depending on the test and were all empirically determined by visual inspection to give feasible results.

The first test statistic for testing the deviation from zero mean and corresponding quantile are

$$H_0 : E\left\{\frac{1}{n_{T_l}}\sum_{i=1}^{n_{T_l}} \mathcal{R}_i\right\} = 0 \text{ vs. } H_A : E\left\{\frac{1}{n_{T_l}}\sum_{i=1}^{n_{T_l}} \mathcal{R}_i\right\} \neq 0$$

$$\mathcal{T}_{\text{mean}} = \frac{1}{n_{T_l}}\sum_{j=1}^{n_{T_l}} \mathcal{R}_j \sim N(0, \tilde{s}^2_{\text{MAD}}/n_{T_l}) \tag{12}$$

$$t_{\text{mean},i} = \frac{1}{n_{T_l}}\left|\sum_{j=i}^{i+n_{T_l}-1} r_j\right| \leq \frac{1}{\sqrt{n_{T_l}}}\, k^{N(0,1)}_{1-\alpha/2}\, \tilde{s}_{\text{MAD}}. \tag{13}$$

We test for changes in the variance by taking the ratio of the sum of squared residuals $\mathcal{R}^T\mathcal{R}$ in the test window (1/50, 1/60 and 1/70 part of the orbit period) and the global robust variance estimator $\tilde{s}^2_{\text{MAD}}$. The quantile value is taken from the χ^2 distribution with the number of data points in the test area as degrees of freedom, i.e.

$$H_0 : E\left\{\frac{1}{n_{T_l}}\sum_{j=1}^{n_{T_l}} \mathcal{R}_i^2\right\} = \tilde{s}^2_{\text{MAD}} \text{ vs.}$$

$$H_A : E\left\{\frac{1}{n_{T_l}}\sum_{j=1}^{n_{T_l}} \mathcal{R}_i^2\right\} > \tilde{s}^2_{\text{MAD}}$$

$$\mathcal{T}_{\text{var}} = \frac{\sum_{j=1}^{n_{T_l}} \mathcal{R}_i^2}{\tilde{s}^2_{\text{MAD}}} \sim \chi^2_{n_{T_l}} \tag{14}$$

$$t_{\text{var},i} = \frac{\sum_{j=i}^{i+n_{T_l}-1} r_j^2}{\tilde{s}^2_{\text{MAD}}} \leq k^{\chi^2_{n_{T_l}}}_{1-\alpha}. \tag{15}$$

The sign test was pioneered by Helmert (1905). Following his work, the difference in the amount of the signs $|\#\mathcal{X}_{\text{pos}} - \#\mathcal{X}_{\text{neg}}|$ of a random variable $\mathcal{X}$ needs to be shifted by population size n_{T_l} and divided by 2 to follow a symmetric binomial distribution. Thus, we can test the sign imbalance occurring within the length of the test window n_{T_l} (1/10, 1/20 and 1/30 part of the orbit period) using the realization $t_{\text{sign},i}$. Formally the test then reads

$$H_0 : E\left\{\sum_{i=1}^{n_{T_l}} \text{sign}(\mathcal{R}_i)\right\} = 0 \text{ vs.}$$

$$H_A : E\left\{\sum_{i=1}^{n_{T_l}} \text{sign}(\mathcal{R}_i)\right\} \neq 0$$

$$\mathcal{T}_{\text{sign}} = \frac{1}{2}\left(n_{T_l} + \sum_{j=1}^{n_{T_l}} \text{sign}(\mathcal{R}_j)\right) \sim B(n_{T_l}, 0.5). \tag{16}$$

$$t_{\text{sign},i} = \left|\sum_{j=i}^{i+n_{T_l}-1} \text{sign}(r_j)\right| \leq 2k^{B(n_{T_l},0.5)}_{1-\alpha/2} - n_{T_l}. \tag{17}$$

Furthermore, we account for a natural occurrence of sign changes. In the sense of Helmert (1905), the cases of no sign changes $(+++++$ and $-----)$ and continuously alternating sign $(+-+-+$ and $-+-+-)$ are equally improbable. This relation is achieved by the sum over the product of consecutive signs. Again, a shift and scale by 2 lead to a binomially distributed test statistic from which the test criterion (applied to 1/20, 1/30, 1/40 part of the orbit) is formulated as

$$H_0 : E\left\{\sum_{i=1}^{n_{T_l}-1} \text{sign}(\mathcal{R}_i)\,\text{sign}(\mathcal{R}_{i+1})\right\} = 0 \text{ vs.}$$

$$H_A : E\left\{\sum_{i=1}^{n_{T_l}-1} \text{sign}(\mathcal{R}_i)\,\text{sign}(\mathcal{R}_{i+1})\right\} \neq 0$$

$$\mathcal{T}_{\text{siCh}} = \frac{1}{2}\left(n_{T_l} - 1 + \sum_{j=1}^{n_{T_l}-1} \text{sign}(\mathcal{R}_j)\,\text{sign}(\mathcal{R}_{j+1})\right) \sim B(n_{T_l} - 1, 0.5) \tag{18}$$

$$t_{\text{siCh},i} = \left|\sum_{j=i}^{i+n_{T_l}-2} \text{sign}(r_j)\,\text{sign}(r_{j+1})\right| \leq 2k^{B(n_{T_l}-1,0.5)}_{1-\alpha/2} - (n_{T_l} - 1). \tag{19}$$

2.5 Further Considerations

Gaps between outliers that are shorter than the filter length are additionally flagged as identified suspicious data to avoid numerical problems. For the same reason, areas with many suspicious data identified are specially handled. Thus, if

60% of the filter length are outliers the area is flagged completely. For the estimation and later when using the filter for decorrelation, special attention has to be paid to data points at the begin of the time series, as they are affected by the filter warmup. Either the observation equations for the filter coefficients can not be constructed as the observations from the past are missing, or – due to the same reason – they cannot be fully decorrelated. Consequently, the first $p = 800$ data points (after a gap) have to be treated as if they were flagged as suspicious. Within this study, no effort was spend on minimizing the effect, as due to the high amount of data, compared to the filter length of $p = 800$, the additional data loss can be neglected. But, from a theoretical point of view, methods exist to minimize the data loss (e.g. Siemes 2008, Chap. 4), which are applicable in this context as well.

All tests summarized above are applied in moving windows and for different area lengths which were empirically determined specific to the tests. The priority of the test criteria is applied in the above order with a decreasing area length. Consequently, 13 tests are performed to each data point. Following this, outlier flags will appear on multiple tests but they are attributed to the first positive test in the order of testing.

2.6 The Data-Fitting Approach

The algorithm scheme at hand is an iterated weighted least squares (IWLS) in the sense of Kleiner et al. (1979). After estimation of the autoregressive parameters using a least squares adjustment, cf. Eq. (7), we apply the screening methods to the residuals $\boldsymbol{r}$. With the acquired outlier flags from hypothesis testing, the least squares adjustment is recomputed without the identified data by eliminating the rows in the design matrix. This corresponds to applying the weighting function

$$w_i = \frac{\psi_R(r_i)}{r_i} = \begin{cases} 1 & \text{accepted} \\ 0 & \text{flagged as suspicious} \end{cases} \qquad (20)$$

of the robust estimator, cf. Eq. (9), to $\boldsymbol{w}$. The procedure is iterated until convergence.

3 Numerical Example: Application to GOCE Gravity Gradients

3.1 GOCE Data Used

We have gravity gradient observations collected by the GOCE satellite mission (Floberghagen et al. 2011; Rummel et al. 2011). Figure 1 shows a short part of the time series comparing the old (imperfect calibration) and the new data (updated calibration from Siemes 2018). Visually, the reduction of the systematic disturbances from the old data to the new data is obvious. In the following, different solutions are generated. To discriminate them the colors as indicated by Table 1 are used.

Two segments of the GOCE time series are used for the analysis of this paper as a proof of concept, cf. Table 2. They are part of the extended mission phase, the data are captured at a lower orbit with −8 and −30 km orbit reduction.

GOCE data contains some inhomogeneities and non-stationarities in the shape of extremely large disturbances, systematic oscillations or data gaps. As these were more

Table 1 Data description: both the non-robust solution from EGM_TIM_RL05 setup (**RL05**) as well as the robust filter solutions denoted as **old** use the old data

ID	Description
RL05	Solution from EGM_TIM_RL05 setup
Old	Official data, robust filters
New	New reprocessed data, robust filters

The new robust filters (**new**) utilize the reprocessed data

Table 2 Data segments used for the numerical tests

ID	Start	End	Orbit info
1	09/Nov/2012	04/Feb/2013	−8 km
2	30/May/2013	31/Jul/2013	−30 km

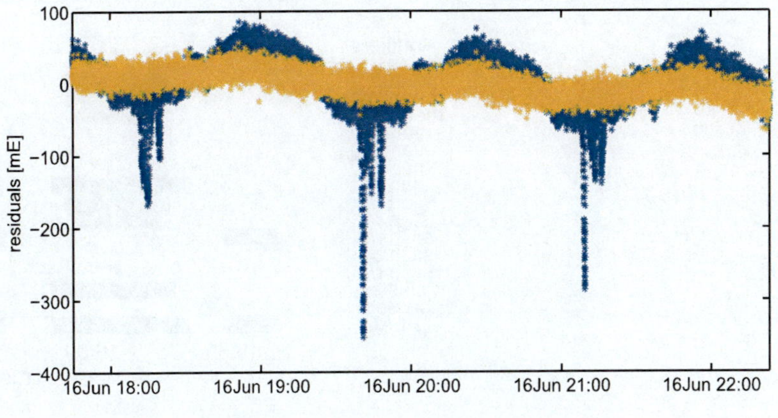

Fig. 1 Comparison of the official data (blue) and the newly processed data (yellow)

prominent in the old, i.e. original data, segments from which decorrelation filters were estimated had to be kept relatively short. With the newly processed level 1B GOCE gravity gradients and the robustified method we can manage to estimate stable decorrelation filters from longer segments. Consequently, the computational effort is reduced.

3.2 Analysis of Suspicious Data

Figure 2 shows possible scenarios of detected and identified suspicious data in the GOCE time series. Within the shown zoom, the systematic disturbances identified by the testing procedures (cf. Sect. 2.4) are obvious. The color code represents the hypothesis test which identified the suspicious data points. However, some effects are not necessarily visible and cannot be visually identified as suspicious due to the high amount of data points. Note, that the data detected by the mean test is typically also identified by the sign test (cf. Fig. 2, bottom). We furthermore see that the remaining time series has mostly the characteristic of white noise.

Comparable to the shown example, the procedure of Sect. 2 produces a log-file, providing the testing results for each individual data point and for each individual test. Within this analysis, an observation is not used for process estimation as well as for gravity field estimation as soon as a single test indicates a suspicious observation.

The numbers of identified outliers are presented in Tables 3 and 4. Using the same data, but improved screening methods, the number of outliers could be significantly decreased ('RL05' vs. 'old') for all components except V_{YY}. For V_{YY}, most affected by the imperfect calibration, more suspicious data were identified. The improved screening detects more contaminated data points and selects only the highest quality data. Using the improved screening, for the old and the new data ('old' vs. 'new'), a significant decrease of detected suspicious data is observed. This reduction is attributed to the improved quality of the new reprocessed data set. The most significant decrease appears in the YY-component, which was affected most. Note, that the generally lower noise level of the reprocessed data can also lead to the detection of smaller systematic effects, such that outlier percentages increase.

Table 3 Percentage of outliers for segment 1 (09/Nov/2012–30/May/2013, number of data: 7.5 million)

	V_{XX}	V_{YY}	V_{ZZ}	V_{XZ}	Total
RL05	10.0%	9.3%	9.5%	10.3%	9.8%
Old	3.3%	**15.5%**	2.5%	2.5%	6.0%
New	3.8%	**3.4%**	3.8%	4.1%	**3.8%**

Bold values indicate notable decreases of outlier percentages

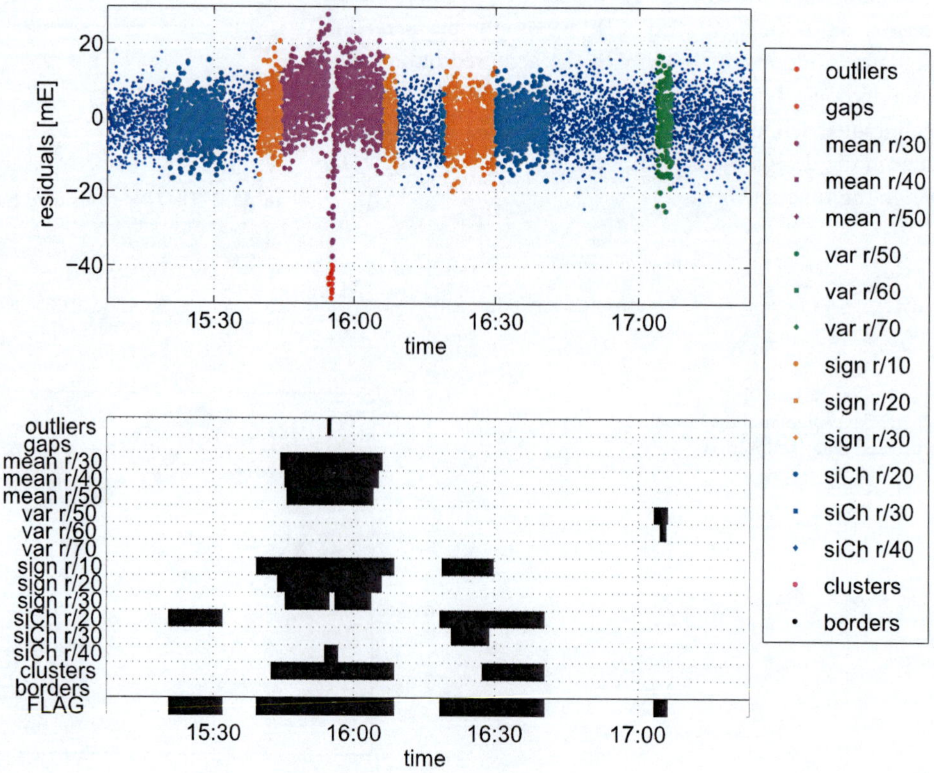

Fig. 2 Example of screening results with time series $\mathcal{R}$ (above) and identified flags (below) for all tests

Table 4 Percentage of outliers for segment 2 (30/May/2013–31/Jul/2013, number of data: 5.5 million)

	V_{XX}	V_{YY}	V_{ZZ}	V_{XZ}	Total
RL05	3.1%	6.2%	1.2%	3.2%	3.4%
Old	1.6%	**6.2%**	1.4%	1.5%	2.7%
New	0.9%	**0.4%**	0.7%	0.5%	**0.6%**

Bold values indicate notable decreases of outlier percentages

3.3 Analysis of the Estimated Processes

With the estimated AR-processes at hand we evaluate the characteristics of the noise model in the spectral domain. Figure 3 presents the power spectral density (PSD) representation of the AR-filter for the V_{YY} gravity gradient. From the design of the gradiometer, we expect white noise, i.e. a flat spectrum in the measurement band (vertical dashed lines in Fig. 3).

The black lines indicate the used filters in the RL05 processing (no robust estimation applied). Due to the non-stationarity of the data, four different filters were required to model the data characteristics. In contrast to that, the blue line shows the robust estimate from the same input data. As the suspicious data are identified, a single filter is used to model the data characteristics. The resulting spectrum is flatter. Using the new data and the robust method improves the filter solutions on a much larger scale (yellow line in Fig. 3). Due to the better calibration, the data quality improves and stationarity becomes a more realistic assumption. This can be shown by the evaluation of subsegments resulting in very similar filter estimates, which is not the case for the old data. Consequently, for the same reasons as stated in Sect. 3.1 the number of segments can be substantially reduced. The noise level decreases for the entire spectrum, even below the measurement band. With the new data, even during the extended mission, the data have characteristics close to white noise in the measurement band. Note, that all other components as well have enhanced data quality and filters whilst the shown V_{YY} component has the overall best improvement.

4 Consequences for Gravity Field Recovery

Output from the analysis of Sect. 3 are robustly estimated AR-processes, which can be used as decorrelation filters within the estimation of a gradiometer only gravity field in a spherical harmonic analysis (Brockmann et al. 2014; Brockmann 2014, Sect. 6.3.3). In addition to the filters per gradient component c and segment s, the suspicious data as identified by the robust estimator is used as outlier flag information in the estimation of the spherical harmonic parameters. The observation equations are set up for that data to keep the segments gap-less and equidistant. After the decorrelation of the observation equations, the identified observations are removed from the decorrelated observation equations. Individual normal equations

$$\mathbf{N}_{s,c}\mathbf{x}_{s,c} = \mathbf{n}_{s,c} \tag{21}$$

for the each segment s and component c are set up and, if positive definite, i.e. enough observations remained, (21) is solved without regularization for each c and s. Results are unconstrained spherical harmonic coefficients $\mathbf{x}_{s,c}$ and its covariance matrix $\Sigma_{\mathbf{x}_{s,c}\mathbf{x}_{s,c}} = \mathbf{N}_{s,c}^{-1}$.

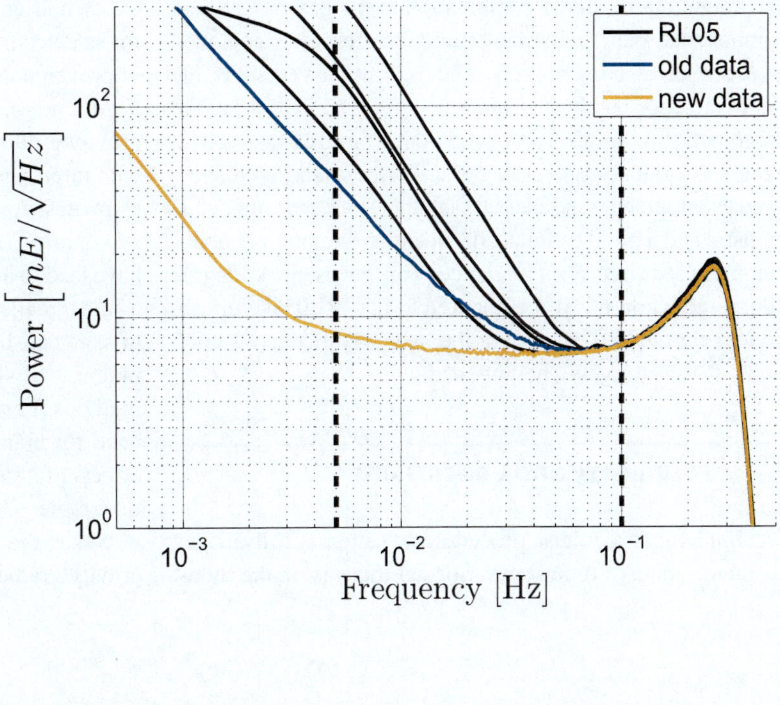

Fig. 3 AR-filter PSD for V_{YY} gravity gradient for the second segment cf. Table 2

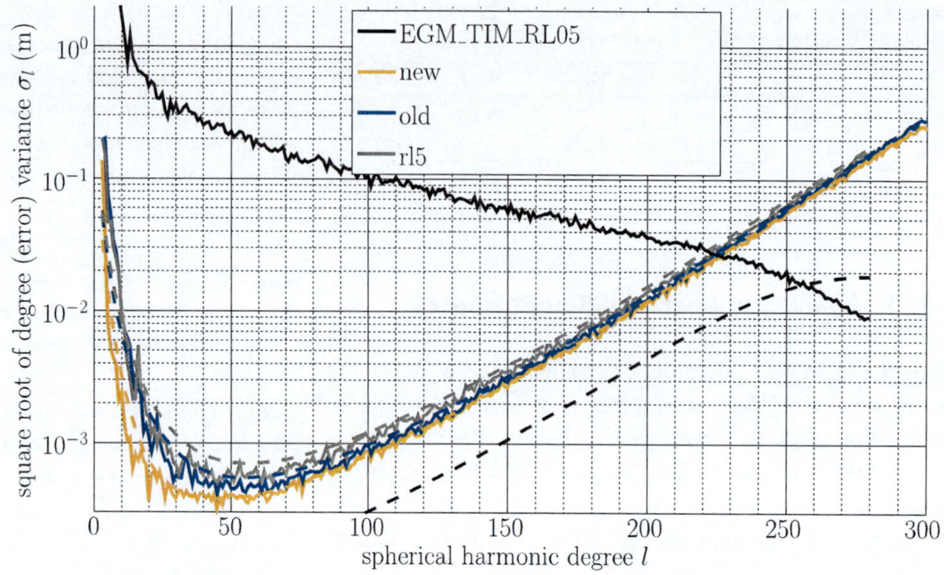

Fig. 4 Degree (error) variances of the gravity field solution using V_{XX}, V_{YY}, V_{ZZ} and V_{XZ} data of the first segment. In solid: empirical from difference to EGM_TIM_RL05. In dashed: formal from covariance matrix. In black: Signal (solid) and error (dashed) degree variances of the whole mission dataset which is used as reference

Figure 4 shows the degree (error) variances of the spherical harmonics gravity field solution for the first segment from Table 2, already combining the V_{XX}, V_{XZ}, V_{YY} and V_{ZZ} gravity gradients. As no regularization is applied, the near-zonal coefficients are excluded, to suppress the polar gap. All three setups (cf. Table 1) are compared using the formal errors (dashed lines) and empirical errors with respect to EGM_TIM_RL05 (solid lines).

The improvement resulting from the robustified AR-process estimation is shown from the improvement from the gray curves to the blue ones. Mainly, the formal errors as well as empirical errors reduce for all degrees, indicating that the robust approach has a positive effect for the gravity field determination, although less data were used in the estimation. Especially the stochastic model improves, which results in a more realistic covariance matrix. As its inverse is the weight matrix in the combination of the gravity gradient components, their individual strengths are better exploited in the relative combination. The yellow curves show the improvements resulting from the improved input data in combination with the robust method. Empirical as well as formal errors decrease even more. Furthermore, both show a nice agreement, indicating that the stochastic model is realistic and a good estimate. To quantify the improvements, cumulative degree error variances can be used. At degree 200, a reduction of 10% compared to the 'RL05' setup can be seen for the 'old' setup. For the 'new' setup, the reduction is about 18% for geoid height errors.

5 Summary and Conclusions

We implement a robust procedure to automatically detect suspicious data in time series. Special focus is on the robust estimation of AR-processes from residual time series. The resulting process can be used as a stochastic model, modeling the error characteristics of the time series. A sequence of statistical hypothesis tests is established, which are used to classify the data into good data points and suspicious data. They are applied in moving windows to the data, such that very long time series can be handled by the hypothesis tests. A wide range of tests is already used, testing for instance the significance of the mean value or for changes in variance or occurrence of signs. Implemented as an iterative procedure, the test decisions are used in a rejection estimator to robustly refine the estimate for the underlying AR-process.

Within this contribution, the procedure was applied to highly correlated GOCE gravity gradient residuals to estimate an AR-processes as a decorrelation filter. This decorrelation filter is used as stochastic model for the gravity gradients within gravity field recovery. Within the numerical tests for selected periods of GOCE observations, it was shown that improvements with respect to the RL05 solution are possible. Furthermore, we applied the method to reprocessed L1B gravity gradients of improved quality. It was shown that the data improves and with the robust method, very stable filter estimates are possible which we attribute to an improved stationarity. The advanced robustified decorrelation filters have lead to an enhanced identification of suspicious data and a more realistic and complete error description. Following this, it can be assumed that the presented tests cover the majority of systematic errors of GOCE data. The screening methods presented here work automatically without further need for manual snooping of outliers and detect systematic effects that may be visible in the time series only at a small time scale.

Since the reprocessing has significantly improved the gravity gradient data the prospect towards the processing of the entire mission dataset for the sixth release is promising. The robust approach will be applied in the processing and used to determine the decorrelation filters for a time-wise RL06 global gravity field model.

Although applied to GOCE here, the proposed method and procedure is generally suitable and applicable for long time series and can be applied to different datasets. Furthermore, the set of hypothesis tests can be continuously extended, thus, it is a flexible and general procedure.

Acknowledgements We thank the anonymous reviewers for their valuable comments which helped improving the manuscript. The authors gratefully acknowledge the Gauss Centre for Supercomputing e.V. (www.gauss-centre.eu) for funding this project by providing computing time through the John von Neumann Institute for Computing (NIC) on the GCS Supercomputer JURECA/JUWELS at Jülich Supercomputing Centre (JSC). Some computations were performed on the cluster at the University of Bonn financed via a DFG Forschungsgroßgeräteantrag (INST 217/749-1 FUGG). This work was financially supported by the ESA GOCE HPF project (main contract No. 18308/04/NL/MM).

References

Baarda W (1968) A testing procedure for use in geodetic networks, vol 5. Netherlands Geodetic Commission, Delft

Brockmann JM (2014) On high performance computing in geodesy – applications in global gravity field determination. Promotionsschrift, Landwirtschaftliche Fakultät der Universität Bonn, Bonn. http://nbn-resolving.de/urn:nbn:de:hbz:5n-38608

Brockmann JM, Zehentner N, Höck E, Pail R, Loth I, Mayer-Gürr T, Schuh WD (2014) EGM_TIM_RL05: An independent geoid with centimeter accuracy purely based on the GOCE mission. Geophys Res Lett 41(22):8089–8099. https://doi.org/10.1002/2014GL061904

Chang I, Tiao GC, Chen C (1988) Estimation of time series parameters in the presence of outliers. Technometrics 30(2):193–204. https://doi.org/10.2307/1270165

Farahani HH, Slobbe DC, Klees R, Seitz K (2017) Impact of accounting for coloured noise in radar altimetry data on a regional quasi-geoid model. Journal of Geodesy 91(1):97–112. https://doi.org/10.1007/s00190-016-0941-6

Floberghagen R, Fehringer M, Lamarre D, Muzi D, Frommknecht B, Steiger C, Piñeiro J, da Costa A (2011) Mission design, operation and exploitation of the gravity field and steady-state ocean circulation explorer mission. Journal of Geodesy 85(11):749–758. https://doi.org/10.1007/s00190-011-0498-3

Fox AJ (1972) Outliers in time series. J R Stat Soc Ser B (Methodol) 34(3):350–363. https://doi.org/10.1111/j.2517-6161.1972.tb00912.x

Helmert FR (1905) Über die Genauigkeit der Kriterien des Zufalls bei Beobachtungsreihen. In: Sitzung der physikalisch-mathematischen Classe vom 25. Mai 1905, Sitzungsberichte der Königlich Preussischen Akademie der Wissenschaften zu Berlin, vol XXVIII, pp 594–612. https://archive.org/details/sitzungsberichte1905deutsch/page/594

Huber PJ (1981) Robust statistics. Wiley, New York. https://doi.org/10.1002/0471725250

Kargoll B, Omidalizarandi M, Alkhatib H, Schuh WD (2018a) Further results on a modified EM algorithm for parameter estimation in linear models with time-dependent autoregressive and t-distributed errors. In: Rojas I, Pomares H, Valenzuela O (eds) Time Series Analysis and Forecasting, Springer International Publishing, Contributions to Statistics, pp 323–337. https://doi.org/10.1007/978-3-319-96944-2_22

Kargoll B, Omidalizarandi M, Loth I, Paffenholz JA, Alkhatib H (2018b) An iteratively reweighted least-squares approach to adaptive robust adjustment of parameters in linear regression models with autoregressive and t-distributed deviations. J Geod 92(3):271–297. https://doi.org/10.1007/s00190-017-1062-6

Kay S, Marple S (1981) Spectrum analysis - a modern perspective. Proc IEEE 69(11):1380–1419. https://doi.org/10.1109/PROC.1981.12184

Klees R, Ditmar P, Broersen P (2003) How to handle colored observation noise in large least-squares problems. J Geod 76(11):629–640. https://doi.org/10.1007/s00190-002-0291-4

Kleiner B, Martin RD, Thomson DJ (1979) Robust estimation of power spectra. J R Stat Soc Ser B (Methodol) 41(3):313–351. https://doi.org/10.1111/j.2517-6161.1979.tb01087.x

Krasbutter I, Brockmann JM, Kargoll B, Schuh WD (2014) Adjustment of digital filters for decorrelation of GOCE SGG data. In: Flechtner F, Sneeuw N, Schuh WD (eds) Observation of the system earth from space - CHAMP, GRACE, GOCE and future missions. Advanced technologies in earth sciences, GEOTECHNOLOGIEN Science Report, vol 20. Springer, Berlin, pp 109–114. https://doi.org/10.1007/978-3-642-32135-1_14

Pail R, Bruinsma S, Migliaccio F, Förste C, Goiginger H, Schuh WD, Höck E, Reguzzoni M, Brockmann JM, Abrikosov O, Veicherts M, Fecher T, Mayrhofer R, Krasbutter I, Sansò F, Tscherning CC (2011) First GOCE gravity field models derived by three different approaches. J Geod 85(11):819. https://doi.org/10.1007/s00190-011-0467-x

Rummel R, Yi W, Stummer C (2011) GOCE gravitational gradiometry. J Geod 85(11):777. https://doi.org/10.1007/s00190-011-0500-0

Schuh WD (1996) Tailored numerical solution strategies for the global determination of the earth's gravity field, Mitteilungen der Geodätischen Institute, vol 81. Technische Universität Graz (TUG), Graz. http://skylab.itg.uni-bonn.de/schuh/Separata/schuh_96.pdf

Schuh WD (2003) The processing of band-limited measurements; filtering techniques in the least squares context and in the presence of data gaps. Space Sci Rev 108(1–2):67–78. https://doi.org/10.1023/A:1026121814042

Schuh WD, Brockmann JM (2018) The numerical treatment of covariance stationary processes in least squares collocation. In: Freeden W, Rummel R (eds) Handbuch der Geodäsie: 6 Bände. Springer Reference Naturwissenschaften. Springer, Berlin/Heidelberg, pp 1–36. https://doi.org/10.1007/978-3-662-46900-2_95-1

Siemes C (2008) Digital filtering algorithms for decorrelation within large least squares problems. Promotionsschrift, Landwirtschaftliche Fakultät der Universität Bonn, Bonn. http://nbn-resolving.de/urn:nbn:de:hbz:5N-13749

Siemes C (2018) Improving GOCE cross-track gravity gradients. J Geod 92(1):33–45. https://doi.org/10.1007/s00190-017-1042-x

Quality and Distribution of Terrestrial Gravity Data for Precise Regional Geoid Modeling: A Generalized Setup

Christian Gerlach, Vegard Ophaug, Ove Christian Dahl Omang, and Martina Idžanović

Abstract

Based on the success of the satellite mission GOCE in providing information on the global gravity field with high quality and spectral resolution, the realization of the 1 cm-geoid is at reach, leading to an increased interest in regional geoid modeling. It is therefore necessary to review theoretical and numerical aspects of regional geoid modeling, including availability of adequate data. In this study, we deal with the latter aspect, specifically the representation error implied by the available gravity data.

We use least-squares collocation to derive formal errors of block mean gravity anomalies and geoid heights for given distributions of scattered gravity stations. By comparison with independent error measures, we validate a generalized procedure in which we do not base the solution on an empirical covariance function of a specific test area, but rather use band-pass filtered global functions. This implies that the procedure is applicable beyond our specific test-bed and can be used to give general error measures, e.g., for network design in poorly surveyed regions.

The computations are carried out in a medium size test area along the Norwegian coast, where the national gravity basis network had been densified in recent years. This allows to show the gain in geoid accuracy that can be expected from adding the new gravity data. We show that the signal variance of the regional gravity field corresponds well with the one derived from the global covariance function, thus validating our generalized procedure. In previous studies, the accuracy of gravity anomalies and geoid heights in Norway were estimated to be (on average) around 2 mGal and 3 cm, respectively. We find good agreement of the formal gravity anomaly error with the empirical measure. By adding the new data, the gravity anomaly error can be reduced to almost 1 mGal. The formal geoid error can be reduced from around 1.7 to 1.3 cm (on average). The discrepancy between the formal error and the empirical measure of 3 cm is probably due to contributions from GNSS and leveling errors, which are not considered in our formal estimate. The results presented here show larger errors over ocean areas, because the computations are restricted to land data. Available airborne and marine gravity will be considered in the future.

Keywords

Geoid · Least-squares collocation · Wenzel modification

C. Gerlach (✉)
Bavarian Academy of Sciences and Humanities, Geodesy and Glaciology, Munich, Germany

Faculty of Science and Technology (RealTek), Norwegian University of Life Sciences (NMBU), Ås, Norway
e-mail: gerlach@badw.de

V. Ophaug · M. Idžanović
Faculty of Science and Technology (RealTek), Norwegian University of Life Sciences (NMBU), Ås, Norway
e-mail: vegard.ophaug@nmbu.no; maid@nmbu.no

O. C. D. Omang
Geodetic Institute, Norwegian Mapping Authority, Hønefoss, Norway
e-mail: ove.omang@kartverket.no

P. Novák et al. (eds.), *IX Hotine-Marussi Symposium on Mathematical Geodesy*,
International Association of Geodesy Symposia 151, https://doi.org/10.1007/1345_2019_71

1 Introduction

The satellite gravity mission GOCE (ESA 1999) has considerably increased our knowledge about the global static gravity field. One of the main geodetic applications of the GOCE mission is the realization of a global vertical datum. GOCE-based global potential models (GPM) allow modeling the geoid with centimeter accuracy for spatial resolutions of around 100 km (Gerlach and Ophaug 2017). Because short-scale signal contributions, on global average, reach values of around 30–40 cm RMS, the GPMs must be combined with terrestrial data in the region of interest (Rummel 2012; Gatti et al. 2013).

In regional geoid modeling, several different data sources are combined. Satellite data provides information on the long-to-medium wavelengths features of the gravity field, while terrestrial data (land, marine, airborne) contribute to the medium-to-short wavelengths. Short-to-ultra-short contributions can be reconstructed from digital terrain models (DTM) (Sansò and Sideris 2013). The classical combination schema makes use of the remove-restore concept (Denker 2013). Thereby, terrestrial data is band-pass filtered by removing signal contributions provided by a satellite-based GPM and a DTM. The remaining residual signal is used for field transformation from gravity to geoid. Thereafter, the GPM- and DTM-contributions to the geoid are restored. Field transformation may be performed by Stokes integration, least-squares collocation (LSC) or employing alternative representation methods like spherical radial base functions (Sansò and Sideris 2013; Ophaug and Gerlach 2017).

The accuracy of a regional geoid depends, among other error sources, on the density and distribution of available gravity data. The ability of gravity values to represent the gravity field in between them is called the representation error, in other words: how well can the continuous gravity field be represented by a finite number of scattered points? The representation error decreases with increasing point density. It is usually larger than the contribution of observational noise of the gravity data and it is of particular interest for gravity network design given some final requirements for the geoid accuracy. Comparison of the representation error with empirical errors also allows to set bounds on the magnitude of other error sources, e.g., from observational noise of input and validation data or from errors in data reductions.

In this study, we derive formal estimates of the representation error in a test area along the Norwegian coast and compare them to empirical errors derived in Ågren et al. (2016) and Gerlach and Ophaug (2017). Test area and point distribution are presented in Sect. 2.

The error estimates are derived by LSC. A realistic description of the signal characteristics is essential for the interpretation of the results. In standard applications, this is achieved by deriving an empirical covariance function from local/regional data. Here, we aim at a generalized procedure, which is not limited to our specific test area. Therefore, our covariance function is derived from band-pass filtered global degree variance models. Band-pass filtering the signal covariance function is applied to reflect the standard remove-restore method. The computational steps are described in Sect. 3.1, the definition of the signal covariance function in Sect. 3.2.

The error estimates for block mean values of (band-pass filtered) residual gravity anomalies as well as for residual geoid heights are validated against empirical error measures in Sect. 4. Section 5 summarizes the main findings and gives an outlook on further research.

2 Test Area and Data Sets

The Norwegian gravity basis network (comprising about 280 stations) as well as its regional densifications (almost 10,000 points) originate from the 1970ies. The number of stations implies an average point distance of about 6 km throughout the country. The data is the basis of all currently available regional geoid models. Country-wide comparison of these regional geoid models with GNSS-leveling data in Norway shows a misfit of geometric and gravimetric geoid heights of about 3 cm RMS and implies an accuracy of about 2 mGal (1 mGal $= 1 \cdot 10^{-5}$ ms^{-2}) for 1 km $\times$ 1 km block mean gravity anomalies, see Ågren et al. (2016) or Gerlach and Ophaug (2017).

In order to investigate the realization of the 1-cm geoid covering land and ocean areas, the Norwegian Mapping Authority (NMA) has set up the Sunnmøre test field. It is about 250 km $\times$ 250 km in extension, covering roughly the county of Møre og Romsdal in western Norway between Sognefjord to the south and the city of Kristiansund in the north. The distribution of gravity stations from the 1970ies in this area is shown in the left panel of Fig. 1. Between 2012 and 2017 NMA acquired about 3,800 additional gravity stations (of which around 2,000 fall into our computation area), leading to an average point distance of only about 2–3 km. The distribution of the new data is shown on the right panel of Fig. 1. Note, that we have limited the computations to the inner part of the area in which most of the new data are located. Therefore, Fig. 1 does not show all of the collected data.

In this study, we derive error estimates from either the old data set, or from the combination of old and new data. We

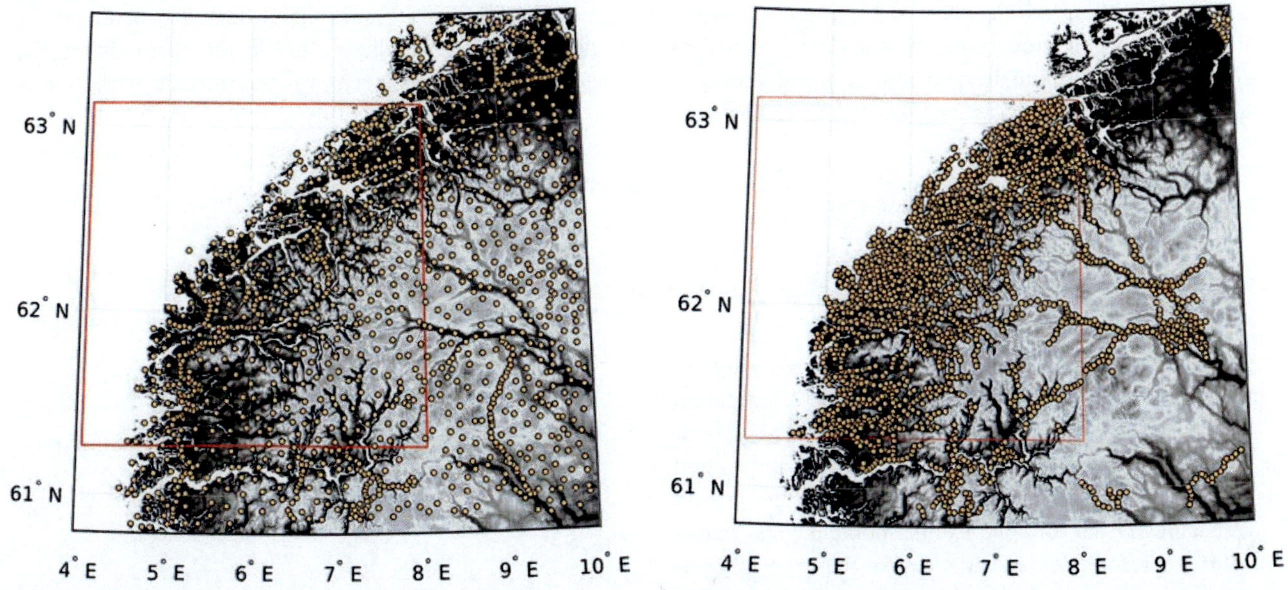

Fig. 1 Distribution of land gravity data available in the Sunnmøre test field before 2012 (left) and of data newly acquired between 2012 and 2017 (right). The red boxes show the computation area of this study

will refer to the old dataset as g_1 and to the combination of old and new data as g_2.

3 Methodology

Our aim is to derive formal error estimates for the accuracy of geoid heights as derived from datasets g_1 and g_2. This will give an indication of the gain in geoid accuracy one can expect from incorporation of the new data. The result will depend on the increased density of terrestrial gravity data (representation error) and also on our assumptions on signal characteristics. Validation of these assumptions is performed by comparison with empirical error measures in Sect. 4.

3.1 Error Propagation by Least-Squares Collocation

We apply LSC to derive formal error estimates for geoid heights and block mean values of terrestrial gravity anomalies. In both cases, gravity anomalies at scattered observation points Q serve as input data, while the computation points P are given on a regular geographic grid with a spacing of $1.2' \times 2.4'$ (around 2 km $\times$ 2 km). The corresponding equations read

$$\Sigma_{PP}^{\bar{g}\bar{g}} = C_{PP}^{\bar{g}\bar{g}} - C_{PQ}^{\bar{g}g}\left(C_{QQ}^{gg} + D_{QQ}^{gg}\right)^{-1} C_{QP}^{\bar{g}g}. \quad (1)$$

$$\Sigma_{PP}^{NN} = C_{PP}^{NN} - C_{PQ}^{Ng}\left(C_{QQ}^{gg} + D_{QQ}^{gg}\right)^{-1} C_{QP}^{Ng}. \quad (2)$$

Matrices Σ and C represent variance-covariance matrices of estimation error and gravity field signal, respectively. Matrix D_{QQ}^{gg} contains the error of the input data. Since we are interested in the representation error, we might neglect the D-matrix. However, we had to add observational white noise in the order of some tens of mGal to overcome numerical problems in the inversion of the C_{QQ}-matrix. Matrix C_{QP} is the transpose of matrix C_{PQ}. Overbar in the superindex indicates block averaging, i.e., $C^{\bar{g}\bar{g}}$ represents auto-covariances of block mean values of gravity and $C^{\bar{g}g}$ are cross-covariances between block mean and point values.

3.2 Definition of the Signal Covariance Function

The aim is to derive an estimate for the contribution of terrestrial gravity anomalies to the geoid in the general remove-restore frame. This implies that the signal covariance function only represents the residual medium-wavelength signal, after reduction of (1) long-wavelength global information (low-pass filtering) as well as (2) short-scale topographic effects (high-pass filtering).

Our residual (band-pass filtered) signal covariance function is based on the combination of two degree variance models:

(1) the model by Tscherning and Rapp (1974) is used to represent the large to medium scale features of the gravity field (global contribution); low-pass filtering is applied to this model explicitly

(2) the model by Flury (2006), is used to represent medium to small-scale features; the model describes RTM-reduced gravity anomalies, i.e., high-pass filtering is already applied implicitly and no further filtering is required.

Degree variances c_l of Flury's model are given by

$$c_l(\Delta g_{RTM}) = \frac{6.8 \cdot 10^{-7}\ \text{mGal}^2}{(l+0.5)^{3.09}}, \tag{3}$$

where l is the spherical harmonic degree and Δg_{RTM} are terrestrial gravity anomalies reduced for the contribution of short-scale topography (residual terrain model, RTM). The model was derived from several regional datasets and therefore is not valid for the global features of the gravity field, specifically not for spherical harmonic degrees below $l \approx 400$. Therefore, degree variances for the lower degrees are taken from the model of Tscherning and Rapp. The same combination of degree-variance models is applied and described in Gerlach and Pettersen (2010).

Finally, the combined degree-variance model needs to be high-pass filtered, reflecting the reduction of global information provided by a satellite-only GPM. The filter coefficients (spectral weights) were derived in Gerlach and Ophaug (2017) from an optimal combination of GOCE and terrestrial information in Norway using the spectral combination approach of Wenzel (1982). Thereby, error degree variances $\sigma^2_{l,\text{GPM}}$ of the GPM are combined with error degree variances $\sigma^2_{l,\text{terr.}}$ of the terrestrial data, yielding

$$w_{l,\text{terr.}} = \frac{\sigma^2_{l,\text{GPM}}}{\sigma^2_{l,\text{GPM}} + \sigma^2_{l,\text{terr.}}}, \tag{4}$$

the spectral weights for the terrestrial data. The complementary weights for the GPM are given by

$$w_{l,\text{GPM}} = 1 - w_{l,\text{terr.}}. \tag{5}$$

The $\sigma^2_{l,\text{GPM}}$ are based on the error covariance matrix of GOCO05s (Mayer-Gürr et al. 2015). The $\sigma^2_{l,\text{terr.}}$ are based on the error covariance function provided by Denker (2013). This function describes the average error behaviour in Europe. In Gerlach and Ophaug (2017) this function was rescaled such, that the combined geoid model minimized the fit to GNSS-levelling data in Norway. This way, the error variance of terrestrial gravity data in Norway was found to be around (2 mGal)^2. The present study is meant to validate whether this error measure reflects the representation error of terrestrial gravity data in our study area, see Eq. (1).

The signal covariance functions required for the computational steps in Eqs. (1) and (2) are based on the combined (Tscherning/Rapp and Flury) signal degree variances $c_{l,\text{TR/F}}$ weighted by $w^2_{l,\text{terr.}}$. In addition, we consider that reducing the GPM-contribution in the remove-step introduces the errors $\sigma_{l,\text{GPM}}$ of the GPM (weighted by $w_{l,\text{GPM}}$) in the low degrees. Therefore, the final (dimensionless) degree variances are given by

$$c_l = w^2_{l,\text{terr.}}\, c_{l,\text{TR/F}} + w^2_{l,\text{GPM}}\, \sigma^2_{l,\text{GPM}} \tag{6}$$

and the signal covariance functions by

$$C^{NN}(\psi) = R^2 \sum_{l=2}^{L} c_l\, P_l(\cos\psi) \tag{7}$$

$$C^{Ng}(\psi) = R\gamma \sum_{l=2}^{L} (l-1)\, c_l\, P_l(\cos\psi) \tag{8}$$

$$C^{gg}(\psi) = \gamma^2 \sum_{l=2}^{L} (l-1)^2\, c_l\, P_l(\cos\psi). \tag{9}$$

If block averaging is applied, Pellinen's smoothing coefficients β_l (see Sjöberg (1980) and references therein) are implemented on the right-hand side of Eq. (9) in either linear or quadratic form, yielding $C^{\bar{g}g}$ or $C^{\bar{g}\bar{g}}$, respectively. The other quantities in Eqs. (7)–(9) are earth radius R (6,378,137 km), normal gravity γ (GM/R^2), spherical distance ψ and maximum spherical harmonic degree L (here chosen to be 5,000).

Figure 2 shows the residual signal covariance functions of geoid heights and gravity anomalies. The expected variability of the signals accounts for around 40 cm and 18 mGal, respectively. The correlation length of residual geoid heights and gravity is 0.31° (around 35 km) and 0.23° (around 25 km), respectively. We note that the standard deviation of the residual (after removing the contribution from GOCO05s) geoid model NMA2014 in our study area is 40.6 cm. This fits very well to the formal signal standard deviation of 39.5 cm and validates our generalized procedure. The slightly lower formal value might reflect the fact that it represents the band-pass filtered signal, while the larger empirical value represents the low-pass filtered signal. The discrepancy between both signals is the contribution of the short-scale topography, which, according to Hirt et al. (2010), accounts for 1–2 cm RMS.

4 Results and Discussion

The error estimates for block mean values of gravity and geoid heights are shown in Fig. 3. The top row contains error estimates for gravity in units of mGal, while the bottom row shows geoid error estimates in units of meter. Error estimates

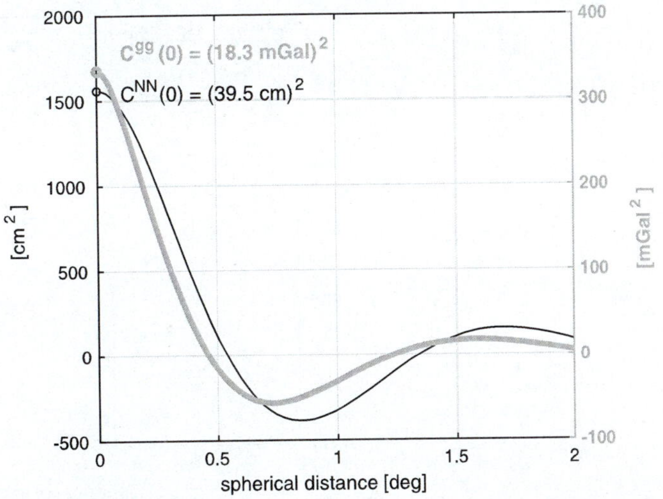

Fig. 2 Residual signal covariance functions for geoid heights (black, left axis) and gravity anomalies (gray, right axis)

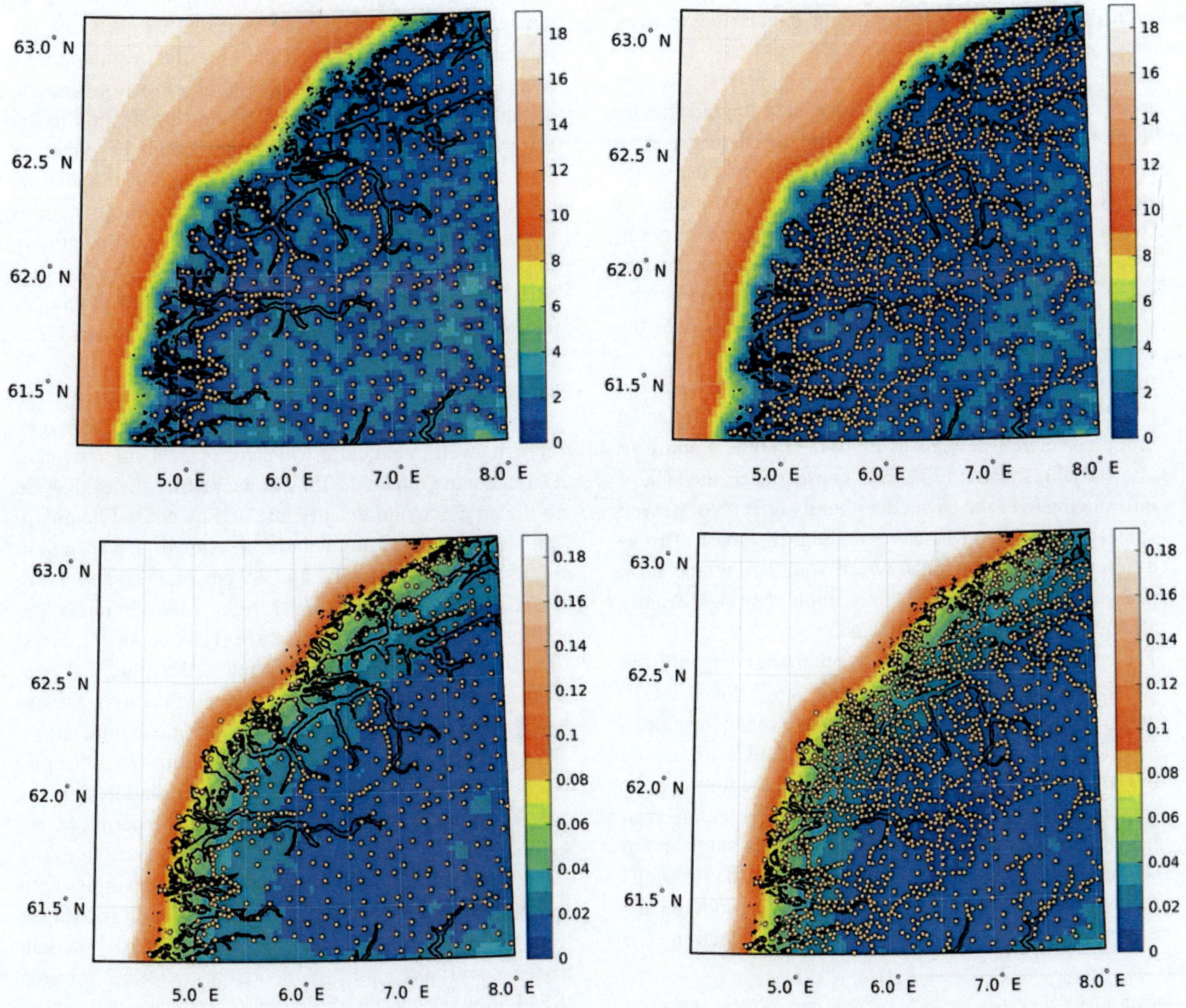

Fig. 3 Formal error estimate for block mean gravity anomalies (top row, units are [mGal]) and geoid heights (bottom row, units are [m]) derived from datasets g_1 (left column) and g_2 (right column). In case of geoid heights, the color bar is restricted to 18 cm in order to make the error amplitudes on land visible

Fig. 4 Formal geoid error covariances along NW-SE profiles with origin at an off-shore station (black line, left scale) and a station at the coast (thick gray line, right scale). Vertical lines indicate correlation length. Units are in [cm^2]

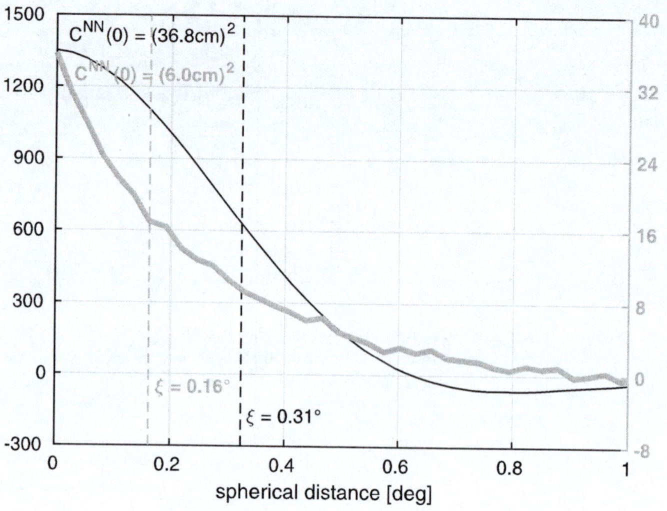

in the left column are based on dataset g_1, those in the right column on dataset g_2. The following conclusions can be drawn from the figures:

- In all cases, restriction of the input data to land gravity leads to large errors over ocean areas. There the error reaches up to about 18 mGal in case of gravity anomalies (top row) and about 37 cm in case of geoid heights (bottom row).[1] This maximum error amplitude over ocean areas corresponds to the standard deviation of the residual signal (see Fig. 2), reflecting the fact, that residual gravity on land hardly affects gravity field estimation over the ocean.
- In general, the geoid signal is dominated by longer, the gravity signal by shorter wavelengths. This is reflected in the correlation lengths of C^{NN} ($\xi = 0.31°$) and C^{gg} ($\xi = 0.23°$), shown in Fig. 2. In consequence, the lack of information over the ocean does hardly affect block mean values of gravity on land, even along the coast. This is different for geoid heights, where boundary effects have a larger impact on land: in our example they significantly affect areas to about 50 km inland.
- The errors of block mean gravity anomalies derived from dataset g_1 (upper left panel in Fig. 3) are below 4 mGal, on average 1.7 mGal (on land). This is close to the empirical error estimate for terrestrial data of 2 mGal as derived in Gerlach and Ophaug (2017). Thereby, the formal error estimate of 1.7 mGal reflects the representation error, i.e., the error arising from the fact that a single point observation is not necessarily a good representative for its neighborhood. The representation error depends on the correlation length of the signal covariance function. The longer the correlation length, i.e., the smoother the signal, the smaller is the representation error. The fact that the formal error is smaller than the empirical estimate may reflect that the signal covariance function is too smooth, or that other error sources contribute significantly to the empirical error. Such error sources could be errors in GNSS or leveling networks, or errors in the computation of gravity anomalies, like errors in topographic reductions or propagation of errors in the coordinates of the gravity stations (mainly the height component).
- On average, densification of the gravity network reduces the formal error of block mean gravity values from 1.7 to 1.1 mGal.
- On average, the error of geoid heights derived from datasets g_1 and g_2 is around 2.9 and 2.6 cm, respectively. This fits well to the empirical error of 2.94 cm derived in Gerlach and Ophaug (2017). If we restrict the evaluation to the area not significantly affected by coastal boundary effects, we find from dataset g_1 (lower left panel in Fig. 3) values between 1.1 and 2.9 cm, on average 1.7 cm; employing dataset g_2 (lower right panel) reduces the errors to 0.5–2.1 cm, on average 1.3 cm. In all cases, the formal error estimates are below the empirical one, indicating that the signal covariance functions are too smooth, or that additional error sources are significant.
- The geoid errors are neither homogeneous nor isotropic, showing varying amplitudes and correlation lengths across the test area, in particular when comparing points over ocean, at the coast, and inland. Figure 4 shows error covariance functions for two selected examples with origin (1) in the NW-corner of the computation grid (black line, left axis) and (2) at the coast (gray, right axis). In both cases, covariances are provided from the origin towards SE. Following the interpretation of stepwise collocation (Moritz 1980), both functions correspond to the original signal covariance function C^{NN}, see Fig. 2, reduced for

[1]Note that the colorbar in the bottom row of Fig. 3 (geoid heights) is restricted to a maximum of 18 cm to increase visibility of the error distribution on land. In consequence, the maximum errors of 37 cm are not represented in the figure.

the contribution of the input data. The off-shore station is hardly affected by data on land, while signal amplitude and correlation length are significantly reduced for the coastal station.

5 Summary and Outlook

We have derived formal error estimates for residual block mean gravity anomalies and geoid heights from two sets of input data in a test area along the Norwegian coast. The formal error estimates prove that the newly acquired data is capable of reducing the geoid error from around 2 to 1 cm, in some parts even to 0.5 cm. Error estimates for block-mean gravity anomalies validate the value of around 2 mGal which was empirically found in Gerlach and Ophaug (2017).

The formal errors are slightly smaller than the empirical ones found by comparison with GNSS-leveling in Gerlach and Ophaug (2017) or Ågren et al. (2016). The level of agreement between formal and empirical errors verifies the basic assumptions on the spectral signal characteristics and shows, that the largest error contribution is the representation error. Remaining discrepancies indicate that there are also other relevant error sources (e.g., computational errors in gravity reduction, or data errors in GNSS or levelling networks) or that the basic signal covariance function may be slightly too smooth. The residual signal covariance function will be compared to empirical covariance functions of the residual gravity field in future investigations.

Data gaps over the oceans reduce geoid quality off-shore and as far as about 50 km inland. Therefore, further studies will concentrate on the integration of available shipborne and airborne data, thus reducing the error variance over ocean areas and along the coast.

Acknowledgement The study is part of the GOCODYN project, funded by the Norwegian Research Council under project number 231017.

References

Ågren J, Strykowski G, Bilker-Koivula M, Omang O, Märdla S, Forsberg R, Ellmann A, Oja T, Liepinš I, Paršeliūnas E, Kaminskis J, Sjöberg LE, Valsson G (2016) The NKG2015 gravimetric geoid model for the Nordic-Baltic region. In: 1st Joint Commission 2 and IGFS Meeting International Symposium on Gravity, Geoid and Height Systems, Thessaloniki, Greece, 19–23 September

Denker H (2013) Regional gravity field modeling: theory and practical results. In: Xu G (ed) Sciences of Geodesy - II. Springer, Berlin, pp 185–291

ESA (1999) Gravity Field and Steady-State Ocean Circulation Mission. Report for mission selection of the four candidate earth explorer core missions. Technical Report ESA SP - 1233(1), ESA

Flury J (2006) Short-wavelength spectral properties of the gravity field from a range of regional data sets. J Geod 79(10–11):624–640. https://doi.org/10.1007/s00190-005-0011-y

Gatti AM, Reguzzoni G (2013) Venuti: the height datum problem and the role of satellite gravity models. J Geod 87(1):15–22. https://doi.org/10.1007/s00190-012-0574-3

Gerlach C, Ophaug V (2017) Accuracy of regional geoid modelling with GOCE. In: International Association of Geodesy Symposia. Springer, Berlin, pp 1–7. https://doi.org/10.1007/1345_2017_6

Gerlach C, Pettersen B (2010) Validation of GOCE with terrestrial gravity data in Norway. In: SP-686. ESA, Bergen, Norway

Hirt C, Featherstone WE, Marti U (2010) Combining EGM2008 and SRTM/DTM2006.0 Residual terrain model data to improve quasi-geoid computations in mountainous areas devoid of gravity data. J Geod **84**(9), 557–67. https://doi.org/10.1007/s00190-010-0395-1

Mayer-Gürr T, Kvas A, Klinger B, Rieser D, Zehentner N, Pail R, Gruber T, Fecher T, Rexer M, Schuh WD, Kusche J, Brockmann JM, Loth I, Müller S, Eicker A, Schall J, Baur O, Höck E, Krauss S, Jäggi A, Meyer U, Prange L, Maier A (2015) The New Combined Satellite Only Model GOCO05s. Geophysical Research Abstracts Vol. 17, EGU2015-12364. https://doi.org/10.13140/RG.2.1.4688.6807

Moritz H (1980) Advanced physical geodesy. Wichmann, Karlsruhe

Ophaug V, Gerlach C (2017) On the equivalence of spherical splines with least-squares collocation and Stokes's Formula for regional geoid computation. J Geod 91(11):1367–1382. https://doi.org/10.1007/s00190-017-1030-1

Rummel R (2012) Height unification using GOCE. J Geod Sci 2(4):355–362

Sansò F, Sideris MG (2013) Geoid determination - theory and methods. Lecture notes in earth system sciences, vol 110. Springer, Berlin

Sjöberg L (1980) A recurrence relation for the bn-function. Bull Géod 54(1):69–72. https://doi.org/10.1007/BF02521097

Tscherning CC, Rapp RH (1974) Closed covariance expressions for gravity anomalies, geoid undulations, and deflections of the vertical implied by anomaly degree variance models. Technical report, Ohio State University

Wenzel HG (1982) Geoid computation by least squares spectral combination using integral formulas. In: Proceedings IAG general meeting, Tokyo, 7–15 May 1982, pp 438–453

Part VI

Interactions of Geodesy and Mathematics

Geodesy and Mathematics: Interactions, Acquisitions, and Open Problems

Willi Freeden and Fernando Sansò

Abstract

The paper highlights arguments that, coming from Mathematics, have fostered the advancement of Geodesy, as well as those that, generated by geodetic problems, have contributed to the enhancement of different branches in Mathematics. Furthermore, not only examples of success are examined, but also open questions that can constitute stimulating challenges for geodesists and mathematicians.

Keywords

Collocation · Generalized random fields · Geodetic boundary · Geodetic inverse Ill-posed problems · Integer estimation · Multiscale mollifier · Multiscale tikhonov and truncated singular value frequency regularization · Space regularization · Value problems

There is no branch of Mathematics, however abstract, which may not some day be applied to phenomena of the real world.

Nikolai Ivanovich Lobachevsky

1 Introduction

The paper highlights arguments that, coming from Mathematics, have fostered the advancement of Geodesy, as well as those that, generated by geodetic problems, have contributed to the enhancement of different branches in Mathematics. Furthermore, not only examples of success are examined, but also open questions that can constitute stimulating challenges for geodesists and mathematicians.

We perform a general overview, without any pretence of completeness, of areas like geometry of the gravity field (GF), boundary value problems (BVP) for the Laplace operator, Runge approximation (RA), probability theory (in particular, Generalized Random Fields), and statistics (in particular integer parameters estimation and rank deficient problems). In Sects. 6–8 we turn the attention to novel applications to Geodesy in the context of multiscale approximation (MA). In fact, multiscale reconstruction and decorrelation methods are a research field originated in geophysics for, e.g., earthquake modeling some decades ago, in which today's Geodesy and Mathematics show mutual influences, especially on the subject of spectral and space data sampling.

Then we focus the attention on the inverse problems of Geodesy and their regularization strategies. Two examples are studied in more detail: Downward continuation of gravitational information to the Earth's surface via satellite gravitational gradiometry (SGG) is seen to be adequately realized in the tensorial frequency framework of non-bandlimited Tykhonov and bandlimited truncated singular value regularization. The inverse gravimetry (IG) problem is shown to be appropriately regularized by use of space multiscale mollifiers, to detect fine particulars of geological relevance.

Finally, the authors want to stress again that the choice of arguments by no means can cover the whole area; it rather reflects the background of the authors and has to be taken as illustrative of a general process of interaction between sciences. In addition, neither extreme depth to explain all facets of the geodetic observational situation nor penetrative handling of mathematical obligations

W. Freeden
University of Kaiserslautern, Kaiserslautern, Germany
e-mail: freeden@rhrk.uni-kl.de

F. Sansò (✉)
Politecnico di Milano, DICA, Milan, Italy
e-mail: fernando.sanso@polimi.it

P. Novák et al. (eds.), *IX Hotine-Marussi Symposium on Mathematical Geodesy*,
International Association of Geodesy Symposia 151, https://doi.org/10.1007/1345_2019_87

and technicalities can be expected. The paper is just an "appetizer" served to enjoy the tasty meal "Mathematical Geodesy Today" to be shared by geodesists and mathematicians.

This notwithstanding, the reader should be warned that a complete understanding of all sections of the paper requires a considerable mathematical background, beyond a general knowledge of many branches of Geodesy.

2 Interactions of Mathematics and Geodesy

Seen from the historic point of view, one of the most important contributions to Mathematics is *geometry* (from the Ancient Greek *γεωμετρία* geo- "Earth", -metron "measurement"). The earliest recorded beginnings of geometry can be traced to ancient Mesopotamia and Egypt in the second millennium BC. Early geometry was a collection of empirically discovered principles concerning lengths, angles, areas, and volumes, which were developed to meet some practical need in surveying and various crafts.

By the third century BC, geometry was put into an axiomatic form by Euclid, whose treatment, *Euclid's Elements*, set a standard for many centuries to follow. It brought the heritage of Mathematics from the antiquity to our time. Some centuries later, the Greeks themselves replaced the term "Geometry", which had meanwhile lost the original meaning of "Earth's measuring" by "Geodesy" as a new meaning of an abstract theory of the "Earth's shape", while geometry now reflected the mathematical rigor through its axiomatic method. In fact, it is the earliest example of the format still used in Mathematics today, that of definition, axiom, theorem, and proof. Although most of the contents of the "Elements" were already known, Euclid arranged them into a single, coherent logical framework.

By the early seventeenth century, geometry had been put on a solid analytic footing by mathematicians such as R. Descartes (1596–1650) and P. de Fermat (1607–1665). Since then, and into modern times, geometry has expanded into non-Euclidean geometry and manifolds, describing spaces that lie beyond the normal range of human experience. Hence, while geometry has evolved significantly throughout the years, there are some general concepts that are more or less fundamental. These include the concepts of points, lines, planes, surfaces, angles, and curves, as well as the more advanced notions of manifolds and topology or metric.

Historians very often use the term "modern" for the period starting with the Italian Renaissance. (Geo)scientists probably call I. Newton (1643–1727) the first modern scientist. The emphasis on a systematic study of the laws such as the Newtonian approach in *"Philosophiae Naturalis Principia Mathematica"* may be regarded as the most obvious characteristic of modern times. It also initiated a new branch of Mathematics, called *analysis*, which from its very beginning is so closely related to Physics and Physical Geodesy, for example in the field of differential equations, that some people finds it difficult to say, where Mathematics starts and Physics ends (see Freeden and Schreiner 2018 for a more detailed study). So, it may be concluded that one great contribution of Mathematics to modern (geo)scientific progress is the setting of the pattern for the theories describing the laws of the real world ("reality").

Apart from the mathematical aspect of the value of a theory, there is a practical side to which it is even more important. Once the theory is recognized to be in accordance with an experience, it renders the experiment unnecessary. This, of course, is very important where experiments are not possible. The best example of a field in which one could not carry out experiments is possibly Astronomy. So it was not surprising that a new approach to the features of the "real world" had its striking success when it was possible to deduce Kepler's empirical law on the movements of planets from Newton's general law of mechanics. The same conceptual development also applies to today's space research, and nowadays satellite technology is a field in which experiments are usual practice. One of the best examples to show the difference in the emphasis between Physics, Mathematics, and Geoengineering is gravitational theory which played a major part in the history of sciences from every angle. In fact, it is well-suited for investigating different trends in thinking as represented in various fields of geoscience.

Geodetically reflected potential determination was in a fortunate position, as the Newtonian theory of gravitation had already developed a mathematical status, which could be naturally extended so as to give a complete description of the laws of this field. As a consequence, potential theory as the scientific collection of ideas, concepts, and structures involving Laplace's equation gained new aspects, and it was challenged with new problems, of which the boundary value problems probably are the best known. Potential theory actually guarantees that if certain values of a potential under specific consideration are given on the boundary of a closed body, the potential is determined via the boundary value problem in the interior or in the exterior (when an additional regularity condition at infinity is supposed to hold true). This assertion, of course, has been checked in many experiments also under geodetic auspices, but naturally it cannot be verified experimentally in the generality in which it can be stated mathematically. More concretely, at a stage, where the theory is regarded as satisfactory from the geophysical point of view, it is a system of fundamental laws, definitions and problems, of which some, under certain conditions, have been solved mathematically. The problems in their full generality, however, are given to Mathematics as conjectures, in a sense, to be proved. They become the object of a study of the well-posedness, i.e., existence, uniqueness, stability proofs, which therefore aim at establishing the consistency of the general physical theory. Altogether, the aspects of potential theory have changed considerably when constituents could be described by means of Laplace's equation, just as scientific tasks arose from the theory of stationary flow, that indeed uses the same differential equation. It could thus be observed how new physical applications developed new aspects of potential theory and the theory of partial differential equations, primarily originated on geodetic gravitational developments.

3 Geometry of the Gravity Field

If we think that the beginning of modern geometry could be identified with the researches of C.F. Gauss (1777–1855) on curvature of surfaces (*Disquisitiones generales circa superficies curvas*) and subsequently with the development of tensor calculus and related differential forms, there is little doubt that this history is intermingled with the development of the concept *shape of the Earth* (and its cartographic representation) which is central in Physical Geodesy and, as we know, fundamentally related to the form of the equipotential surfaces of the gravity field.

If Gauss was thinking of Geodesy and mapping as prominent examples of this geometric theories, Geodesy in turn has been developing its own geometric tools to describe the gravity field and its observable quantities. Besides matters concerning the ellipsoidal geometry and related gravity models (Grossman 1974; Pizzetti 1910; Somigliana 1929), we could mention the work by Stokes (1867), Bruns (1878), Helmert (1884), leading to the fundamental definition of *anomalies* with respect to the normal gravity field and several equations, which populate the first chapters of textbooks on Geodesy (e.g. Heiskanen and Moritz 1967; Sansò and Sideris 2013; Vaniček and Krakiwsky 1986), like the behaviour of the gravity vector along a plumbline

$$\nabla g(P) = (-2gJ + 4\pi G\rho - 2\omega^2)\boldsymbol{n} + gk\boldsymbol{n}_1 \quad (1)$$

(g modulus of gravity at P, J mean curvature of the equipotential surface, ρ mass density, ω the Earth angular velocity, k first curvature of the plumbline, $\boldsymbol{n}$ tangent to the plumbline, $\boldsymbol{n}_1$ principal normal of the plumbline) and the linearized equation, also known as *fundamental equation of Physical Geodesy*

$$\Delta g = -\frac{\partial T}{\partial h} + \frac{1}{\gamma}\frac{\partial \gamma}{\partial h}T \quad (2)$$

(Δg free-air gravity anomaly, h ellipsoidal height, γ modulus of normal gravity, T anomalous gravity potential).

The work has got a mature formulation in books like "Mathematical Geodesy" by Hotine (1985), and "Intrinsic Geodesy" by Marussi (1985).

To quote T. Krarup (see Krarup 2006b, p. 53) "The notion itself of Mathematical Geodesy goes back to Martin Hotine, who launched it as the title of a formidable book [Hotine 1985] But the concept was born in 1951 with Marussi's beautiful paper *Fondamenti di Geodesia Intrinseca*". These authors clearly highlighted that the main role of Geometric Geodesy was to find suitable coordinates to represent physical entities related to the gravity field, possibly observable, and ultimately to formulate the physical laws that can in principle uniquely determine the unknown potential W on the Earth's surface and in the outer space. Certainly belongs to this area the work done by T. Krarup (see Krarup 1969) and his definition of isozenithals to arrive at a rigorous definition of the so-called *Vector Molodensky Problem* (see Sansò and Sideris 2013) in a linearized form.

A similar concept applies too, to the work by F. Sansò on the *gravity space* approach to Molodensky's problem (Sansò 1977).

More modern geometric tools have then been introduced into Geodesy by authors like Grossman (1974), Zund (2012), and Grafarend (1986, 1975). Maybe it is worth mentioning here the discussion, raised by E.W. Grafarend, on the holonomity of certain coordinates, specially height coordinates, so important in applied surveying. One possible solution has been found in a rigorous application of the definition of coordinates, contrary to the interpretation of the observation equation of levelling in terms of a differential form, Sansò and Vaniček (2006).

Summarizing, we believe that the main role of geometric methods in Geodesy is in trying to find suitable coordinates to express observation equations as well as physical laws, in particular supporting the analysis of geodetic boundary value problems.

In such a process one important feature is the capability of approximating the exact equations with a linearized version. This always leaves open the question of the degree of approximation so obtained and of the limits of validity of the solutions developed when a priori bounds on the errors are imposed. Just as examples take two classical questions:

- how large can be a leveling network if we treat leveling increments as orthometric height differentials

$$\delta n \cong dH\,, \quad (3)$$

before the systematic effects implicit in such an equation exceed the measurement noise propagated into a compensation practice?

- how accurate is the fundamental equation of Physical Geodesy (2) if we use it to estimate local geoid undulations?

Questions like these are often met in geodetic literature (Seitz and Heck 1993; Sansò and Sideris 2013; Vaniček and Krakiwsky 1986), but they are often treated by a rough (sometimes very rough) estimate of the errors, rather than with a systematic use of simulations and numerical analysis.

So we can formulate the first open question:

OQ1) The aspect of, so to say, numerical geometry is still an area requiring further investigation by geodesists. In particular, it would be nice to know, in different geological settings, how large can be the deflection of the vertical, the curvature of plumblines and equipotential surfaces on the Earth surface and inside the topographic masses, by a systematic simulation work.

4 Geodetic Boundary Value Problems (GBVP)

The history of GBVP's starts in the middle of nineteenth century with the miracle of the formulation first and the solution then of a problem relating gravity anomalies to the anomalous potential as proposed by Stokes (1867).

The miracle is that, notwithstanding the many approximations implied by moving the gravity values observed on the Earth surface down to the geoid, approximating the geoid with a sphere and the direction of the vertical with the radial direction, one can arrive at something physically meaningful, i.e., to estimate the geoid undulation with an error of the order of 1–10 m, over 100 m of the original signal.

After one century of chewing forth and back corrections due to topographic masses, more or less complicated reductions, with the notable contribution by Helmert (1884), we arrive at the middle of the twentieth century with the cornerstone work by Molodensky et al. (1960), where the gravimetric GBVP was recognized as a free-boundary BVP and an approach to its linearization was attempted.

The exact linearization, in the modern meaning of the Fréchet differential of a non-linear problem, has arrived with T. Krarup in his famous letters on Molodensky's problem (Krarup 2006a).

The first analysis of the fully non-linear Molodensky problem has been performed by the great mathematician (Hörmander 1976). Other more advanced results can be found in the review paper (Sansò 2018). What is important here is to underline that meanwhile it was understood that with an excess of carelessness two different problems were treated in geodetic literature as if they were the same, namely the vector and the scalar Molodensky problem (Sansò 2018). In the former the data are the potential $W|_S$, and the gravity vector $\boldsymbol{g}|_S$, given on the unknown surface of the Earth and the unknowns are S itself (i.e., the three Cartesian coordinates of its points) as well as the potential W in the outer space. In the latter data are $W|_S$ and $g|_S$, with g the modulus of the vector $\boldsymbol{g}$ and unknowns are again W, but S is modelled only as height h above points on the ellipsoid, of known (horizontal) coordinates.

Nowadays it is the scalar Molodensky problem that is recognized as "the" GBVP and its linearized form as the linearized Molodensky problem.

The conceptual role of these (and other) BVP's in Geodesy, is in that we need a clear theoretical frame, defining a limit to which we want that our approximate estimates of the gravity field would ultimately tend.

The non-linear and the linear problems need to be analyzed with different tools, so the regularity results obtained are different. Basically we obtain existence uniqueness and stability of the non-linear problem for data belonging to Hölder–Banach spaces, $W \in H_{1,\lambda}, g \in H_{1,\lambda}$, (see F. Sansò in Sansò and Rummel 1997 and Sansò 2018), and in this case the unknown boundary S belongs to $H_{1,\lambda}$, too.

It is maybe worth mentioning already here that results about the linearized Molodensky problem hold under much less restrictive conditions. Yet the theorem of existence, uniqueness and stability of the solution of the non-linear problem guarantees that under the above conditions of Hölder regularity a converging iterative solution of the non-linear problem could be constructed by a sequence of solutions of the linearized problem.

As for the linear Molodensky problem, we need to be more specific. For our purposes the following formulation is convenient. Assume a suitable approximate surface $\tilde{S}$ to the Earth's surface S is given, e.g. the Marussi telluroid (cf. Sansò and Sideris 2013); S and $\tilde{S}$ uniquely separate the Euclidean space $\mathbb{R}^3$ into the inner space B and $\tilde{B}$ and the outer space $\mathbb{R}^3\backslash(B \cup S)$, and $\mathbb{R}^3\backslash(\tilde{B} \cup \tilde{S})$, respectively. As usual, we let $\overline{B} = B \cup S, S = \partial B$. Moreover, we assume that $\tilde{S} = \partial\tilde{B}$ is a star-shaped surface with equation

$$x = \underbrace{R_{\tilde{S}}(\sigma)}_{=|x|}\,\sigma, \quad \sigma \in \mathbb{S}^2, \ \mathbb{S}^2 \text{ unit sphere in } \mathbb{R}^3. \tag{4}$$

Now we are looking for the anomalous potential T in the outer space $\mathbb{R}^3\backslash(\tilde{B} \cup \tilde{S})$ of $\tilde{S}$ satisfying the following BVP

$$\begin{cases} \Delta T = 0 & \text{in } \mathbb{R}^3\backslash(\tilde{B} \cup \tilde{S}), \\ B(T) = -\frac{\partial T}{\partial h} + \frac{1}{\gamma}\frac{\partial \gamma}{\partial h} T = \Delta g + & \\ + \sum\limits_{j=0}^{L} \sum\limits_{k=1}^{2j+1} a_{jk} \overline{\Psi}_{jk} = b & \text{on } \tilde{S} \\ T(x) = O\left(\frac{1}{|x|^{L+2}}\right), & |x| \to \infty. \end{cases} \tag{5}$$

The third equation in (5) says that we assume to know within the asymptotic development of T in terms of spherical harmonics, all the harmonic coefficients up to degree L; the second equation in (5) introduces suitable functions $\overline{\psi}_{jk}$, exactly to subtract to the free-air anomaly data, $\Delta g|_S$, the components of the harmonics up to degree L.

The problem (5) is then transformed into a perturbative form noticing that the boundary condition can be written as

$$rT' + 2T + (r\boldsymbol{\varepsilon}\cdot\nabla T - \eta T) = -R_{\tilde{S}}\,\Delta g + \sum_{j=0}^{L}\sum_{k=1}^{2j+1} a_{jk}\psi_{jk}, \tag{6}$$

where $R_{\tilde{S}}$ is the function defined in Eq. (4) and we have denoted with T' the radial derivative $\frac{\partial T}{\partial r}$. Furthermore,

$$\boldsymbol{\varepsilon} = \boldsymbol{\nu} - \boldsymbol{e}_r \qquad (\boldsymbol{\nu} \text{ ellipsoidal normal, } \boldsymbol{e}_r \text{ radial unit vector})$$

$$\eta = r\frac{1}{\gamma}\frac{\partial\gamma}{\partial h} + 2,$$

$$\psi_{jk} = -r\overline{\psi}_{jk}, \quad r = |\boldsymbol{x}|.$$

Equation (6) is in a perturbative form because, calling $\varepsilon_+ = \max|\boldsymbol{\varepsilon}| \sim e^2 (\cong 6.7\cdot 10^{-3})$, $\eta_+ = \max|\eta| \sim 2e^2$, one has

$$|DT| = |r\boldsymbol{\varepsilon}\cdot\nabla T - \eta T| \le r\varepsilon_+|\nabla T| + \eta_+|T|, \tag{7}$$

which is a "small" operator when T is assumed (as we do) to be at least in the Sobolev space $H^1(\mathbb{S}^2)$ on the boundary, namely, using an unconventional norm definition,

$$\| T \|^2_{H^1(\mathbb{S}^2)} = \int_{\mathbb{S}^2} (|\nabla T|^2_S)(\sigma)\, dS(\sigma) \qquad < \infty. \tag{8}$$

where dS is the surface element. We would like to find a solution $T \in H^1(\mathbb{S}^2)$ under the assumption that the data $f = R_{\tilde{S}}\,\Delta g$ belong to $L^2(\mathbb{S}^2) = H^0(\mathbb{S}^2)$, i.e.,

$$\| f \|^2_{H^0} = \int_{\mathbb{S}^2} |f(\sigma)|^2\, dS(\sigma) \qquad < \infty. \tag{9}$$

To specify completely the problem we need to say what are the functions $\{\psi_{jk}\}$. They are defined as follows: let u be any harmonic function in $\mathbb{R}^3\backslash(\tilde{B}\cup\tilde{S})$, $u \in H^0$; moreover, let $\overline{R}$ be such that the sphere of radius $\overline{R}$, i.e., $\mathbb{S}^2_{\overline{R}}$, is totally contained in $\mathbb{R}^3\backslash(\tilde{B}\cup\tilde{S})$

$$\mathbb{S}^2_{\overline{R}} \subset \mathbb{R}^3\backslash(\tilde{B}\cup\tilde{S}), \tag{10}$$

then we can define the harmonic coefficients

$$\overline{u}_{jk} = \frac{1}{4\pi}\int_{\mathbb{S}^2} u(\overline{R},\sigma)\, Y_{jk}(\sigma)\, dS(\sigma), \tag{11}$$

where $\{Y_{jk}\}$ constitutes an L^2-orthonormal system of spherical harmonics with respect to $\mathbb{S}^2$. It is obvious that $\overline{u}_{jk}$ define linear functionals of $u \in H^0$ and it is not difficult to prove that they are bounded linear functionals (Sansò 2018).

So, by the Riesz theorem, there must be $\{\psi_{jk}\}$ such that

$$\langle \psi_{jk}, u\rangle_{H^0} = \overline{u}_{jk}. \tag{12}$$

One can easily prove that $\{\psi_{jk}\}$ are linearly independent and, in fact, they form a bi-orthogonal system with the solid (outer) spherical harmonics

$$\left\{\frac{1}{4\pi}\left(\frac{\overline{R}}{r}\right)^{j+1} Y_{jk}\right\}, \tag{13}$$

namely

$$\frac{1}{4\pi}\left\langle \psi_{jk}, \left(\frac{\overline{R}}{R_{\tilde{S}}}\right)^{\ell+1} Y_{\ell m}\right\rangle_{H^0} = \delta_{j\ell}\,\delta_{km}, \tag{14}$$

as one can see by combining (11) and (12).

At this point one can proceed with the analysis by a perturbative approach, namely first neglecting the term DT in (6), what leaves us with the so-called *simple Molodensky problem*, and then showing that, if the perturbation is small enough, then also the original problem has one and only one solution in $H^1(\mathbb{S}^2)$. So we are left with the following Proposition, the proof of which can be found in Sansò (2018).

Proposition 1 *Let $\tilde{S}$ be star-shaped with $R_{\tilde{S}}$ be a Lipschitz function given by (4), and $f = R_{\tilde{S}}\Delta g \in H^0(\mathbb{S}^2)$, then there is one and only one solution of (5) if the maximum inclination of $\tilde{S}$ with respect to $\boldsymbol{e}_r$, $I_+ (\cos I = \boldsymbol{e}_r\cdot\boldsymbol{n}$, with $\boldsymbol{n}$ normal to $\tilde{S}$), satisfies the inequality*

$$I_+ \le \arccos\left(\frac{3.0662(0.0044 + \frac{4}{2L+5})}{-0.0068 + \sqrt{0.0068^2 + 3.0662(0.0044 + \frac{4}{2L+5})}}\right). \tag{15}$$

In (15) we have used the estimates

$$\varepsilon_+ \sim 0.0067, \quad \eta_+ = 0.0136,$$

and

$$\frac{\max R}{\min R} \sim 1.0043.$$

The result seems to be acceptable, if we consider that already fixing $L = 24$, (15) says that the maximum inclination I_+ should not be larger than $\sim 60°$. As a matter of fact, we know quite well the harmonic coefficients of T up to degree 24 and so the answer given by the above proposition is at least realistic. Yet it is not satisfactory on a theoretical ground. In fact, it is the L^2-theory developed that forces us to put the constraint (15) on the maximum inclination I_+. On the other hand it would be much nicer if we could develop an L^p-theory in such a way that a constraint of the form

$$\frac{1}{4\pi}\int_{\mathbb{S}^2}\frac{1}{(\cos I(\sigma))^\alpha}\, dS(\sigma) = O(1) \tag{16}$$

could represent for some $\alpha > 1$ a sufficient condition for existence and uniqueness of the solution. In fact, we know that $\cos I$ can become small in rugged parts of $\tilde{S}$, which however cover a small area of the globe.

So the following open question becomes natural:

OQ2) Is it possible to develop an L^p quantitative theory of the linearized Molodensky problem in such a way that existence and uniqueness of the solution would depend on a sufficient condition of the form (16), for some exponent $\alpha > 1$?

Another interesting remark is as follows. In the very successful estimation of global gravity models, in the form of a truncated series of spherical harmonics, (see Pavlis 2103), complete up to a high degree (at present 2159, but probably up to a higher degree in next future), an original method has been devised which has been called the *change of boundary method* and formalized in Sansò and Sideris (2013).

In short, the method consists in pullbacking the data from the present boundary to an internal surface, e.g. a (Runge) Bjerhammar sphere or the Earth's ellipsoid; solving the BVP for such a surface and computing the residuals (errors) at the level of the current surface; pullbacking the residual to the internal surface and so forth, iteratively.

The method is not standard in Mathematics and so it requires some analysis as for the question of its convergence. A perturbative argument has been recently presented in Sansò and Sideris (2017), yet the question in its general form is still open:

OQ3) Is the change of boundary method convergent, at least for models of finite maximum degree?

In the geodetic approach to boundary value problems, as in other scientific fields, the purely mathematical aspect is met by the geodesist's need for sufficiently handling a great number of applications, which are generally more complicated than the typical cases in which the physicists verified this theory. It is, for instance, certainly not enough for geodetic boundary value problems to be able to calculate the "geopotential" for spheres and ellipsoids, but more complicated bodies (e.g., the geoid, telluroid, and/or the real Earth provided by GNSS-technology) must be taken into account. The boundary value problem, which provides a mathematical apparatus of these questions, therefore, is not only a purely mathematical subtlety, but also a problem of great practical importance.

As an example, in the theory of harmonic functions related to "potato-like" surfaces (such as S or $\tilde{S}$), a result first motivated by Runge (1885) in complex analysis and later generalized, e.g., by Walsh (1929), Vekua (1953), and Hörmander (1976) to potential theory in three-dimensional Euclidean space $\mathbb{R}^3$ is of basic interest for geodetically relevant questions see, e.g., Bjerhammar (1962), Krarup (1969), Moritz (1977, 1989, 2015), Sansò (1982), Grafarend (2015), and the references therein. In fact, the "Runge approach" as proposed in Freeden (1980), Freeden and Mayer (2006) may be used to solve the linear Molodensky problem in the following way:

Let $\{H_n\}$ be a system of harmonic functions in the outer space of $\mathbb{S}^2_R$ with L^2-completeness property of $\{H_n|_{\mathbb{S}^2_R}\}$ on a Bjerhammar (Runge) sphere $\mathbb{S}^2_R$ (i.e., the outer space of $\mathbb{S}^2_R$ totally contains $\mathbb{R}^3\backslash(\tilde{B}\cup\tilde{S})$ and $\text{dist}(\mathbb{S}^2_R, \tilde{S}) > 0$). As examples, conventional solid (outer) spherical harmonics, outer ellipsoidal harmonics, certain buried mass point systems, harmonic spline and spline-wavelet systems may be chosen (for more details see, e.g., Freeden and Gerhards 2013). By orthonormalizing $\{BH_n\}$ on the boundary surface $\tilde{S} = \partial\tilde{B}$ we obtain a new system $\{H_n^*\}$ of harmonic functions in the outer space of $\mathbb{S}^2_R$ with $\{BH_n^*\}$ complete and orthonormal on the boundary surface $\tilde{S} = \partial\tilde{B}$. As a consequence, we get

$$\lim_{N\to\infty}\|\sum_{n=0}^{N}\langle BH_n^*, b\rangle_{L^2(\tilde{S})}BH_n^* - b\|_{L^2(\tilde{S})} = 0 \tag{17}$$

and

$$\lim_{N\to\infty}\|\sum_{n=0}^{N}\langle BH_n^*, b\rangle_{L^2(\tilde{S})}H_n^* - T\|_{C^{(0)}(\Gamma)} = 0 \tag{18}$$

for every $\Gamma \subset \mathbb{R}^3\backslash(\tilde{B}\cup\tilde{S})$ with positive distance of $\partial\Gamma$ from $\tilde{S}$. In other words, L^2-convergence to the boundary values on the boundary values on the boundary $\tilde{S}$ implies uniform convergence to the solution, in particular on compact subdomains of the outer space $\mathbb{R}^3\backslash(\tilde{B}\cup\tilde{S})$.

It is remarkable that the Runge approach allows the calculation of gravitational quantities such as the disturbing potential T with respect to arbitrary telluroids, just by suitably operating with spherically based equipment such as multipoles (i.e., solid (outer) spherical harmonics). We are able to avoid, e.g., ellipsoidal outer harmonics, which are much more difficult to handle numerically. The price to be paid are multivariate definite integration procedures over surfaces $\tilde{S}$ (cf. Freeden and Gutting 2017) and an orthonormalization process, for which, however, powerful rules are available today (see, e.g., Freeden and Gutting 2017 and the references therein). Nonetheless, the following open questions arise naturally:

OQ4) Is it possible to get a better numerical efficiency by combining appropriate trial systems, for example, to use outer spherical harmonics for a trend solution and to continue with mass points, harmonic splines, harmonic spline-wavelets for the modeling of more detailed signatures? What is the right upper limit N to stop the whole solution process?

Many other items in the geodetic BVP theory would merit attention, yet it is time to pass to another argument.

5 Probability Theory and Statistics

This is a very large branch of Mathematics, with an origin again historically related to Geodesy. May it be sufficient to mention the Gaussian distribution and its relation to the error theory and the subsequent development of the least squares method. Nor, we could refrain, in this context to mention the name of Markov regarding the problem of optimal estimation and all the large geodetic literature that has followed its trail, from the germinal paper by H. Moritz on "least squares" (Moritz 1972), to the comprehensive book of Grafarend and Awange (2012) on linear and non-linear models.

Given the widespread of this matter, we will concentrate on three arguments only, leading to interesting open questions. The three items are:

(a) *generalized random fields and BVP's,*
(b) *collocation theory and non-isotropic random fields,*
(c) *integer estimation theory.*

(a) Generalized Random Fields and BVP's We have uniquely identified the gravity field, exterior to the Earth's surface S, by means of observations performed on it, e.g. gravity and potential observations. This has led us to the formulation of BVP's, where boundary data are derived from measurements and therefore they must be considered as samples from a random field (RF). When the errors are small, we can work in the regime of the linearized theory and in general we have to solve a problem of the form

$$Bu = v \quad \text{on } \tilde{S} \tag{19}$$

for instance with $v \in L^2(S)$ and $u \in H^1(S)$ (u harmonic in the outer space of S) as for the linear Molodensky problem treated in Sect. 4. Notice that we skip the part of boundary equation depending on the unknown constants $\{a_{ik}\}$ because they are finite in number and clearly inessential to the present reasoning.

Now the point is that if instead of v we have an "observed" RF

$$v_0 = v + \nu \tag{20}$$

with ν a field of random disturbances, the problem (19) has to be substituted by

$$B\widehat{u} = v_0 = v + \nu \tag{21}$$

and its solution, whatever it is, will be a random estimator of u.

When ν is a sufficiently smooth RF, so that its samples belong to H^0 almost surely, then we know what it means to solve (21). In fact, if we call M the Molodensky solver, namely

$$M = B^{-1}\,;\; H^0(S) \to H^1(S) \tag{22}$$

we can simply put

$$\widehat{u} = M v_0 = M v + M \nu = u + \varepsilon \tag{23}$$

and all that has a clear meaning.

In particular, we can sample ω from an abstract probability space $(\Omega, \mathcal{A}, P)$ and build the chain

$$\begin{array}{ccccc} \omega & \to & \nu(\omega) & \to & \widehat{u} = u + M\nu(\omega) \\ \text{sample} & & \text{sample noise} & & \text{sample solution} \end{array}$$

and from the statement $\nu(\omega) \in H^0(S)$ with probability 1, we deduce that $\widehat{u} \in H^1(S)$ with probability 1, namely it is a sample from a RF with realization in $H^1(S)$.

All that is elementary, yet the reasoning is not able to cover maybe the most important case of noise, namely the Wiener white noise (WWN). In other words, how can we define a solution of (8) when ν is a WWN, which is well-known to have sample fields that do not belong to $H^0(S)$?

A natural idea, when we have to deal with an equation with a very irregular known term, is to put it in a weak form, so that the irregularity can be let off in the coupling with a smooth test function.

These are basically the ideas that led to the definition in particular of the Wiener integral (Lamperti 1977), and more generally to the concept of generalized random field; on this argument our reference is the work of Rozanov (1998), a great mathematician of Kolmogorov's school that has contributed to geodetic literature, too (based on these concepts, the field of stochastic differential equations has also seen tremendous progress in Mathematics during the last years, see, e.g., Grothaus and Raskop 2018 as a point of entry).

Definition 1 Let X be Hilbert space with elements $\{x\}$ and norm $\| x \|_X$; let $\mathcal{P} = (\Omega, \mathcal{A}, P)$ be a probability space and $\mathcal{L}^2(\Omega)$ the Hilbert space of random variables $\{f(\omega)\}$, defined on $\mathcal{P}$ with norm

$$\| f \|^2_{\mathcal{L}^2(\Omega)} = E\{f(\omega)^2\}\,; \tag{24}$$

a generalized random field F on X is a bounded linear operator from X into $\mathcal{L}^2(\Omega)$

$$F(\omega; x) = f(\omega) \tag{25}$$

$$\| F(\omega; x) \|_{\mathcal{L}^2(\Omega)} = \| f(\omega) \|_{\mathcal{L}^2(\Omega)} \leq c \| x \|_X ; \tag{26}$$

to sightly simplify matters, we will assume that each of the random variables $\{F(\omega; x)\}$ has zero mean.

From Definition 1 the following proposition can be easily derived.

Proposition 2 *There is a bounded, positive, selfadjoint operator $C : X \to X$, such that*

$$\| F(\omega; x) \|^2_{\mathcal{L}(\Omega)} = E\{F(\omega; x)^2\} = \langle x, Cx\rangle_X ; \tag{27}$$

We shall always assume that C is strictly positive so that the statement that F is into $\mathcal{L}^2(\Omega)$ in the above Definition is justified; in fact, in this case from $F(\omega, x) = 0$, we deduce that $x = 0$.

Remark 1 The GRF, F, in general cannot be onto $\mathcal{L}^2(\Omega)$; rather we can have that the set

$$\mathcal{L}^2_X = \{F(\omega; x),\ x \in X\} \tag{28}$$

is a closed subspace of $\mathcal{L}^2(\Omega)$ if $C \geq c_0 I > 0$, if suitable conditions are satisfied.

Example 1 A WWN on S is a GRF, $N(x)$, on $L^2(S)$ with covariance operator

$$C = \sigma^2 I ; \tag{29}$$

namely $N(x)$ is a WWN if

$$E\{N(x)^2\} = \sigma^2 \| x \|^2_{L^2(S)} = \sigma^2 \int_{t\in S} x^2(t)\, dS(t)\,, \tag{30}$$

with t some coordinate system on S for the purpose of integration.

Notation 1 *Although in general there is no element $F(\omega) \in X$, that can represent (25) as a scalar product with x, yet by analogy with integral kernels, we write*

$$F(\omega; x) = \langle F(\omega), x\rangle_X ; \tag{31}$$

we stress that (31) is just a symbol and does not mean that $F(\omega) \in X$.

For instance, in case of a WWN on S, we write

$$N(\omega; x) = \langle N(\omega), x\rangle_{L^2(S)} = \int_{t\in S} N(\omega; t)\, x(t) dS(t). \tag{32}$$

The first integral in (32) is precisely a Wiener integral, what does not imply that $N(\omega; \cdot) \in L^2(S)$.

Remark 2 Let us notice that, as proved by the example of WWN, the GRF, $F(\omega; x)$, in general has not realizations in X, in which case (31) could be written in the ordinary sense, sample by sample. Nevertheless, such a circumstance is not impossible. One can prove for instance that, if

$$Tr\ C < \infty \tag{33}$$

then the realizations of $F(\omega, t)$ are in X with $P = 1$, so that (31) becomes almost surely a veritable scalar product, and we can write

$$F(\omega; x) = \langle F(\omega, t), x(t)\rangle_X\,. \tag{34}$$

In other words, (31) is a real generalization to GRF's of (34), which holds for ordinary RF's with realizations in X.

We follow the same scheme of the above remark, to generalize the notion of the application of an operator $A : X \to Y$ to a GRF on X.

So let A be a given continuous linear operator, $X \to Y$, which is into and onto, so that there is a continuous linear operator $A^{-1} : Y \to X$ (see Yosida 1980). This implies that the equation

$$y = Ax\,,\ y \in Y \tag{35}$$

has one and only one solution $x \in X$,

$$x = A^{-1} y\,,\ x \in X\,. \tag{36}$$

Now assume that a GRF, U, is given on X, which is also an ordinary RF with realizations in X; this means that $\| u(\omega) \|^2_X$ is bounded a.s., and we further assume that its average is bounded too, which is equivalent to assume that, for U, (33) holds. Then we can define ω by ω

$$V(\omega) = AU(\omega)\,,\ V \in Y \text{ a.s.} \tag{37}$$

Notice that we can subsequently write for all $y \in Y$

$$\langle V(\omega), y\rangle_Y = \langle AU(\omega), y\rangle_Y = \langle U(\omega), A^T y\rangle_X ; \tag{38}$$

since A^T is a bounded operator $Y \to X$, we see that in fact (37) generates a GRF, V, on Y because

$$E\{\langle V(\omega), y\rangle^2_Y\} = E\{\langle U(\omega), A^T y\rangle^2_X\} \leq \tag{39}$$
$$\leq E\{\| U(\omega) \|^2_X\} \| A^T y \|^2_X \leq c \| y \|^2_Y\,.$$

Therefore, with such smoother RF's, we can rewrite (37) in the form

$$V(\omega; y) = \langle V(\omega), y\rangle_Y = \langle U(\omega), A^T y\rangle_X = U(\omega; A^T y)\,. \tag{40}$$

We use (40) to define the application of the operator A to a GRF, V.

Definition 2 We say that, given U, a GRF on X, the GRF on Y, AU is defined by

$$V = AU \Leftrightarrow V(\omega; y) = U(\omega; A^T y) \tag{41}$$

for all $y \in Y$.

This allows to prove a Lemma of general equivalence between deterministic linear equations and corresponding stochastic linear equations, in the framework of GRF's theory.

Lemma 1 (The General Equivalence Lemma) *Let a linear deterministic equation be established in the form (35) with the operator A sending X into and onto Y; then, given any GRF, V on Y, we can find one and only one GRF, U on X, such that*

$$V = AU\,; \tag{42}$$

in fact, $U(\omega; x)$ *is given explicitly by*

$$U(\omega; x) = V(\omega; (A^T)^{-1}x)\,. \tag{43}$$

Proof The proof is elementary and amounts to verifying that (43) is equivalent to (41). □

At this point we have available a general rule that gives a meaning for instance to the solution of stochastic BVP's even when boundary data contain a WWN. This approach, thanks to its generality supersedes previous ad hoc analyses like in Rozanov and Sansò (1997), and Sansò and Rozanov (2003).

So, we could say that in this branch Geodesy and Mathematics have been working hand in hand, with good results. But this is just the beginning of the story, because our problems are mostly non-linear and, even more, our GBVP is a free BVP. Therefore, a hard difficult theoretical open question is now in front of us.

OQ5) Is it possible to define a stochastic BVP with random boundary? How can we find solutions?

Just to show that such a question is difficult, but not impossible, we build a very elementary example.

Example 2 We assume that the family of possible boundaries is just that of spheres $\mathbb{S}^2_R$, with R a random variable, distributed on the positive axis. Furthermore, we consider a family of purely radial potentials, namely

$$u = \frac{M}{r}$$

with M again a positive RV, independent of R. The "geodetic" BVP in this case would be: Given $W = \frac{M}{R}$ and $g = \frac{M}{R^2}$ (obviously two RV's), to find R and the potential u agreeing with W on $\mathbb{S}^2_R$. In such an elementary context the solution is just algebraic, namely

$$R = \frac{W}{g}\,,$$

$$M = \frac{W^2}{g}\,,$$

$$u = \frac{W^2}{g}\frac{1}{r}\,.$$

Notice that, due to the overwhelming simplicity of the example, one characteristic of the solution is lost. Namely, the third of the above relations should be more correctly written in a conditional form, as

$$u|_{r \geq R} = \frac{W^2}{g}\frac{1}{r}\,;$$

in fact the u so found has to be harmonic only outside the random $\mathbb{S}^2_R$. In this particular case $r \mapsto \frac{1}{r}$ is in any way harmonic down to the origin and so the above statement has little relevance, nevertheless in a realistic case where the surface S is not a sphere, and u cannot be continued to the origin, the RF u should be harmonic conditional on S. In other words, in general there is not a unique linear space to which u has to belong.

(b) Collocation Theory and Non-isotropic RF's Collocation theory is certainly a notable contribution of Geodesy to mathematical approximation theory, in the framework of harmonic fields. This theory has been developed from the Sixties of the last century, stretching from two different approaches: one stochastic, by Moritz (1989), the other deterministic, by Krarup (1969). In some way the two approaches reflect respectively the approach of Kolmogorov (stochastic) and that of N. Wiener (deterministic) to the problem of filtering and prediction of stochastic processes, with the due differences because in collocation theory we treat fields and not functions of time, and in addition our fields have to possess the harmonic property in the set where measurements can be performed and predictions are required.

This is the reason why in Geodesy we often speak of the Wiener–Kolmogorov theory. The two approaches have been proved to be equivalent on a rigorous mathematical ground (see Sansò 1986). The basic hypothesis is that the anomalous potential T is a GRF on a Hilbert space, $H(O)$, of functions harmonic in the outer domain O; T is assumed to have zero mean, $E\{T\} = 0$. This means that, for all $h \in H$, which we can view as a linear functional on H itself by the Riesz representation formula

$$L(k) = \langle h, k\rangle_H , \tag{44}$$

we can as well define a RV $\in \mathcal{L}^2(\Omega)$ by (recall the discussion on notation and Eq. (31))

$$L(T) = \langle T, h\rangle_H = T(h) \in \mathcal{L}^2(\Omega) , \tag{45}$$

and a covariance operator C such that

$$E\{L(T)^2\} = E\{T(h)^2\} = < h, Ch >_H . \tag{46}$$

The problem we have to solve is then as follows: Given a certain number of input data with (linear) observation equations

$$\begin{aligned} &V = \boldsymbol{L}(T) = T(\boldsymbol{h}) = \langle T, \boldsymbol{h}\rangle_H , \\ &(\boldsymbol{L} = \{L_i\},\ \boldsymbol{h} = \{h_i\},\ i = 1, \ldots, N, \\ &\text{observation functionals}) \end{aligned} \tag{47}$$

we want to predict another functional

$$U = L_0(T) = T(h_0) = \langle T, h_0\rangle_H . \tag{48}$$

The RV are all in $\mathcal{L}^2(\Omega)$ and their covariance–crosscovariance structure is given by

$$\begin{aligned} &C_{UU} = \langle h_0, Ch_0\rangle_H \\ &C_{UV} = \langle h_0, C\boldsymbol{h}^T\rangle_H ,\ C_{VV} = \langle \boldsymbol{h}, C\boldsymbol{h}^T\rangle_H . \end{aligned} \tag{49}$$

The optimal linear predictor of U given V is just the $\mathcal{L}^2(\Omega)$ orthogonal projection of U onto the linear space spanned by the vector V, i.e. $\{\boldsymbol{\lambda}^T V ; \boldsymbol{\lambda} \in R^N\}$. This is well-known from the theory of ordinary linear regression and it is given by

$$\widehat{U} = C_{UV} C_{VV}^{-1} V \tag{50}$$

with a prediction error $\mathcal{E}^2$

$$\mathcal{E}^2 = C_{UU} - C_{UV} C_{VV}^{-1} C_{VU} . \tag{51}$$

Now assume that $H(O)$ has a reproducing kernel K, namely that for all $h \in H(O)$

$$h(t) = \langle K(t, \cdot), h\rangle_H ; \tag{52}$$

this certainly happens with Hilbert spaces of harmonic functions for all $t \in O$. Yet, if the traces on S of functions in $H(O)$ are regular enough, (52) has to hold for all $t \in S$. Let us recall that by definition a reproducing kernel $K(t, \cdot) \in H(O)$, for all $t \in O$, and it is bounded, namely

$$\begin{aligned} |K(t, t')|^2 &= |\langle K(t, \cdot), K(t', \cdot)\rangle_H|^2 \leq \| K(t, \cdot) \|_H^2 \cdot \| K(t', \cdot) \|_H^2 \\ &= K(t, t)\, K(t', t') \leq a^2 . \end{aligned} \tag{53}$$

In this case $T(\omega; h)$ has pointwise "values", namely

$$T(\omega; K(t, \cdot)) = T(\omega; t) \in \mathcal{L}^2(\Omega) \tag{54}$$

because

$$\begin{aligned} \| T(\omega; K(t, \cdot)) \|_{\mathcal{L}^2}^2 &= \langle K(t, \cdot) , CK(t, \cdot)\rangle_{H^2(O)} \\ &\leq c\ \langle K(t, \cdot) , K(t, \cdot)\rangle = c\ K(t, t) ; \end{aligned} \tag{55}$$

given (54) any sample from $T(\omega; t)$ is almost surely finite, and so we could say that T is an ordinary RF. When this happens, we can compute a very important function which is called the *covariance function of the RF*, T; namely

$$\begin{aligned} C(t, t') &= E\{T(\omega; t) T(\omega; t')\} \\ &= E\{T(\omega; K(t, \cdot)) T(\omega; K(t', \cdot)\} \\ &= \langle K(t, \cdot), CK(t', \cdot)\rangle_H . \end{aligned} \tag{56}$$

It is clear that $C(t, t')$ uniquely identifies the operator C. In particular, with the help of the covariance function we can conveniently rewrite (50), adopting Krarup's notation; namely for any two functionals L_i, L_k we write

$$\begin{aligned} (L_i)_t\{(L_k)_{t'} C(t, t')\} &= E\{T(h_i) T(h_k)\} \\ &= E\{L_i(T) L_k(T)\} = C(L_i, L_k) . \end{aligned} \tag{57}$$

Then we have

$$\widehat{U} = \widehat{L_0(T)} = \sum_{i,k=1}^{N} C(L_0, L_i)\{C(L_i L_k)\}^{(-1)} L_k(T) ; \tag{58}$$

similarly we can rewrite (51) as

$$\mathcal{E}^2 = C(L_0, L_0) - \sum_{i,k=1}^{N} C(L_0, L_i)\{C(L_i, L_k)\}^{(-1)} C(L_k, L_0)$$

The above formulas are suitably modified when we have observational models containing a measurement noise (see Sansò and Sideris 2013), which is analogous to smoothing methods in spline theory (see. e.g., Freeden 1981).

Up to here the linear prediction problem seems to be solved by a computational algorithm; yet a crucial point is that we assumed to know C, namely the covariance function $C(t,t')$. We know that $C(t,t')$ has to satisfy certain properties: namely it has to be continuous, symmetric and positive definite, i.e. all matrices of the kind $\{C(L_i, L_k)\}$ have to be positive definite for all choices of (admissible) functionals $\{L_i,\ i = 1 \ldots N\}$ and for all N.

Furthermore, as one can see from the definition (56), $C(t,t')$ has to be harmonic in $t, t' \in \Omega$. But the specific shape of C depends on T, of which we have at most one realization, in fact even less, only a certain finite number of observations. We are here in front of the same dilemma that in signal analysis for stochastic processes has been solved by the so-called hypothesis of stationarity of the signal. This is basically the assumption that the signal is self-similar under the group of translations of the time axis. The analogous assumption used in Geodesy is that T has to be self-similar under the group of rotations; namely the stochastic parameter ω in this case is just constituted by proper rotations of R^3 and the probability model is a uniform distribution on the rotation group (see Moritz 1989 and Sansò 1986).

This forces the harmonicity domain to be the exterior of a sphere, because the domain has to be invariant too under rotations. Furthermore, it is not difficult to see that $C(t,t')$ has to assume a peculiar form, namely

$$C(t,t') = C(\psi_{t,t'}, r_t, r_{t'}) \tag{59}$$

where $r_t, r_{t'}$ are the distances of t, t' from the origin and $\psi_{t,t'}$ is the spherical angle between the directions of t, t'. All such hypotheses can be summarized by saying that we assume T to be an *isotropic* RF.

When we have enough data reduced to the same sphere, we have practical formulas to estimate discrete values of what is called the *empirical covariance function*, which is then interpolated by a suitable model taken from a family of positive definite functions (Tscherning and Rapp 1974; Tscherning 2013).

The fit of this interpolation is usually not very tight, nevertheless we have a theorem showing that there is a low sensitivity of the collocation solution (58) to estimation errors in the covariance, so the method works and has a wide range of applications, for instance in estimating geoid undulations from gravity anomalies (see Tscherning 2013). Nevertheless, the actual anomalous gravity field of the Earth is not isotropic. There are areas like the Himalaya, the Andes, the Alps or oceanic subduction zones that have a heavy gravitational signature, drastically different from surroundings, so that the idea of a statistical similarity moving them randomly on the sphere does not carry at all. Therefore the above theory is usually applied only on reduced parts of the sphere and by first applying a number of reductions to the data, for instance subtracting the influence of topographic masses (see Sansò and Sideris 2013).

Nevertheless, take the case of two neighbouring areas of size 10° × 10°, such that for each of them a reasonable isotropic covariance function can be estimated, so that the above collocation formulas give good results, but the two covariances are different from one another. At this point, the prediction of the field along the border will display an ugly signature.

Even more, if we have many areas close one to the other, for which individual covariances are worked out, the prediction of the field will have a very patchy pattern which is clearly an artifact of the algorithm rather than a physical property.

So the question arises as to whether it is possible to create a unique covariance model, keeping the fundamental positive definiteness property, that however is changing form area to area. In particular, the amplitude and the width (correlation length) of the covariance function are slowly but continuously changing, and the problem arises on how to perform the interpolation for such models.

Some work has been done for general RF in $\mathbb{R}^1$ and $\mathbb{R}^2$ (see Migliaccio et al. 1998; Darbeheshti and Featherstone 2009), but disregarding completely the harmonicity property.

This is an interesting open question, concerning non-isotropic harmonic RF's.

OQ6) Is it possible to find a non-isotropic covariance model with slowly varying amplitude and correlation length, still keeping the harmonicity of the covariance? Is it possible that maybe the case of multiresolution functions (spherical wavelets, as presented in Freeden et al. 2018a) could solve this problem?

(c) Least Squares with Integer Parameters Statistics is a scientific area where Mathematics and Geodesy have got a parallel and entangled development, with Geodesy from times to times surpassing Mathematics. Indeed this is not surprising, given that Geodesy is a science so heavily based on measurements.

This statement certainly applies to least squares theory, in particular regarding the rank deficiency problem, with the implied concept of generalized inverse of matrices, proposed by Bjerhammar (1973) (on the basis of the work by Moore 1935 and Penrose 1955), leading to an impressive mathematical monograph by Nashed (1976). Something similar could be maintained for the introduction of statistical inference in Geodesy, by the work of Baarda (1967).

Similarly one has to mention the research on rank deficient estimation problems which has formally engaged statisticians (Rao 1965), but is central to Geodesy where we want to estimate point coordinates and on the same time define a reference system (Grafarend and Awange 2012; Dermanis and Sansò 2018).

One particular field that Geodesy shares with e.m. signals transmission theory is the problem of estimating integer parameters that has become very important after the advent of GNSS tools for point positioning. In reality the problem was present much before in Geodesy in the application of e.m. equipments to measure 3D distances. Although that problem was solved by a satisfactory approach to the treatment of phase observations, yet the underlying idea has been inherited in the era of GNSS for the combination of code and phase observations, in a way that has generated a certain confusion that only in recent years is getting a clear theoretical setting.

We start with an example that can be used to show what is the effect of "fixing" an integer parameter on an estimation problem; the example can then be adapted to two different cases: that we are measuring a distance by a distance meter working with two wavelengths, or that we are measuring a pseudorange by a GNSS receiver, with one code and one phase observation.

Example 3 The mathematical model we have in mind is done of two observation equations

$$\begin{cases} Y_1 = x \quad\quad\quad +\varepsilon_1, \\ Y_2 = x - N\Lambda + \varepsilon_2 \end{cases} \tag{60}$$

by hypothesis we assume $\boldsymbol{\varepsilon} = [\varepsilon_1 \varepsilon_2]^T$ to be a Gaussian random variate, the error part, with zero mean and covariance

$$C_{\boldsymbol{\varepsilon}} = \begin{vmatrix} \sigma_1^2 & 0 \\ 0 & \sigma_2^2 \end{vmatrix} . \tag{61}$$

The model (60) seems to be linear in the couple (x, N) and in fact it is, if you assume $(x, N) \in \mathbb{R}^2$; nevertheless, if you restrict the domain of parameters to $\mathbb{R} \times \mathbb{Z}$ ($\mathbb{Z}$ the set of integers) the problem of estimating (x, N) is not anymore linear and it requires some stricter analysis. In fact, if one thinks of using a Maximum Likelihood (ML) criterion, due to the hypothesis of normality of $\boldsymbol{\varepsilon}$, the problem is translated into a least squares form, namely

$$\begin{cases} (\widehat{x}, \widehat{N}) = \arg\min Q(x, N\,;\, Y_1, Y_2), \\ Q(x, N\,;\, Y_1, Y_2) = \frac{(Y_1 - x)^2}{\sigma_1^2} + \frac{(Y_2 - x + N\Lambda)^2}{\sigma_2^2} . \end{cases} \tag{62}$$

If you consider for one moment the same problem as (62) on $\mathbb{R}^2$, you know that the solution is given by

$$\begin{cases} x_{\mathrm{fl}} = Y_1, \\ N_{\mathrm{fl}} = \frac{1}{\Lambda}(Y_1 - Y_2) \end{cases} \tag{63}$$

the so-called *floating* solution.

The covariance of this solution is given by

$$C_{\mathrm{fl}} = \begin{vmatrix} \sigma_1^2 & \frac{\sigma_1^2}{\Lambda} \\ \frac{\sigma_1^2}{\Lambda} & \frac{1}{\Lambda^2}(\sigma_1^2 + \sigma_2^2) \end{vmatrix} . \tag{64}$$

If we put

$$\xi = x_{\mathrm{fl}} - x,$$
$$\nu = N_{\mathrm{fl}} - N$$

and substitute in (62), we find after some boring but elementary algebra, that (62) can be rewritten as

$$Q(x, N\,;\, Y_1, Y_2) = Q(x_{\mathrm{fl}}, N_{\mathrm{fl}}\,;\, Y_1, Y_2) + \overline{Q}(\xi, \nu\,;\, Y_1, Y_2) \tag{65}$$

with

$$\overline{Q}(\xi, \nu\,;\, Y_1, Y_2) = [\xi \;\; \nu]\, C_{\mathrm{fl}}^{-1} \begin{bmatrix} \xi \\ \nu \end{bmatrix} . \tag{66}$$

In our example

$$Q(x_{\mathrm{fl}}, N_{\mathrm{fl}}\,;\, Y_1, Y_2) = 0$$

because $(x_{\mathrm{fl}}, N_{\mathrm{fl}})$ satisfy exactly (60), which is not the general case when we start from an overdetermined problem. Moreover, (65) is verified by no means by chance, but it is just a consequence of a well-known theorem of orthogonal decomposition of the residuals in least squares theory (Koch 1988). So our problem now becomes to find

$$\min_{\xi \in \mathbb{R}\,,\, \nu \in N_{\mathrm{fl}} + \mathbb{Z}} \overline{Q}(\xi, \nu\,;\, Y_1, Y_2) . \tag{67}$$

This can be achieved in two steps: First we find the value ξ_ν, which solves the minimum problem

$$\min_{\xi \in \mathbb{R}} \overline{Q}(\xi, \nu\,;\, Y_1, Y_2) ,$$

then we substitute ξ_ν in $\overline{Q}$ and we look for the minimum of this new function of ν. It results

$$\overline{Q}(\xi_\nu, \nu\,;\, Y_1, Y_2) = \frac{\Lambda^2}{\sigma_1^2 + \sigma_3^2} \nu^2 = \frac{(N_{\mathrm{fl}} - N)^2}{\sigma_{N\mathrm{fl}}^2} . \tag{68}$$

Now it is clear that the ML estimator, i.e. the minimizer of $\overline{Q}$, is just the integer closest to N_{fl}, we could call it the round-off of N_{fl}, namely

$$N_{ML} = N_r = [N_{\mathrm{fl}}] . \tag{69}$$

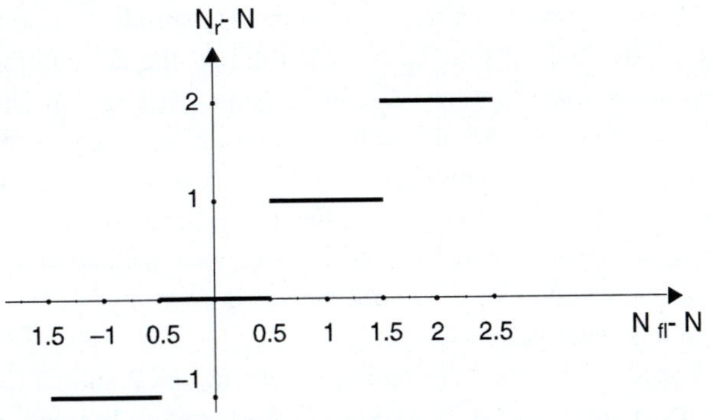

Fig. 1 The plot of the function $N_r(N_{f\ell} - N)$; note that $N_r(N_{\text{fl}}) = N_r(N_{\text{fl}} - N) + N$

To understand the result it is important to notice that N_{fl} is a real Gaussian variate with mean N and variance $\sigma^2_{N_{\text{fl}}}$, i.e. $N_{\text{fl}} = N + \sigma_{\text{fl}} Z$ with Z a standard normal; hence N_r is a discrete variable, the distribution of which can be easily derived considering that $N_r(N_{\text{fl}})$ is the staircase function represented in Fig. 1

In fact, if $N - 0.5 < N_{\text{fl}} < N + 0.5$ we have $N_r = N$, if $N + 0.5 < N_{\text{fl}} < N + 1.5$ we have $N_r = N + 1$, and so forth.

Accordingly, we have

$$P\{N_r = N\} = P\left\{|Z| < \frac{0.5}{\sigma_{\text{fl}}}\right\} = P_0 ,$$

$$P\{N_r = N \pm 1\} = P\left\{\frac{0.5}{\sigma_{\text{fl}}} < Z < \frac{1.5}{\sigma_{\text{fl}}}\right\} = P_1 .$$

As we see, the distribution of N_r is symmetric around N, i.e. N is its mean and the variance will be written as

$$\sigma^2_{N_r} = 2\sum_{k=1}^{+\infty} k^2 P_k = 2(1 \cdot P_1 + 4 \cdot P_2 + \ldots) \tag{70}$$

To conclude the example it is necessary to compute the estimator x_{ML} as well; this is more easily done by going back to (60), estimate $\widehat{x}(N)$ for every N and then substitute N with N_r. The result is

$$x_{ML} = \frac{\sigma_2^2 Y_1 + \sigma_1^2 (Y_2 + N_r \Lambda)}{\sigma_1^2 + \sigma_2^2} , \tag{71}$$

namely the weighted average of Y_1 and $Y_2 + N_r\Lambda$. It is easy to see that $E\{x_{ML}\} = x$ so that x_{ML} is an unbiased estimator of x; however, to compute the variance of x_{ML} is a more difficult exercise due to the non-linearity of the function $N_r(N_{\text{fl}}) = N_r(Y_1, Y_2)$, specially when P_0 is close to 1 but not so close as to decide that one can put $P_1 = P_2 = \ldots = 0$, i.e. ultimately $\sigma_{N_r} = 0$ (see (70)).

If we take this last choice we have the so-called "fixed" solution N_{fix}, where N now becomes just a constant so that

$$\sigma(Y_2 + N_{\text{fix}}\Lambda) = \sigma(Y_2) = \sigma_2 . \tag{72}$$

Before we apply the above reasoning to different numerical and physical situations we notice that, when $\sigma_2 \ll \sigma_1$, as we shall assume, (71) simplifies to

$$x_{ML} \sim Y_2 + N_r \Lambda . \tag{73}$$

Now we apply what we have learnt to two different contexts: distance meters, GNSS positioning.

Distance Meters Assume that by combining different wavelengths we can produce a first wave with wavelength longer than 1 km and a second one with wavelength $\Lambda = 10$ m. Then, if we know a priori that $x < 1$ km, the first equation of (60) can represent (with a very simplified model) the observation of the distance with

$$\sigma_1 = 10 \text{ cm} ,$$

while the second of (60) can represent the observation of the distance, with a hypothesized standard deviation

$$\sigma_2 = 1 \text{ mm} .$$

In this case (64) tells that

$$\sigma^2(N_{\text{fl}}) = \frac{1}{\Lambda^2}(\sigma_1^2 + \sigma_2^2) \cong \frac{\sigma_1^2}{\Lambda^2} = 10^{-4} ,$$

i.e., $\sigma(N_{\text{fl}}) = 10^{-2}$ and

$$P_0 = \{|Z| < \frac{0.5}{\sigma(N_{\text{fl}})}\} = P\{|Z| < 50\} .$$

By exploiting the approximation

$$P(Z > L) < e^{-\frac{L^2}{2}} P(Z > 0) = \frac{1}{2} e^{-\frac{L^2}{2}} \tag{74}$$

one immediately sees that

$$2(P_1 + P_2 + \ldots) < e^{-1250} \sim 10^{-542} ,$$

namely such a small number that justifies the assumption

$$P_0 \cong 1 , \ \sigma(N_{\text{fix}}\Lambda) \cong 0 .$$

The statement holds too with a more precise justification from (70). In this case the use of $N_r = N_{\text{fix}}$ and the conclusion

$$\sigma(x_{ML}) \cong \sigma(Y_2 + N_{\text{fix}}\Lambda) = 1 \text{ mm}$$

is quite reasonable.

However, if one repeats the same reasoning with the same σ_1 and σ_2 as above, but $\Lambda = 1$ m, one finds $\sigma_{N_{\text{fl}}} = 10^{-1}$ and

$$P\{|Z| > \frac{0.5}{N_{\text{fl}}}\} = P\{Z > 5\} < e^{-12.5} \cong 4 \cdot 10^{-6} .$$

With the help of (70) and $P_1 \sim 4 \cdot 10^{-6}$ one finds

$$\sigma(N_{\text{fl}}\Lambda) \sim 2 \text{ mm}$$

so that the simple relation $\sigma(x_{ML}) \sim \sigma_2$ cannot carry any more.

GNSS Positioning Again with a lot of simplification, one can take the first of (60) as a code observation, with for instance

$$\sigma_1 = 15 \text{ cm} ,$$

and the second as a phase observation, with

$$\sigma_2 = 1 \text{ mm} ;$$

moreover, for an L_1 career, $\Lambda \cong 20$ cm. In this case

$$\sigma(N_{\text{fl}}) = \frac{\sigma_1}{\Lambda} \cong 0.75$$

and one can more precisely compute

$$P_0 = 0.4972 ,$$

$$P_1 = 0.2286 ,$$

$$P_2 = 0.0200 ,$$

$$P_3 = 0.0013 ;$$

the remaining probabilities can be safely put to zero. In this case one reckons

$$\sigma(N_r\Lambda) \sim 18 \text{ cm}$$

showing that the term $N_r\Lambda$ is giving by far the largest contribution to $\sigma(x_{ML})$.

Even if we assume to have a very precise code, like the nowadays available L2C code, with an sqm of say $\sigma_1 \sim 33$ mm, by repeating the above computation one gets

$$\sigma(N_r\Lambda) \sim 93 \text{ mm} ,$$

showing that one cannot take N_r as constant and ignore the term $N_r\Lambda$ in the variance propagation.

The lesson we learn from this extremely simplistic example, is that specially in GNSS positioning, the assumption that the estimated integer N can be considered as constant is very dangerous and misleading, as for the accuracy of the estimate x_{ML}. Ultimately this is the effect of the choice of using a test (yes or no) to fix N; this leads, depending on the significance parameter α, to a correct conclusion many times, as for the estimate of the coordinates, but to a large error in them in a few cases.

The analysis of a more realistic, yet simplified, model for GNSS positioning can be written in the form of observation equations as

$$\boldsymbol{Y} = A\boldsymbol{x} + B\boldsymbol{N} + \boldsymbol{\varepsilon} , \tag{75}$$

with $\boldsymbol{x} \in \mathbb{R}^n$ a vector of coordinates and other parameters and $\boldsymbol{N} \in \mathbb{Z}^N$ a vector of integers. This case runs exactly along the same lines as in the example. We first compute a floating solution $(\boldsymbol{x}_{\text{fl}} , \boldsymbol{N}_{\text{fl}})$ by minimizing the quadratic form

$$Q(\boldsymbol{x}, \boldsymbol{N}; \boldsymbol{Y}) = (\boldsymbol{Y} - A\boldsymbol{x} - B\boldsymbol{N})^T \Sigma_\varepsilon^{-1} (\boldsymbol{Y} - A\boldsymbol{x} - B\boldsymbol{N}) \tag{76}$$

considering $\boldsymbol{N}$ just a continuous vector in $\mathbb{R}^N$. Then the floating estimates $(\boldsymbol{x}_{\text{fl}}, \boldsymbol{N}_{\text{fl}})$ will have a covariance matrix, C_{fl}, given by

$$C_{\text{fl}} = \begin{bmatrix} A^T W A & A^T W B \\ B^T W A & B^T W B \end{bmatrix}^{-1} , W = \Sigma_\varepsilon^{-1} . \tag{77}$$

By using the formula for the partitioned inverse of a matrix, we know then that

$$C_{N_{\text{fl}}} = [B^T W B - B^T W A (A^T W A)^{-1} A^T W B]^{-1} = \Gamma^{-1}$$

or

$$C_{N_{\text{fl}}}^{-1} = B^T W B - B^T W A (A^T W A)^{-1} A^T W B = \Gamma , \tag{78}$$

which is a known, easily computable matrix.

If we return now to the minimum problem

$$\min_{\substack{x \in \mathbb{R}^n \\ N \in \mathbb{Z}^N}} Q(\boldsymbol{x}, \boldsymbol{N}; \boldsymbol{Y}) \tag{79}$$

we see that first of all we can use the decomposition

$$\begin{cases} Q(\boldsymbol{x}, \boldsymbol{N}; \boldsymbol{Y}) = Q(\boldsymbol{x}_{\text{fl}}, \boldsymbol{N}_{\text{fl}}; \boldsymbol{Y}) + \overline{Q}(\boldsymbol{x}_{\text{fl}} - x, \boldsymbol{N}_{\text{fl}} - N), \\ \overline{Q}(\boldsymbol{x}_{\text{fl}} - \boldsymbol{x}, \boldsymbol{N}_{\text{fl}} - N) = [(\boldsymbol{x}_{\text{fl}} - \boldsymbol{x})^T (\boldsymbol{N}_{\text{fl}} - \boldsymbol{N})^T] C_{\text{fl}}^{-1} \begin{bmatrix} \boldsymbol{x}_{\text{fl}} - \boldsymbol{x} \\ \boldsymbol{N}_{\text{fl}} - \boldsymbol{N} \end{bmatrix} \end{cases} \tag{80}$$

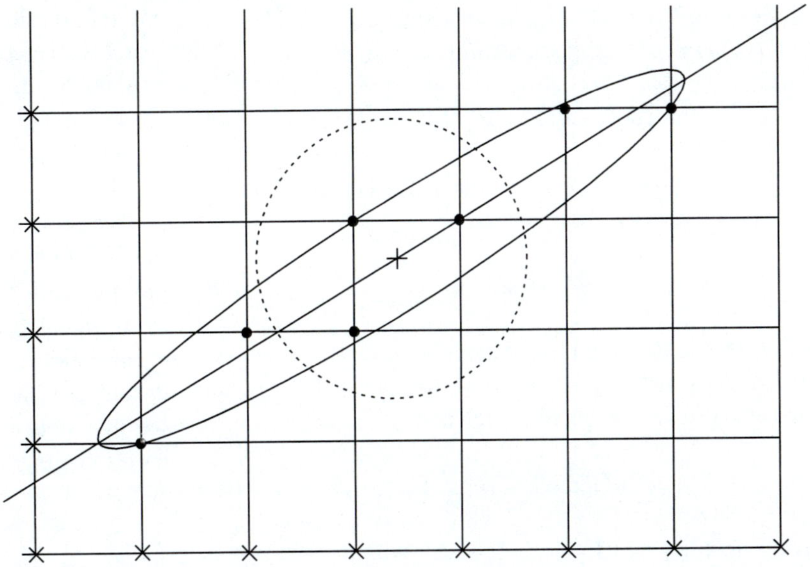

Fig. 2 A plot of the search space for N in 2D; + N_{fl} point, • • • points falling in the ellipsoid (see (87)), grid points falling in the rectangular envelope of the ellipsoid, a sphere with the same volume of the ellipsoid and its reduced search space

reducing the minimization of (78) to that of $\overline{Q}$. Then by minimizing $\overline{Q}$ with respect to $\boldsymbol{x}$, given $\boldsymbol{N}$, we find the "conditional" solution

$$\boldsymbol{x}_N = \boldsymbol{x}_{fl} + (A^T WA)^{-1} A^T WB(N_{fl} - N) . \quad (81)$$

With the use of (81) and some algebra, we achieve the further decomposition

$$\begin{aligned}\overline{Q}(\boldsymbol{x}_{fl} - \boldsymbol{x}, \boldsymbol{N}_{fl} - \boldsymbol{N}) &= (\boldsymbol{x}_N - \boldsymbol{x})^T (A^T WA)(\boldsymbol{x}_N - \boldsymbol{x}) + \\ &\quad + (\boldsymbol{n}_{fl} - \boldsymbol{N})^T \Gamma (\boldsymbol{n}_{fl} - \boldsymbol{N}) \\ &= (\boldsymbol{x}_N - x)^T (A^T WA)(\boldsymbol{x}_N - x) + \overline{\overline{Q}}(\boldsymbol{N}) . \end{aligned} \quad (82)$$

But then the minimum of $\overline{Q}$ on $\boldsymbol{x} \in \mathbb{R}^n$ and $\boldsymbol{N} \in \mathbb{Z}^N$ can be obtained in two separate steps; first we find

$$\boldsymbol{N}_{ML} = \arg \min_{N \in \mathbb{Z}^N} (\boldsymbol{N}_{fl} - \boldsymbol{N})^T \Gamma (\boldsymbol{N}_{fl} - \boldsymbol{N}) , \quad (83)$$

then we can simply put

$$\boldsymbol{x}_{ML} = \boldsymbol{x}_{N_{ML}} , \quad (84)$$

because this annihilates the first term in the right hand side of (82).

So essentially we are left with the problem of finding in the lattice $\mathbb{Z}^N$ the point $\boldsymbol{N}_{ML}$ nearest to the point $\boldsymbol{N}_{fl}$ in the metric induced by (83).

Let us imagine for a moment that we are so lucky that the matrix Γ results to be diagonal. In this case the quadratic form in (83) is decomposed as

$$(\boldsymbol{N}_{fl} - \boldsymbol{N})^T \Gamma (\boldsymbol{N}_{fl} - \boldsymbol{N}) = \sum_{i=1}^{N} \gamma_i (N_{fli} - N_i)^2 \quad (85)$$

and the minimum is immediately found by rounding, namely (recalling (69))

$$N_{ML_i} = [N_{fli}] . \quad (86)$$

This however is not what is found in practice, specially when we process GNSS data for a short time. Typically, what happens is that the ellipsoid

$$(\boldsymbol{N}_{fl} - \boldsymbol{N})^T \Gamma (\boldsymbol{N}_{fl} - \boldsymbol{N}) \le c , \quad (87)$$

with c chosen by using a χ^2 distribution, in a way that it bears a probability close to 1, has in fact quite an elongated form. An example is displayed in Fig. 2 when $\boldsymbol{N}$ has only two components

If we just consider this example we immediately perceive that the points at which the form (83) needs to be computed are only 7, while if we search on the whole rectangle we have to compute (83) 48 times.

It is clear that, when the dimension of the vector $\boldsymbol{N}$ increases, the number of computations bursts exponentially. For instance, if one has to probe a cube of side 5 in 12 dimensions one has to make more than $244 \cdot 10^6$ computations; not a very practical solution.

This problem has found a large number of tricks and recipes to get a quick solution, however a rational settlement of the argument has come with the so-called Λ-method (see Teunissen 2017).

Here we apologize for using the symbol Z for different objects, however, we prefer to keep the original notation of the author (Teunissen 2017).

The idea of the method works on the basis of the following proposition.

Proposition 3 *Consider the set of matrices $\mathcal{Z} \in \mathbb{R}^N \otimes \mathbb{R}^N$, with integer entries and unimodular, namely*

$$Z \in \mathcal{IU} = \{Z; Z \in \mathcal{Z}\,,\ |Z| = \pm 1\}\,, \tag{88}$$

where $|Z|$ denotes the determinant of Z. Then $\mathcal{IU}$ is invariant under inversion, namely

$$Z \in \mathcal{IU} \Leftrightarrow Z^{-1} \in \mathcal{IU}\,. \tag{89}$$

Proof If $Z \in \mathcal{IU}$, thinking of the inverse Z^{-1} as computed by Kramer's rule, it is obvious that the entries $(Z^{-1})_{ik}$ are integers; also on account of the relation

$$|Z||Z^{-1}| = |I| = 1\,, \tag{90}$$

we see that $|Z| = \pm 1 \Rightarrow |Z^{-1}| = \pm 1$, with the same order. The inverse relation is trivial because $(Z^{-1})^{-1} = Z$.

We note that, by using (90), it is elementary to show that (87) is equivalent to the other condition

$$Z\,,\ Z^{-1} \in \mathcal{IU}\,,\ \text{i.e.}\ Z\,,\ Z^{-1} \in \mathcal{IU} \Rightarrow |Z| = |Z^{-1}| = \pm 1$$

□

Given that, we return to the integer estimation problem of finding the minimum of

$$\boldsymbol{N} \in \mathbb{Z}^N\,,\ \overline{\overline{Q}}_{\Gamma}(\boldsymbol{N}) = (\boldsymbol{N}_{\text{fl}} - \boldsymbol{N})^T \Gamma (\boldsymbol{N}_{\text{fl}} - \boldsymbol{N})\,, \tag{91}$$

with $\Gamma = C_{\text{fl}}^{-1}$, given by (78). Now let us transform by means of a $Z \in \mathcal{IU}$ the unknowns, according to

$$\boldsymbol{M} = Z^{-1}\boldsymbol{N}\,,\ \boldsymbol{M}_{\text{fl}} = Z^{-1}\boldsymbol{N}_{\text{fl}}\,. \tag{92}$$

Accordingly, (91) is transformed into the search of the minimum of

$$\boldsymbol{M} \in \mathbb{Z}^N\,,\ \overline{\overline{Q}}_{\overline{\Gamma}}(\boldsymbol{M}) = (\boldsymbol{M}_{\text{fl}} - \boldsymbol{M})^T \overline{\Gamma} (\boldsymbol{M}_{\text{fl}} - \boldsymbol{M})\,, \tag{93}$$

with

$$\overline{\Gamma} = Z^T \Gamma Z\,. \tag{94}$$

We note that, since

$$|\overline{\Gamma}| = |Z^T||\Gamma||Z| = |\Gamma|\,, \tag{95}$$

the volume of the ellipsoid $\overline{\overline{Q}}_{\Gamma}(N) \le c$ is the same as that of the ellipsoid $\overline{\overline{Q}}_{\overline{\Gamma}}(\boldsymbol{M}) \le c$; so the point now is whether we can take advantage of the Z transform so that the new ellipsoid becomes as close as possible to one referred to its axis. Under this circumstance in fact a search space constituted by a parallelepiped parallel to the axis in the $\boldsymbol{M}$ space, has much smaller volume and the corresponding numerical burden of computing $\overline{\overline{Q}}_{\overline{\Gamma}}(\boldsymbol{M})$, in this smaller search space, is strongly reduced (see Fig. 2).

Ideally, we would like to use (94) to make $\overline{\Gamma}$ diagonal, but this is not possible in general due to the discrete character of the entries of Z. We note that diagonalizing $\overline{\Gamma}$ is equivalent to diagonalizing $C_{M_{\text{fl}}} = \overline{\Gamma}^{-1}$ and this last target has a clearer statistical meaning, when we say that we want to decorrelate the components of $\boldsymbol{M}_{\text{fl}}$. To achieve this result we first notice that, given any covariance matrix C, the corresponding correlation matrix R is obtained by

$$R = D(C)^{-1/2} C D(C)^{-1/2}\,, \tag{96}$$

where $D(C)$ is a diagonal matrix with the same diagonal as C. Then our purpose is to make the matrix R as close as possible to the identity; since

$$|R| \le 1, \quad |R| = 1 \Leftrightarrow R = I\,, \tag{97}$$

decorrelation can be translated into trying to maximalize the $|R|$, i.e.

$$|R| = \frac{|C|}{|D(C)|} = \frac{|C|}{\prod_{i=1}^N C_{ii}}\,. \tag{98}$$

In our case $C = C_{M\text{fl}}$ and we know from (95) that

$$|C_{M_{\text{fl}}}| = |\overline{\Gamma}^{-1}| = |\Gamma^{-1}| = |C_{N_{\text{fl}}}|\,,$$

so that (98) writes

$$|R_{M_{\text{fl}}}| = \frac{|C_{N_{\text{fl}}}|}{\prod_{i=1}^N C_{M_{\text{fl}}ii}}\,; \tag{99}$$

therefore

$$\max |R_{M\text{fl}}| \Leftrightarrow \min \prod_{i=1}^N C_{M_{\text{fl}}ii}\,. \tag{100}$$

Since it is clear that the product of two Z transforms is again a Z transform, the algorithm used for decorrelation works on a couple of variables at time and then takes the product of the so found Z transforms. So, following Teunissen (2017), we concentrate on the algorithm for a two-dimensional case. Also in this reduced example we act iteratively, namely we fix one of two components, say $M_1 = N_1$, and we try to minimize the variance of M_2. Once this is achieved, we fix the second component and we try to minimize the variance of the first with a new Z transform, and so forth. At the end we multiply all the Z transforms above defined. Then, starting from

$$C_{fl} = \begin{vmatrix} C_1 & C_{12} \\ C_{21} & C_2 \end{vmatrix} , \; (C_{12} = C_{21}) , \qquad (101)$$

and applying a Z transform of the form

$$Z = \begin{vmatrix} 1 & 0 \\ \alpha & 1 \end{vmatrix} ,$$

where α for the moment is left just real, we see that $\boldsymbol{M}_{fl} = Z\boldsymbol{N}_{fl}$ will have variances

$$\sigma^2_{M_{fl}1} = \sigma^2_{N_{fl}1} = C_1 , \; \sigma^2_{M_{fl}2} = C_1\alpha^2 + 2C_{12}\alpha + C_2 . \quad (102)$$

So the real α minimizing $\sigma^2_{M_{fl}2}$ is given by

$$\alpha = -\frac{C_{12}}{C_1} ; \qquad (103)$$

indeed this value is in general real and not an integer. However, we can round it to the nearest integer $-\left[\frac{C_{12}}{C_1}\right]$. Since from (102) $\sigma^2_{M_{fl}}$ can be written as

$$\sigma^2_{M_{fl}2} = C_2 - C_1 \left[\left(\frac{C_{12}}{C_1} \right)^2 - \left(\alpha + \frac{C_{12}}{C_1} \right)^2 \right] , \qquad (104)$$

we see that indeed α given by (103) can only decrease $\sigma^2_{M_{fl}2}$, but for the trivial case that $\alpha = 0$, in which instance $\sigma^2_{M_{fl}2} = C_2 = \sigma^2_{N_{fl}2}$.

So we have found an integer $\alpha = -\left[\frac{C_{12}}{C_2}\right]$ that decreases $\sigma^2_{M_{fl}2}$ and then we can proceed as described above.

Once the numerical problem of finding $\boldsymbol{N}_{ML}$ has been solved, we still have to try to give a reasonable method to compute the covariance of $\boldsymbol{x}_{ML}$. This however implies that we are able to compute the covariance of $\boldsymbol{N}_{ML}$, which indeed is not an easy task. However, as our elementary example has revealed, in general assuming that $\boldsymbol{N}_{ML}$ is a fixed vector, with zero covariance, easily leads to an unrealistic conclusion. In this case in fact we have an $\boldsymbol{N}_{fix}$ solution and the corresponding $\boldsymbol{x}_{fix}$ solution has obviously the covariance $(A^T WA)^{-1}$; this often is simply too small to be true.

Now that we have visited so to say the ground zero of the integer estimation problem, at least focussing the main drawbacks of too simplistic solutions, we have to account for an important mathematical construction in estimation theory which is essentially due to P. Teunissen and his school (Teunissen 2001, 2003, 2007).

The analysis starts with a more precise definition of class of integer estimator $\widehat{\boldsymbol{N}}_I$ (see Teunissen 2001), of which the Least Squares Estimator, discussed above, is just one member. To be precise, $\widehat{\boldsymbol{N}}_I$ is defined first by a partition of R^N into regions $S_N(N \in \mathcal{Z}^N)$ that are obtained by shifting the neighbor of the origin S_0, namely

$$S_N = S_0 + N \qquad (105)$$

in such a way that the two conditions

$$\begin{cases} \bigcup_{S_N} = R^N \\ \text{int}\,(S_N) \cap \text{int}\,(S_N) = \emptyset \, , \; N \neq N' \end{cases} \qquad (106)$$

are satisfied. Since the origin $\mathbf{0}$ has to belong to S_0, we have as well $N \in S_N$. The set S_N is called the *pull-in region* and it defines the integer estimator $\widehat{\boldsymbol{N}}_I$ by the rule

$$\widehat{\boldsymbol{N}}_{fl} \in S_N \;\Rightarrow\; \widehat{\boldsymbol{N}}_I = N . \qquad (107)$$

This can be cast into an analytic form by the use of characteristic functions

$$\chi_N(\widehat{\boldsymbol{N}}_{fl}) = \begin{cases} 1 \; \widehat{\boldsymbol{N}}_{fl} \in S_N \\ 0 \text{ otherwise} , \end{cases} \qquad (108)$$

namely

$$\widehat{\boldsymbol{N}}_I = \sum_{N \in \mathcal{Z}^N} N \chi_N(\widehat{\boldsymbol{N}}_{fl}) . \qquad (109)$$

Clearly the choice of any S_0 satisfying (108) determines the specific integer estimator under consideration.

For instance, the simple *round off* estimator $\widehat{\boldsymbol{N}}_R$ has pull-in regions determined by

$$S_0 = \{\xi \in R^N \, , \; |\xi_i| \le \frac{1}{2} \quad i = 1 \ldots N\} . \qquad (110)$$

A little more elaborated, but more effective, specially if preceded by a step of a quasi-decorrelation, is the so-called *bootstrap integer estimator*, $\widehat{\boldsymbol{N}}_B$. This is defined according to the concept of sequential conditional Least Squares Adjustment, going through the following steps.

We start with N_1 and fix it as in the round off method,

$$\widehat{N}_{1B} = [\widehat{N}_{1fl}] ,$$

then the rest of the vector $\widehat{\boldsymbol{N}}_{fl}$ (i.e. $\widehat{N}_{2fl}, \widehat{N}_{3fl} \ldots \widehat{N}_{Nfl}$) is corrected by the effect of N_{1B}, now fixed, due to the correlation of these components with $\widehat{N}_{1fl} - \widehat{N}_{1B}$. We obtain $(\widehat{N}_{2|1fl}, \widehat{N}_{3|1fl} \ldots)$ and we proceed by rounding $\widehat{N}_{2|1fl}$, getting

$$\widehat{N}_{2B} = [\widehat{N}_{2|1fl}] .$$

At this point we subtract from $(\widehat{N}_{3fl} \dots \widehat{N}_{Nfl})$ the influence of $(\widehat{N}_{1B}, \widehat{N}_{2B})$ due to correlation and we can now fix $\widehat{N}$ and so forth, since all components are fixed. Then one can prove that, performing the Cholesky decomposition

$$C_{N_{fl}} = LDL^T \ ,$$

with L a lower triangular matrix with all units on the main diagonal, the pull-in region for the bootstrap estimator is given by

$$S_{0B} = \{\boldsymbol{\xi} \in R^N \ ; \ |\boldsymbol{e}_i^T L^{-1} \xi| \leq \frac{1}{2}, \ i = 1, 2 \dots N\} \tag{111}$$

$$(\boldsymbol{e}_i^T = (0, 0 \dots 1 \dots 0), i \text{ in the } i\text{-th entry}) \ .$$

Finally, the integer Least Squares estimator $\widehat{N}_{IL}$ has a pull-in region that is described by the formula

$$S_{OL} = \{\boldsymbol{\xi} \in R^N \ ; \ |\boldsymbol{c}^T C_{N_{fl}}^{-1} \xi| \leq \frac{1}{2} \boldsymbol{c}^T C_{N fl}^{-1} \boldsymbol{c} \ , \ \forall \boldsymbol{c} \in \mathcal{Z}^N\} \ . \tag{112}$$

The importance of this systematic approach to integer estimation is in that we can in principle describe the (discrete) probability distribution of $\widehat{N}_I$ as function of $\widehat{N}_{fl}$. In fact, if $P_{fl}(N)$ is the probability density of $\widehat{N}_{fl}$, often taken as Gaussian $\widehat{N}_{fl} \sim \mathcal{N}(\overline{N}, C_{N_{fl}})$, we have

$$P(\widehat{N}_{fl} \in S_N) = \int_{S_N} P_{fl}(\boldsymbol{\xi}) d\xi = \int_{S_0} P_{fl}(N + \boldsymbol{\xi}) d\xi \tag{113}$$

This distribution can be explicitly computed for the bootstrap estimator (see Teunissen 2001); otherwise upper and lower bounds for (113) can be given for the other cases. Indeed other numerical methods, for instance a Monte-Carlo method, could be applied too.

In any event, knowledge of the distribution of $\widehat{N}_I$ allows one to define an important index of quality for a specific integer estimator, namely the success rate, i.e. the probability that $\widehat{N}_I = \overline{N}$ when $\overline{N}$ is the true value of the variable N in the model (75), namely

$$P(\widehat{N}_I = \overline{N}) = P(\widehat{N}_{fl} \in S_{\overline{N}}) = P(\widehat{N}_{fl} - \overline{N} \in S_0) = \\ = \int_{S_0} P_{fl}(\overline{N} + \boldsymbol{\xi}) d\xi \ . \tag{114}$$

In addition all that permits to study the sensitivity of the estimator $\widehat{N}_I$ to the presence of biases, accounted by the model (75). This is discussed in Teunissen (2001), but on the same time it opens the problem that an integer estimator $\widehat{N}_I$ has always as output a vector of integer variates by its very definition. This however is not always the best solution, exactly because the presence of some biases can shift the $\widehat{N}_{fl}$ "far" from a knot of $\mathcal{Z}^N$, sometimes even with a peaked distribution. To explain, take the case that $N \in R^1$ (one ambiguity only) and $\widehat{N}_{fl} = 0.4$ while $\sigma(\widehat{N}_{fl}) = 0.01$; if we are forced to chose an integer, we will certainly take $N_I = 0$, yet any test with a reasonable value of the significance parameter α will tell us that the hypothesis that $\overline{N} = 0$ has to be rejected.

This has suggested to define a new class of estimators (see Teunissen 2004), called by the author "integer aperture" estimators, $\widehat{N}_{IA}$. The basic idea is to define a pull-in region $\Omega_0 \subset S_0$ and then propagate it to the full design of $\{S_N\}$ by setting

$$\Omega_N = N + \Omega_0 \subset S_N \ . \tag{115}$$

Then the following rule is established

$$\widehat{N}_{IA} = \begin{cases} N & \text{if } \widehat{N}_{fl} \in \Omega_N \\ \widehat{N}_{fl} & \text{if } \widehat{N}_{fl} \in R^N \setminus \bigcup_N \Omega_N \ . \end{cases} \tag{116}$$

In other words, $\widehat{N}_{IA}$ is discrete when $\widehat{N}_{fl}$ falls in the region $\Omega = \bigcup_N \Omega_N$, and continuous when $\widehat{N}_{fl}$ falls in the complementary region, Ω^c.

With the help of the characteristic function $\chi_{\Omega_N}(\boldsymbol{\xi})$ we have then the IA estimator defined as

$$\widehat{N}_{IA} = N_{fl} + \sum_{N \in \mathcal{Z}^N} (N - \widehat{N}_{fl}) \chi_{\Omega_N}(\widehat{N}_{fl}) \ . \tag{117}$$

In Teunissen (2004) it is shown that some rules commonly used to fix ambiguities, like for example the ratio test criterion, give rise to estimators falling in the IA class. Connatural to IA estimators is the definition of three probabilities

$$P_S = P(\widehat{N}_{IA} = \overline{N} | N = \overline{N}) = \boldsymbol{P}(\widehat{N}_{fl} \in \Omega_{\overline{N}})$$
$$P_F = P(\widehat{N}_{IA} = N | N \neq \overline{N}) = \boldsymbol{P}(\widehat{N}_{fl} \in \Omega \setminus \Omega_{\overline{N}})$$
$$P_U = P(\widehat{N}_{IA} = \widehat{N}_{fl}) = \boldsymbol{P}(\widehat{N}_{fl} \in \Omega^c)$$

called *success, failure and undecided rates*.

If one attaches to the three events S, F, U some penalty figures, one is then conducted to find minimal penalty estimators. The solution of this problem can be found in Teunissen (2004).

Finally, an even larger class of estimators has been defined in Teunissen (2003), namely the Integer Equivariant (IE) estimators, defined as follows.

Let ϑ be any linear function of parameters in (81), namely

$$\vartheta = C_N^T N + C_x^T \boldsymbol{x} \ ; \tag{118}$$

then an estimator

$$\widehat{\vartheta} = g_{\vartheta}(Y) \tag{119}$$

belongs to IE if it satisfies the relations

$$g_{\vartheta}(Y + Bz) = g_{\vartheta}(Y) + C_N^T z \quad \forall z \in \mathcal{Z}^N \tag{120}$$

$$g_{\vartheta}(Y + A\xi) = g_{\vartheta}(Y) + C_x^T \xi\,, \quad \forall \xi \in R^N\,. \tag{121}$$

It has to be noted that the class of linear unbiased estimators, LU, is included in IE. In fact, if

$$\widehat{\vartheta}_{\mathrm{LU}} = \boldsymbol{g}_{\vartheta}^T Y = \boldsymbol{g}_{\vartheta}^T BN + \boldsymbol{g}_{\vartheta}^T A\boldsymbol{x} + \boldsymbol{g}_{\vartheta}^T \boldsymbol{\nu}\,, \tag{122}$$

for some $\boldsymbol{g}_{\vartheta} \in R^m$, and $\widehat{\vartheta}_{\mathrm{LU}}$ has to be unbiased, one has to have

$$E\{\widehat{\vartheta}_{\mathrm{LU}}\} = \boldsymbol{g}_{\vartheta}^T BN + \boldsymbol{g}_{\vartheta}^T A\boldsymbol{x} \equiv C_N^T N + C_x^T \boldsymbol{x} \quad \forall N \in \mathcal{Z}^N,\ \forall \boldsymbol{x} \in R^n.$$

Therefore it is enough to choose $\boldsymbol{g}_{\vartheta}$ such that

$$\begin{cases} B^T \boldsymbol{g}_{\vartheta} = C_N \\ A^T g_{\vartheta} = C_x \end{cases} \tag{123}$$

to see that (122) is an IE estimator too.

One can prove that $LU \subset IA$ strictly.

Since $\begin{bmatrix} B^T \\ A^T \end{bmatrix}$ is a matrix with less rows than columns, roughly there are many solutions of (123) and therefore the idea of finding an "optimal" IE estimator is appropriate.

The problem of finding the g_{ϑ} estimator with minimum mean square error is analyzed and solved in Teunissen (2003).

Such best solution can be summarized by the formulas

$$\widehat{\vartheta}_{B|E} = C_N^T \widehat{\boldsymbol{N}}_{B|E} + C_x^T \widehat{\boldsymbol{x}}_{B|E}\,, \tag{124}$$

$$\boldsymbol{N}_{B|E} = \sum_{N \in \mathcal{Z}^N} N \frac{P_{\hat{\text{n}}}(N_{\hat{\text{n}}} - N)}{\sum_{z \in \mathcal{Z}^N} P_{\hat{\text{n}}}(N_{\hat{\text{n}}} - z)}\,, \tag{125}$$

$$\boldsymbol{x}_{B|E} = \boldsymbol{x}_{\hat{\text{n}}} - C_{\widehat{x}_{\hat{\text{n}}} \widehat{N}_{\hat{\text{n}}}} C_{\widehat{N}_{\hat{\text{n}}}}^{-1} (N_{\hat{\text{n}}} - \boldsymbol{N}_{B|E})\,. \tag{126}$$

An interesting feature of this solution is that it comes to one and the same form as the mean of the a posteriori distribution of estimators of $\boldsymbol{N}$ and $\boldsymbol{x}$, based on a Bayesian approach (Betti et al. 1993). So in this case too a Bayesian approach, with a suitable prior distribution, and a frequentist approach give similar results.

It has to be noted that, while the class of Integer estimators has been considered in mathematical statistics, the other two broader classes constitute an original contribution from Geodesy.

Finally, it is worth at least mentioning that not only the estimation of integer/continuous unknowns has been studied, but also the problem of prediction of mixed integer/continuous random variables has been analyzed in Teunissen (2007), arriving at closely related results.

Moreover, the analysis of a problem "dual" to the integer estimation described in this section, namely the so-called *search of the closest point in a lattice*, has been developed in the 1980s (see Pohst 1981; Kannan 1983) suggested by informatic problems in coding and cryptography.

Once more Mathematics and Geodesy have attacked similar problems, until recently they started to interact (see Jazaeri et al. 2014). Yet it seems not by chance that Geodesy has approached the integer estimation problem from a typical statistical point of view, as a particular aspect of the general least squares theory.

6 Inverse Problems and Regularization

Next we are interested in another important area of Mathematical Geodesy, namely the theory of inverse problems. A geodetically reflected survey about inverse problems is given, e.g., in Freeden and Nashed (2018b). In the approach presented here we also follow this contribution, however, only considering such parts, that are necessary prerequisites for a novel regularization strategy, i.e., multiscale regularization in Hilbert spaces. The preparatory material taken almost literally from Freeden and Nashed (2018b) is used to study two essential problems of Geodesy, namely satellite gravitational gradiometry (SGG) and inverse gravimetry (IG).

We start our preparatory consideration by assuming that X and Y are two Hilbert spaces with inner products $\langle\cdot,\cdot\rangle_X$ and $\langle\cdot,\cdot\rangle_Y$, respectively. Let

$$A : X \longrightarrow Y \tag{127}$$

be a linear bounded operator. For a given element $y \in Y$ we are looking for a solution of

$$Ax = y. \tag{128}$$

In accordance to the *classification due to Hadamard* (see Hadamard 1902), we call such a problem (128) *well-posed*, if the following properties are valid:

- *For all data, a solution exists (existence).*
- *For all data, the solution is unique (uniqueness).*
- *The solution depends continuously on the data (stability).*

In the language of functional analysis, these requirements can be translated (see, e.g., Freeden and Nashed 2018b) into the properties:

- *A is surjective, i.e., the range $\mathcal{R}(A)$ is equal to Y.*
- *A is injective, i.e., the null-space $\mathcal{N}(A)$ only consists of $\{0\}$.*
- *A^{-1} is continuous (hence, bounded).*

By convention, if one of the three conditions is not fulfilled, the problem (128) is called *ill-posed* (for more details the reader is referred to, e.g., Freeden and Nashed 2018b).

Let us point out the consequences of the violations of the above requirements for the well-posedness of (128). The lack of injectivity of A is perhaps the easiest problem. The space X can be replaced by the orthogonal complement $\mathcal{N}(A)^{\perp}$, and the restriction of the operator A to $\mathcal{N}(A)^{\perp}$ leads to an injective problem.

In geoscientific practice, one is very often confronted with the problem that $\mathcal{R}(A) \neq Y$, since the right side is given by measurements and is, therefore, disturbed by errors. In this case, instead of $y \in \mathcal{R}(A)$, we have to consider a perturbed right side y^{ε} (see, e.g., Nashed 1987a,b). We suppose that

$$\|y - y^{\varepsilon}\|_Y < \varepsilon. \tag{129}$$

The aim now is to find a solution x^{ε} of the equation

$$Ax^{\varepsilon} = y^{\varepsilon}. \tag{130}$$

Since y^{ε} might not be in $\mathcal{R}(A)$, the solution of this equation might not exist, and we have to generalize what is meant by a solution (cf. Nashed 1976). x^{ε} is called *least-squares solution* of (130), if

$$\|Ax^{\varepsilon} - y^{\varepsilon}\|_Y = \inf\{\|Az - y^{\varepsilon}\|_Y : z \in X\}. \tag{131}$$

The solution of (131) might not be unique, and therefore one looks for the solution of (131) with minimal norm. x^{ε} is called *best approximate solution* of $Ax^{\varepsilon} = y^{\varepsilon}$, if x^{ε} is a least-squares solution and the condition

$$\|x^{\varepsilon}\|_X = \inf\{\|z\|_X : z \text{ is a least-squares solution of } Az = y^{\varepsilon}\} \tag{132}$$

holds true. The notion of a best-approximate solution is closely related to the Moore-Penrose (generalized) inverse of A (for more details on the theory and applications of generalized inverse operators, see, e.g., Nashed 1987a). We let

$$\breve{A} : \mathcal{N}(A)^{\perp} \longrightarrow \mathcal{R}(A) \tag{133}$$

by

$$\breve{A} = A|_{\mathcal{N}(A)^{\perp}} \tag{134}$$

and define the *Moore-Penrose (generalized) inverse* $A^{\dagger}$ to be the unique linear extension of $\breve{A}^{-1}$ to

$$\mathcal{D}(A^{\dagger}) = \mathcal{R}(A) + \mathcal{R}(A)^{\perp}, \tag{135}$$

where

$$\mathcal{N}(A^{\dagger}) = \mathcal{R}(A)^{\perp}. \tag{136}$$

A standard result (see, e.g., Nashed 1976, 1987a) is provided by the following statement:

If $y \in \mathcal{D}(A^{\dagger})$, then $Ax = y$ has a unique best-approximate solution which is given by

$$x^{\dagger} = A^{\dagger} y. \tag{137}$$

The set of all least-squares solutions is $x^{\dagger} + \mathcal{N}(A)$.

A serious problem for ill-posed problems occurs when A^{-1} or $A^{\dagger}$ are not continuous. This means that small errors in the data or even small numerical noise can cause large errors in the solution. In fact, in most cases the application of an unbounded A^{-1} or $A^{\dagger}$ does not make any sense in computations. The usual strategy to overcome this difficulty is to substitute the unbounded inverse operator

$$A^{-1} : \mathcal{R}(A) \longrightarrow X \tag{138}$$

by a suitable bounded approximation

$$R : Y \longrightarrow X. \tag{139}$$

The operator R is not chosen to be fixed, but dependent on a *regularization parameter* α. According to the usual approach in inverse theory we are led to introduce the following setting.

Definition 3 A regularization strategy is a family of linear bounded operators

$$R_{\alpha} : Y \longrightarrow X, \quad \alpha > 0, \tag{140}$$

so that

$$\lim_{\alpha \to 0} R_{\alpha} Ax = x \text{ for all } x \in X, \tag{141}$$

i.e., the operators $R_{\alpha} A$ converge pointwise to the identity.

From the theory of inverse problems (see, e.g., Engl et al. 1996) it is also clear that if $A : X \to Y$ is compact and Y has an infinite dimension (as it is the case of all satellite applications we have in mind), then the operators R_{α} are not uniformly bounded, i.e., there exists a sequence $\{\alpha_j\}$ with $\lim_{j\to\infty} \alpha_j = 0$ such that

$$\|R_{\alpha_j}\|_{L(Y,X)} \to \infty, \qquad j \to \infty. \tag{142}$$

Here we have used the standard convention of the operator norm

$$\|L\|_{L(Y,X)} = \inf_{\|y\|_Y \neq 0} \frac{\|Ly\|_X}{\|y\|_Y}. \tag{143}$$

Note that the convergence of $R_\alpha Ax$ in Definition 3 is based on $y = Ax$, i.e., on unperturbed data. In practice, the right side is affected by errors and no convergence is achieved. Instead, one is (or has to be) satisfied with an approximate solution based on a certain choice of the regularization parameter.

Let us discuss the error of the solution: We let $y \in \mathcal{R}(A)$ be the (unknown) exact right-hand side and $y^\varepsilon \in Y$ be the measured data with

$$\|y - y^\varepsilon\|_Y < \varepsilon. \tag{144}$$

For a fixed $\alpha > 0$, we let

$$x^{\alpha,\varepsilon} = R_\alpha y^\varepsilon, \tag{145}$$

and look at $x^{\alpha,\varepsilon}$ as an approximation of the solution x of $Ax = y$. Then the error can be split in standard way (see, e.g., Engl et al. 1996) as follows:

$$\begin{aligned} \|x^{\alpha,\varepsilon} - x\|_X &= \|R_\alpha y^\varepsilon - x\|_X \\ &\le \|R_\alpha y^\varepsilon - R_\alpha y\|_X + \|R_\alpha y - x\|_X \\ &\le \|R_\alpha\|_{L(Y,X)} \, \|y^\varepsilon - y\|_Y + \|R_\alpha y - x\|_X, \end{aligned} \tag{146}$$

such that

$$\|x^{\alpha,\varepsilon} - x\|_X \le \|R_\alpha\|_{L(Y,X)}\varepsilon + \|R_\alpha Ax - x\|_X. \tag{147}$$

We see that the error between the exact and the approximate solution consists of two parts: The first term is the product of the bound for the error in the data and the norm of the regularization operator R_α. This term will usually tend to infinity for $\alpha \to 0$ if the inverse A^{-1} is unbounded (for example, if A is compact). The second term denotes the approximation error $\|(R_\alpha - A^{-1})y\|_X$ for the exact right-hand side $y = Ax$. This error tends to zero as $\alpha \to 0$ by the definition of a regularization strategy. Thus, both parts of the error show a diametrically reflected behavior. A typical picture of the errors in dependence on the regularization parameter α is sketched in Fig. 3. Thus, a strategy is needed to choose α dependent an ε in order to keep the error as small as possible, i.e., we would like to minimize

$$\varepsilon\|R_\alpha\|_{L(Y,X)} + \|R_\alpha Ax - x\|_X. \tag{148}$$

In principle, we distinguish two classes of parameter choice rules: If $\alpha = \alpha(\varepsilon)$ does not depend on ε, we call $\alpha = \alpha(\varepsilon)$ an *a-priori* parameter choice rule. Otherwise α

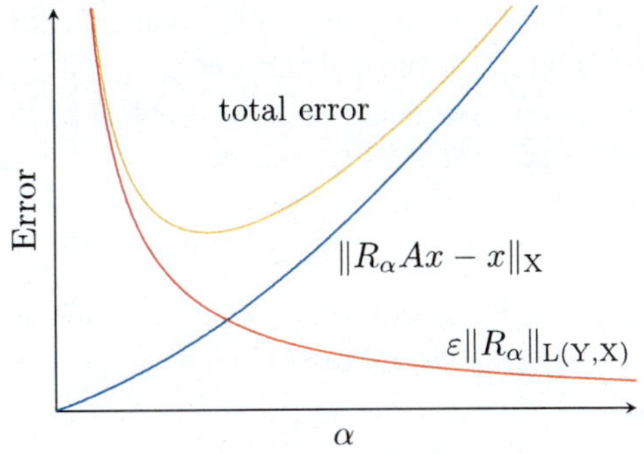

Fig. 3 Typical behavior of the total error in a regularization process

depends also on y^ε and we call $\alpha = \alpha(\varepsilon, y^\varepsilon)$ an *a-posteriori* parameter choice rule. It is conventional to say a parameter choice rule is convergent, if for $\varepsilon \to 0$ the rule is such that

$$\lim_{\varepsilon \to 0} \sup\{\|R_{\alpha(\varepsilon,y^\varepsilon)}y^\varepsilon - A^\dagger y\|_X \ : y^\varepsilon \in Y,\ \|y^\varepsilon - y\|_Y \le \varepsilon\} = 0 \tag{149}$$

and

$$\lim_{\varepsilon \to 0} \sup\{\alpha(\varepsilon, y^\varepsilon) \ : y^\varepsilon \in Y, \|y - y^\varepsilon\|_Y \le \varepsilon\} = 0. \tag{150}$$

All in all, the rationale in most methods for resolution (approximate solvability) of ill-posed inverse problems is to construct a "solution" that is acceptable physically as a meaningful approximation and is sufficiently stable from the computational standpoint, hence, an emphasis is put on the distinction between "solution" and "resolution". The main dilemma of modeling of ill-posed problems (IPP) is that the closer the mathematical model describes the IPP, the worse is the "condition number" of the associated computational problem (i.e., the more sensitive to errors). For ill-posed problems, the difficulty is to bring additional information about the desired solution, compromises, or new outlooks as aids to the resolution of IPP. It is conventional to use the phrase "regularization of an ill-posed problem" to refer to various approaches to circumvent the lack of continuous dependence (as well as to bring about existence and uniqueness if necessary). As already pointed out, this entails an analysis of an IPP via an analysis of associated well-posed problems, i.e., a system (usually a sequence or a family) of well-posed "regularizations", yielding meaningful answers to the IPP.

The *strategy of resolution and reconstruction of ill-posed problems* involves one or more of the following intuitive ideas (cf. Nashed 1987a,b; Freeden and Nashed 2018b):

- change the notion of what is meant by a solution (e.g., ε-approximate solution: $\|A\tilde{x} - y\| \le \varepsilon$, where $\varepsilon > 0$ is prescribed; quasi-solution: $\|A\tilde{x} - y\| \le \|Ax - y\|$ for all

$x \in \mathcal{M}$, a prescribed subset of the domain of A; least-squares solution of minimal norm, etc),
- modify the operator equation or the problem itself,
- change the spaces and/or topologies,
- specify the type of involved noise ("strong" or "weak" noise as discussed, e.g., in Eggermont et al. 2015).

From the standpoint of mathematical and numerical analysis one can roughly group "regularization methods" into six categories:

1. Regularization methods in function spaces is one category. This includes Tikhonov-type regularization, the method of quasi-reversibility, the use for certain function spaces such as scale spaces in multi-resolutions, the method of generalized inverses (pseudoinverses) in reproducing kernel Hilbert spaces, and multiscale wavelet regularization.
2. Resolution of ill-posed problems by "control of dimensionality" is another category. This includes projection methods, moment-discretization schemes. The success of these methods hinges on the possibility of obtaining approximate solution while keeping the dimensionality of the finite dimensional problem within the "range of numerical stability". It also hinges on deriving error estimates for the approximate solutions that is crucial to the control of the dimensionality.
3. A third category is iterative methods which can be applied either to the problem in function spaces or to a discrete version of it. The crucial ingredient in iterative methods is to stop the iteration before instability creeps into the process. Thus iterative methods have to be modified or accelerated so as to provide a desirable accuracy by the time a stopping rule is applied.
4. A fourth category is filter methods. Filter methods refer to procedures where, for example, values producing highly oscillatory solutions are eliminated. Various "low pass" filters can, of course, be used. They are also crucial for determination of a stopping rule. Mollifiers are known in filtering as smooth functions with special properties to create sequences of smooth functions approximating a non-smooth function or a singular function.
5. The original idea of a mollifier method (see, e.g., Freeden and Nashed 2018b and the references therein) is interested in the solution of an operator equation, but we realize that the problem is "too ill-posed" for being able to determine the (pseudo)inverse accurately. Mollifiers are known as smooth functions with special properties to create sequences of smooth functions approximating a non-smooth function. Thus, we compromise by changing the problem into a more well-posed one, namely that of trying to determine a mollified version of the solution. The heuristic motivation is that the trouble usually comes from high frequency components of the data and of the solution, which are damped out by mollification.
6. The root of the Backus-Gilbert method (BG-method) was geophysical (cf. Backus and Gilbert 1967, 1968, 1970). The characterization involved in the model is known as moment problem in the mathematical literature. The BG-method can be thought of as resulting from discretizing an integral equation of the first kind. Where other regularization methods, such as the frequently used Tikhonov regularization method (see, e.g., Freeden and Nashed 2018b and the references therein), seek to impose smoothness constraints on the solution, the BG-method instead realizes stability constraints. As a consequence, the solution is varying as little as possible if the input data were resampled multiple times. The common feature between mollification and the BG-method is that an approximate inverse is determined independently from the right hand side of the equation.

OQ7) The philosophy of resolution leads to the use of algebraic methods versus function space methods, statistical versus deterministic approaches, strong versus weak noise, etc. A *regularization-approximation scheme* refers to a variety of methods such as Tikhonov's regularization, projection methods, multiscale methods, iterative approximation, etc., that can be applied to ill-posed problems. These schemes turn into algorithms once a resolution strategy can be effectively implemented. Unfortunately, this requires a determination of a suitable value of a certain parameter associated with the scheme (e.g., regularization parameter, mesh size, dimension of subspace in the projection scheme, specification of the level of a scale space, classification of noise, etc.). This is not a trivial problem since it involves a trade-off between accuracy and numerical stability, a situation that does not usually arise in well-posed problems. Are there more appropriate concepts of regularization stopping rules in future?

In Geodesy we are confronted with two essential types of problems, for which the following examples in the structural manifestations of Fredholm integral equations of the first kind may serve as prototypes for a large class of ill-posed inverse problems in Mathematics:

- *satellite gravitational gradiometry* (SGG) problem, i.e., "downward continuation" of gravitational satellite data to the Earth's surface to obtain gravitational information: A prototype is the SGG problem of determining the (anomalous) potential T on the Earth's surface S from values of the Hesse tensor $\mathbf{T}(x) = (\nabla \otimes \nabla)T(x)$ on the satellite orbit. A viable way to model SGG may be based on the Runge assumption that there exists a potential $\tilde{T}$ outside a Bjerhammar (Runge) sphere $\mathbb{S}^2_\beta$ in ε-accuracy ($\varepsilon > 0$, arbitrarily small) with the potential T outside and on the Earth's surface S. This leads to the integral equation of type

$$\mathbf{T}(x) \cong A(\tilde{T})(x) = \int_{\mathbb{S}^2_\beta} \underbrace{(\nabla_x \otimes \nabla_x) \frac{1}{4\pi\beta} \frac{x^2 - \beta^2}{|x-y|^3}}_{=P(x,y)} \tilde{T}(y)\, dS(y),$$

$$\tilde{T}|_{\mathbb{S}^2_\beta} \in L^2(\mathbb{S}^2_\beta), \tag{151}$$

for points x on the satellite orbit, where $\mathbf{T}(x) = (\nabla \otimes \nabla)T(x)$ are the tensorial satellite input data, P is the Poisson kernel, and the scalar function $\tilde{T}|_S$ obtained by "upward continuation" from $\tilde{T}_{\mathbb{S}^2_\beta}$ represents the required SGG-solution, and A is the operator of the Abel-Poisson-type integral (151). So, the inverse problem consists of finding $\tilde{T}|_{\mathbb{S}^2_\beta}$ from the Hesse tensor $(\nabla \otimes \nabla)\tilde{T}$ on the satellite orbit. In the jargon of Mathematics (see, e.g., Engl et al. 1996), this is an exponentially ill-posed problem which inextricably needs regularization because of the unboundedness of the existing inverse A^{-1} (note that A given by (151) is a compact operator).

In principle, all aforementioned regularization techniques of Mathematics are applicable to SGG. However, because of the nature of ill-posedness, there are good reasons to restrict our discussion to multiscale regularization methods developed during the last decades based on an IPP-modification (Freeden and Schneider 1998) of the wavelet philosophy (as known from Chui 1992, Daubechies 1992 in classical one-dimensional theory and Freeden and Windheuser (1997) in the spherical case).

– *inverse gravimetry* (IG) problem, i.e., "downward computation" of gravitational terrestrial and/or external data to the Earth's interior to obtain density information: A prototype is the IG problem based on Newton's integral equation

$$V(x) = \int_B G(x-y)\, F(y)\, dy = A(F)(x) \quad (152)$$

where B represents the Earth's body (or a part of it), V is the potential generated by the mass density anomaly F distributed on B, G is the fundamental solution of the Laplace equation, i.e., the Newton kernel

$$G(x-y) = \frac{1}{4\pi}\frac{1}{|x-y|}, \quad x \neq y, \quad (153)$$

and the operator A here stands for the Newton integral (152).

Contrary to the case of $L^2(\partial B)$ (see, e.g., Freeden 1980 for more details), the class $L^2(B)$ of square-integrable functions is not obtainable only by the L^2-completion of a countable harmonic function system such as solid (outer) spherical harmonics, outer ellipsoidal harmonics, certain mass point systems, harmonic spline and spline-wavelet systems. In addition, we have to take into account a so-called "anharmonic function system" (see, e.g., Weck 1972, Ballani and Stromeyer 1982) forming the "ghosts" of IG. This observation explains the ill-posedness of the inverse gravimetry problem. In fact, in accordance with the famous *Hadamard classification* (cf. Hadamard 1902), the IG problem with $F \in L^2(B)$ violates all criteria, viz. uniqueness, existence, and stability.

According to Freeden and Nashed (2018a) IG is "too inverse" to apply a conventional regularization technique, instead they propose multiscale mollifier methods for its regularization. Moreover, Freeden and Nashed (2018a) studies the case (important at least in exploration), where $V(x)$ is also known for some internal points $x \in B$. In fact, Blick et al. (2018) validate for representative data sets that additional internal data points can be used to stabilize the numerical solution process of IG.

7 The SGG Problem and Its Frequency-Multiscale Analysis

Tensor spherical harmonics of type (1,1) allow to express the Hesse tensor applied to solid (outer) harmonics in the form (see Freeden and Schreiner 2009 for more details on tensorial harmonics)

$$\nabla \otimes \nabla\, H^\beta_{-n-1,k} = \sqrt{\tilde{\mu}_n^{(1,1)}}\; \mathbf{h}^{(1,1);\beta}_{-n-1,k},$$

where we have used the abbreviation

$$\tilde{\mu}_n^{(1,1)} = (n+2)(n+2)(2n-3)(2n-1).$$

Consequently, we find the *tensor-isotropic SGG-pseudodifferential equation*

$$A(\tilde{T}) = \sum_{n=2}^{\infty}\sum_{k=1}^{2n+1} \tilde{T}^{\wedge_{L^2(\mathbb{S}^2_\beta)}}(n,k)\sqrt{\tilde{\mu}_n^{(1,1)}}\; \mathbf{h}^{(1,1);\beta}_{-n-1,m} = (\nabla\otimes\nabla)\tilde{T} \quad (154)$$

with

$$\tilde{T}^{\wedge_{L^2(\mathbb{S}^2_\beta)}}(n,k) = \langle \tilde{T}, H^\beta_{-n-1,k}\rangle_{L^2(\mathbb{S}^2_\beta)}$$

and

$$\begin{aligned} A(H^\beta_{-p-1,q}) &= A^\wedge(p)\,\mathbf{h}^{(1,1);\beta}_{-p-1,q} \\ &= \sqrt{\tilde{\mu}_p^{(1,1)}}\;\mathbf{h}^{(1,1);\beta}_{-p-1,q} = \nabla\otimes\nabla\, H^\beta_{-p-1,q} \end{aligned} \quad (155)$$

as spectral (frequency) representation to be used for the inversion of the SGG-integral equation (151). Equivalently,

$$\begin{aligned} A(\tilde{T})(x) &= \int_{L^2(\mathbb{S}^2_\beta)} \mathbf{K}(x,y)\,\tilde{T}(y)\,dS(y) \\ &= (\nabla\otimes\nabla)\tilde{T}(x) \cong \mathbf{T}(x), \end{aligned} \quad (156)$$

where the tensorial kernel $\mathbf{K}(\cdot,\cdot)$ is given by

$$\mathbf{K}(x,y) = \sum_{n=2}^{\infty}\sum_{k=1}^{2n+1} \sqrt{\tilde{\mu}_n^{(1,1)}}\, \mathbf{h}^{(1,1);\beta}_{-n-1,m}(x)\, H^{\beta}_{-n-1,k}(y). \tag{157}$$

By the completeness of the system $\{H^{\beta}_{-n-1,k}\}$ this enables us to conclude in the framework of $\mathbf{L}^2(\mathbb{S}^2_\beta)$ of square-integrable tensor fields on $\mathbb{S}^2_\beta$ that

$$\begin{aligned}\langle(\nabla\otimes\nabla)\tilde{T}, \mathbf{h}^{(1,1);S}_{-p-1,q}\rangle_{\mathbf{L}^2(\mathbb{S}^2_\beta)} &= \int_{L^2(\mathbb{S}^2_\beta)} (\nabla\otimes\nabla)\tilde{T}(y)\cdot \mathbf{h}^{(1,1);\beta}_{-p-1,q}(y)\,dS(y) \qquad (158)\\ &= \tilde{T}^{\wedge L^2(\mathbb{S}^2_\beta)}(p,q)\sqrt{\tilde{\mu}_n^{(1,1)}}.\end{aligned}$$

As an immediate result, we obtain the following expansion for the potential $\tilde{T}$

$$\begin{aligned}\tilde{T}(x) &= A^{-1}((\nabla\otimes\nabla)\tilde{T})(x) \qquad (159)\\ &= \sum_{n=2}^{\infty}\sum_{k=1}^{2n+1}\langle(\nabla\otimes\nabla)\tilde{T}, \mathbf{h}^{(1,1);\beta}_{-n-1,k}\rangle_{\mathbf{L}^2(\mathbb{S}^2_\beta)}(\tilde{\mu}_n^{(1,1)})^{-1/2}H^{\beta}_{-n-1,k}(x)\end{aligned}$$

for all points x of the outer space of $\mathbb{S}^2_\beta$. Equivalently, if all input (satellite) tensor data points $(x, (\nabla\otimes\nabla)T(x))$ satisfy the canonical assumption $|x| > \gamma \gg \beta$, i.e., $\mathbb{S}^2_\gamma$ is a "Bjerhammar (Runge) sphere" for the satellite orbit, then

$$\begin{aligned}\tilde{T}(x) &= A^{-1}((\nabla\otimes\nabla)\tilde{T})(x) \qquad (160)\\ &= \sum_{n=2}^{\infty}\sum_{k=1}^{2n+1}\langle(\nabla\otimes\nabla)\tilde{T}, \mathbf{h}^{(1,1);\gamma}_{-n-1,k}\rangle_{\mathbf{L}^2(\mathbb{S}^2_\gamma)}\\ &\quad\times(\tilde{\mu}_n^{(1,1)})^{-1/2}\left(\frac{\gamma}{\beta}\right)^{n+2} H^{\beta}_{-n-1,k}(x).\end{aligned}$$

This formula expresses the (approximation of the) anomalous potential $\tilde{T}$ in terms of the tensor $\mathbf{T} \cong (\nabla\otimes\nabla)\tilde{T}$ related to $\mathbb{S}^2_\gamma$.

A is a linear bounded injective compact operator so that the SGG-problem is ill-posed because of the unboundedness of A^{-1}.

As described earlier in Definition 3, a *regularization strategy for the SGG-problem* is a family of linear bounded pseudodifferential operators R_j, so that $\lim_{j\to\infty} R_j A(\tilde{T}) = \tilde{T}$ in the outer space of $\mathbb{S}^2_\beta$, i.e., the operators $R_j A$ converge in pointwise sense to the identity operator.

In what follows, we are only interested in two *multiscale regularization strategies,* particularly relevant in geodetic SGG-application (for a more general and deeper explanation see Freeden and Nutz (2011), Freeden et al. (2018a); the publication (Freeden et al. 2018b) is based on Freeden and Nutz (2011), Freeden et al. (2018a) and on this work, it provides more general material in frequency as well as space domain):

Tikhonov Regularization Strategy It uses the (non-bandlimited) isotropic Tikhonov-kernels (scaling functions) Φ_j, $j = 0, 1, \ldots$, given by

$$\Phi_j(x,y) = \sum_{n=2}^{\infty}\sum_{k=1}^{2n+1}(\Phi_j)^\wedge(n) H^{\beta}_{-n-1,k}(x) H^{\beta}_{-n-1,k}(y) \tag{161}$$

with

$$(\Phi_j)^\wedge(n) = \frac{(A^\wedge(n))^2}{(A^\wedge(n))^2 + \mu_j^2}, \quad n = 2, 3, \ldots, \; j = 0, 1, \ldots, \tag{162}$$

where $\{\mu_j\}$, $j = 0, 1, \ldots$, is a sequence of real numbers satisfying $\lim_{j\to\infty}\mu_j = 0$. Then the operators $R_j = \Phi_j * A^{-1}$ constitute a multiscale regularization strategy:

$$\begin{aligned}T_j(x) & \qquad (163)\\ &= R_j A(\tilde{T})(x)\\ &= \sum_{n=2}^{\infty}\sum_{k=1}^{2n+1}(\Phi_j)^\wedge(n)\langle(\nabla\otimes\nabla)\tilde{T}, \mathbf{h}^{(1,1);\beta}_{-n-1,k}\rangle_{\mathbf{L}^2(\mathbb{S}^2_\beta)}\\ &\qquad\times(\tilde{\mu}_n^{(1,1)})^{-1/2}H^{\beta}_{-n-1,k}(x)\\ &= \sum_{n=2}^{\infty}\sum_{k=1}^{2n+1}(\Phi_j)^\wedge(n)\langle(\nabla\otimes\nabla)\tilde{T}, \mathbf{h}^{(1,1);\gamma}_{-n-1,k}\rangle_{\mathbf{L}^2(\mathbb{S}^2_\gamma)}\\ &\qquad\times(\tilde{\mu}_n^{(1,1)})^{-1/2}\left(\frac{\gamma}{\beta}\right)^{n+2} H^{\beta}_{-n-1,k}(x).\end{aligned}$$

Truncated Singular Value Regularization Strategy It starts from a one-dimensional function $\varphi_0 : [0,\infty) \to \mathbb{R}$ with the following properties:

1. $\varphi_0(0) = 1$,
2. φ_0 is monotonically decreasing,
3. φ_0 is continuous at 0.
4. $\varphi_0 : [0,\infty) \to \mathbb{R}$ has a local support, i.e., supp $\varphi_0 \subset [0, 1]$.
5. The generators $\psi_0, \tilde{\psi}_0 : [0,\infty) \to \mathbb{R}$ of the mother wavelet and the dual mother wavelet, respectively, also possess a local support, i.e.,

$$\text{supp}\ \psi_0 \subset [0, 1], \tag{164}$$

$$\text{supp}\ \tilde{\psi}_0 \subset [0, 1]. \tag{165}$$

Accordingly, we are led to the isotropic scaling functions $\Phi_j,\ j = 0, 1, \ldots$, given by

$$\Phi_j(x, y) = \sum_{n=2}^{\infty} \sum_{k=1}^{2n+1} \varphi_0(2^{-j} n)\ H^{\beta}_{-n-1,k}(x) H^{\beta}_{-n-1,k}(y), \tag{166}$$

where

$$0 \le (\Phi_j)^\wedge(n) = \varphi_0(2^{-j} n) \le 1, \qquad n = 0, 1, \ldots \tag{167}$$

and

$$\lim_{j\to\infty} (\Phi_j)^\wedge(n) = \lim_{j\to\infty} \varphi_0(2^{-j} n) = 1, \qquad n = 0, 1, \ldots . \tag{168}$$

In accordance with our construction, the compact support of φ_0 implies that only finitely many $(\Phi_j)^\wedge(n)$ are different from 0. Obviously, it follows that

$$\begin{aligned} \operatorname{supp} \varphi_j &\subset [0, 2^j], \\ \operatorname{supp} \psi_j &\subset [0, 2^j], \\ \operatorname{supp} \tilde{\psi}_j &\subset [0, 2^j]. \end{aligned}$$

So, the bandlimited kernels $\Phi_j,\ j = 0, 1, \ldots$, defined via a generator φ_0 satisfying the properties (i)–(iv) as stated above define operators

$$R_j = \Phi_j * A^{-1}, \tag{169}$$

which constitute a regularization strategy in the following sense:

1. R_j is bounded,
2. the limit relation

$$\lim_{j\to\infty} R_j A(\tilde{T}) = \tilde{T} \tag{170}$$

holds true in the outer space of $\mathbb{S}^2_\beta$.

For more non-bandlimited as well as bandlimited regularization strategies and multiscale realizations in form of tree algorithms the reader is referred to Freeden et al. (2018a).

OQ9) As in collocational and spline theory (see, e.g., Sansò 1980, Sansò et al. 1968, Tscherning 1977), the choice of the kernel, i.e. regularization scaling function, is an important problem in SGG. Mathematically, all regularization strategies are equivalent, however, what is the right computational compromise between mathematical rigor and geodetic relevance in respect to geometry of orbit, data width, and accuracy?

The recently published approach (Freeden et al. 2018b) also contains a novel space-based multiscale regularization of SGG.

8 The IG Problem and Its Space-Multiscale Analysis

Let us begin with some preparatory remarks:

1. Sometimes instead of Newton's integral equation (152) we consider a different integral relation between V and F by applying a linear(ized) operator L to it, transforming V into some other field, e.g., a gravity anomaly field, namely

$$LV(x) = \int_B \{L_x G(x - y)\} F(y) dy . \tag{171}$$

For simplicity, however, we will work with (152) and the IG problem, which is just to derive F from V (instead of deriving F from certain linear functionals $LV(x)$ of V).

2. We formally know that (cf. Miranda 1970; Sansò and Sideris 2013)

$$-\Delta_x G(x - y) = \delta(x - y), \tag{172}$$

where the Dirac distribution δ is so to say the "density", in the sense of measure/distribution theory, of a point mass at y. The formal inversion of (152) is therefore well known and consists in applying $-\Delta$ to both members of this equation in B. However, we must recall that the study of the integral operator A and its inverse depends significantly on the spaces, where we choose that domain and image of the operator A are embedded. Note that here the operator is the same as that defined in (152).

3. In classical potential theory (see, e.g., Miranda 1970), if the domain of A is constituted by Hölder continuous functions F on $\overline{B}$, then the Poisson equation

$$\Delta V(x) = -\alpha(x) F(x),\ x \in \mathbb{R}^3, \tag{173}$$

yields an inverse equation to (152), where $\alpha(x)$ is the solid angle, divided by 4π, under which one "sees" the surface S from x, so that $\alpha = 1$ in B and $\alpha = 0$ in $\mathbb{R}^3 \setminus \overline{B}$. On the boundary $S = \partial B$, the value of α depends on the smoothness of S. For a continuously differentiable surface S we have $\alpha = \frac{1}{2}$. A rigorous mathematical proof of (173) is given in Freeden and Gerhards (2013), which also serves as a preparation for the multiscale mollifier method.

In what follows, we only provide a sketch of the inversion process, if we take as the domain of A the space $L^2(B)$. In this case, the image

$$Y = A(L^2(B)) , \tag{174}$$

is a closed subspace of $H^{2,2}_{\text{loc}}(\mathbb{R}^3)$, where "loc" has to be added since potentials in Y may go to zero at infinity as of the order $O\left(|\boldsymbol{x}|^{-1}\right)$, which does not imply square-integrability. Moreover, the solid angle α on S is not of importance, in fact, F in the integral equation (152) may be understood as a function in $L^2(\mathbb{R}^3)$ with $F|_{\mathbb{R}^3\setminus\overline{B}} = 0, \overline{B} = B \cup \partial B$, so that it has no specific values on a set of zero 3D-measure like S. A potential theoretic description of Y can be provided by saying that $V \in Y$ means that

a) $V \in H^{2,2}(B)$
b) $V \in H^{2,2}_{\text{loc}}(\mathbb{R}^3\setminus\overline{B})$, $\Delta V=0$ in $\mathbb{R}^3\setminus\overline{B}$; $|V(\boldsymbol{x})| \to 0$ for $|\boldsymbol{x}| \to \infty$,
c) the traces of V and $\frac{\partial V}{\partial n}$ on $S=\partial B$ from inside and outside, coincide, i.e.,

$$V_+ = V_-, \qquad \frac{\partial V}{\partial n_+} = \frac{\partial V}{\partial n_-} . \tag{175}$$

(note that another characterization of Y will be given later on).

The reason why we call

$$F = -\Delta V, \quad V \in Y = A(L^2(B)) \tag{176}$$

a "formal inverse" of (152) is that, since V is given as a datum, it might be polluted by errors and therefore suitable regularization/approximation procedures have to be applied before (173) can become effective. This is the viewpoint of a general review paper (Freeden and Nashed 2018b) on regularization methods, that we will follow in the sequel, with the specific target of showing briefly how multiscale methods can enter into this process.

Remark 3 In reality, the most relevant problem from the geophysical point of view is when V is known only in $\overline{B} = B \cup \partial B$, and therefore also on S due to the trace property c) above. In this case, however, we cannot expect uniqueness of the solution of (152) because, according to (176), the Laplacian applied to any harmonic function that continues V in B respecting the conditions (175) generates the same outer potential.

Letting formally $\tilde{V}_B$ be a member of class $H^{2,2}(B)$ satisfying

$$\tilde{V}_B|_S = (V|_{(\mathbb{R}^3\setminus\overline{B})})|_S \, , \quad \frac{\partial \tilde{V}_B}{\partial n}|_S = \frac{\partial (V|_{(\mathbb{R}^3\setminus\overline{B})})}{\partial n}|_S \tag{177}$$

we are able to show that the class of all solutions of IG consists of functions $\tilde{F}$ of the form

$$\tilde{F} = -\Delta V = -\Delta\left(V|_{(\mathbb{R}^3\setminus\overline{B})} + \tilde{V}_B\right), \tag{178}$$

where the condition (175), in fact, guarantees that $V = V|_{(\mathbb{R}^3\setminus\overline{B})} + \tilde{V}_B$ is still a member of $H^{2,2}_{\text{loc}}(R^3)$.

Indeed, the family $\{\tilde{V}_B\}$, which is clearly closed in $H^{2,2}(B)$, can be completely described by the equation

$$\tilde{V}_B = \tilde{V}_0 + \varphi, \quad \varphi \in H^{2,2}_0(B) \, , \text{ i.e.}, \varphi|_S = \left.\frac{\partial \varphi}{\partial n}\right|_S = 0, \tag{179}$$

where $\tilde{V}_0$ is any function of $H^{2,2}(B)$ satisfying the boundary condition (175). The problem how to find such $\tilde{V}_0$ is well-known from the literature (see, e.g. Miranda 1970; McLean 2000).

So, continuing φ with 0 outside B, we have found the class of all solutions of (152), namely

$$\tilde{F} = -\Delta\left(V|_{(\mathbb{R}^3\setminus\overline{B})} + \tilde{V}_0\right) - \Delta\varphi \, , \; \varphi \in H^{2,2}_0(B) \, , \tag{180}$$

and in particular we can say that the kernel of A, which is a closed subspace of $L^2(B)$ is given by the null-space

$$\mathcal{N}(A) = -\Delta\left(H^{2,2}_0(B)\right) . \tag{181}$$

The elements of $\mathcal{N}(A)$ are called *anharmonic functions* (see, e.g., Ballani and Stromeyer 1982). If we further observe that the orthogonal complement in $L^2(B)$ of $\mathcal{N}(A)$ is just

$$HL^2(B) = \{f \in L^2(B) \, : \, \Delta f = 0 \text{ in } B\} \, , \tag{182}$$

and we restrict (180) to B, where $\tilde{F}$ is supported, we arrive at a further fact specifying that the general solution of (152) can be written as

$$\tilde{F} = f - \Delta\varphi \, , \; f \in HL^2(B) \, , \; \varphi \in H^{2,2}_0(B) \tag{183}$$

with f the orthogonal projection of $-\Delta\tilde{V}_0$ onto $HL^2(B)$.

Incidentally, the harmonic f can as well be characterized as the minimum $L^2(B)$-norm solution of (152), because (183) is nothing but the decomposition of $\tilde{F}$ along two complementary orthogonal subspaces. Up to here what we can call the classical analysis of the inverse gravimetric problem in geophysical sense, as described by several authors, among which Freeden and Nashed (2018a), Ballani and Stromeyer (1982), and others. Such a theory can be significantly generalized to different couples of spaces as shown in Sansò (2014).

Remark 4 As promised, we are able to provide an alternative characterization of $Y = A(L^2(B))$. We first introduce on Y the topology which is the image of the $L^2(B)$-topology through A, namely

$$V = A(F) \quad \Rightarrow \quad \| V \|_Y^2 = \| F \|_{L^2(B)}^2 \, . \tag{184}$$

This makes Y a Hilbert space with the scalar product

$$V = A(F)\,,\ V' = A(F') \quad \Rightarrow \quad \langle V, V' \rangle_Y = \langle F, F' \rangle_{L^2(B)} \, . \tag{185}$$

We notice explicitly that, since we can view Y as a closed subspace of $H^{2,2}_{\text{loc}}(R^3)$, we know that the new topology of Y introduced by (184) should be equivalent to that of $H^{2,2}_{\text{loc}}$; in particular, if $V_n \to V$ in Y, this would imply as well that $F_n = -\Delta V_n$ tends to $F = -\Delta V$ in L^2-sense, i.e.,

$$\| \Delta V_n - \Delta V \|_{L^2(B)} \to 0 \, . \tag{186}$$

Now we first maintain that the evaluation functional is bounded in Y as a consequence of the fact that B is bounded. In fact, calling D the diameter of B we have by the Cauchy–Schwartz inequality (cf. Freeden and Nashed 2018a)

$$|V(\boldsymbol{x})|^2 \le \int_B |G(\boldsymbol{x}-\boldsymbol{y})|^2 dy \int_B |F(\boldsymbol{y})|^2 dy \tag{187}$$

and

$$\int_B |G(\boldsymbol{x}-\boldsymbol{y})|^2 \, dy = \left(\frac{1}{4\pi}\right)^2 \int_B \frac{1}{|\boldsymbol{x}-\boldsymbol{y}|^2} \, dy \le \frac{D}{4\pi} \, . \tag{188}$$

for <u>all</u> points $\boldsymbol{x} \in \mathbb{R}^3$.

Therefore, Y must have a reproducing kernel K, according to the Aronszajn-Bergman theorem (see, e.g., Yosida 1980), which completely characterizes it. It follows that

$$K(\boldsymbol{x}, \boldsymbol{y}) = \int_B G(\boldsymbol{x}-z) G(z-\boldsymbol{y}) \, dy \, . \tag{189}$$

In fact, since $G(\cdot - \boldsymbol{y})$ for every fixed $\boldsymbol{y}$ belongs to $L^2(B)$, we see that $K(\cdot, \boldsymbol{y}) \in Y$. Moreover, we can write (189) in the form

$$K(\cdot - \boldsymbol{y}) = A(G(\cdot - \boldsymbol{y})), \tag{190}$$

so that we can easily conclude from (185) that

$$\langle K(\boldsymbol{x}, \cdot), V \rangle_Y = \langle G(\boldsymbol{x} - \cdot), F \rangle_{L^2(B)} = V(\boldsymbol{x}) \tag{191}$$

holds true for all $V \in Y, V = A(F),\ F \in L^2$, so that all properties of a reproducing Hilbert space framework for Y are satisfied.

Armed with this tool we can think of a first regularization of (173), namely the projection of (173) on a finite-dimensional subspace by interpolating V with a finite number of *splines*. The use of the reprokernel K as a constituting function for splines then becomes expedient from the following procedure (we restrict ourselves to an intuitive description of the geometric situation, a more detailed consideration can be found in Freeden and Michel 2004):

Let $L_\rho = \{\boldsymbol{x}_i,\ i = 1 \ldots N_\rho\}$ be a "regular network" of points in $B \cup S$, with mesh side of length ρ. By halving ρ k times and putting $\rho_k = \rho/2^k$ we can construct finer and finer nets, so that

$$\bigcup_{k=0}^{\infty} L_{\rho_k} = \bigcup_{k=0}^{\infty} \{\boldsymbol{x}_i,\ i = 1 \ldots N_{\rho_k}\}$$

becomes a subset of B, being dense in it. Then we put

$$Y_{N_k} = \text{span}\{K(\boldsymbol{x}_i, \cdot) \ : \ \boldsymbol{x}_i \in L_{\rho_k}\}. \tag{192}$$

We notice that any V orthogonal (in Y) to Y_{N_k} satisfies

$$\langle K(\boldsymbol{x}_i, \cdot), V \rangle_Y = V(\boldsymbol{x}_i) = 0 \qquad i = 1, \ldots N_k \, . \tag{193}$$

Therefore it is clear that the linear space of "splines", i.e., the set of linear combinations of $K(\boldsymbol{x}_i, \cdot)$, $i = 1, 2 \ldots$

$$\text{span}\{K(\boldsymbol{x}_i, \cdot),\ i = 1, 2 \ldots\} = \bigcup_{k=0}^{\infty} Y_{N_k}$$

is dense in Y and any V orthogonal to it has to be zero. In fact, if V satisfies (193) on a set of point dense in $\overline{B}$, since V is continuous it has to be identically zero in $\overline{B}$, and the same will then be true for its Laplacian. As a consequence, $\Delta V = 0$ in the whole $\mathbb{R}^3$, implying that (remember that V has to be regular at infinity) V is identically zero.

Calling P_{N_k} the orthogonal projection in Y on Y_{N_k} we are able to conclude that

$$V_{N_k} = P_{N_k} V \to V \text{ in } Y \, , \tag{194}$$

implies

$$\Delta V_{N_k} = F_{N_k} \underset{L^2(B)}{\rightarrow} \Delta V = F\,. \tag{195}$$

The orthogonal projector P_{N_k} can be represented as usual, leading to the spline function V_{N_k} of the form

$$\begin{cases} V_{N_k} = \sum_{i=1}^{N_k} \lambda_i K(\boldsymbol{x}_i, \cdot) \\ \lambda_i = \sum_{i=j}^{N_k} \{K(\boldsymbol{x}_i, \boldsymbol{x}_j)\}^{-1} V(\boldsymbol{x}_j)\,, \end{cases} \tag{196}$$

where $\{K(\boldsymbol{x}_i, \boldsymbol{x}_j)\}^{-1}$ represent the (i, j) entry of the matrix inverse to $\{K(\boldsymbol{x}_i, \boldsymbol{x}_j)\}$. Since Y is a Hilbert space, (196) provides even the *minimum norm solution under the interpolatory conditions*

$$V_{N_k}(\boldsymbol{x}_i) = V(\boldsymbol{x}_i), \boldsymbol{x}_i \in L_{\rho_k}, \tag{197}$$

among all spline functions, i.e., all members of $Y_{N_k} = \text{span}\{K(\boldsymbol{x}_i, \cdot) : \boldsymbol{x}_i \in L_{\rho_k}\}$.

Furthermore, by explicitating (196) and interchanging integral and sum we immediately see that

$$V_{N_k} = \int_B \sum_{i=1}^{N_k} \lambda_i G(x_i - z) G(z - \cdot)\, dz\,, \tag{198}$$

showing that the corresponding F_{N_k} is given by

$$F_{N_k}(z) = \sum_{i=1}^{N_k} \lambda_i G(x_i - z)\,. \tag{199}$$

In other words, the approximate density distributions F_{N_k} are just potentials of N_k point masses placed at the nodes of L_{ρ_k}.

The orthogonal projection is always a stable operation. However, the algebraic computation of the coefficients λ_i becomes more and more ill-conditioned when $k \to \infty$. This is a well-known problem in regularization theory, and its handling leads to the so-called *Morozov criterion* to fix an optimal N_k. In order to go deeper into such questions of inverse theory one can read, for instance, Freeden and Nashed (2018b).

Remark 5 Let us observe that V would be known only on S, so that a point network replacing L_ρ should be chosen on S itself, because only there we can satisfy (197). Subsequently, the approximate density F_ρ, (199), becomes a function harmonic in B. One can readily prove that, for $\rho_k \to 0$, F^{ρ_k} tends to the harmonic solution of the inverse problem, already mentioned in Remark 4.

In any way we know that the mentioned instability comes from the increasing "correlation" between the "constituting trial kernel functions" $K(\boldsymbol{x}_i, \cdot)$. In more detail, if one computes the cosine of the angle between two of them, as a correlation coefficient ρ_{corr},

$$\rho_{\text{corr}} = \frac{\langle K(\boldsymbol{x}_i, \cdot),\, K(\boldsymbol{x}_i, \cdot)\rangle_Y}{\| K(\boldsymbol{x}_i, \cdot) \|_Y \| K(\boldsymbol{x}_j, \cdot) \|_Y} = \frac{K(\boldsymbol{x}_i, \boldsymbol{x}_j)}{\sqrt{K(\boldsymbol{x}_i, \boldsymbol{x}_i) K(\boldsymbol{x}_j, \boldsymbol{x}_j)}}. \tag{200}$$

One can see that $\rho_{\text{corr}} \to 1$ when $\boldsymbol{x}_i, \boldsymbol{x}_j$ become closer and closer. To implement bases that reduce, or even annihilate, such correlation is exactly the idea underlying the construction of a wavelet scheme.

In fact, the critical point in the aforementioned spline approach is the solution of the linear system

$$V_{N_k}(\boldsymbol{x}_j) = \sum_{i=1}^{N_k} \lambda_i\, K(\boldsymbol{x}_i, \boldsymbol{x}_j), \quad j = 1, \ldots, N_k. \tag{201}$$

Each coefficient $K(\boldsymbol{x}_i, \boldsymbol{x}_j)$ must be determined by numerical integration and the coefficient matrix $\{K(\boldsymbol{x}_i, \boldsymbol{x}_j)\}$ is full-sized. These numerical calamities led Freeden and Nashed (2018a) either to decorrelate the kernel K by Gaussian bell functions or to replace K by mollifier spline-wavelets. The forthcoming study can be regarded as basic background material for mollifier spline-wavelets, however, only for the special case of spacelimited Haar mollifiers (for more general mollifiers see Freeden and Nashed 2018a).

In fact, a simple mollifier, useful in our context, is the *moving average* or *Haar mollifier* defined by

$$M_\rho(f) = \frac{1}{\|B_\rho\|} \int_{B_\rho} H_\rho(x - y) f(y) dy \tag{202}$$

where H is the spherically symmetric Heaviside function

$$H_\rho(x - y) = \begin{cases} 1, & |x - y| < \rho, \\ 0, & |x - y| \geq \rho. \end{cases} \tag{203}$$

and $\|B_\rho\| = \frac{4}{3}\pi\rho^3$. In the nomenclature of convolutions, (202) can be written in the form

$$M_\rho(f) = \frac{1}{\|B_\rho\|} H_\rho * f\,, \tag{204}$$

where $*$ denotes the convolution product.

In addition, Newton's equation allows a representation as convolution, namely

$$V = G * F\,. \tag{205}$$

Thus we obtain

$$V^\rho = M_\rho(V) = \frac{1}{\|B_\rho\|} H_\rho * V = \frac{1}{\|B_\rho\|} H_\rho * G * F \quad (206)$$

$$= G * \frac{1}{\|B_\rho\|} H_\rho * F = G * F_\rho .$$

After denoting

$$G^\rho = \frac{1}{\|B_\rho\|} H_\rho * G \quad (207)$$

we are able to rewrite (206) in the form

$$V^\rho = G^\rho * F = G * F^\rho \,. \quad (208)$$

From the second identity in (208) we derive

$$-\Delta V^\rho = F^\rho \quad (209)$$

and since $F^\rho \underset{L^2(B)}{\rightarrow} F$ when $\rho \rightarrow 0$, thanks to a well-known theorem by Lebesgue, we have

$$-\Delta V^\rho = F_\rho \underset{L^2(B)}{\rightarrow} -\Delta V = F \,. \quad (210)$$

It is easy to find the explicit form of G^ρ if one considers that (207) can be expressed in the form

$$G^\rho = G * H_\rho, \quad H^\rho = \frac{1}{\|B_\rho\|} H_\rho \,, \quad (211)$$

showing that G^ρ is nothing but the potential of a uniform distribution with mass density equal to $\frac{1}{B_\rho}$. So G^ρ can be read out of literature (e.g. Sansò and Sideris 2013) and is given by

$$G^\rho(r) = \frac{1}{\|B_\rho\|} \begin{cases} \frac{1}{2}(\rho^2 - \frac{1}{3} r^2), & r < \rho, \\ \frac{1}{3}\frac{\rho^3}{r}, & r > \rho. \end{cases} \quad (212)$$

Obviously, G^ρ satisfies the transmission conditions (175) at $r = \rho$.

Often in numerics, a disadvantage of the Haar based mollifier approach is the discontinuity of the Haar kernel. Alternative continuous mollifier kernels are available from the theory of singular integrals (for more details see Freeden and Nashed 2018a).

Following here the Haar concept as in Freeden and Gerhards (2013) and its extensions in Freeden and Nashed (2018a), the functions (208) can be regarded as scaling functions in a multiscale context. Moreover, we can go over from scale continuous scaling functions $\{V^\rho\}, \{F^\rho\}$ to scale discrete scaling functions $\{V^{\rho_k}\}, \{F^{\rho_k}\}$.

By discretization, with suitable cubature weights w_i and nodes $\boldsymbol{x}_i$ in a ρ_k-point system L_{ρ_k} $\left(\rho_k = \frac{\rho}{2^k}\right)$ we then obtain

$$F^{\rho_k} \cong F^k = \sum_{i=1}^{N_k} w_i F(\boldsymbol{x}_i) H^{\rho_k}(\cdot - \boldsymbol{x}_i) \quad (213)$$

$$= \sum_{i=1}^{N_k} \mu_i H^{\rho_k}(\cdot - \boldsymbol{x}_i), \qquad \mu_i = w_i F(\boldsymbol{x}_i),$$

so that

$$V^k = G * F^k$$

$$= \sum_{i=1}^{N_k} \lambda_i G^{\rho_k}(\cdot - \boldsymbol{x}_i) \,. \quad (214)$$

Finally, evaluating (214) at the ρ^k-network nodes one obtains a linear system in the unknowns λ_i

$$V^k(\boldsymbol{x}_j) = \sum_{i=1}^{N_k} \lambda_i G^{\rho_k}(\boldsymbol{x}_j - \boldsymbol{x}_i) \,. \quad (215)$$

In other words, the solution of the system (215) enables us to evaluate the values $\{\lambda_i\}$, hence, the functional values $F(\boldsymbol{x}_i)$ are available, since the cubature weights are assumed to be known. As a consequence, we are able to reconstruct from (213) the approximate mass density F^k.

It is important here to underline that the coefficients $\{\lambda_i\}$ in (215) and (213) do depend on k, too. They change when we change the approximation scale ρ_k.

The convergence of V^k to V and of F^k to F for $k \rightarrow \infty$ is guaranteed by the reasoning already developed above. However, we are not in a very different position from that where we were with (196) and (199), because again the scale function G^{ρ_k} presents a degree of smoothness, and then of correlation.

Freeden and Nashed (2018a) introduced potential wavelets $(WV)^k$ and density wavelets, $(WF)^k$, according to

$$(WV)^k = V^{k+1} - V^k, \quad (216)$$

$$(WF)^k = F^{k+1} - F^k \,. \quad (217)$$

So, sparse techniques from numerics become applicable, since the wavelets have local supports, which become smaller with increasing scale.

It is readily seen then that (215) and (213), respectively, lead to the following linear systems

$$(WV)^k(x_j) = \sum_{i=1}^{N_k} \lambda_i \Psi_V^k(x_j - x_i) \tag{218}$$

and

$$(WF)^k(x_j) = \sum_{i=1}^{N_k} \lambda_i \Psi_H^k(x_j - x_i)\,, \tag{219}$$

where

$$\Psi_V^k(x_j - x_i) = G^{\rho_{k+1}}(x_j - x_i) - G^{\rho_k}(x_j - x_i) \tag{220}$$

and

$$\Psi_H^k(x_j - x) = H^{\rho_{k+1}}(x_j - x_i) - H^{\rho_k}(x_j - x_i)\,. \tag{221}$$

The problem now is to calculate from (218) the coefficients λ_i (as already observed they depend on k, too) of the development of V in terms of $\Psi_V^k(x_j - x_i)$ and then use them to reconstruct the mass density expressions $(WP)^k(x_j)$ by (219) in terms of $\Psi_H^k(x_j - x_i)$.

It is readily seen that the reconstruction formula

$$F = F^0 + \sum_{k=1}^{\infty} (WF)^k \tag{222}$$

holds true.

The key point in the present step is that $\Psi_V^k(x_j - x_i)$, although not strictly orthogonal, are in fact much less "correlated" than $G^{\rho_k}(x_i - x_i)$ and so the development (218), which is originated in wavelet theory, can be easily performed at least up to some approximation degree $k_{\max}$. The exact value of the maximum approximation degree is again a matter of regularization theory and respond to criteria like that of the Morozov principle or some other truncation criterion as in Freeden and Maier (2002).

A numerical investigation of the above method applied to a structural model proposed by the Rice University, Houston Texas (2002), the so-called *Marmousi model*, can be found in Blick et al. (2018).

Nevertheless one point is non-standard and worth of further investigations, both in theoretical and numerical way:

OQ8) When we restrict the IG problem to the geophysically relevant case, namely that V or some linear functional such as the gravity anomaly Δg are given on S, what is a good criterion to stop the maximum approximation index $k_{\max}$ and therefore the resolution at which we can perform a reconstruction of F?

9 Summary and Outlook

It is clear that a subject like the relation between Geodesy and Mathematics cannot have a "conclusion" (hopefully!). As mentioned in the Introduction, we have not tried to be complete, but only to bring the attention to some relevant interactions, examples, and open questions (sometimes in a more formal and accessible language avoiding larger mathematical technicalities).

In Sects. 2–3 we have examined two subjects where the relation has a historical character. In Sects. 4–8 we have come to more recent framework and resulting arguments.

In all of the items presented the purpose was to show the fruitful interaction between the two disciplines along the way of a general enhancement of scientific knowledge. May this last for long!

Acknowledgements The authors would like to thank the reviewers for their useful work that has certainly contributed to improve the paper.

References

Baarda W (1967) Statistical concepts in geodesy. New series V2 N4. Netherland Geod Comm Pub Geodesy, Delft

Backus GE, Gilbert F (1967) Numerical applications of a formalism for geophysical inverse problems. Geophys J R Astron Soc 13:247–276

Backus GE, Gilbert F (1968) The resolving power of gross Earth data. Geophys J R Astron Soc 16:169–205

Backus GE, Gilbert F (1970) Uniqueness in the inversion of inaccurate gross Earth data. Philos Trans R Soc Lond A 266:123–192

Ballani L, Stromeyer D (1982) The inverse gravimetric problem: a Hilbert space approach. In: Holota P (ed) Proceedings international symposium figure of the Earth, Moon, and other planets, Prague, pp 359–373

Betti B, Crespi M, Sansò F (1993) A geometric illustration of ambiguity resolution in GPS theory and a Bayesian approach. Man Geod 18(5):318–330

Bjerhammar A (1962) Gravity reduction to an internal sphere. Division of Geodesy, Stockholm

Bjerhammar A (1973) Theory of errors and generalized matrix inverses. Elsevier, Amsterdam

Blick C, Freeden W, Nutz H (2018) Gravimetry and exploration. In: Freeden W, Nashed MZ (eds) Handbook of Mathematical Geodesy: functional analytic and potential methods. Geosystems Mathematics. Birkhäuser/Springer, Basel/New York/Heidelberg, pp 687–752

Bruns H (1878) Die Figur der Erde "Ein Beitrag zur europäischen Gradmessung". P. Stankiewicz, Berlin

Chui CK (1992) An introduction to wavelets. Texas A&M University, College Station, Texas

Darbeheshti N, Featherstone WE (2009) Non-stationary covariance function modelling in 2D least-squares collocation. J Geod 83:495–508

Daubechies I (1992) Ten lectures on wavelets. CBMS-NSF regional conference. Series in applied mathematics, vol 61. Capital City Press, Montpelier, VT

Dermanis A, Sansò F (2018) Different equivalent approaches to the geodetic reference system. In: Satellite positioning for geosciences. Rend Lincei Sc Fis Nat 29:11–22

Eggermont PN, LaRiccia V, Nashed, MZ (2015) Noise models for ill-posed problems. In: Freeden W, Nashed, MZ, Sonar T (eds) Handbook of Geomathematics, 2nd edn, vol 2. Springer, New York, pp 1633–1658

Engl, HW, Hanke, M, Neubauer, A (1996) Regularization of inverse problems. Kluwer, Dordrecht

Freeden W (1980) On the approximation of the external gravitational potential with closed systems of (trial) functions. Bull Géod 54:1–20

Freeden W (1981) On approximation by harmonic splines. Manuscr Geod 6:193–244

Freeden W, Gerhards C (2013) Geomathematically oriented potential theory. CRC Press, Taylor & Francis Group, Boca Raton

Freeden W, Gutting M (2017) Integration and cubature methods. CRC Press, Taylor & Francis Group, Boca Raton

Freeden W, Maier T (2002) On multiscale denoising of spherical functions: basic theory and numerical aspects. Electron Trans Numer Anal (ETNA) 14:40–62

Freeden W, Mayer C (2006) Multiscale solution for the Molodensky problem on regular telluroidal surfaces. Acta Geod Geophys Hung 41:55–86

Freeden W, Michel V (2004) Multiscale potential theory (with applications to geoscience). Birkhäuser, Boston

Freeden W, Nashed MZ (2018a) Inverse gravimetry: background material and multiscale mollifier approaches. GEM Int J Geomath 9:199–264

Freeden W, Nashed MZ (2018b) Ill-posed problems: Operator methodologies of resolution and regularization. In: Freeden W, Nashed MZ (eds) Handbook of Mathematical Geodesy: functional analytic and potential methods. Geosystem Mathematics. Birkhäuser/Springer International Publishing, Basel/New York/Heidelberg, pp 201–314

Freeden W, Nutz H (2011) Satellite gravity gradiometry as tensorial inverse problem. GEM Int J Geomath 2:177–218

Freeden W, Schneider F (1998) Regularization wavelets and multiresolution. Inv Probl 14: 493–515

Freeden W, Schreiner M (2009) Spherical functions of mathematical geosciences: a scalar, vectorial, and tensorial setup. Springer, Berlin/Heidelberg

Freeden W, Schreiner M (2018) Mathematical geodesy: its role, its potential and its perspective. In: Freeden W, Rummel R (eds) Handbuch der Geodäsie. Mathematical Geodesy, Springer Reference Naturwissenschaften

Freeden W, Windheuser U (1997) Combined spherical harmonic and wavelet expansion. A future concept in Earth's gravitational determination. Appl Comput Harmon Anal 4:1–37

Freeden W, Nashed MZ, Schreiner M (2018a) Spherical sampling. Geosystems Mathematics. Springer, Basel/New York/Heidelberg

Freeden W, Nutz H, Rummel R, Schreiner M (2018b) Satellite gravity gradiometry (SGG): Methodological foundation and geomathematical advances. In: Freeden W, Rummel R (eds) Handbuch der Geodäsie. Mathematical Geodesy. Springer Reference Naturwissenschaften

Grafarend EW (1975) Cartan frames and a foundation of physical Geodesy. In: Brosowski B, Martensen E (eds) Methoden und Verfahren der Mathematischen Physik, Band 12. BI-Verlag, Mathematical Geodesy, Mannheim, pp 179–208

Grafarend EW (1986) Differential geometry of the gravity field. Proc First Hotine Marussi Symposium, Roma, 3–6 June 1985. Politecnico di Milano

Grafarend EW (2015) The reference figure of the rotating Earth in geometry and gravity space and an attempt to generalize the celebrated Runge-Walsh approximation theorem for irregular surfaces. GEM Int J Geomath 6:101–140

Grafarend EW, Awange JL (2012) Applications of linear and non-linear models. Springer, Berlin/Heidelberg

Grossman N (1974) Holonomic measurables in geodesy. J Geophys Res 79:689–694

Grothaus M, Raskop T (2018) Oblique stochastic boundary value problems. In: Freeden W, Nashed MZ (eds) Handbook of Mathematical Geodesy: functional analytic and potential methods. Geosystems Mathematics. Birkhäuser/Springer, Basel/New York/Heidelberg, pp 491–516

Hadamard J (1902) Sur les problémes aux dérivés partielles et leur signification physique. Princeton Univ Bull 13:49–52

Heiskanen WA, Moritz H (1967) Physical geodesy. Freeman and Co, S. Francisco

Helmert FR (1884) Die Mathematischen und Physikalischen Theorien der Höheren Geodäsie. Band 1+ 2, Teubner, Leipzig

Hörmander L (1976) The boundary problems of physical geodesy. The Royal Institute of Technology, Division of Geodesy, Stockholm, Report 9, 1975 (also: Arch Rat Mech Anal, 62:1–52, 1976)

Hotine M (1985) Mathematical geodesy. ESSA monographs. U.S. Dept. of Commerce, Rockville

Jazaeri S, Amri-Simkooei A, Sharifi MA (2014) On lattice reduction algorithms for solving weighted integer least square problem: comparative study. GPS Solutions 18:105–114

Kannan R (1983) Improved algorithms for integer programming and related lattice problems. Proc ACM symposium on theory of computing, Boston, MA, pp 193–206

Koch KR (1988) Parameter estimation and hypothesis testing in linear models. Springer, Berlin/Heidelberg

Krarup T (1969) A contribution to the mathematical foundation of physical geodesy. Meddelelse no 44, Geodätisk Inst Köbenhavn

Krarup T (2006a) Letters on Molodensky's problem: III. A mathematical formulation of Molodensky's problem. In: Borre K (ed) Mathematical foundation of Geodesy. Springer, Berlin/Heidelberg

Krarup T (2006b) In: Borre K (ed) Mathematical foundation of Geodesy. Springer, Berlin/Heidelberg

Lamperti J (1977) Stochastic processes: a survey of mathematical theory. Springer, New York

Marussi A (1985) Intrinsic geodesy. Springer, Berlin/Heidelberg

McLean W (2000) Strongly elliptic systems and boundary integral equations. Cambridge University press, Cambridge

Migliaccio F, Sansò F, Tornatore V (1998) Cluster and probabilistic models for a refined estimation theory. Boll Geod Sc Aff 57:241–256

Miranda C (1970) Partial differential equations of elliptic type. Springer, Berlin

Molodensky MS, Eremeev VF, Yurkina MI (1960) Methods for the study of the gravitational field of the Earth. Transl Russian Israel Program for Scient Trans, Jerusalem

Moore EH (1935) General analysis. Mem Am Math Soc 1:197–209

Moritz H (1972) Advanced least squares estimation. Scientific Report, 175, The Ohio State University, Department of Geodetic Science, Columbus, Ohio

Moritz H (1977) Recent developments in the geodetic boundary value problem. Scientif Report, 266, The Ohio State University, Department of Geodetic Science, Columbus, Ohio

Moritz H (1989) Advanced physical geodesy, 2nd edn. Wichmann Verlag, Karlsruhe

Moritz H (2015) Classical physical geodesy. In: Freeden W, Nashed MZ, Sonar T (eds) Handbook of Geomathematics, 2nd edn, vol 1. Springer, New York, pp 253–289

Nashed MZ (1976) Generalized inverses and applications. Academic, New York

Nashed MZ (1987a) A new approach to classification and regularization of ill-posed operator equations. In: Engl H, Groetsch, CW (eds) Inverse and ill-posed problems. Notes and reports in Mathematics and Science and Engineering, vol 4. Academic, Boston

Nashed MZ (1987b) Inner, outer, and generalized inverses in Banach and Hilbert spaces. Numer Funct Anal Optim 9:261–326

Pavlis NK (2103) Global gravitational models. In: Sansò F, Sideris M (eds) Geoid determination: theory and methods, Chapter 6, Part II. Springer, Berlin/Heidelberg

Penrose R (1955) A generalized inverse for matrices. Proc Camb Philol Soc 51:406–413

Pizzetti P (1910) Intorno alle possibili distribuzioni di massa nell'interno della Terra. Analisi di Mat, Milano, XVII, 225–258

Pohst M (1981) On the computation of lattice vectors of minimal length, successive minima and reduced bases with applications. ACM SIGSAM Bull 15: 37–44

Rao CR (1965) Linear statistical inference and its applications. Wiley, New York

Rozanov Y (1998) Random fields and stochastic partial differential equations. Kluwer, Doordrecht

Rozanov Y, Sansò F (1997) Boundary value problems for harmonic random fields. In: Geodetic boundary value problems in view of the one centimeter geoid. Lecture notes in Earth Sciences, vol 65. Springer, Heidelberg

Runge C (1885) Zur Theorie der eindeutigen analytischen Funktionen. Acta Math 6:229–234

Sansò F (1977) The geodetic boundary value problem in gravity space. Memorie Acc Lincei 14, S8, n3

Sansò F (1980) Internal collocation. Atti Della Academia Nazionale Dei Lincei 16:4–52

Sansò F (1982) A note on density problems and the Runge-Krarup theorem. Boll Geod Sci Affini 41:422–477

Sansò F (1986) Statistical methods in physical geodesy. In: H. Sünkel (ed) Mathematical and numerical techniques in physical geodesy. Lecture notes in Earth Sciences, vol 7. Springer, Berlin, pp 49–155

Sansò F (2014) On the regular decomposition of the inverse gravimetric problem in non-L^2 spaces. GEM Int J Geomath 5(1):33–61

Sansò F (2018) The analysis of the geodetic boundary value problem: state and perspectives. In: Freeden W, Nashed MZ (eds) Handbook of Mathematical Geodesy. Geosystems Mathematics. Birkhäuser, Springer International Publishing, Basel/New York/Heidelberg, pp 459–489

Sansò F, Rozanov Y (2003) The analysis of the Neumann and oblique derivative problem. Weak theory. In: Grafarend EW, Krumm FW, Schwarze VS (eds) Geodesy. The challenge of the third millennium. Springer, Berlin/Heidelberg

Sansò F, Rummel R (eds) (1997) Geodetic boundary value problems in view of the one centimeter geoid. Lecture notes in Earth Sciences, vol 65. Springer, Berlin/Heidelberg

Sansò F, Sideris M (2013) Geoid determination: theory and methods. Springer, Berlin/Heidelberg

Sansò F, Sideris M (2017) Geodetic boundary value problem: the equivalence between Molodensky's and Helmert's solutions. In: Springer briefs in Earth Sciences. Springer, Cham

Sansò F, Vaniček P (2006) The orthometric height and the holonomity problem. J of Geod 80:225–232

Sansò F, Barzaghi R, Tscherning CC (1968) Choice of norm for the density distribution of the Earth. Geophys J R Astr Soc 87:123–141

Seitz K, Heck B (1993) Effects of non-linearity in the geodetic boundary value problems. Deutsche Geodätische Kommission, Reihe A, Heft Nr. 109, Verlag des Bayerische Akad der Wissenschaften, Munich

Somigliana C (1929) Teoria generale del campo gravitazionale dell'ellissoide di rotazione. Mem Soc Astron It 54

Stokes GG (1867) On the internal distribution of matter which shall produce a given potential at the surface of a gravitating mass. Proc R Soc 482–486

Teunissen PJ (2001) Integer estimation in the presence of biases. J Geod 75:399–407

Teunissen PJ (2003) Theory of integer equivariant estimation with application to GNSS. J Geod 77:402–410

Teunissen PJ (2004) Penalized GNSS ambiguity resolution. J Geod 78:235–244

Teunissen PJ (2007) Best prediction in linear models with mixed integer/real unknowns: theory and applications. J Geod 81:759–780

Teunissen PJ (2017) Carrier phase integer ambiguity resolution. In: Teunissen PJ, Montenbrook O (eds) Springer handbook of global navigation systems. Springer, New York

Tscherning CC (1977) A note of the choice of norm when using collocation for the computation of approximations to the anomalous potential. Bull Géod 51:137–147

Tscherning CC (2013) Geoid determination by 3D least squares collocation. In: Geoid determination: theory and methods. Springer, Berlin/Heidelberg

Tscherning CC, Rapp RH (1974) Closed covariance expressions for gravity anomalies, geoid undulations, and deflections of the vertical implied by anomaly degree-variance models. Scientif Report, 208, The Ohio State University, Department of Geodetic Science, Columbus, Ohio

Vaniček P, Krakiwsky EJ (1986) Geodesy: the concepts. Elsevier, Amsterdam

Vekua IN (1953) Über die Vollständigkeit des Systems harmonischer Polynome im Raum. Dokl Akad Nauk 90:495–498

Walsh JL (1929) The approximation of harmonic functions by harmonic polynomials and by harmonic rational functions. Bull Am Math Soc 35:499–544

Weck N (1972) Zwei inverse Probleme in der Potentialtheorie. Mitt Inst Theor Geod Univ Bonn 4:27–36

Yosida K (1980) Functional analysis. Classics in mathematics. Springer, Berlin/Heidelberg

Zund J (2012) Foundations of differential geodesy. Springer, Berlin/Heidelberg

List of Reviewers

Aleš Bezděk
Alfonso Vitti
Ali Reza Amiri-Simkooei
Anna Klos
Anno Löcher
Athanasios Dermanis
Balaji Devaraju
Battista Benciolini
Bofeng Li
Boris Kargoll
Brigitte Gundlich
Dimitrios Rossikopoulos
Dimitrios Tsoulis
Elena Mazurova
Erricos C. Pavlis
Felix Norman Teferle
Fernando Sansò
Gilad Even-Tzur
Giovanna Venuti
Hamza Alkhatib
Hana Krasná
Hans Neuner
Heiner Denker
Holger-soren Steffen
Jan Douša
Jianqing Cai
Jianzhong Zhang
Johannes Bureick
Khosro Ghobadi-Far
Karl-Rudolf Koch
Lars E. Sjöberg
Ludovico Biagi
Maria Clara de Lacy
Martin Pitoňák
Martin Vermeer
Michal Šprlák
Mirko Reguzzoni
Pavel Novák
Peter J.G. Teunissen
Petr Holota
Riccardo Barzaghi
Robert Čunderlík
Roberto Devoti
Stefano Federico
Tanghua Li
Torsten Mayer-Guerr
Toshio Fukushima
WenBin Shen
Wieslaw Kosek
Willfried Schwarz
Wolf-Dieter Schuh
Xing Fang
Zuheir Altamimi

P. Novák et al. (eds.), *IX Hotine-Marussi Symposium on Mathematical Geodesy*,
International Association of Geodesy Symposia 151, https://doi.org/10.1007/978-3-030-54267-2

Author Index

P. Novák et al. (eds.), *IX Hotine-Marussi Symposium on Mathematical Geodesy*,
International Association of Geodesy Symposia 151, https://doi.org/10.1007/978-3-030-54267-2

Subject Index

P. Novák et al. (eds.), *IX Hotine-Marussi Symposium on Mathematical Geodesy*,
International Association of Geodesy Symposia 151, https://doi.org/10.1007/978-3-030-54267-2